VOLUME ONE HUNDRED AND EIGHTEEN

Advances in
Cancer Research

VOLUME ONE HUNDRED AND EIGHTEEN

Advances in
CANCER RESEARCH

Edited by

KENNETH D. TEW

*Professor and Chairman,
Department of Cell and Molecular Pharmacology,
John C. West Chair of Cancer Research,
Medical University of South Carolina,
South Carolina, USA*

PAUL B. FISHER

*Professor and Chair,
Department of Human & Molecular Genetics,
Director, VCU Institute of Molecular Medicine,
Thelma Newmeyer Corman Chair in Cancer Research,
VCU Massey Cancer Center,
Virginia Commonwealth University, School of Medicine,
Richmond, Virginia, USA*

AMSTERDAM • BOSTON • HEIDELBERG • LONDON
NEW YORK • OXFORD • PARIS • SAN DIEGO
SAN FRANCISCO • SINGAPORE • SYDNEY • TOKYO

Academic Press is an imprint of Elsevier

Academic Press is an imprint of Elsevier
525 B Street, Suite 1800, San Diego, CA 92101-4495, USA
225 Wyman Street, Waltham, MA 02451, USA
32 Jamestown Road, London, NW1 7BY, UK
The Boulevard, Langford Lane, Kidlington, Oxford, OX5 1GB, UK
Radarweg 29, PO Box 211, 1000 AE Amsterdam, The Netherlands

First edition 2013

Copyright © 2013 Elsevier Inc. All rights reserved.

No part of this publication may be reproduced, stored in a retrieval system or transmitted in any form or by any means electronic, mechanical, photocopying, recording or otherwise without the prior written permission of the Publisher.

Permissions may be sought directly from Elsevier's Science & Technology Rights Department in Oxford, UK: phone (+44) (0) 1865 843830; fax (+44) (0) 1865 853333; email: permissions@elsevier.com. Alternatively you can submit your request online by visiting the Elsevier website at http://elsevier.com/locate/permissions, and selecting *Obtaining permission to use Elsevier material*.

Notice
No responsibility is assumed by the publisher for any injury and/or damage to persons or property as a matter of products liability, negligence or otherwise, or from any use or operation of any methods, products, instructions or ideas contained in the material herein. Because of rapid advances in the medical sciences, in particular, independent verification of diagnoses and drug dosages should be made.

ISBN: 978-0-12-407173-5
ISSN: 0065-230X

For information on all Academic Press publications
visit our website at store.elsevier.com

Printed and bound in USA

13 14 15 16   12 11 10 9 8 7 6 5 4 3 2 1

CONTENTS

Contributors vii

1. **Bioengineering Strategies for Designing Targeted Cancer Therapies** 1
   Angela A. Alexander-Bryant, Wendy S. Vanden Berg-Foels, and Xuejun Wen

   1. Introduction 2
   2. Cancer-Targeting Mechanisms 4
   3. Design Criteria for Cancer Targeting 9
   4. Delivery Vehicles 13
   5. Targeting Moieties 23
   6. Drug Release 33
   7. Future Directions in Cancer Targeting 39
   8. Concluding Remarks 42
   Acknowledgments 44
   References 44

2. **Autophagy: Cancer's Friend or Foe?** 61
   Sujit K. Bhutia, Subhadip Mukhopadhyay, Niharika Sinha, Durgesh Nandini Das, Prashanta Kumar Panda, Samir K. Patra, Tapas K. Maiti, Mahitosh Mandal, Paul Dent, Xiang-Yang Wang, Swadesh K. Das, Devanand Sarkar, and Paul B. Fisher

   1. Introduction 62
   2. Autophagy and Autophagic Death 63
   3. Autophagy in Tumor Initiation and Development 70
   4. Autophagy in Tumor Progression and Metastasis 73
   5. Autophagy in Tumor Dormancy 79
   6. Autophagy in Cancer Initiating/Stem Cells 81
   7. Autophagy in Cancer Therapy 82
   8. Conclusions 86
   Acknowledgments 87
   References 87

3. **The Transcription Factor FOXM1 (Forkhead box M1): Proliferation-Specific Expression, Transcription Factor Function, Target Genes, Mouse Models, and Normal Biological Roles**   97
   Inken Wierstra

   1. FOXM1 is an Activating Transcription Factor   99
   2. The Four Known FOXM1 Splice Variants   102
   3. FOXM1 Target Genes   110
   4. Molecular Mechanisms of FOXM1 for Gene Regulation   164
   5. The Molecular Function of FOXM1 as a Conventional Transcription Factor   167
   6. The FOXM1 Expression   197
   7. FOXM1 Mouse Models   266
   8. Biological Functions of FOXM1   283
   References   343

Index   *399*

CONTRIBUTORS

Angela A. Alexander-Bryant
Department of Bioengineering, Clemson University, Clemson, and Department of Craniofacial Biology, Medical University of South Carolina, Charleston, South Carolina, USA

Sujit K. Bhutia
Department of Life Science, National Institute of Technology Rourkela, Rourkela, Odisha, India

Durgesh Nandini Das
Department of Life Science, National Institute of Technology Rourkela, Rourkela, Odisha, India

Swadesh K. Das
Department of Human and Molecular Genetics, and VCU Institute of Molecular Medicine, Virginia Commonwealth University School of Medicine, Richmond, Virginia, USA

Paul Dent
Department of Neurosurgery; VCU Institute of Molecular Medicine, and VCU Massey Cancer Center, Virginia Commonwealth University School of Medicine, Richmond, Virginia, USA

Paul B. Fisher
VCU Institute of Molecular Medicine; VCU Massey Cancer Center, and Department of Human and Molecular Genetics, Virginia Commonwealth University School of Medicine, Richmond, Virginia, USA

Tapas K. Maiti
Department of Biotechnology, Indian Institute of Technology Kharagpur, Kharagpur, West Bengal, India

Mahitosh Mandal
School of Medical Science and Technology, Indian Institute of Technology Kharagpur, Kharagpur, West Bengal, India

Subhadip Mukhopadhyay
Department of Life Science, National Institute of Technology Rourkela, Rourkela, Odisha, India

Prashanta Kumar Panda
Department of Life Science, National Institute of Technology Rourkela, Rourkela, Odisha, India

Samir K. Patra
Department of Life Science, National Institute of Technology Rourkela, Rourkela, Odisha, India

Devanand Sarkar
VCU Institute of Molecular Medicine; VCU Massey Cancer Center, and Department of Human and Molecular Genetics, Virginia Commonwealth University School of Medicine, Richmond, Virginia, USA

Niharika Sinha
Department of Life Science, National Institute of Technology Rourkela, Rourkela, Odisha, India

Wendy S. Vanden Berg-Foels
Department of Bioengineering, Clemson University, Clemson; Department of Craniofacial Biology, Medical University of South Carolina, Charleston, South Carolina, and Department of Chemical and Life Science Engineering, Virginia Commonwealth University, Richmond, Virgina, USA

Xiang-Yang Wang
VCU Institute of Molecular Medicine; VCU Massey Cancer Center, and Department of Human and Molecular Genetics, Virginia Commonwealth University School of Medicine, Richmond, Virginia, USA

Xuejun Wen
Department of Bioengineering, Clemson University, Clemson; Department of Craniofacial Biology; Department of Chemical and Life Science Engineering, Virginia Commonwealth University, Richmond, Virginia; Department of Regenerative Medicine and Cell Biology; Department of Orthopedic Surgery; Institute for Biomedical Engineering and Nanotechnology, Tongji University School of Medicine, Shanghai, China; Hollings Cancer Center, and College of Dental Medicine, Medical University of South Carolina, Charleston, South Carolina, USA

Inken Wierstra
Wiβmannstr. 17, D–30173 Hannover, Germany

CHAPTER ONE

Bioengineering Strategies for Designing Targeted Cancer Therapies

Angela A. Alexander-Bryant[*,†,1], Wendy S. Vanden Berg-Foels[*,†,‡,1], Xuejun Wen[*,†,‡,§,¶,||,#,**,2]

[*]Department of Bioengineering, Clemson University, Clemson, South Carolina, USA
[†]Department of Craniofacial Biology, Medical University of South Carolina, Charleston, South Carolina, USA
[‡]Department of Chemical and Life Science Engineering, Virginia Commonwealth University, Richmond, Virginia, USA
[§]Department of Regenerative Medicine and Cell Biology, Medical University of South Carolina, Charleston, South Carolina, USA
[¶]Department of Orthopedic Surgery, Medical University of South Carolina, Charleston, South Carolina, USA
[||]Institute for Biomedical Engineering and Nanotechnology, Tongji University School of Medicine, Shanghai, China
[#]Hollings Cancer Center, Medical University of South Carolina, Charleston, South Carolina, USA
[**]College of Dental Medicine, Medical University of South Carolina, Charleston, South Carolina, USA
[1]These authors contributed equally to this work.
[2]Corresponding author: e-mail address: xuejun@musc.edu; xjwen@clemson.edu; xwen@vcu.edu

Contents

1. Introduction   2
2. Cancer-Targeting Mechanisms   4
   2.1 Passive targeting   5
   2.2 Active targeting   7
3. Design Criteria for Cancer Targeting   9
   3.1 Opsonization and the mononuclear phagocytic system   9
   3.2 Extravasation   10
   3.3 Drug uptake and release   11
   3.4 Multidrug resistance   12
   3.5 Particle elimination   12
4. Delivery Vehicles   13
   4.1 Liposomes   13
   4.2 Micelles   15
   4.3 Dendrimers   17
   4.4 Polymeric particles   20
   4.5 Carbon nanotubes   22
5. Targeting Moieties   23
   5.1 Folate   24
   5.2 Transferrin   25
   5.3 Epidermal growth factor   26
   5.4 Peptides   27
   5.5 Aptamers   30

5.6 Monoclonal antibodies	31
5.7 Vascular endothelial growth factor	32
6. Drug Release	33
6.1 pH-responsive drug release	34
6.2 Enzyme-responsive drug release	36
6.3 Thermoresponsive drug release	37
7. Future Directions in Cancer Targeting	39
7.1 Multifunctional delivery systems	39
7.2 Dual ligand targeting	41
8. Concluding Remarks	42
Acknowledgments	44
References	44

Abstract

The goals of bioengineering strategies for targeted cancer therapies are (1) to deliver a high dose of an anticancer drug directly to a cancer tumor, (2) to enhance drug uptake by malignant cells, and (3) to minimize drug uptake by nonmalignant cells. Effective cancer-targeting therapies will require both passive- and active-targeting strategies and a thorough understanding of physiologic barriers to targeted drug delivery. Designing a targeted therapy includes the selection and optimization of a nanoparticle delivery vehicle for passive accumulation in tumors, a targeting moiety for active receptor-mediated uptake, and stimuli-responsive polymers for control of drug release. The future direction of cancer targeting is a combinatorial approach, in which targeting therapies are designed to use multiple-targeting strategies. The combinatorial approach will enable combination therapy for delivery of multiple drugs and dual ligand targeting to improve targeting specificity. Targeted cancer treatments in development and the new combinatorial approaches show promise for improving targeted anticancer drug delivery and improving treatment outcomes.

1. INTRODUCTION

Cancer is the leading cause of death worldwide (Lovett, Liang, & Mackey, 2012) and the second leading cause of death in the United States, accounting for one in every four deaths (Siegel, Naishadham, & Jemal, 2012). The American Cancer Society reported in 2012 that over one million new cancer diagnoses and half a million cancer deaths are recorded each year. The National Institutes of Health estimated that $103.8 billion was spent on direct health care costs for cancer treatment in 2007 (ACS, 2012). Advances in cancer detection and treatment have led to a decline in cancer deaths by one percent per year over the past decade; however, survival rates for several types of cancer remain low. The lowest survival rates have been recorded for cancer of the esophagus (17%), liver (14%), lung and bronchus (16%), stomach (26%), brain (35%), and pancreas (6%) (Howlader et al., 2011). Survival rates for head and neck cancer, which are currently 40–50%, have not significantly improved over the past few decades (Leemans, Braakhuis, & Brakenhoff, 2011). Patients diagnosed with these cancer types may benefit from new, targeted approaches to cancer therapy.

Radiation and chemotherapy are standards of care for cancer treatment; however, traditional radiation and chemotherapy have many limitations. Although radiation therapy is focused on the cancer tumor, this therapy risks severe damage to nonmalignant tissues that are in the path of the radiation beam (Shepard, Ferris, Olivera, & Mackie, 1999). Radiation also has limited effectiveness in treating metastasized cancers because it requires the detection and treatment of each tumor. Chemotherapy is a systemic treatment that typically targets highly proliferative cells. Systemic delivery exposes all cells to the drug. This lack of specificity also results in damage to highly proliferative nonmalignant cells, such as bone marrow, gonads, gastrointestinal mucosa, and hair follicles (Corrie, 2008), resulting in acute complications and systemic toxicity (Liu, Miyoshi, & Nakamura, 2007; Sahoo & Labhasetwar, 2003). Further, nonspecific uptake of the chemotherapy drug by nonmalignant cells reduces the dose delivered to the target malignant cells, and as a result, higher doses of the cytotoxic drugs must be administered systemically to achieve treatment efficacy (Yotsumoto et al., 2009). Traditional chemotherapy is also ineffective in overcoming multidrug resistance, a condition in which cancer cells become resistant to anticancer drugs. Although the traditional radiation and chemotherapy treatments can successfully fight cancer, there is an urgent need for a more targeted approach that will increase treatment efficacy and reduce treatment side effects.

The goal of targeted cancer therapy is (1) to deliver a high dose of an anticancer drug directly to the site of a tumor, (2) to enhance drug uptake by malignant cells, and (3) to minimize drug uptake by nonmalignant cells. The general approach for designing targeted cancer therapies is to design the drug delivery system to exploit the features that are unique to tumor cells and tumor tissues. Targeted delivery research has focused on unique features of the tumor microenvironment, such as leaky vasculature, overexpressed cell surface receptors, and intratumoral pH differences, as well as features of the cell uptake process, such as endosomal pH. Innovation in micro- and nanotechnology has led to the development of micro- and nanoparticles, such as liposomes and micelles which are particle shaped carriers that can encapsulate and deliver drugs (Egusquiaguirre, Igartua, Hernández, & Pedraz, 2012; Gong, Chen, Zheng, Wang, & Wang, 2012; Kedar, Phutane, Shidhaye, & Kadam, 2010; Malam, Loizidou, & Seifalian, 2009). Nanoparticles are typically defined as particles that are less than 100 nm in size. Some delivery vehicle types discussed may be fabricated at both micro- and nano-scales. Due to the inherent physiologic barriers, particles used for targeted cancer therapy are predominantly nanoscale. We have used the term nanoparticle in reference to any particle type or particle research that is at the nanoscale. These particles have been shown to enhance drug delivery to malignant cells. Advances in cancer research have identified receptors that are overexpressed in various types of cancer. Ligands that bind overexpressed receptors have been successfully used to target malignant cells (Byrne, Betancourt, & Brannon-Peppas, 2008; Das, Mohanty, & Sahoo, 2009; Yu, Tai, Xue, Lee, & Lee, 2010). Stimuli-responsive polymers have also been developed to control release of chemotherapy drugs in response to environmental triggers (Cabane, Zhang, Langowska, Palivan, & Meier, 2012).

Advances in cancer research in combination with advances in biomaterials and nanotechnology have enabled the development of targeted anticancer drug delivery and a more tailored approach to treating individual cancer types. A multidisciplinary approach that includes cancer biology, biomaterials, and nanotechnology has the potential to improve treatment outcomes while minimizing harmful side effects. The design of an effective targeted therapy will require optimization of therapeutic particles, cancer cell targeting, and drug release mechanisms. Both passive- and active-targeting mechanisms may be utilized to enhance targeted delivery. Physical properties of particles can be modified to decrease toxicity to nonmalignant cells and increase circulation time. Targeting moieties, ligands that bind to receptors overexpressed on malignant cells, can be conjugated to particles to increase cellular uptake, and as a result, enhance treatment efficacy. Environmentally responsive polymers can be used to achieve controlled release under defined conditions. This review discusses targeted cancer therapies currently under development and focuses on the design and optimization of individual components required to achieve effective cancer treatment.

2. CANCER-TARGETING MECHANISMS

Cancer targeting can be divided into two general categories: signaling and delivery. Traditional chemotherapy drugs can be thought of as signal transduction therapeutics because they target and inhibit signal transduction essential for tumor survival. Targets for signaling inhibitors include cell proliferation, cell survival, angiogenesis, and nuclear factors (Klein & Levitzki, 2007). These drugs target cellular processes; they do not specifically target malignant cells. A more recent therapeutic approach to cancer targeting is gene therapy, which involves the delivery of DNA or RNA to cancer cells (Ali, Urbinati, Raouane, & Massaad-Massade, 2012; Gao, Xiao, et al., 2010). There are several strategies for cancer treatment using gene therapy, such as blocking expression of oncogenes or insertion of a suicide gene or suppressor gene in tumor cells. However, similar to chemotherapy, delivery of RNA or DNA alone does not specifically target malignant cells, and cell transfection is inefficient. Targeted delivery, on the other hand, is the process by which carriers specifically deliver a drug to malignant cells, while avoiding delivery to nonmalignant cells. In this review, the terms drugs and anticancer drugs are broadly defined to include chemotherapy agents, DNA or RNA for gene therapy, recombinant proteins, or any other therapeutic that can be delivered by particles for cancer therapy. Targeted delivery is essential to improve cancer treatment efficacy and reduce side effects of anticancer drugs. Particles, ligands, and controlled release mechanisms can be used in combination to design a drug–particle complex that achieves targeted delivery. To design an effective targeted cancer therapy, it is essential to understand the two primary mechanisms of cancer targeting: passive targeting and active targeting.

2.1. Passive targeting

Passive targeting, first described in 1986, takes advantage of the greater vascular permeability and poor lymphatic drainage of tumors that result in the accumulation of micro- and nano-particles in tumor tissue (Matsumura & Maeda, 1986). The particles accumulate through passive diffusion, a phenomenon known as the enhanced permeability and retention (EPR) effect (Maeda, 2001; Maeda, Bharate, & Daruwalla, 2009; Maeda, Sawa, & Konno, 2001).

Enhanced permeability of the EPR effect is the result of the leaky vasculature in tumor tissue. Vessels in tumors are irregularly shaped, leaky, and dilated due to rapid growth and abnormal blood flow (Iyer, Khaled, Fang, & Maeda, 2006). Endothelial junctions, gaps in the endothelium that mediate passage of macromolecules from the blood to tissue, vary between nonmalignant and malignant tissue. In normal vasculature, endothelial junctions between cells are narrow, ranging from 5 to 10 nm in width (Haley & Frenkel, 2008). However, in tumor tissue, these junctions range from 100 to 780 nm depending on the tumor type (Hobbs et al., 1998; Rubin & Casarett, 1966; Shubik, 1982). These large gaps allow extravasation of particles out of circulation and into the tumor tissue (Fig. 1.1A).

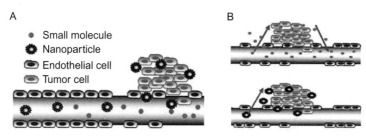

Figure 1.1 Passive targeting. (A) Nanoparticles accumulate in tumor tissue due to the enhanced permeability and retention (EPR) effect. Enhanced permeability is due to large endothelial gaps that result in leaky vasculature. Enhanced retention is due to poor lymphatic drainage. (B) Smaller particles are able to quickly enter and exit tumor tissue through large endothelial gaps, while larger particles diffuse more slowly, resulting in the accumulation of a greater number of particles and drug. *Figure modified and used with permission from Danhier, Feron, and Preat (2010).*

Enhanced retention of the EPR effect is the result of poor lymphatic drainage of cancer tumors. Lymphatic vessels are responsible for draining fluid from the tissue and returning it to the vascular system. However, tumor tissues either lack lymphatic vessels or the vessels they contain are nonfunctional, resulting in the retention of macromolecules in the tumor interstitium (Iyer et al., 2006). Poor lymphatic drainage allows particles that extravasate into tumor vasculature to be retained in the tissue. Thus, the EPR effect is a passive mechanism that enables the accumulation of particle-encapsulated drugs in tumor tissue.

Particle size must be tailored to enable permeation of particles into tumor tissue and their retention within the tissue. Very small nanoparticles (<20 nm) are able to quickly diffuse out of the vasculature through the large endothelial gaps. However, very small nanoparticles are also less likely to be retained because they can just as easily diffuse out of the tumor tissue (Perrault, Walkey, Jennings, Fischer, & Chan, 2009). Larger nanoparticles (50–100 nm) have longer circulation times, allowing sufficient time for the nanoparticles to travel to, and diffuse into, the tumor tissue (Liu, Mori, & Huang, 1992; Perrault et al., 2009). Larger particles diffuse out of tumor tissue very slowly, and as a result, they accumulate in tumors (Fig. 1.1B).

The EPR effect has become a gold standard in the design of cancer therapeutics (Maeda, 2001) because particles can be specifically designed to take advantage of passive targeting. By using passive targeting, anticancer drugs encapsulated in particles will accumulate in tumor tissue, reaching concentrations 10–100 times higher than that of free drug alone (Sinha, Kim,

Nie, & Shin, 2006). The EPR effect also promotes prolonged drug retention (Maeda et al., 2009) due to poor lymphatic drainage.

Although passive targeting promotes drug accumulation in cancer tumors, there are also limitations to this approach. Results of passive targeting are not consistent because tumor vascularization and angiogenesis vary for different cancer types. Consequently, particle concentrations achieved will vary with tumor type and site. For example, the EPR effect is not observed in hypovascular tumors, which are common in prostate and pancreatic cancers (Maeda et al., 2009); therefore, passive targeting alone will not promote particle accumulation in these tumors. The EPR effect can be enhanced in hypovascular tumors by using chemotherapeutic agents, such as nitroglycerin (Seki, Fang, & Maeda, 2009) and S-1, a prodrug of 5-fluorouracil (Nakamura et al., 2011), to alter the tumor microenvironment (Asai, 2012). Particle accumulation in tumors is also influenced by particle size. The EPR effect favors accumulation and retention of larger particles, and particles greater than 100 nm in size (Danhier et al., 2010); smaller particles in nanoscale will be retained to a lesser degree by passive targeting, which may result in nonspecific delivery of anticancer drugs to nonmalignant cells. Passive targeting results in higher concentrations of particles in tumor tissue; however, it does not eliminate nonspecific uptake of particles by nonmalignant tissue. When nonspecific uptake occurs, higher doses of particles must be administered systemically to achieve a therapeutic dose within the tumor.

2.2. Active targeting

Active targeting uses ligands to specifically target receptors that are overexpressed on malignant cells. Ligands are molecules, such as folate, transferrin, epidermal growth factor (EGF), and aptamers, which bind to receptors on the surface of a cell. Ligands are conjugated to anticancer drugs or particle-encapsulated drugs to target malignant cells or tumor endothelium. Conjugation is the physical or chemical attachment of a ligand directly to an anticancer drug or attachment to a particle encapsulating an anticancer drug. Ligand candidates for cancer treatment target receptors that are overexpressed on malignant cells. The folate receptor (FR) and the epidermal growth factor receptor (EGFR) are two examples of receptors that are overexpressed on many types of malignant cells. Therefore, conjugation of these ligands to drugs or particles will result in receptor-mediated active targeting and higher drug or particle concentration in malignant cells than in nonmalignant cells.

Active targeting promotes internalization of ligand-conjugated drug carriers into a cell via receptor-mediated endocytosis (Fig. 1.2). The drug may be released either at the surface of the cell or upon internalization. The ligand-conjugated particle and receptor are first internalized via invagination, and then an endosome is formed. The anticancer drug must escape the endosome before it fuses with the lysosome to avoid being damaged or destroyed by lysosomal enzymes. After release of the drug and receptor from the endosome, some receptors are recycled back to the surface of the cell where they will be available for another cycle of endocytosis (Mohanty, Das, Kanwar, & Sahoo, 2011).

The active-targeting approach addresses many of the goals for improving cancer therapies. Ligands conjugated to cancer drugs often help protect the cancer drug from degradation and enhance the physical and chemical stability of the drug (Mohanty et al., 2011). Ligand binding also increases the drug dose delivered to malignant cells, which permits systemic administration of smaller doses. Particle internalization that occurs by active targeting has been shown to enhance therapeutic effects (Iinuma et al., 2002; Kobayashi et al., 2007), an important advantage over passive targeting. However, active targeting alone will not achieve optimal results. If an anticancer drug is delivered systemically, the ligand-conjugated drug or drug–particle complex must first reach the cancer tumor before the advantages of active targeting can be realized. Passive-targeting mechanisms using the EPR effect are still necessary for extravasation and drug or particle accumulation in tumor tissue. Therefore, it is necessary to use a combination of active and passive targeting in designing drug carriers to improve targeted delivery of cancer therapeutics.

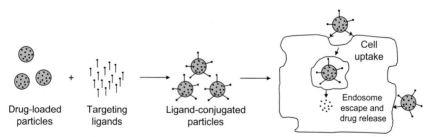

Figure 1.2 Active targeting. Active targeting of particles is achieved by conjugation of ligands that target overexpressed receptors on the surface of cancer cells. Ligands bind to overexpressed receptors on malignant cells and are endocytosed. Anticancer drugs must then escape the endosome to avoid degradation and to perform their function within the cell.

3. DESIGN CRITERIA FOR CANCER TARGETING

The design goals for effective cancer-targeting therapeutics include (1) selective targeting of malignant cells, (2) release of anticancer drugs at the target site, and (3) elimination of cytotoxicity to nonmalignant tissue. When cancer drugs enter the body, they face many physiological barriers that can prevent them from reaching the target site and achieving these design goals. These physiological barriers dictate the design parameters for targeted cancer treatments. A thorough understanding of these barriers is necessary for the development of effective targeting. Particles provide a flexible platform for targeted treatment development. Particle selection, size, surface modification, ligand conjugation, and controlled release mechanisms all contribute to overcoming physiological barriers to targeted delivery and to enhancing therapeutic outcomes (Table 1.1).

3.1. Opsonization and the mononuclear phagocytic system

When cancer drugs are introduced into the circulatory system, they first encounter the blood. Proteins in the blood called opsonins are absorbed onto the surface of the drug or particle in a process called opsonization. Opsonized drugs or particles are recognized by macrophages of the mononuclear phagocytic system (MPS), which trigger removal of drugs or particles from circulation, preventing them from reaching the cancer tissue (Owens & Peppas, 2006). An advantage of the drug–particle complex over unconjugated free drug is that it can be designed to reduce elimination by the MPS.

Surface hydrophobicity determines the amount of opsonins absorbed onto the surface of a drug or drug–particle complex, which will ultimately determine the fate of the drug. Hydrophilic surfaces repel plasma proteins that cause opsonization, thus preventing particle elimination by the MPS (Brigger, Dubernet, & Couvreur, 2002; Moghimi, Hunter, & Murray, 2005). "Stealth" particles, created by pegylation through covalent attachment of polyethylene glycol (PEG), a nonionic hydrophilic biocompatible polymer, are used to improve biocompatibility and prevent elimination by the MPS (Storm, Belliot, Daemen, & Lasic, 1995). Hydrophilic polymers can also be used to coat the surface of cytotoxic materials to reduce their toxicity (Jokerst, Lobovkina, Zare, & Gambhir, 2011). Hydrophobic particle surfaces are not beneficial for drug delivery because they promote particle agglomeration, resulting in rapid removal by the MPS (Veiseh, Gunn, & Zhang, 2010).

Table 1.1 Bioengineering design strategies to overcome physiologic barriers to targeted delivery

Barrier	Design feature	Strategy
Opsonization and MPS	Hydrophobicity	Increase hydrophilicity
	Size	Decrease particle size (<200 nm)
Extravasation	Surface charge	Use cationic particles to increase extravasation and uptake
	Size	Tailor particle size to endothelial junction size of tumor tissue (<200 nm, with enhanced tissue penetration and retention at ~50–100 nm)
Drug uptake and release	Ligand conjugation	Conjugate ligands to particles that target overexpressed receptors on malignant cells. Use multiple ligands and oblique particles to increase binding avidity
	Stimuli-responsive element	Use pH, temperature, or enzyme-responsive polymers or linkers to trigger release
Multidrug resistance	Particle encapsulation	Encapsulate therapeutics using particles such as micelles, liposomes, polymeric particles, dendrimers, or carbon nanotubes
Particle elimination	Biodegradable	Use biodegradable materials to avoid accumulation in the liver and spleen
	Size	Particles <20 nm are excreted, while larger, nondegradable particles accumulate in the liver and spleen

Particles must be small enough to avoid uptake by the MPS (Cho, Wang, Nie, Chen, & Shin, 2008). Large particles (>200 nm) are quickly recognized by the MPS and are removed from circulation (Albanese, Tang, & Chan, 2012). Small (<100 nm) hydrophilic nanoparticles have less interaction with the MPS and reach their cellular targets in higher numbers (Dong & Feng, 2004; Park, Lee, & Lee, 2005). A particle will have a longer circulation time, and thus a higher probability of reaching its target, if it avoids removal by the MPS (Kumari, Yadav, & Yadav, 2009).

3.2. Extravasation

After a drug–particle complex successfully avoids uptake by the MPS, it must then extravasate out of circulation into the target tissue. However, to enter the tissue, it must first traverse endothelial junctions, as described in Section 2.1. While particles must be small enough to traverse the endothelial junctions, they must also be large enough to be retained in the leaky tumor vasculature (Cho et al., 2008). There are also size limitations specific to the type of target site. Endothelial junctions in tumor vasculature range from 100 to 780 nm (Hobbs et al., 1998; Rubin & Casarett, 1966; Shubik, 1982), while nanoparticles with a hydrodynamic size (size in solution accounting for movement and viscosity) of 15–50 nm are able to cross the blood–brain barrier (Sonavane, Tomoda, & Makino, 2008). Diffusion across the blood–brain barrier requires a low-molecular weight lipid-soluble carrier (Koo et al., 2006; Pardridge, 2002).

Surface charge influences particle–cell interactions and particle cytotoxicity. The plasma membrane of cells is naturally negatively charged. Therefore, cationic carriers have more electrostatic interaction with the cell membrane, which results in increased extravasation and cell uptake (Nigavekar et al., 2004). However, this increased interaction also results in greater toxicity and nonspecific uptake by nonmalignant cells (Merdan, Kopecek, & Kissel, 2002). Nonspecific accumulation of cationic particles is typically found in the liver, kidney, spleen, and pancreas (Roberts, Bhalgat, & Zera, 1996). Electrostatic interactions of cationic particles with endothelial cells also shorten the particle circulation time (Lee, MacKay, Frechet, & Szoka, 2005; Malik et al., 2000; Wijagkanalan, Kawakami, & Hashida, 2011). Anionic and neutral particles are nontoxic to cells (Fant et al., 2010). However, anionic and neutral particles have reduced electrostatic interactions with cell membranes, which results in lower cellular uptake, decreased ability to bind to vascular walls (Boas & Heegaard, 2004), and increased circulation time (Kukowska-Latallo et al., 2005).

3.3. Drug uptake and release

After the therapeutic agent reaches the target tissue, receptor binding/uptake and drug release are required to achieve a therapeutic effect. As described in Section 2.2, cell binding and uptake are facilitated through receptors on the cell surface. The level of overexpression of individual receptors on malignant cells will determine which ligand(s) are best suited for conjugation to the particle for active targeting. Ligand conjugation also decreases toxicity by preventing binding/uptake of particles by cells in nonmalignant tissues where the cancer drug has been nonspecifically distributed through passive mechanisms (Vega-Villa et al., 2008). Ligands and other moieties are discussed in detail in Section 5.

Particle shape influences ligand-binding avidity, which is the bond strength of multiple ligand–receptor interactions. Ligand-conjugated oblique particles, which have a flatter surface, have enhanced binding avidity relative to spherical particles (Decuzzi & Ferrari, 2006) due to increased multivalent interactions. A flatter particle surface is able to present a greater number of ligands to the malignant cell surface, increasing interaction and enhancing binding strength.

Drug release mechanisms of particles can be designed to decrease toxicity. If the particle is stimuli-responsive, nonmalignant cells can be protected from exposure. For example, polymers can be designed to release drug when exposed to either the acidic intratumoral environment or the acidic intracellular environment of the endosome. Thermoresponsive polymers such as poly(N-isopropyl acrylamide) (pNIPAM) can be used to trigger release when heat is applied to the tumor region (Peng et al., 2011). Drug release mechanisms are discussed in detail in Section 6.

3.4. Multidrug resistance

A major barrier to effective cancer chemotherapy is multidrug resistance, in which malignant cells become resistant to chemotherapy drugs. The mechanism of multidrug resistance is associated with at least two proteins found in the cell membrane of malignant cells, p-glycoprotein and multidrug resistance protein (Leslie, Deeley, & Cole, 2005). These proteins function as transmembrane transporters, or efflux pumps, that recognize and bind to drugs as they enter the cell membrane. Drug binding activates an adenosine triphosphate (ATP) binding domain, and hydrolysis of ATP results in a shape change of p-glycoprotein and release of the drug into the extracellular space (Gottesman, 2002). One of the key advantages of particle encapsulation of chemotherapy drugs is the ability to combat multidrug resistance. Particle-encapsulated drugs accumulate in cells because they are not recognized by the drug-expelling pumps (Cho et al., 2008), thus avoiding multidrug resistance and improving treatment efficacy. Particle-encapsulated drugs are transported into the cell in endosomal vesicles where they are released out of the reach of the drug efflux pumps, increasing the likelihood that they will reach their target (Hu & Zhang, 2009).

3.5. Particle elimination

The mechanisms of particle elimination from the body also affect particle toxicity. Biodegradable particles do not accumulate and disrupt cellular processes because they are degraded and metabolized (Haley & Frenkel, 2008). If the particle is not degradable, particle size affects the way it is eliminated. Nanoparticles with a molecular weight of <5000 (Owens & Peppas, 2006) and a size of <20 nm (Banerjee, Mitra, Kumar Singh, Kumar Sharma, & Maitra, 2002) are excreted by the renal system. Larger particles that are nondegradable accumulate in the spleen and liver where they are eliminated by the MPS (Albanese et al., 2012). Particle elimination is also affected by particle shape. Anisotropically shaped (nonspherical) particles have a lower bioelimination rate, possibly due to decreased interaction with phagocytes (Geng et al., 2007).

4. DELIVERY VEHICLES

Particle delivery vehicles are an important and promising tool for the design of targeted cancer therapies. Particle delivery vehicles aid in drug delivery by protecting the surrounding environment from nonspecific effects. Particle features, including size, shape, surface charge, and hydrophobicity, affect drug delivery and vary among the particle types. Each of these features must be considered when selecting and optimizing a particle for targeted delivery of cancer drugs. There are already several FDA-approved cancer drugs that use particle drug encapsulation technology. Particles discussed in this section include liposomes, micelles, dendrimers, polymeric particles, and carbon nanotubes (CNTs).

4.1. Liposomes

Liposomes are composed of a bilayer structure of either natural or synthetic phospholipids. Phospholipids are a key structural component of the cell membrane. They are amphiphilic in nature, possessing both hydrophilic and hydrophobic regions. Amphiphilic phospholipids self-assemble into bilayers (Malam et al., 2009) by arranging their hydrophilic groups outward to interact with aqueous environments and arranging their lipophilic groups toward the center of the bilayer (Fig. 1.3). Liposomes can be unilamellar (Fig. 1.3A) or multilamellar (Fig. 1.3B) depending on the number of lipid bilayers formed (Haley & Frenkel, 2008).

Liposome surface properties can be easily manipulated for drug delivery. Surface charge can be modified by adjusting the lipid composition to add either more neutral or cationic lipids, which directly influences liposome interactions with the negatively charged cell membrane. Neutral liposomes have no significant cell membrane interaction, and the neutral charge results

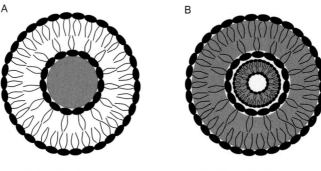

Figure 1.3 Structure of liposomes. (A) Unilamellar liposomes contain a large aqueous core for storage of water-soluble drugs. (B) Multilamellar liposomes are composed of multiple layers of phospholipid groups for storage of lipid-soluble drugs between the hydrophobic tail groups in each layer.

in liposome aggregation (Sharma & Sharma, 1997). Aggregation is a key issue for drug delivery because particle aggregates are rapidly cleared by the MPS, which greatly reduces drug delivery. Anionic liposomes are internalized through clathrin-mediated endocytosis (Straubinger, Hong, Friend, & Papahadjopoulos, 1983), while cationic liposomes deliver their contents by membrane fusion (Felgner et al., 1994) and by endocytosis. However, at high doses, cationic liposomes cause toxicity and nonspecific uptake, which harms nonmalignant cells.

Liposomes can be used to carry water- or lipid-soluble drugs. Unilamellar liposomes have an aqueous core that is used to carry water-soluble drugs, while multilamellar liposomes have lipophilic layers between hydrophobic tail groups that are used to carry lipid-soluble drugs. There are currently four liposome-based cancer drugs available on the market, Doxil, DaunoXome, LipoDox, and Myocet, and many others are in clinical trials (Chang & Yeh, 2012). The four available drug formulations are all unilamellar liposomes containing water-soluble drugs. Doxil and its second generation drug, LipoDox, both contain doxorubicin encapsulated within a liposome with a PEG-modified surface. Doxil has been proven effective in treating drug-resistant tumors in clinical trials (Chou et al., 2006). Both Doxil and LipoDox have been used successfully to treat many cancers including Kaposi's sarcoma, ovarian cancer, and metastatic breast cancer (Tejada-Berges, Granai, Gordinier, & Gajewski, 2002). Myocet is an unpegylated liposomal doxorubicin approved for use in Canada and Europe to treat metastatic breast cancer (Lorusso, Manzione, & Silvestris, 2007). DaunoXome, which is a pegylated liposomal daunorubicin, has been approved to treat blood tumors (Chang & Yeh, 2012).

Liposomes have advantages and disadvantages to consider when developing targeted drug delivery therapies. Liposomes can be used to encapsulate hydrophilic and hydrophobic drugs. Liposomes also have the potential to treat multidrug resistant cancers. However, liposomes have poor control over drug release, often dispersing the loaded drug in a rapid burst release. Liposomes also exhibit low-drug encapsulation efficiency due to poor solubility of many drugs in solution (Das et al., 2009), and as a result, they are only useful for very potent drugs (Alexis, Pridgen, Langer, & Farokhzad, 2010). Liposomes are physically unstable; the bilayer structure can quickly disintegrate in response to hydrophobic, electrostatic, and van der Waals forces (Haley & Frenkel, 2008), resulting in particle aggregation, drug leakage, and a reduced shelf life (Sharma & Sharma, 1997). Liposomes measure up to 400 nm in size, and as a result, the larger sized liposomes are rapidly cleared by the MPS (Malam et al., 2009). Pegylation of the liposome surface reduces opsonization and clearance by the MPS. Liposomes also cause severe side effects due to nonspecific uptake in skin tissue.

4.2. Micelles

Micelles are composed of amphiphilic macromolecules that contain both hydrophilic and hydrophobic segments (Sutton, Nasongkla, Blanco, & Gao, 2007). Depending on the size of these segments, micelles with various morphologies, including spheres, rods, tubules, lamellae, and vesicles, can be created (Choucair & Eisenberg, 2003). A micelle consists of a core and a shell, where hydrophobic end groups form the core and hydrophilic head groups form the outer shell (Fig. 1.4), or vice versa (Haley & Frenkel, 2008). Amphiphilic micelles are formed through self-assembly of unimers with hydrophilic and hydrophobic segments in solution. In aqueous solutions, water-insoluble (hydrophobic) drugs are encapsulated during self-assembly (Kedar et al., 2010). While in nonaqueous solutions, water-soluble (hydrophilic) drug molecules, including proteins, peptides, and nucleic acids, are encapsulated during assembly (Allen, 1998; Momekova et al., 2007). Micelle nanoparticles range in size from 5 to 100 nm (Oerlemans et al., 2010). Many different polymers can be used to create micelles; however, the selection is limited for drug delivery applications because the micelle must be biocompatible and biodegradable. PEG is commonly used to fabricate micelles because it is neutral, nontoxic, and water soluble. Other

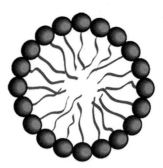

Figure 1.4 Micelle structure. Micelles are composed of phospholipids, with hydrophilic head groups forming the outer shell. Micelles encapsulate water-insoluble drugs in their hydrophobic cores. *Figure modified and used with permission from Husseini and Pitt (2008).*

hydrophilic polymers used include poly(N-vinyl pyrrolidone) (PVP) and pNIPAM (Chung, Yokoyama, Aoyagi, Sakurai, & Okano, 1998; Chung, Yokoyama, & Okano, 2000; Chung et al., 1999). Degradable hydrophobic polyesters are commonly used to make the hydrophobic segment of the amphiphilic macromolecule (Sutton et al., 2007). Micelles are intrinsically stealth particles when formed with a hydrophilic outer shell, and they are able to avoid uptake by the MPS without further modification.

Passive and active targeting can be achieved using micelles. Drug-encapsulated micelle systems increase targeting of tumors due to passive targeting via the EPR effect (Kwon et al., 1994). Polymeric micelles that are not ligand-conjugated accumulate in tumor tissue, and release occurs intratumorally, in the tumor tissue outside of the malignant cells. Ligands can be attached to micelles for active targeting and drug release within the cell. Ligand-conjugated micelles are internalized via pinocytosis (Shuai, Merdan, Schaper, Xi, & Kissel, 2004).

Micelles can be used to deliver water-insoluble drugs in their hydrophobic core (Moghimi et al., 2005). One-third of all drugs used in cancer therapy are hydrophobic (Lee et al., 2012). Amphiphilic copolymers used to fabricate micelles are ideal for delivery of water-insoluble drugs because the outer shell creates a hydrophilic corona to stabilize and protect the hydrophobic drug. Polymeric micelles can increase water solubility of drugs by 10- to 500-fold (Savić, Eisenberg, & Maysinger, 2006), which enables the intravenous injection of micelle-encapsulated hydrophobic drugs. For example, paclitaxel is a water-insoluble drug; however, when encapsulated in a micelle, its water solubility is significantly enhanced (Soga et al., 2005).

Drug loading in micelles is achieved by physical entrapment or chemical conjugation; drug release is dependent on the loading method (Kedar et al., 2010). Drugs loaded by chemical conjugation are released by bulk degradation or surface erosion of the polymer, while drugs loaded by physical entrapment are released by diffusion. Release is also influenced by the extent of cross-linking in the core of the micelle, with cross-linked micelles resulting in slower release, which results in longer release times (Liu, Hsieh, Fan, Yang, & Chung, 2007). Three micelle-based cancer drugs have been approved for clinical trials for treatment of solid tumors: one that delivers doxorubin and two others that deliver paclitaxel (Chen, Khemtong, Yang, Chang, & Gao, 2011).

A key advantage of micelles over other particles is the core–shell structure. The hydrophobic core enables solubilization of hydrophobic drugs in water, and the hydrophilic shell, or corona, protects the drug by preventing elimination by the MPS (Kedar et al., 2010), which enables prolonged circulation (Sutton et al., 2007). Another advantage of micelles is that they generally possess low toxicity and can be eliminated through renal filtration if the molecular weight of the micelle polymer chains is below the critical value for renal filtration (Yokoyama, 2011). Disadvantages of micelles include poor drug-loading efficiency, poor physical stability *in vivo*, and insufficient cellular interaction of neutral micelles with malignant cells for uptake (Kim, Shi, Kim, Park, & Cheng, 2010). Additionally, micelles have a smaller drug-loading capacity than liposomes.

4.3. Dendrimers

Dendrimers are the smallest of the nanoparticles (Svenson & Tomalia, 2005), with a radius ranging from 2.5 to 8 nm (Kaminskas, Boyd, & Porter, 2011). Dendrimers are spherical, with a three-dimensional branched structure, which contains a core, branching repeat units arranged radially from the core, and outer terminal functional groups (Roberts et al., 1996) (Fig. 1.5). Branches can be created by polymerization of a monomer that diverges from the core or converges from the periphery. The size of the dendrimer is easily controlled via polymerization of repeat layers, with each layer making up a generation. Dendrimers can be made cationic, anionic, or neutral depending on their terminal groups. For example, amine-, carboxyl-, or hydroxyl-terminated dendrimers would create cationic, anionic, and neutral surface charges, respectively. Poly(amido amine) (PAMAM) dendrimers are the most widely studied. PAMAM is described as a "starburst dendrimer" because its branched

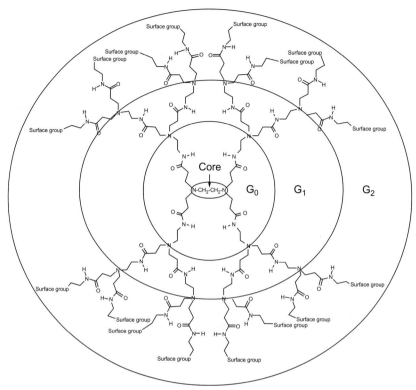

Figure 1.5 Dendrimer branch structure. Dendrimers are created by polymerization of a core monomer to create a branched polymer structure. Each polymerized layer radiates from the core and correlates to the generation of the dendrimer. Various surface functional groups may be used to modify the surface charge of the dendrimer. *Figure used with permission from Kaczorowska and Cooper (2008).*

structure has a star-like pattern (Labieniec et al., 2009). PAMAM dendrimers contain an initiator core of ethane-1,2-diamine with radially extending repeat branching layers. Full generation PAMAM dendrimers are amine-terminated, while half-generations are carboxylic acid-terminated.

Dendrimer size and surface charge directly affect their drug delivery properties. As discussed in Section 3, surface charge influences carrier uptake. Cationic dendrimers tend to be rapidly cleared from circulation due to electrostatic binding to tissue and vasculature, while anionic and neutral dendrimers have longer circulation times. Smaller dendrimers (<25 kDa) are also rapidly cleared from circulation through urinary excretion (Kaminskas et al., 2011). Dendrimers can be pegylated to increase their size, resulting in increased circulation time and decreased clearance by the

MPS. Pegylation or acetylation of cationic dendrimers also reduces their cytotoxicity (Chen, Neerman, Parrish, & Simanek, 2004; Kolhatkar, Kitchens, Swaan, & Ghandehari, 2007). Acetylation, which is the chemical attachment of acetyl groups, reduces the surface charge by shielding cationic groups to create a more neutral dendrimer; the resulting surface charge depends on the degree of acetylation.

Dendrimers have desirable properties for drug conjugation or encapsulation. Drugs can be attached to functional groups on branch ends or encapsulated in dendritic channels within the branches (Hughes, 2005; Moghimi et al., 2005). Incorporating degradable linkages between the drug and the dendrimer controls drug release. Drug loading in the dendritic channels can be manipulated by varying the dendrimer generation. Drug loading is performed by physical entrapment or through interactions between drug molecules and dendritic groups (D'Emanuele & Attwood, 2005). Dendrimers synthesized with hydrophobic cores are able to encapsulate hydrophobic drugs (Frechet, 2002; Patri, Majoros, & Baker, 2002), and electrostatic interactions between the dendrimer and the drug increase drug solubility at high pH (Milhem, Myles, McKeown, Attwood, & D'Emanuele, 2000). Encapsulation of hydrophobic drugs is achieved through hydrophobic interactions between the drug and hydrophobic regions of the polymer when the drug and polymer are mixed. Currently, there are no FDA-approved dendrimer cancer drugs; however, they are promising nanoparticles that are receiving much attention for future anticancer drug delivery.

One of the key advantages of dendrimers is that they are highly customizable. They can be uniformly reproduced with monodisperse sizes during synthesis by controlling the amount of monomer used. End groups can also be tailored to produce neutral, cationic, or anionic dendrimers. Dendrimers have a high drug-loading capacity and may encapsulate both hydrophilic and hydrophobic drugs (Sampathkumar & Yarema, 2007). Further, dendrimers can be functionalized and conjugated with many different therapeutic molecules. Their small size may also help them avoid recognition by the MPS. A possible disadvantage of dendrimers is that steric crowding of the branching arms occurs at higher generation numbers, which limits growth to a larger size (Medina & El-Sayed, 2009) and results in decreased drug encapsulation. A key issue for cationic dendrimer drug delivery systems is the balance between function and surface charge. The positively charged surface promotes cellular uptake and endosome escape, but it also results in toxicity to nonmalignant cells. Surface charge must be balanced to optimize drug delivery and minimize cytotoxicity.

4.4. Polymeric particles

Polymeric particles can be fabricated as nanospheres or nanocapsules (Fig. 1.6). Nanospheres are solid spherical structures composed of a polymer matrix. Drugs can be conjugated to the nanosphere surface or absorbed within the bulk of the polymer matrix. Nanocapsules are vesicular hollow polymers that encapsulate a drug solution within a polymeric membrane. Polymeric particles consist of a polymeric backbone that is typically composed of a biodegradable monomer (Malam et al., 2009). Polymer particles provide flexibility in design because they can either be biodegradable or nonbiodegradable, natural or synthetic (Alexis et al., 2010). Typical natural polymers used to make polymer particles are alginate, albumin, dextran, and chitosan. Commonly used synthetic polymers include PLA/PLGA and PEG block copolymers, which create stealth with longer circulation times (Gref et al., 1994) because they inhibit opsonization and elimination by the MPS.

Conjugation, functionalization, and targeted delivery can be achieved using polymeric particles. The surface of polymeric particles is typically sterically stabilized by using grafting, conjugation, or pegylation to increase repulsion between particles, which prevents their aggregation (Gref et al., 2000). Steric stabilization results in reduced elimination by the MPS and longer circulation times (Gref et al., 2000; Peracchia et al., 1999). Ligands can be conjugated to polymeric particles for active targeting. Polymeric particles can also be coated with other polymers to enhance drug delivery. Hydrophobic particles can be coated with PEG to create stealth particles. Polymeric nanoparticles may also be tailored for delivery across the

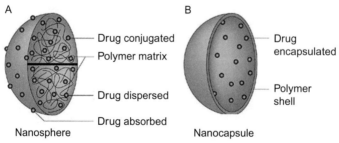

Figure 1.6 Structure of polymeric nanoparticles. (A) Polymer nanospheres are solid structures composed of a polymer matrix. Drugs can be loaded into the matrix by absorption or chemically conjugated to the surface. (B) Polymer nanocapsules are hollow structures containing an outer polymer shell. Drug can be encapsulated in the core of nanocapsules during formation or absorbed into the polymer shell. *Figure modified and used with permission from Griffiths, Nystrom, Sable, and Khuller (2010).*

blood–brain barrier by coating them with polysorbate 80 (Kreuter, 2001). Absorption of apolipoprotein E onto the surface produces a coating that mimics low-density lipoprotein (LDL) and results in transport, via LDL receptors, across the blood–brain barrier (Hans & Lowman, 2002).

Drugs can be incorporated into polymeric particles via conjugation, absorption, and encapsulation. Anticancer drugs can be chemically conjugated to the surface of both nanocapsules and nanospheres using ester or amide bonds (Malam et al., 2009). The drug is released *in vivo* by the hydrolysis of these bonds (Malam et al., 2009). The anticancer drug does not become active until the bond has been hydrolyzed. Drugs can also be absorbed into solid nanospheres and entrapped in the inner polymer matrix. Nanocapsules containing a core and shell can either have drug encapsulated in the core or absorbed into the polymer matrix of the outer shell (Panyam & Labhasetwar, 2003; Parveen, Misra, & Sahoo, 2012). Encapsulation is performed either during particle formation (for capsules) or via adsorption following particle synthesis (for both capsules and spheres).

The mechanism of encapsulation affects the rate of drug release. Drugs loaded using adsorption have a burst effect, in which the bulk of the drug is quickly released, while drugs that are chemically conjugated have a slower controlled release (Fresta et al., 1995). There are a few polymeric particle anticancer formulations in clinical trials for cancer treatment. A cyclodextrin–PEG nanoparticle conjugated with camptothecin, a broad-spectrum anticancer drug, has exhibited significant antitumor activity (Svenson, Wolfgang, Hwang, Ryan, & Eliasof, 2011). A PEG–PLGA particle encapsulating docetaxel, BIND-014, has exhibited antitumor activity in mouse models (Farokhzad, Karp, & Langer, 2006), and it has been approved for clinical trials in patients with advanced or metastatic cancer (Service, 2010).

Polymeric particles have some unique advantages for drug delivery. Particles composed of natural polymers have greater biocompatibility than synthetic polymers and low recognition by the MPS. Like dendrimers, polymeric particles are highly customizable; polymer composition, conjugation, and drug release can each be modified to optimize targeted delivery for a specific cancer type. Biodegradable particles can be used to control drug release and avoid issues with elimination. Polymeric particles also have potential for large-scale production for drug delivery applications using several fabrication techniques (Fang & Zhang, 2011). Maintaining purity and sterility during manufacturing will be imperative for commercialization of polymeric particles for cancer drug delivery (Fang & Zhang, 2011).

4.5. Carbon nanotubes

CNTs are hollow tubes formed from rolled sheets of graphene rings, which are composed of sp2 hybridized carbons (Prakash, Malhotra, Shao, Tomaro-Duchesneau, & Abbasi, 2011; Fig. 1.7). There are two types of CNTs: single-walled carbon nanotubes (SWNTs) and multiwalled carbon nanotubes (MWNTs). SWNTs are made of one layer of graphene and have a diameter of 1–2 nm, while MWNTs are made from multiple coaxially arranged layers of graphene and have diameters from 5 to 100 nm (Liang & Chen, 2010).

CNTs are naturally hydrophobic; however, they can be functionalized via adsorption, electrostatic interaction, or covalent bonding to create a hydrophilic surface (Prakash et al., 2011). CNTs are functionalized through linkages created with the sidewall carbons or ends of the nanotubes. The tips or ends of the CNTs are typically more reactive than the sidewalls (Tasis, Tagmatarchis, Bianco, & Prato, 2006). CNTs can be reacted with acids to purify them and create carboxylic acid groups for further functionalization (Bianco, Kostarelos, & Prato, 2005). Hydrophilically functionalized CNTs exhibit increased tumor tissue accumulation via passive targeting compared to nonfunctionalized CNTs (Pantarotto, Briand, Prato, & Bianco, 2004). CNTs can be conjugated with ligands for active targeting. Water-soluble CNTs can be conjugated with peptides, proteins, nucleic acids, and other water-soluble therapeutic agents for active targeting and drug delivery (Bianco et al., 2005).

CNTs exhibit effective drug loading and release. Binding and release of anticancer drugs from SWNTs can be regulated by pH. CNTs have a high drug-loading capacity, a CNT with a diameter of 80 nm has been shown to contain approximately five million drug molecules (Kam & Dai, 2005).

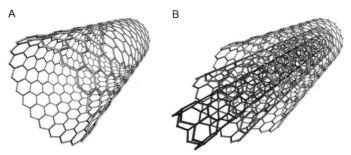

Figure 1.7 Carbon nanotube structure. Carbon nanotubes (CNTs) are hollow tubes composed of graphene sheets. CNTs can be either (A) single-walled or (B) multiwalled.

Drug-loading capacity decreases as pH decreases, and drug release is triggered in acidic conditions; both effects are due to increased hydrophilicity and solubility of the CNT (Liu, Sun, Nakayama-Ratchford, & Dai, 2007). CNTs have not yet been approved for use in clinical trials, but they have shown promising results in drug delivery studies. SWNTs functionalized with platinum(IV)–PEG conjugates increased uptake and toxicity in testicular carcinoma cells *in vitro* compared to free platinum(IV) (Feazell, Nakayama-Ratchford, Dai, & Lippard, 2007). Paclitaxel-conjugated SWNTs exhibited enhanced uptake by malignant cells compared to free drug in a mouse model (Liu et al., 2008).

CNTs have properties that are advantageous for cancer targeting. CNTs have a high drug-loading capacity. Additionally, CNTs are triggered to release their drug load in acidic environments. However, one of the main disadvantages of CNTs is that they may cause inflammation and toxicity even after functionalization (Roda et al., 2011). The CNT purification process is also laborious, and commercially available CNTs are contaminated with particles that make them nonbiocompatible. Noncovalent bonding of drugs can result in inefficient attachment and release of the drug before it reaches its therapeutic target (Gottesman, Fojo, & Bates, 2002; Shi Kam, Jessop, Wender, & Dai, 2004). These limitations must be addressed before CNT-based cancer therapies can advance to clinical trials.

5. TARGETING MOIETIES

Targeting moieties are ligands that bind to receptors that are overexpressed on cancer cells. Conjugating targeting moieties to the surface of particles promotes uptake and intracellular retention of particles by malignant cells, both of which enhance therapeutic efficacy (Sahoo & Labhasetwar, 2005). Active targeting of malignant cells using ligands promotes receptor-mediated endocytosis. After a ligand binds to its corresponding cell surface receptor, the receptor–ligand–particle complex is endocytosed. Active targeting promotes direct cell kill and enhances cytotoxicity of anticancer drugs against malignant cells, while passive targeting promotes accumulation of particles in tumor tissue (Pastorino et al., 2006). Targeting moieties discussed here include folate, transferrin, monoclonal antibodies (MAbs), peptides, EGF, and aptamers.

Ideally, unique cell surface antigens would be expressed exclusively on, and homogeneously among, cancer cells (Danhier et al., 2010). However, receptors overexpressed on cancer cells are also expressed on nonmalignant cells in lower numbers. Therefore, ligands should have high affinity and specificity for overexpressed cell surface receptors (Brumlik et al., 2008). Additional considerations for ligand selection include whether or not the ligand promotes internalization of the anticancer drug or drug–particle complex into malignant cells and the ease of ligand conjugation to the anticancer drug or particle.

5.1. Folate

Folate is the most extensively researched receptor for cancer targeting. Folate, or folic acid, is an essential vitamin necessary for cell function and synthesis of purine and pyrimidine for DNA. The FR is typically overexpressed on malignant cells, and may be upregulated on cancer cells by up to two orders of magnitude (Low & Antony, 2004). FR is overexpressed in lung, head, neck, brain, ovarian, uterine, bone, breast, and renal cancer (Elnakat & Ratnam, 2004; Shmeeda et al., 2006; Vasir & Labhasetwar, 2005). FR is present in a reduced form on most nonmalignant cells as a low-affinity carrier that selectively transports reduced forms of folate into the cell (Hilgenbrink & Low, 2005). Folate uptake on most cells is mediated by either the reduced folate carrier or a proton-coupled folate transporter, neither of which display affinity for folate conjugates (Low & Kularatne, 2009). In contrast, a high-affinity form of FR is overexpressed on many cancer cells and has limited expression on nonmalignant cells. There are two FR isoforms known to be overexpressed in cancer: FR-α is found in epithelial cancer, and FR-β is found in myeloid leukemia and macrophages of chronic inflammatory disease. FR-α is found in low levels in nonmalignant epithelia (Lu & Low, 2002), with the exception of the placenta and the choroid plexus (Sudimack & Lee, 2000). However, FR-α is found on the apical surface of the nonmalignant epithelial, and it is therefore not accessible to folate conjugates on the surface of cancer-targeted delivery vehicles (Low & Kularatne, 2009). FR-β is generally found in low levels in nonepithelial and hematopoietic cells (Lu & Low, 2002).

Folate is easily conjugated to particles, and it has a high-binding affinity for the FR (Low & Antony, 2004; Shmeeda et al., 2006; Stella et al., 2000). PAMAM–folate–methotrexate dendrimers have been shown to decrease toxicity and enhance efficacy 10-fold relative to free methotrexate in mice with human KB tumors (Kukowska-Latallo et al., 2005). KB cells are a keratin-producing cell line established by HeLa cell contamination (Lacroix, 2008). Additionally, liposomes encapsulating doxorubicin and coated with a folate-conjugated poly(L-lysine) polymer exhibited increased toxicity in both human lung and nasopharyngeal cancer cells *in vitro* compared to nontargeted doxorubicin-encapsulated liposomes (Watanabe, Kaneko, & Maitani, 2012). Folate-conjugated particles are also used to target various malignancies for tumor imaging in mouse models (Chen, Li, et al., 2011; Hou et al., 2011).

Folate has many advantages as a targeting ligand. Folate is inexpensive, nonimmunogenic, available in large quantities, and easily conjugated to particles and anticancer drugs (Lee & Low, 1994; Sudimack & Lee, 2000). Folate is small in size and has a high affinity for its receptor. Some folate conjugates are released from the FR in the endosome during acidification (Lee, Wang, & Low, 1996; Wileman, Harding, & Stahl, 1985). Folate-conjugated particles have demonstrated endosomal escape through an unknown mechanism (Hilgenbrink & Low, 2005), preventing damage to the anticancer drug by lysosomal enzymes. Further research on this phenomenon is necessary to elucidate its mechanism. It has also been demonstrated that monovalent FR-containing endosomes are only mildly acidic (pH 6.8), whereas endosomes of multivalent FR-conjugates are more acidic (pH 5) (Lee et al., 1996; Yang, Chen, Vlahov, Cheng, & Low, 2007). The pathway of multivalent folate conjugates can be exploited to trigger drug release using pH-responsive linkers (Low & Kularatne, 2009). FRs also exhibit limited expression on nonmalignant cells. Cells that do express folate are inaccessible to folate–drug conjugates in circulation because they are localized to apical surfaces of polarized epithelia (Hilgenbrink & Low, 2005).

5.2. Transferrin

Transferrin is a glycoprotein that binds iron and transports it into cells via transferrin receptor-mediated endocytosis. Because cancer cells have a high rate of proliferation, they have a greatly increased need for iron to carry out cellular functions, which results in an overexpression of the transferrin receptor (Gatter, Brown, Trowbridge, Woolston, & Mason, 1983). Transferrin receptors are particularly overexpressed on metastasizing and drug-resistant cells. In fact, tumor cells typically overexpress transferrin receptors by 2- to 10-fold compared to nonmalignant cells (Anabousi et al., 2006).

Several transferrin-conjugated drugs have been tested and have shown improved outcomes in clinical trials, including adriamycin, cisplatin, and diphtheria toxin (Faulk, Taylor, Yeh, & McIntyre, 1990; Head, Wang, & Elliott, 1997; Rainov & Soling, 2005). Transferrin–polymer–drug conjugation has been used to treat several types of malignant cells in mouse models (Hu-Lieskovan, Heidel, Bartlett, Davis, & Triche, 2005).

Transferrin-conjugated particles loaded with paclitaxel have exhibited enhanced antiproliferative activity compared to free drug and nontargeted particles due to increased cellular uptake and reduced exocytosis (Sahoo & Labhasetwar, 2005). These particles have proven effective against breast cancer cells *in vitro* (Sahoo & Labhasetwar, 2005) and prostate cancer in mouse models (Sahoo, Ma, & Labhasetwar, 2004). Lipid particles (blend of solid and oil lipid) conjugated with transferrin also exhibited increased uptake in leukemia cells *in vitro* (Khajavinia, Varshosaz, & Dehkordi, 2012). Transferrin-conjugated particles can cross the blood–brain barrier and target rat brain tumor glioma cells in *in vitro* models (Pulkkinen et al., 2008).

Transferrin has several advantages for targeted delivery to cancer cells. Human transferrin is nonimmunogenic; therefore, it can be safely delivered without causing toxicity. Transferrin-conjugated drugs have been shown to prevent cardiotoxicity and drug resistance (Singh, 1999). Transferrin has been used to successfully deliver many cancer drugs. Additionally, the likelihood of transferrin binding to tumor cells is higher compared to nonmalignant cells because transferrin receptors on tumor cells recycle back to the surface more quickly (Singh, 1999). Transferrin can also be easily obtained from human sources. However, there are some disadvantages to using transferrin as a targeting ligand. Delivery of transferrin-conjugated drugs in high doses may result in damage to other cells that express low levels of transferrin receptor (Laske, Youle, & Oldfield, 1997). Transferrin ligands may target liver cells and highly proliferative nonmalignant cells (Daniels, Delgado, Rodriguez, Helguera, & Penichet, 2006). Further, free transferrin in the blood may competitively bind to malignant cell receptors, preventing uptake and limiting efficacy (Xu, Pirollo, & Chang, 1997; Xu et al., 2001).

5.3. Epidermal growth factor

EGFR is a transmembrane receptor that contains an extracellular-binding domain (Tai, Mahato, & Cheng, 2010). EGFR has been found to play a role in the progression of many malignancies (Danhier et al., 2010), and it is indicative of the metastatic capability of a malignant tumor (Radinsky et al., 1995; Verbeek, Adriaansen-Slot, Vroom, Beckers, & Rijksen, 1998). EGFR is overexpressed in many cancer types, including colorectal, lung, head and neck, ovarian, kidney, prostate, and pancreatic cancer (Lurje & Lenz, 2009). EGFR is overexpressed by 100-fold in some cancer cell types (Salomon, Brandt, Ciardiello, & Normanno, 1995). EGFR is actually a family of receptors that includes EGFR, human epidermal growth factor receptor-2 (HER-2), HER-3, and HER-4 (Zwick, Bange, & Ullrich, 2001). HER-2 is overexpressed in some invasive breast cancers and correlates with a poor prognosis (Sakai et al., 1986). EGFR and HER-2 have been the most extensively studied for cancer-targeting applications because their activation leads to rapid growth, differentiation, migration, and survival of cancer cells (Kaptain, Tan, & Chen, 2001; Laskin & Sandler, 2004).

EGF and transforming growth factor-α are both ligands for EGFR (Byrne et al., 2008). MAbs can also be used as ligands to target EGFR, and they are discussed in Section 5.6. EGF-conjugated particles with encapsulated gemcitabine exhibited greater uptake and antiproliferative activity in human breast cancer cells compared to nontargeted particles in mouse tumor models (Sandoval et al., 2012). EGF-conjugated liposomes have been shown to overcome low specificity and inefficient drug accumulation in an *in vitro* model using human epithelial cancer cells and neuronal glioblastoma cells (Carlsson et al., 2003; Kullberg, Nestor, & Gedda, 2003). Additionally, EGF-conjugated gold nanoparticles (GNPs) have been shown to target glioma brain tumors, increasing uptake by 10-fold over untargeted nanoparticles in mice (Cheng et al., 2011).

There are several advantages to using EGF as a targeting ligand. EGF binds to EGFR with high affinity. A key advantage of EGF relative to other ligands is that it enables drug delivery to the cell nucleus. EGFR is able to translocate from the cell surface into the nucleus through an endocytic pathway in highly proliferative cells (Lin et al., 2001; Torrisi et al., 1999; Wang et al., 2001). Internalization of EGF also occurs more quickly than antibody-conjugated anticancer drugs (Fan, Lu, Wu, & Mendelsohn, 1994). A disadvantage of EGF is that it is mitogenic (Kitchens, Snyder, & Gottlieb, 1994; Norman et al., 1987) and therefore may promote cell division.

5.4. Peptides

Peptides are molecules consisting of 2–50 amino acids linked by peptide bonds (Reubi, 2003). Some peptide receptors overexpressed in tumor cells include somatostatin, vasoactive intestinal peptide (VIP), luteinizing hormone-releasing hormone (LHRH), cholecystokinin (CCK), gastrin-releasing peptide, and bombesin receptor. Somatostatin peptides regulate the endocrine system and bind to G-coupled peptide receptors with high affinity (Graff et al., 2005). Somatostatin receptors are overexpressed on many tumor cells and in tumor blood vessels (Honer et al., 2011). Somatostatin is the most commonly used radiopeptide, a radioactively labeled peptide that can be used for diagnosis and treatment of malignances (Graham & Menda, 2011). VIP is a vasodilator that is secreted from some tumors of the pancreas and nervous system (Das et al., 2009). LHRH is a key regulator of reproduction, and it is overexpressed in many cancer types including breast, prostate, endometrial, and ovarian (Kakar et al., 2008). CCK is a digestive hormone, and CCK receptors bind CCK or gastrin peptides. CCK subtype two receptor (CCK-2R) is overexpressed in medullary thyroid carcinomas, neuroendocrine tumors, small cell lung carcinomas, and colorectal cancers (Sosabowski et al., 2009). Bombesin peptides stimulate gastrin release and are overexpressed in small cell carcinoma of the lung, gastric cancer, and neuroblastoma (Ohlsson, Rehfeld, & Axelson, 1999).

Peptides have been used in many tumor-targeting studies. *In vitro*, LHRH peptide-conjugated particles enhanced internalization in cancer cells overexpressing LHRH receptors when compared to nontargeted particles (Taheri et al., 2011). Divalent gastrin peptides, which are composed of two peptides, have been used to target CCK-2Rs. Contrast agent-conjugated gastrin peptide increased binding affinity and tumor uptake of the peptide in mice (Sosabowski et al., 2009). VIP-conjugated micelles loaded with anticancer drug had significantly higher toxicity than non-targeted micelles when targeting VIP receptor-expressing breast cancer cells *in vivo* (Onyuksel, Mohanty, & Rubinstein, 2009). VIP-conjugated liposomes and micelles exhibited increased accumulation in malignant cells compared to nontargeted particles in breast cancer *in situ* tumor models in rats (Dagar, Krishnadas, Rubinstein, Blend, & Onyuksel, 2003; Onyuksel et al., 2009). Somatostatin has been conjugated to many anticancer drugs, and it has displayed significant somatostatin receptor-selective antitumor ability in mouse models (Sun & Coy, 2011). The bombesin analog, a man-made peptide comparable to bombesin, also exhibits specific targeting *in vivo*. The bombesin analog has been used to target gastrin-releasing peptide receptors to treat prostate carcinoma in a mouse model (Honer et al., 2011). A peptide that has been used to treat hypovascular tumors is the Ala-Pro-Arg-Pro-Gly (APRPG) peptide. APRPG targets angiogenic vessels; it is not specific to tumor cells. However, these peptides are beneficial for treating hypovascular tumors because they interact with endothelial cells facing the circulating blood, and APRPG peptides damage angiogenic vessels without the need for particle extravasation (Asai, 2012). Anticancer drug-loaded liposomes (Yonezawa, Asai, & Oku, 2007) and micelles (Saito, Yasunaga, Kuroda, Koga, & Matsumura, 2007) exhibited increased tumor regression relative to nontargeted particles in a mouse model of a hypovascular orthotopic pancreatic tumor.

In addition to targeting peptide receptors, peptides are commonly used as ligands to target integrin. Integrin, a transmembrane receptor that mediates adhesion between cells and surrounding tissue, is typically overexpressed in tumor neovasculature (Arias, 2011). There are many types of integrin; however, $\alpha_v\beta_3$ is the most commonly targeted integrin receptor for cancer drug delivery. $\alpha_v\beta_3$ integrin is a receptor on endothelial cells that binds extracellular matrix proteins that contain the arginine–glycine–aspartic acid (RGD) sequence (Byrne et al., 2008). The RGD peptide has high-binding affinity for $\alpha_v\beta_3$, which is expressed on angiogenic endothelium in malignant tissue (Brooks et al., 1994). RGD peptides are commonly used to target $\alpha_v\beta_3$ integrin (Byrne et al., 2008). The $\alpha_v\beta_3$ ligand has been used for *in vivo* gene delivery to cancer cells using cationic polymerized liposomes (Hood et al., 2002). RGD10, a high-affinity peptide conjugated to doxorubicin-loaded liposomes, resulted in improved uptake compared to free doxorubicin and untargeted liposomes in a mouse colon cancer model (Hölig et al., 2004). Additionally, cyclic-RGD peptide targeted $\alpha_v\beta_3$ positive cells in ovarian carcinoma tumors in a xenograft mouse model (Dijkgraaf et al., 2007). Doxorubicin encapsulated in RGD-conjugated particles resulted in selective apoptosis of $\alpha_v\beta_3$-positive cells and a 15-fold increase in anticancer activity in a mouse melanoma model (Murphy et al., 2008).

There are several advantages to using peptides as ligands for cancer targeting. Generally, peptides have low immunogenicity, are small in size, and are excreted via the liver or kidney (Dijkgraaf et al., 2007). Peptides have high-binding affinity for their receptors and are designed to duplicate the target region of the physiological receptor, resulting in high specificity. RGD peptides have high-chemical stability *in vivo*, and they are easy to fabricate (Gu et al., 2007). Additionally, RGD peptides for integrin targeting have strong avidity, due to the formation of multiple ligand–receptor bonds, and extended blood circulation half-lives compared to other peptides (Montet, Funovics, Montet-Abou, Weissleder, & Josephson, 2006). Diagnostic imaging has been achieved in clinical trials using radiopeptides to target malignant cells and also to administer treatment using radiotherapy (Graham & Menda, 2011). However, peptides are rapidly degraded by peptidases, enzymes that catalyze hydrolysis of peptide linkages (Reubi, 2003); therefore, stable analog peptides will be required for clinical applications.

5.5. Aptamers

Aptamers are DNA or RNA oligonucleotide sequences that are folded into unique 3D conformations and bind to target antigens (Pestourie, Tavitian, & Duconge, 2005). Aptamers have potential as cancer-targeting ligands because they can facilitate delivery of particles to tumor antigens on the surface of malignant cells (Smith et al., 2007). Nucleic acid aptamers can be engineered using *in vitro* selection or SELEX (systemic evolution of ligands by exponential enrichment) (Huang et al., 2009). SELEX is used to select aptamers that bind to their targets with high affinity (Ellington & Szostak, 1990; Tuerk & Gold, 1990). Aptamers from a DNA or RNA pool can be selected and enriched via repetitive binding to their target molecule. Aptamers can target proteins, small molecules, nucleic acids, cells, and tissues (Huang et al., 2009).

Aptamers can be conjugated directly to anticancer drugs or particles, and they can be used to target receptors on the cell surface or ECM molecules expressed by tumors (Das et al., 2009). Covalent conjugation of aptamers to particles can be achieved using succinimidyl ester amine chemistry that produces stable amide linkages or through maleimide thiol chemistry (Farokhzad et al., 2004) to form thioether bonds. Noncovalent conjugation involves affinity interactions and metal coordination (Farokhzad et al., 2006, 2005). PLGA–PEG particles conjugated with prostate-specific membrane antigen aptamer have been engineered to deliver an increased drug dose to prostate cancer cells compared to nontargeted particles in xenograft mouse models (Dhar, Kolishetti, Lippard, & Farokhzad, 2011).

There are several advantages to using aptamers as ligands for cancer drug delivery. The advantages include high specificity and binding affinity for the target antigen, minimal immunogenicity, and easy modification (Drolet et al., 2000; Torchilin, 2005). Aptamers are stable over wide temperature and pH ranges, and they are stable in organic solvents (Nimjee, Rusconi, Harrington, & Sullenger, 2005; Wilson & Szostak, 1998). Additionally, aptamers are able to bind to proteins, molecules, nucleic acids, cells, or tissues (Burgstaller, Jenne, & Blind, 2002). Aptamers also exhibit effective tissue penetration (Hicke & Stephens, 2000). A key advantage of aptamers is their ease of synthesis and capability for large-scale production. However, aptamers are susceptible to nuclease degradation, have very short circulation half-lives, and can be quickly cleared by the kidneys (Das et al., 2009). The aptamer circulation half-life can be extended by using pyrimidine or PEG linkages between the aptamer and particle (Ulrich, Martins, & Pesquero, 2005) to prevent opsonization.

5.6. Monoclonal antibodies

MAbs are produced by clones of a single B lymphocyte and bind to a single epitope of a target antigen. MAbs are used to target specific receptors and interfere with signal transduction, which disrupts cancer cell proliferation (Danhier et al., 2010). Examples of MAbs include trastuzumab, bevacizumab, and etaracizumab, which are antibodies for ERBB2, vascular endothelial growth factor (VEGF), and $\alpha_v\beta_3$, respectively.

MAbs can be conjugated to liposomes for cancer targeting. MAb-conjugated liposomes are called immunoliposomes. Immunoliposomes can be prepared through conjugation of the MAb to the phospholipid head or PEG tail group (Sapra & Allen, 2003). Immunoliposomes have been used to target HER-2. Anti-HER-2 immunoliposomes encapsulating doxorubicin exhibit enhanced anticancer activity compared to untargeted liposomes in HER-2-overexpressing cancer cells in rat xenograft tumor models (Park et al., 2002, 2001). MAbs have been used in targeted therapy for breast cancer. Cytokeratin is a MAb used to target invasive breast cancer epithelial cells. Drug-loaded PLGA particles conjugated with cytokeratin showed improved efficacy in treating breast cancer cells *in vitro* (Kos, Obermajer, Doljak, Kocbek, & Kristl, 2009). Additionally, HER-2 trastuzumab antibody-conjugated PLGA particles exhibited improved cellular uptake and anticancer activity compared to nontargeted polymers in breast cancer cells *in vitro* (Sun, Ranganathan, & Feng, 2008).

The use of MAbs has also been explored for targeting cancer stem cells (CSCs). CD44, a cell surface ECM receptor, is an established marker of CSCs (Deonarain, Kousparou, & Epenetos, 2009). CD44 may exhibit various signaling functions depending on its protein interaction, including proliferation, apoptosis, survival, migration, and differentiation (Deonarain et al., 2009). Anti-CD44 antibodies have been used for targeted anti-CSC therapy. H90, a MAb specific to CD44, mediated eradication of acute myeloid leukemic stem cells in nonobese diabetic/severe combined immunodeficient mice (Jin, Hope, Zhai, Smadja-Joffe, & Dick, 2006). A patented humanized version of H460-16-2, an anti-CD44 MAb, reduced tumor growth in pancreatic cancer xenografts in mice (Young et al., 2007).

There are advantages and disadvantages to using MAbs as targeting ligands. MAbs have high affinity and specificity for their target receptor, and they are able to achieve targeted delivery across the blood–brain barrier. OX26 is a MAb that targets the transferrin receptor. Drug–OX26 conjugates have been shown to cross the blood–brain barrier using receptor-mediated transcytosis via brain capillary endothelial transferrin receptors (Bickel,

Yoshikawa, & Pardridge, 2001). However, MAbs are also immunogenic and have short half-lives. The high-molecular weight of MAbs also hinders extravasation, resulting in slow diffusion of MAb-conjugated particles into the target tissue (Dijkgraaf et al., 2007).

5.7. Vascular endothelial growth factor

VEGF mediates tumor angiogenesis, a common target for cancer drugs. Angiogenesis enables growth of new blood vessels, which are necessary to provide nutrients to a rapidly growing malignant tumor. Inhibiting angiogenesis inhibits tumor growth. VEGF is upregulated in neoplastic tumor cells due to hypoxia in the tumor tissue and oncogene expression, which results in overexpression of VEGF receptors-1 and -2 on tumor endothelial cells (Kremer, Breier, Risau, & Plate, 1997; Shweiki, Itin, Soffer, & Keshet, 1992). VEGF expression correlates to the degree of vascularization and to the stage of the cancer (Brown et al., 1993; Jain et al., 1998). VEGF is upregulated in bone marrow of patients with hematological tumors (Malik & Gerber, 2003). VEGF receptors (VEGFRs) are commonly expressed on endothelial cells of tumor vasculature, in solid tumors, leukemias, and other hematological tumors (Bozec et al., 2008). VEGFR-2 is typically targeted in therapeutic application because it binds VEGF and is overexpressed on malignant cells and endothelial cells of tumor vasculature.

VEGF-targeting strategies are different than those of other receptors. VEGF and VEGFR ligands are used therapeutically for their inhibitory functions, and not simply for targeting a receptor or antigen. Targeting angiogenesis is typically executed using two approaches: targeting VEGFRs to decrease binding of VEGF or targeting VEGF to inhibit its binding to VEGFRs (Backer et al., 2005; Veikkola, Karkkainen, Claesson-Welsh, & Alitalo, 2000). The most-studied approaches include inhibition of VEGF and VEGFRs by MAbs to inhibit receptor signaling (Malik & Gerber, 2003). Bevacizumab, an antibody that binds to VEGF and inhibits tumor angiogenesis, has been FDA approved to treat colorectal cancer (Ferrara, 2005). Bevacizumab-conjugated polymer microspheres loaded with paclitaxel have been demonstrated to target angiogenesis resulting in growth inhibition of prostate tumors in mice (Lu, Jackson, Gleave, & Burt, 2008). Further, boronated dendrimers conjugated with human recombinant VEGF$_{121}$, a splice variant of VEGF containing 121 amino acids, have the ability to selectively target VEGFR-2 receptors on breast cancer tumor cells and inhibit tumor growth through receptor binding and endocytosis of anticancer drugs in a mouse model (Backer et al., 2005). Additionally, pegylated liposomal doxorubicin effectively reduced breast tumor growth when conjugated with anti-VEGFR-2 MAbs in various cancer types using a mouse model (Wicki et al., 2012).

One of the advantages of targeting VEGF and VEGFR is that tumor angiogenesis is directly targeted to inhibit tumor growth. VEGFRs overexpressed both in the tumor endothelium and tumor cells may be targeted. However, ligands that target VEGFR may also bind to other VEGF-related receptors and cause harmful side effects (Ferrara, 2005). Using recombinant VEGF to target VEGFRs is also very costly due to difficulty of manufacture and their short half-lives (Cao, 2001).

6. DRUG RELEASE

Methods of controlled drug release can be categorized as either temporal control or distribution control (Ulrich et al., 2005). The aim of temporal control is to enable release at a specific time and/or to extend duration of release. The goal of distribution control is to release drug at a target site. Distribution control provides additional strategies to enhance cancer targeting by maximizing the drug concentration at the cancer site, thus minimizing side effects to nonmalignant cells. This section will focus on methods for distribution control. The simplest method for achieving distribution control is to inject or implant anticancer drugs at the target site; however, direct delivery is only feasible if the target site is safely and easily accessible, and the drug remains at the target site (Ulrich et al., 2005). An engineering approach for enhancing cancer targeting is to use stimuli-responsive particles to trigger drug release. Particles can be designed to interact with their physiological environment to promote drug release. Stimuli for release may be intrinsic, exploiting factors such as physiological pH or enzymes present, or stimuli may be extrinsic, using application of heat or other applied stimuli to trigger release.

The temporal and spatial drug release profiles that will be required for therapeutic efficacy are fundamental considerations when designing stimuli-responsive particles for targeted cancer therapy. Stimuli-responsive particles should enhance uptake by malignant cells and stimulate intracellular release in order to maximize the drug dose delivered to malignant cells. The particles should also be designed to promote endosomal escape after endocytosis for delivery to intracellular compartments. After a particle has been endocytosed, endosomal trafficking from the endosome to the lysosome occurs within 2–3 min (Liechty & Peppas, 2012). Particles must be designed to escape the endosome to avoid damage to the drug by the lysosomal enzymes. To avoid harming nonmalignant cells, particles must also be strictly designed to release their payloads at the targeted cancer sites and not in nonmalignant cells or tissues. If the pH- or thermoresponse of a particle is not tightly controlled, it may result in drug release to nontargeted cells. The fate of the particle polymer component after drug release is also an important consideration. As reviewed in Section 3.5, polymer particles that can be degraded or excreted by the kidneys will have lower toxicity. Stimuli-responsive particles should also enable attachment of ligands to promote active targeting of the treatment site before the drug release is triggered.

6.1. pH-responsive drug release

Drug release can be triggered by taking advantage of the pH of the intratumoral environment or the intracellular environment. In nonmalignant body tissue, pH is maintained near 7.4. The pH of solid tumors varies in the range of approximately 5.7–7.6 due to accumulation of acidic metabolites that are the result of low blood pressure and hypoxia caused by abnormal tumor vessels (Dellian, Helmlinger, Yuan, & Jain, 1996; Stubbs, McSheehy, & Griffiths, 1999; Tannock & Rotin, 1989). The pH variation is even greater within the cell. Once drug–particle complexes are endocytosed, endocytotic vesicles convert to early endosomes, where the pH range is 5.9–6.8, and then to late endosomes, where the pH range is 5–6 (Liechty & Peppas, 2012). Late endosomes then fuse with lysosomes, in which the protein concentration is 100-fold higher than physiological conditions, and the pH decreases to an average of 5 (Bae, Fukushima, Harada, & Kataoka, 2003).

pH sensitivity can be achieved by conjugating the drug to particles using acid-cleavable linkers or by conjugating a drug to, or encapsulating it within a pH-sensitive polymer. Acid-cleavable linkages can be used to conjugate the drug to the polymer to form a prodrug that will remain inactive until the polymer has been cleaved. Prodrug particles are easily hydrolyzed through cleavage of their chemical bonds with water molecules at lower pH. Common pH-responsive linkages include acetals, anhydrides, *cis*-aconityl, and hydrazones (Duncan, 2003). Dimethylmaleic anhydride has been used as a pH-sensitive linker for polymeric particles (Kamada et al., 2004). The linker binds to amine groups at pH greater than 8 and reverts to anhydride to release the drug at pH less than 7. Hydrazone is another pH-sensitive linker that can be used to create an environmentally reactive cancer drug.

Doxorubicin-conjugated polymer particles linked with hydrazone are stable at physiological pH; however, at pH 5, they release almost 50% of the drug (Etrych, Jelinkova, Rihova, & Ulbrich, 2001). Experiments in a mouse model demonstrated that these pH-sensitive particles slowed T cell lymphoma tumor growth when compared to free doxorubicin (Ulbrich, Etrych, Chytil, Jelinkova, & Rihova, 2003) due to a triggered and controlled release of DOX at the tumor site.

In addition to pH-sensitive linkers, pH-responsive polymers can be used to trigger drug release under acidic conditions. pH sensitivity is found in ionizable groups, such as amine, azo, carboxylic acid, imidazole, pyridine, and thiol groups (Kim, Kim, Jeon, Kwon, & Park, 2009). pH-responsive polymers are triggered to escape the endosome by endosomal acidification and release their payload. Under low pH, pH-responsive particles absorb protons in the endosome, which increases endosome osmotic pressure, resulting in disruption of the endosomal membrane and particle release into the cytoplasm (Gao, Chan, & Farokhzad, 2010). This mechanism is called the "proton sponge" effect (Behr, 1997). The most common pH-sensitive liposomes contain phosphatidylethanolamine (PE) in the membrane bilayer (Simoes, Moreira, Fonseca, Duzgunes, & de Lima, 2004). Poly(ethylene oxide)–poly(propylene oxide) (PEO–PPO) copolymer modified with poly(β-amino ester) (PβEA) creates a pH-sensitive particle. This modified PEO–PPO–PβEA particle is stable at physiological pH and exhibits burst release at pH less than 6.5 (Shenoy, Little, Langer, & Amiji, 2005). When loaded with paclitaxel, these pH-sensitive particles exhibited increased accumulation and antitumor activity in mouse xenograft tumor models of ovarian cancer (Devalapally, Shenoy, Little, Langer, & Amiji, 2007). Acetal polymers can also be used for controlled release in pH-sensitive micelles. The acetal groups, which form the hydrophobic core to encapsulate the drug, are hydrolyzed under acidic conditions (Gillies & Frechet, 2005), resulting in micelle disruption and drug release.

One of the key advantages of pH-responsive particles is that they are able to exploit pH variances to promote endosomal escape. pH-sensitive particles enhance endosomal escape using the proton sponge effect. However, pH-sensitive particles can be toxic to nonmalignant cells if they are not designed to release their drug payload only within a precise pH range (Kim et al., 2009). Few pH-responsive delivery systems for cancer therapy have progressed to clinical trials due to concerns of toxicity to nonmalignant tissues, regulatory issues, and inefficient translation of *in vitro* efficacy to *in vivo* efficacy (Liechty & Peppas, 2012).

6.2. Enzyme-responsive drug release

Enzyme-responsive materials can be used to trigger drug release. After a particle has been endocytosed, it is exposed to pH changes and enzymatic degradation in the lysosome. When an endosome fuses with a lysosome, the particle is exposed to lysosomal enzymes such as glycosidases, proteases, and sulfatases that digest cellular waste. Enzymatically degradable spacers can be conjugated to polymeric particles to produce enzyme-mediated drug release (Hoste, De Winne, & Schacht, 2004). Cancer drugs can be attached to the spacers, and upon lysosomal fusion, hydrolases degrade the spacer and release the drug. Lysosomal enzymes are not only expressed intracellularly, they are often overexpressed intratumorally, making them a promising target for enzyme-controlled drug release. Enzymes including cathepsin B and D and metalloproteinase are overexpressed in tumor tissue and play a role in cancer progression (Gialeli, Theocharis, & Karamanos, 2010; Masson et al., 2011; Podgorski & Sloane, 2003).

The most common type of spacer for enzyme-mediated drug delivery is an oligopeptide, a peptide composed of 2–20 amino acids. The rate of cleavage or enzymatic degradation of the peptide depends on the amino acid composition. Drug release can be modified by adjusting the length and composition of the spacer. Oligopeptide spacers with a C-terminal glycine are degraded more quickly, whereas hydrophobic terminal amino acids, such as leucine and phenylalanine, are degraded slowly (De Marre, Seymour, & Schacht, 1994; De Marre, Soyez, & Schacht, 1994). In an *in vitro* study, polymer-coated GNPs with glycine–phenylalanine–leucine–glycine spacers were designed to effectively control release of cancer drugs upon enzymatic degradation of the spacer (Schneider, Subr, Ulbrich, & Decher, 2009).

Controlled release and delivery of cytotoxic drugs can also be achieved by using antibody-directed enzyme prodrug therapy (ADEPT) (Sharma, Bagshawe, Reddy, & Couvreur, 2010). In ADEPT, enzymes are first delivered to tumors using antibody–enzyme conjugates to target tumor cells. After enzymes bind to receptors on tumor cells and residual enzymes are cleared from circulation, prodrug is administered that is then enzymatically degraded within the tumor by the bound enzymes to create an active drug. ADEPT has been successful in clinical trials for the treatment of malignant tumors (Napier et al., 2000).

Enzyme-responsive drug carriers have advantages for cancer targeting because they minimize toxicity to nonmalignant cells. Through the use of prodrugs, the anticancer drug only becomes active at the target cancer site, limiting exposure of nonmalignant cells to the harmful effects of the drug. The prodrug–enzyme therapy enables delivery of large amounts of drug to the tumor site. However, one of the disadvantages of this therapy is the poor solubility of prodrugs (Hoste et al., 2004). Further, even though ADEPT therapy results in local delivery, some active drug may escape the tumor site and cause damage to nonmalignant cells (Springer, Poon, Sharma, & Bagshawe, 1993).

6.3. Thermoresponsive drug release

Hyperthermia can be used to trigger drug release from particles composed of thermoresponsive polymers. Local hyperthermia can be generated using infrared light, microwaves, hot water bath, ultrasound, or radiofrequency, depending on the location of the tumor (Hosokawa, Sami, Kato, & Hayakawa, 2003). Magnetic particles, in conjunction with an alternating magnetic field, can also be used to generate localized hyperthermia (Kobayashi, 2011). Regional hyperthermia is typically generated via perfusion by heating the blood. The temperature range for hyperthermia is between 37 and 42 °C; at 42 °C, 5 °C above physiologic temperature, cells begin to show signs of thermal damage (Crile, 1962). Thermosensitive particles can be delivered intratumorally through passive accumulation due to the EPR effect. After particle accumulation, heat can be applied locally to stimulate drug release from the thermosensitive particles. Hyperthermia also increases the permeability of the heated tissue and cells, promoting intracellular uptake of drug or particle–drug conjugates. Electroporation, a technique in which an electric field is generated around the target tissue, also increases permeability, and it can be used to ablate tissue through irreversible electroporation (Davalos, Mir, & Rubinsky, 2005).

Poly(N-isopropyl acrylamide) (PNIPAAm) is a commonly used thermoresponsive polymer. The thermosensitivity of PNIPAAm results from a balance between hydrophobic and hydrophilic polymer segments; interactions between these segments cause chain aggregation and cross-linking (Kim et al., 2009). PNIPAAm is soluble at room temperature and gels above its lower critical solution temperature (LCST) (Kaneko et al., 1999; Nakayama et al., 2006), the temperature below which the polymer is miscible. The LCST of PNIPAAm is close to body temperature; therefore, the polymer will gel and trap drug molecules upon injection. However, PNIPAAm is commonly blended with other polymers to create particles or particle coatings. LCST of PNIPAAm is tunable via copolymerization of hydrophobic monomers or adjusting the molecular weight of the polymer. Hydrophobic monomers and high-molecular weight decrease LCST, while hydrophilic monomers increase LCST (Kim et al., 2009). Doxorubicin-encapsulated magnetic particles coated with PNIPAAm exhibit drug release when heated at 42 °C *in vitro* (Purushotham et al., 2009). A PNIPAAm copolymer loaded with camptothecin anticancer drug in conjunction with hyperthermia suppressed tumor growth in a mouse model of colon cancer (Peng et al., 2011).

Thermosensitive polymers that have been used to encapsulate cancer drugs include block copolymers of PEG–PLGA and PEO–PPO (Jeong, Bae, & Kim, 2000; Kabanov, Batrakova, & Alakhov, 2002). A multiblock polymer composed of poly(ε-caprolactone), PEG, and poly(propylene glycol) segments was synthesized to promote controlled release of paclitaxel (Loh, Yee, & Chia, 2012). The copolymer exhibited sustained drug release for more than 2 weeks, and it arrested growth of cervical cancer-derived HeLa cells *in vitro* compared to paclitaxel encapsulated in nonthermosensitive particles. Adriamycin encapsulated in a thermosensitive liposome had greater delivery across the blood–brain barrier and greater antitumor activity in hyperthermia-treated mice with glioma than treatment with free drug or adriamycin encapsulated in nonthermosensitive liposomes (Gong et al., 2011). Preliminary research has shown that permeability of the liposome across an *in vitro* model of the blood–brain barrier increased significantly with elevated temperature. Another liposomal formulation termed "Hyperthermia-activated cytoToxic" (HaT), resulted in greater cell membrane permeability and drug release for mammary cancer cells in mice at elevated temperature (40–41 °C) (Tagami, Ernsting, & Li, 2011). The liposomal HaT increased uptake of doxorubicin into heated tumor tissue by 5.2-fold and reduced delivery to the heart by 15-fold compared to free doxorubicin. Using hyperthermia, administration of a single intravenous treatment of the doxorubicin-encapsulated thermosensitive liposome resulted in greater mammary carcinoma tumor regression compared to free doxorubicin and temperature-sensitive lysolipid (phospholipid containing one carbon chain) liposomes in a mouse model (Tagami et al., 2011).

Thermosensitive drug carriers have several advantages for cancer targeting. Hyperthermia increases accumulation of drug in malignant tissue and cells (Gaber et al., 1996; Kong, Braun, & Dewhirst, 2001) by increasing the permeability of tumor vessels and tumor cells. Hyperthermia also increases cell sensitivity to thermally induced injury (Fajardo & Prionas, 1994), which increases the death rate of malignant cells. Additionally, thermoresponsive polymers can solubilize hydrophobic drugs, and they can be used with poorly soluble drugs (Vukelja et al., 2007; Zentner et al., 2001).

7. FUTURE DIRECTIONS IN CANCER TARGETING

The successes of current research in targeted anticancer drug delivery demonstrate that the future of cancer-targeting therapy is promising. Particle drug encapsulation, passive and active targeting, and controlled release each enhance the targeting capabilities of anticancer drugs. Targeted cancer treatments such as particle encapsulation, targeting moiety conjugation, and stimuli-responsive drug release have already demonstrated improved patient outcomes relative to untargeted systemic drug administration in clinical trials. Research is ongoing to further improve targeted drug delivery for a variety of cancers. However, the tumor microenvironment is very complex, and optimizing targeted delivery requires a sophisticated drug delivery system that utilizes more than one targeting approach, for example, combining targeting moiety conjugation with stimuli-responsive drug release. Recent studies have shown that particle systems designed to have multiple functions further enhance the specificity of targeted drug delivery that is required to improve treatment efficacy and reduce treatment side effects. These systems combine design elements, such as particle encapsulation of the anticancer drug, ligand conjugation, controlled release, and imaging, to improve targeted delivery and to allow greater control over cancer therapy. Multifunctional systems have explored the delivery of therapeutics beyond chemotherapy, including DNA and RNA for cancer gene targeting and silencing. Multidrug therapy has also proven successful in many particle systems (Parhi, Mohanty, & Sahoo, 2012). Additionally, dual ligand targeting has been employed to enhance cancer-targeting specificity to receptors on malignant cells. The goal of the combinatorial approach is to further enhance targeted delivery.

7.1. Multifunctional delivery systems

Recently, multifunctional delivery systems have been designed to provide a comprehensive approach to targeting cancer. Multifunctional delivery systems are particles that have multiple functionalities, including diagnostic imaging, targeted drug delivery, and controlled drug release. These particles enable active monitoring of the cancer during administration of chemotherapy, which allows further control over the treatment process. Multifunctional particles combine multiple-targeting methods in an attempt to achieve optimal therapeutic results. These carriers may also support combination drug therapy by loading the particle with multiple drugs to achieve a better therapeutic outcome. Single drug chemotherapy may sometimes result in poor prognosis due to multidrug resistance, toxicity, and undesirable side effects (Parhi et al., 2012). Particle-based combination therapy provides a synergistic effect that enhances treatment effectiveness and overcomes limitations of single drug therapy. Multifunctional delivery systems can be used to deliver and monitor combination drug therapy to further improve cancer treatment.

There are many design options for multifunctional delivery systems. For example, a poly(propylene imine) dendrimer was designed for multifunctional drug delivery by loading it with both small interfering RNA (siRNA) and superparamagnetic iron oxide (SPIO) nanoparticles and by conjugating it with LHRH peptide (Taratula et al., 2009). SPIOs are used as magnetic resonance imaging (MRI) contrast agents. The combination of the LHRH peptide and the SPIO contrast agent enabled successful targeting and monitoring of siRNA nanoparticles to LHRH receptor-positive ovarian and lung cancer cells, which resulted in tumor suppression in a mouse model. In another recent example, magnetic particles loaded with an anticancer superparamagnetic manganese complex were functionalized with PEG and RGD peptides and doped with fluorescent dye (Chen, Tao, Liu, & Yang, 2012). The magnetic particles enabled *in vitro* imaging and differentiation between malignant and nonmalignant liver cells using MRI. RGD peptides enabled effective targeting and internalization of particles into HeLa cells *in vitro*. Multifunctional degradable polymeric particles have been developed for cancer "theranostics," which combines diagnostic and therapeutic capabilities (Wang et al., 2012). Photosensitive and fluorescent imaging agents were incorporated into a polymer particle matrix with PEG and peptide-targeting ligand conjugated to the surface. The degradable particles efficiently targeted breast cancer cells *in vitro* and enabled fluorescent imaging. The photosensitive drug also caused damage when tumor cells were irradiated with light at the appropriate wavelength.

Microbubbles (MBs), another type of multifunctional theranostic agent, have recently been studied for drug and gene delivery applications. MBs contain a gas core and an outer shell composed of lipids, proteins, or polymers (Sirsi & Borden, 2009). MBs have decreased solubility and diffusivity due to the high-molecular weight gases contained in their core, making the bubble robust and able to recirculate in the blood stream several times (Greco et al., 2010). MBs exhibit useful properties for diagnostic imaging and drug delivery when subjected to ultrasound. MBs produce a backscattered echo under ultrasound that can be used to detect the bubble, making it useful as an imaging contrast agent. A MB also functions as a vehicle for gene and drug delivery. The MB is triggered for local release of its cargo by an ultrasound "destruction" pulse in a technique referred to as ultrasound-targeted MB-destruction (UTMD) (Sirsi & Borden, 2009). The UTMD approach has been used to enhance therapeutic outcomes by delivering cancer terminator viruses (Das et al., 2012). MBs and UTMD have been used for delivery of a combination of adenoviruses that resulted in tumor growth inhibition in a mouse model of prostate cancer (Dash et al., 2011). In another study, ultrasound-guided MBs, complexed with a cancer terminator virus, eradicated targeted therapy-resistant tumors and nontargeted distant tumors in xenograft mouse models of prostate cancer (Greco et al., 2010).

7.2. Dual ligand targeting

The dual ligand-targeting approach involves conjugating more than one type of ligand to the surface of a particle. Dual ligand targeting is based on the idea that cancer cells tend to overexpress multiple types of surface receptors (Li et al., 2011). For example, both HER-2 and EGFR are often overexpressed in breast cancer, while both EGFR and bombesin peptide are often overexpressed in prostate cancer. A key challenge for targeted cancer treatment is that many receptors that are overexpressed on malignant cells are also expressed on nonmalignant cells at lower, normal expression levels. Conjugation of two different ligands to particles to target overexpressed receptors has been shown to result in a higher rate of malignant cell recognition than that of either of the individual ligands (Li et al., 2011) by increasing specificity. Dual ligands increase the binding avidity of the particle complex to the malignant cell by binding two different ligands to two different receptors on the surface of the same cell (Meng et al., 2010). Dual ligands can further improve targeted delivery by enhancing the specificity of malignant cell targeting, thereby reducing toxicity to other cells.

Dual ligand targeting was demonstrated *in vitro* with human KB cells that overexpress both FR and EGFR (Saul, Annapragada, & Bellamkonda, 2006). Both folic acid and MAbs for EGFR were conjugated to liposomal particles. When KB cells expressing both receptors were tested against cells in which one or both of the receptors were blocked, the results showed that the dual ligand improved detection of the KB cells.

Dual ligands have been shown to enhance cancer targeting in several other studies. Dual ligand GNPs have been designed to promote multivalent interactions between GNPs and FR-overexpressing cancer cells (Li et al., 2011). GNPs with folate and glucose conjugation enhanced cellular recognition of FR-overexpressing cells. Cellular uptake was increased by 3.9- and 12.7-fold compared to GNP–folate and GNP–glucose. In another study, α_v integrins and neuropilin-1 (NRP-1), a receptor for VEGF, were chosen as simultaneous tumor targets (Meng et al., 2010). RGD was used to target α_v integrins, and a heptapeptide was used to target NRP-1. Targeting was tested *in vitro* in lung carcinoma cells and human umbilical vein endothelial cells, both of which express α_v integrins and NRP-1. The dual-targeting ligands increased uptake of paclitaxel in both cell lines. Dual targeting of brain glioma was achieved using the TGN peptide, a 12 amino acid peptide that targets the blood–brain barrier, and AS1411 aptamer, a DNA aptamer that binds to nucleolin protein in glioma cells (Gao et al., 2012). The dual ligand polymeric particles not only targeted endothelial and glioma cells, they also penetrated the endothelial monolayers and tumor cells to reach the core of the tumor spheroids *in vitro* (Gao et al., 2012). A similar strategy was used by another group to also successfully target brain glioma using *p*-aminophenyl-α-D-mannopyranoside (MAN) and transferrin-conjugated liposomes to enhance penetration of the blood–brain barrier and targeted delivery to glioma, respectively, in a rat model (Ying et al., 2010).

8. CONCLUDING REMARKS

This review has focused on bioengineering approaches to improve targeted delivery of cancer therapeutics, which include particles, targeting moieties, and stimuli-responsive drug release mechanisms. Design criteria must be understood in order to achieve the goals of cancer targeting, which include (1) selective targeting of cancer cells, (2) efficient release of anticancer drugs at the target site, and (3) elimination of cytotoxicity to noncancerous tissue. To achieve these goals, barriers to targeted delivery, including opsonization, nonspecific uptake by nonmalignant cells, and multidrug resistance, must be addressed when designing targeted anticancer drugs. By combining particles, targeting ligands, and triggered release mechanisms, cancer-targeting drugs can be improved and customized to achieve targeted cancer treatment.

Advances in nanotechnology and biomaterials have enabled the development of enhanced drug carriers for cancer targeting. Particles have been designed to encapsulate drugs to reduce toxicity to nonmalignant cells and to take advantage of passive-targeting mechanisms using the EPR effect.

Additionally, the surface properties of particles can be customized using techniques such as pegylation to enhance biocompatibility and to reduce cytotoxicity. Micelle and liposomal particles can be used to encapsulate hydrophobic drugs for delivery; however, there is a need for improved drug-loading capacity. CNTs have high drug-loading capacity, but efficient purification methods are needed before CNTs can be used commercially as nanocarriers for cancer drugs. Dendrimers and other polymeric particles show promise for many cancer drug delivery applications because they are both highly customizable, and they can be designed with minimal immunogenicity and high drug-loading capacity.

Targeting ligands can be attached to particles to promote active targeting and endocytosis of the drug by the targeted cancer cell. Ligands are selected to target receptors that are overexpressed on the surface of malignant cells. Folate is a widely used targeting ligand because its receptor is overexpressed in many cancer cells, but not on most nonmalignant tissues. Transferrin is nonimmunogenic and can be easily obtained from human sources, but it can also cause toxicity to nontarget tissue due to nonspecific uptake in nonmalignant cells that have low-level expression of transferrin receptors. MAbs have high affinity and specificity for their target receptors but are also immunogenic and diffuse more slowly due to a high MW. Peptides have potential for many cancer-targeting applications due to the large selection of targeting ligands, high-binding affinity and specificity, and low immunogenicity.

Controlled release of the anticancer drug from a particle carrier can be achieved using pH, enzyme, or thermoresponsive particles or conjugates. Stimuli-responsive particles promote release of the drug under the physiological conditions of the cancer tissue or cell. pH-sensitive particles take advantage of the intratumoral or intracellular pH differences to promote intratumoral or intracellular drug release. Enzyme-responsive spacers create prodrugs that are nontoxic until enzymes in the cancer tissue degrade the spacer. Thermosensitive release is achieved by applying external heat to enhance permeability of tumor tissue and cells, which increases the accumulation of particle-encapsulated anticancer drugs in malignant cells. Stimuli-responsive polymers or spacers must be tightly controlled to avoid toxicity to noncancerous tissues.

The future of bioengineering strategies for cancer targeting will be to use a combinatorial approach to customize delivery of anticancer drugs and improve treatment outcomes. Multifunctional drug delivery systems are being designed that achieve imaging, targeted delivery, and controlled release from each particle. In addition, dual ligand targeting to multiple overexpressed receptors on malignant cells enhances specificity of particles for malignant cells while further reducing nonspecific uptake by nonmalignant cells. This combinatorial and customized approach to the design of particles shows promise for optimizing targeted cancer drug delivery to improve efficacy and to minimize side effects associated with drug delivery to nonmalignant cells.

ACKNOWLEDGMENTS

This work was made possible by the MUSC Hollings Cancer Center Translational Research Pilot Funds and the William H. Goodwin Endowment Funds (X. W.). This work was also supported by the AO Foundation (S0955V: W. V. B. F. X. W.) and the National Institutes of Health (T32DE017551: W. V. B. F. A. A. B.; K99DE023123: W. V. B. F.). The authors thank Drs. Andrew G. Jakymiw and Frank Alexis for their critical reviews of this work.

REFERENCES

ACS (2012). *Cancer facts & figures 2012*. Atlanta: American Cancer Society.
Albanese, A., Tang, P. S., & Chan, W. C. (2012). The effect of nanoparticle size, shape, and surface chemistry on biological systems. *Annual Review of Biomedical Engineering, 14,* 1–16.
Alexis, F., Pridgen, E. M., Langer, R., & Farokhzad, O. C. (2010). Nanoparticle technologies for cancer therapy drug delivery. In M. Schäfer-Korting (Ed.), *Handbook of Experimental Pharmacology: Vol. 197*. (pp. 55–86). Berlin, Heidelberg: Springer.
Ali, H. M., Urbinati, G., Raouane, M., & Massaad-Massade, L. (2012). Significance and applications of nanoparticles in siRNA delivery for cancer therapy. *Expert Review of Clinical Pharmacology, 5,* 403–412.
Allen, T. M. (1998). Liposomal drug formulations. Rationale for development and what we can expect for the future. *Drugs, 56,* 747–756.
Anabousi, S., Bakowsky, U., Schneider, M., Huwer, H., Lehr, C. M., & Ehrhardt, C. (2006). In vitro assessment of transferrin-conjugated liposomes as drug delivery systems for inhalation therapy of lung cancer. *European Journal of Pharmaceutical Sciences, 29,* 367–374.
Arias, J. L. (2011). Drug targeting strategies in cancer treatment: An overview. *Mini Reviews in Medicinal Chemistry, 11,* 1–17.
Asai, T. (2012). Nanoparticle-mediated delivery of anticancer agents to tumor angiogenic vessels. *Biological & Pharmaceutical Bulletin, 35,* 1855–1861.
Backer, M. V., Gaynutdinov, T. I., Patel, V., Bandyopadhyaya, A. K., Thirumamagal, B. T., Tjarks, W., et al. (2005). Vascular endothelial growth factor selectively targets boronated dendrimers to tumor vasculature. *Molecular Cancer Therapeutics, 4,* 1423–1429.
Bae, Y., Fukushima, S., Harada, A., & Kataoka, K. (2003). Design of environment-sensitive supramolecular assemblies for intracellular drug delivery: Polymeric micelles that are responsive to intracellular pH change. *Angewandte Chemie (International Ed. in English), 42,* 4640–4643.
Banerjee, T., Mitra, S., Kumar Singh, A., Kumar Sharma, R., & Maitra, A. (2002). Preparation, characterization and biodistribution of ultrafine chitosan nanoparticles. *International Journal of Pharmaceutics, 243,* 93–105.
Behr, J.-P. (1997). The proton sponge: A trick to enter cells the viruses did not exploit. *Chimia (Aarau), 51,* 34–36.
Bianco, A., Kostarelos, K., & Prato, M. (2005). Applications of carbon nanotubes in drug delivery. *Current Opinion in Chemical Biology, 9,* 674–679.

Bickel, U., Yoshikawa, T., & Pardridge, W. M. (2001). Delivery of peptides and proteins through the blood–brain barrier. *Advanced Drug Delivery Reviews, 46*, 247–279.

Boas, U., & Heegaard, P. M. (2004). Dendrimers in drug research. *Chemical Society Reviews, 33*, 43–63.

Bozec, A., Gros, F. X., Penault-Llorca, F., Formento, P., Cayre, A., Dental, C., et al. (2008). Vertical VEGF targeting: A combination of ligand blockade with receptor tyrosine kinase inhibition. *European Journal of Cancer, 44*, 1922–1930.

Brigger, I., Dubernet, C., & Couvreur, P. (2002). Nanoparticles in cancer therapy and diagnosis. *Advanced Drug Delivery Reviews, 54*, 631–651.

Brooks, P. C., Montgomery, A. M., Rosenfeld, M., Reisfeld, R. A., Hu, T., Klier, G., et al. (1994). Integrin alpha v beta 3 antagonists promote tumor regression by inducing apoptosis of angiogenic blood vessels. *Cell, 79*, 1157–1164.

Brown, L. F., Berse, B., Jackman, R. W., Tognazzi, K., Manseau, E. J., Senger, D. R., et al. (1993). Expression of vascular permeability factor (vascular endothelial growth factor) and its receptors in adenocarcinomas of the gastrointestinal tract. *Cancer Research, 53*, 4727–4735.

Brumlik, M. J., Daniel, B. J., Waehler, R., Curiel, D. T., Giles, F. J., & Curiel, T. J. (2008). Trends in immunoconjugate and ligand-receptor based targeting development for cancer therapy. *Expert Opinion on Drug Delivery, 5*, 87–103.

Burgstaller, P., Jenne, A., & Blind, M. (2002). Aptamers and aptazymes: Accelerating small molecule drug discovery. *Current Opinion in Drug Discovery & Development, 5*, 690–700.

Byrne, J. D., Betancourt, T., & Brannon-Peppas, L. (2008). Active targeting schemes for nanoparticle systems in cancer therapeutics. *Advanced Drug Delivery Reviews, 60*, 1615–1626.

Cabane, E., Zhang, X., Langowska, K., Palivan, C., & Meier, W. (2012). Stimuli-responsive polymers and their applications in nanomedicine. *Biointerphases, 7*, 1–27.

Cao, Y. (2001). Endogenous angiogenesis inhibitors and their therapeutic implications. *The International Journal of Biochemistry & Cell Biology, 33*, 357–369.

Carlsson, J., Kullberg, E. B., Capala, J., Sjoberg, S., Edwards, K., & Gedda, L. (2003). Ligand liposomes and boron neutron capture therapy. *Journal of Neuro-Oncology, 62*, 47–59.

Chang, H. I., & Yeh, M. K. (2012). Clinical development of liposome-based drugs: Formulation, characterization, and therapeutic efficacy. *International Journal of Nanomedicine, 7*, 49–60.

Chen, H., Khemtong, C., Yang, X., Chang, X., & Gao, J. (2011). Nanonization strategies for poorly water-soluble drugs. *Drug Discovery Today, 16*, 354–360.

Chen, H., Li, L., Cui, S., Mahounga, D., Zhang, J., & Gu, Y. (2011). Folate conjugated CdHgTe quantum dots with high targeting affinity and sensitivity for *in vivo* early tumor diagnosis. *Journal of Fluorescence, 21*, 793–801.

Chen, H. T., Neerman, M. F., Parrish, A. R., & Simanek, E. E. (2004). Cytotoxicity, hemolysis, and acute *in vivo* toxicity of dendrimers based on melamine, candidate vehicles for drug delivery. *Journal of the American Chemical Society, 126*, 10044–10048.

Chen, Q.-Y., Tao, G.-P., Liu, Y.-Q., & Yang, X. (2012). Synthesis, characterization, cell imaging and anti-tumor activity of multifunctional nanoparticles. *Spectrochimica Acta. Part A, Molecular and Biomolecular Spectroscopy, 96*, 284–288.

Cheng, Y., Meyers, J. D., Agnes, R. S., Doane, T. L., Kenney, M. E., Broome, A. M., et al. (2011). Addressing brain tumors with targeted gold nanoparticles: A new gold standard for hydrophobic drug delivery? *Small, 7*, 2301–2306.

Cho, K., Wang, X., Nie, S., Chen, Z., & Shin, D. M. (2008). Therapeutic nanoparticles for drug delivery in cancer. *Clinical Cancer Research, 14*, 1310–1316.

Chou, H.-H., Wang, K.-L., Chen, C.-A., Wei, L.-H., Lai, C.-H., Hsieh, C.-Y., et al. (2006). Pegylated liposomal doxorubicin (Lipo-Dox®) for platinum-resistant or refractory epithelial ovarian carcinoma: A Taiwanese gynecologic oncology group study with long-term follow-up. *Gynecologic Oncology, 101*, 423–428.

Choucair, A., & Eisenberg, A. (2003). Control of amphiphilic block copolymer morphologies using solution conditions. *The European Physical Journal. E, Soft Matter, 10*, 37–44.

Chung, J. E., Yokoyama, M., Aoyagi, T., Sakurai, Y., & Okano, T. (1998). Effect of molecular architecture of hydrophobically modified poly(N-isopropylacrylamide) on the formation of thermoresponsive core-shell micellar drug carriers. *Journal of Controlled Release, 53*, 119–130.

Chung, J. E., Yokoyama, M., & Okano, T. (2000). Inner core segment design for drug delivery control of thermo-responsive polymeric micelles. *Journal of Controlled Release, 65*, 93–103.

Chung, J. E., Yokoyama, M., Yamato, M., Aoyagi, T., Sakurai, Y., & Okano, T. (1999). Thermo-responsive drug delivery from polymeric micelles constructed using block copolymers of poly(N-isopropylacrylamide) and poly(butylmethacrylate). *Journal of Controlled Release, 62*, 115–127.

Corrie, P. G. (2008). Cytotoxic chemotherapy: Clinical aspects. *Medicine, 36*, 24–28.

Crile, G., Jr. (1962). Selective destruction of cancers after exposure to heat. *Annals of Surgery, 156*, 404–407.

Dagar, S., Krishnadas, A., Rubinstein, I., Blend, M. J., & Onyuksel, H. (2003). VIP grafted sterically stabilized liposomes for targeted imaging of breast cancer: In vivo studies. *Journal of Controlled Release, 91*, 123–133.

Danhier, F., Feron, O., & Preat, V. (2010). To exploit the tumor microenvironment: Passive and active tumor targeting of nanocarriers for anti-cancer drug delivery. *Journal of Controlled Release, 148*, 135–146.

Daniels, T. R., Delgado, T., Rodriguez, J. A., Helguera, G., & Penichet, M. L. (2006). The transferrin receptor part I: Biology and targeting with cytotoxic antibodies for the treatment of cancer. *Clinical Immunology, 121*, 144–158.

Das, M., Mohanty, C., & Sahoo, S. K. (2009). Ligand-based targeted therapy for cancer tissue. *Expert Opinion on Drug Delivery, 6*, 285–304.

Das, S. K., Sarkar, S., Dash, R., Dent, P., Wang, X.-Y., Sarkar, D., et al. (2012). Chapter one—Cancer terminator viruses and approaches for enhancing therapeutic outcomes. In D. T. Curiel & P. B. Fisher (Eds.), *Advances in Cancer Research*, Vol. 115, (pp. 1–38). New York: Academic Press.

Dash, R., Azab, B., Quinn, B. A., Shen, X., Wang, X. Y., Das, S. K., et al. (2011). Apogossypol derivative BI-97C1 (Sabutoclax) targeting Mcl-1 sensitizes prostate cancer cells to mda-7/IL-24-mediated toxicity. *Proceedings of the National Academy of Sciences of the United States of America, 108*, 8785–8790.

Davalos, R. V., Mir, I. L., & Rubinsky, B. (2005). Tissue ablation with irreversible electroporation. *Annals of Biomedical Engineering, 33*, 223–231.

Decuzzi, P., & Ferrari, M. (2006). The adhesive strength of non-spherical particles mediated by specific interactions. *Biomaterials, 27*, 5307–5314.

Dellian, M., Helmlinger, G., Yuan, F., & Jain, R. K. (1996). Fluorescence ratio imaging of interstitial pH in solid tumours: Effect of glucose on spatial and temporal gradients. *British Journal of Cancer, 74*, 1206–1215.

D'Emanuele, A., & Attwood, D. (2005). Dendrimer-drug interactions. *Advanced Drug Delivery Reviews, 57*, 2147–2162.

De Marre, A., Seymour, L. W., & Schacht, E. (1994). Evaluation of the hydrolytic and enzymatic stability of macromolecular Mitomycin C derivatives. *Journal of Controlled Release, 31*, 89–97.

DeMarre, A., Soyez, H., & Schacht, E. (1994). Synthesis of macromolecular Mitomycin C derivatives. *Journal of Controlled Release, 32*, 129–137.

Deonarain, M. P., Kousparou, C. A., & Epenetos, A. A. (2009). Antibodies targeting cancer stem cells: A new paradigm in immunotherapy? *MAbs, 1*, 12–25.

Devalapally, H., Shenoy, D., Little, S., Langer, R., & Amiji, M. (2007). Poly(ethylene oxide)-modified poly(beta-amino ester) nanoparticles as a pH-sensitive system for tumor-targeted delivery of hydrophobic drugs: Part 3. Therapeutic efficacy and safety studies in ovarian cancer xenograft model. *Cancer Chemotherapy and Pharmacology*, *59*, 477–484.

Dhar, S., Kolishetti, N., Lippard, S. J., & Farokhzad, O. C. (2011). Targeted delivery of a cisplatin prodrug for safer and more effective prostate cancer therapy *in vivo*. *Proceedings of the National Academy of Sciences*, *108*, 1850–1855.

Dijkgraaf, I., Kruijtzer, J. A. W., Frielink, C., Corstens, F. H. M., Oyen, W. J. G., Liskamp, R. M. J., et al. (2007). αvβ3 Integrin-targeting of intraperitoneally growing tumors with a radiolabeled RGD peptide. *International Journal of Cancer*, *120*, 605–610.

Dong, Y., & Feng, S.-S. (2004). Methoxy poly(ethylene glycol)-poly(lactide) (MPEG-PLA) nanoparticles for controlled delivery of anticancer drugs. *Biomaterials*, *25*, 2843–2849.

Drolet, D. W., Nelson, J., Tucker, C. E., Zack, P. M., Nixon, K., Bolin, R., et al. (2000). Pharmacokinetics and safety of an anti-vascular endothelial growth factor aptamer (NX1838) following injection into the vitreous humor of rhesus monkeys. *Pharmaceutical Research*, *17*, 1503–1510.

Duncan, R. (2003). The dawning era of polymer therapeutics. *Nature Reviews. Drug Discovery*, *2*, 347–360.

Egusquiaguirre, S., Igartua, M., Hernández, R., & Pedraz, J. (2012). Nanoparticle delivery systems for cancer therapy: Advances in clinical and preclinical research. *Clinical & Translational Oncology*, *14*, 83–93.

Ellington, A. D., & Szostak, J. W. (1990). *In vitro* selection of RNA molecules that bind specific ligands. *Nature*, *346*, 818–822.

Elnakat, H., & Ratnam, M. (2004). Distribution, functionality and gene regulation of folate receptor isoforms: Implications in targeted therapy. *Advanced Drug Delivery Reviews*, *56*, 1067–1084.

Etrych, T., Jelinkova, M., Rihova, B., & Ulbrich, K. (2001). New HPMA copolymers containing doxorubicin bound via pH-sensitive linkage: Synthesis and preliminary *in vitro* and *in vivo* biological properties. *Journal of Controlled Release*, *73*, 89–102.

Fajardo, L. F., & Prionas, S. D. (1994). Endothelial cells and hyperthermia. *International Journal of Hyperthermia*, *10*, 347–353.

Fan, Z., Lu, Y., Wu, X., & Mendelsohn, J. (1994). Antibody-induced epidermal growth factor receptor dimerization mediates inhibition of autocrine proliferation of A431 squamous carcinoma cells. *Journal of Biological Chemistry*, *269*, 27595–27602.

Fang, R. H., & Zhang, L. (2011). Dispersion-based methods for the engineering and manufacture of polymeric nanoparticles for drug delivery applications. *Journal of Nanoengineering and Nanomanufacturing*, *1*, 106–112.

Fant, K., Esbjorner, E. K., Jenkins, A., Grossel, M. C., Lincoln, P., & Norden, B. (2010). Effects of PEGylation and acetylation of PAMAM dendrimers on DNA binding, cytotoxicity and in vitro transfection efficiency. *Molecular Pharmaceutics*, *7*, 1734–1746.

Farokhzad, O. C., Jon, S., Khademhosseini, A., Tran, T.-N. T., LaVan, D. A., & Langer, R. (2004). Nanoparticle-aptamer bioconjugates. *Cancer Research*, *64*, 7668–7672.

Farokhzad, O. C., Karp, J. M., & Langer, R. (2006). Nanoparticle-aptamer bioconjugates for cancer targeting. *Expert Opinion on Drug Delivery*, *3*, 311–324.

Farokhzad, O. C., Khademhosseini, A., Jon, S., Hermmann, A., Cheng, J., Chin, C., et al. (2005). Microfluidic system for studying the interaction of nanoparticles and microparticles with cells. *Analytical Chemistry*, *77*, 5453–5459.

Faulk, W. P., Taylor, C. G., Yeh, C. J., & McIntyre, J. A. (1990). Preliminary clinical study of transferrin-adriamycin conjugate for drug delivery to acute leukemia patients. *Molecular Biotherapy*, *2*, 57–60.

Feazell, R. P., Nakayama-Ratchford, N., Dai, H., & Lippard, S. J. (2007). Soluble single-walled carbon nanotubes as longboat delivery systems for platinum(IV) anticancer drug design. *Journal of the American Chemical Society*, *129*, 8438–8439.

Felgner, J. H., Kumar, R., Sridhar, C. N., Wheeler, C. J., Tsai, Y. J., Border, R., et al. (1994). Enhanced gene delivery and mechanism studies with a novel series of cationic lipid formulations. *Journal of Biological Chemistry, 269*, 2550–2561.

Ferrara, N. (2005). VEGF as a therapeutic target in cancer. *Oncology, 69*(Suppl. 3), 11–16.

Frechet, J. M. (2002). Dendrimers and supramolecular chemistry. *Proceedings of the National Academy of Sciences of the United States of America, 99*, 4782–4787.

Fresta, M., Puglisi, G., Giammona, G., Cavallaro, G., Micali, N., & Furneri, P. M. (1995). Pefloxacine mesilate- and ofloxacin-loaded polyethylcyanoacrylate nanoparticles: Characterization of the colloidal drug carrier formulation. *Journal of Pharmaceutical Sciences, 84*, 895–902.

Gaber, M. H., Wu, N. Z., Hong, K., Huang, S. K., Dewhirst, M. W., & Papahadjopoulos, D. (1996). Thermosensitive liposomes: Extravasation and release of contents in tumor microvascular networks. *International Journal of Radiation Oncology, Biology, Physics, 36*, 1177–1187.

Gao, W., Chan, J. M., & Farokhzad, O. C. (2010). pH-Responsive nanoparticles for drug delivery. *Molecular Pharmaceutics, 7*, 1913–1920.

Gao, H., Qian, J., Cao, S., Yang, Z., Pang, Z., Pan, S., et al. (2012). Precise glioma targeting of and penetration by aptamer and peptide dual-functioned nanoparticles. *Biomaterials, 33*, 5115–5123.

Gao, W., Xiao, Z., Radovic-Moreno, A., Shi, J., Langer, R., & Farokhzad, O. C. (2010). Progress in siRNA delivery using multifunctional nanoparticles. *Methods in Molecular Biology, 629*, 53–67.

Gatter, K. C., Brown, G., Trowbridge, I. S., Woolston, R. E., & Mason, D. Y. (1983). Transferrin receptors in human tissues: Their distribution and possible clinical relevance. *Journal of Clinical Pathology, 36*, 539–545.

Geng, Y., Dalhaimer, P., Cai, S., Tsai, R., Tewari, M., Minko, T., et al. (2007). Shape effects of filaments versus spherical particles in flow and drug delivery. *Nature Nanotechnology, 2*, 249–255.

Gialeli, C., Theocharis, A. D., & Karamanos, N. K. (2010). Roles of matrix metalloproteinases in cancer progression and their pharmacological targeting. *FEBS Journal, 278*, 16–27.

Gillies, E. R., & Frechet, J. M. (2005). pH-Responsive copolymer assemblies for controlled release of doxorubicin. *Bioconjugate Chemistry, 16*, 361–368.

Gong, J., Chen, M., Zheng, Y., Wang, S., & Wang, Y. (2012). Polymeric micelles drug delivery system in oncology. *Journal of Controlled Release, 159*, 312–323.

Gong, W., Wang, Z., Liu, N., Lin, W., Wang, X., Xu, D., et al. (2011). Improving Efficiency of Adriamycin Crossing Blood Brain Barrier by Combination of Thermosensitive Liposomes and Hyperthermia. *Biological & Pharmaceutical Bulletin, 34*, 1058–1064.

Gottesman, M. M. (2002). Mechanisms of cancer drug resistance. *Annual Review of Medicine, 53*, 615–627.

Gottesman, M. M., Fojo, T., & Bates, S. E. (2002). Multidrug resistance in cancer: Role of ATP-dependent transporters. *Nature Reviews. Cancer, 2*, 48–58.

Graff, A., Tropel, D., Raman, S. K., Machaidze, G., Aebi, U., & Burkhard, P. (2005). Peptidic nanoparticles: For cancer diagnosis and therapy. *NanoBiotechnology, 1*, 293–294.

Graham, M. M., & Menda, Y. (2011). Radiopeptide imaging and therapy in the United States. *Journal of Nuclear Medicine, 52*(Suppl. 2), 56S–63S.

Greco, A., Di Benedetto, A., Howard, C. M., Kelly, S., Nande, R., Dementieva, Y., et al. (2010). Eradication of therapy-resistant human prostate tumors using an ultrasound-guided site-specific cancer terminator virus delivery approach. *Molecular Therapy, 18*, 295–306.

Gref, R., Luck, M., Quellec, P., Marchand, M., Dellacherie, E., Harnisch, S., et al. (2000). 'Stealth' corona-core nanoparticles surface modified by polyethylene glycol (PEG): Influences of the corona (PEG chain length and surface density) and of the core composition on phagocytic uptake and plasma protein adsorption. *Colloids and Surfaces. B, Biointerfaces, 18*, 301–313.

Gref, R., Minamitake, Y., Peracchia, M. T., Trubetskoy, V., Torchilin, V., & Langer, R. (1994). Biodegradable long-circulating polymeric nanospheres. *Science, 263*, 1600–1603.

Griffiths, G., Nystrom, B., Sable, S. B., & Khuller, G. K. (2010). Nanobead-based interventions for the treatment and prevention of tuberculosis. *Nature Reviews. Microbiology, 8*, 827–834.

Gu, F. X., Karnik, R., Wang, A. Z., Alexis, F., Levy-Nissenbaum, E., Hong, S., et al. (2007). Targeted nanoparticles for cancer therapy. *Nano Today, 2*, 14–21.

Haley, B., & Frenkel, E. (2008). Nanoparticles for drug delivery in cancer treatment. *Urologic Oncology, 26*, 57–64.

Hans, M. L., & Lowman, A. M. (2002). Biodegradable nanoparticles for drug delivery and targeting. *Current Opinion in Solid State and Materials Science, 6*, 319–327.

Head, J. F., Wang, F., & Elliott, R. L. (1997). Antineoplastic drugs that interfere with iron metabolism in cancer cells. *Advances in Enzyme Regulation, 37*, 147–169.

Hicke, B. J., & Stephens, A. W. (2000). Escort aptamers: A delivery service for diagnosis and therapy. *The Journal of Clinical Investigation, 106*, 923–928.

Hilgenbrink, A. R., & Low, P. S. (2005). Folate receptor-mediated drug targeting: From therapeutics to diagnostics. *Journal of Pharmaceutical Sciences, 94*, 2135–2146.

Hobbs, S. K., Monsky, W. L., Yuan, F., Roberts, W. G., Griffith, L., Torchilin, V. P., et al. (1998). Regulation of transport pathways in tumor vessels: Role of tumor type and microenvironment. *Proceedings of the National Academy of Sciences of the United States of America, 95*, 4607–4612.

Hölig, P., Bach, M., Völkel, T., Nahde, T., Hoffmann, S., Müller, R., et al. (2004). Novel RGD lipopeptides for the targeting of liposomes to integrin-expressing endothelial and melanoma cells. *Protein Engineering, Design & Selection, 17*, 433–441.

Honer, M., Mu, L., Stellfeld, T., Graham, K., Martic, M., Fischer, C. R., et al. (2011). 18F-labeled bombesin analog for specific and effective targeting of prostate tumors expressing gastrin-releasing peptide receptors. *Journal of Nuclear Medicine, 52*, 270–278.

Hood, J. D., Bednarski, M., Frausto, R., Guccione, S., Reisfeld, R. A., Xiang, R., et al. (2002). Tumor regression by targeted gene delivery to the neovasculature. *Science, 296*, 2404–2407.

Hosokawa, T., Sami, M., Kato, Y., & Hayakawa, E. (2003). Alteration in the temperature-dependent content release property of thermosensitive liposomes in plasma. *Chemical & Pharmaceutical Bulletin (Tokyo), 51*, 1227–1232.

Hoste, K., De Winne, K., & Schacht, E. (2004). Polymeric prodrugs. *International Journal of Pharmaceutics, 277*, 119–131.

Hou, J., Zhang, Q., Li, X., Tang, Y., Cao, M. R., Bai, F., et al. (2011). Synthesis of novel folate conjugated fluorescent nanoparticles for tumor imaging. *Journal of Biomedical Materials Research. Part A, 99*, 684–689.

Howlader, N., Noone, A. M., Krapcho, M., Neyman, N., Aminou, R., Waldron, W., et al. (2011). *SEER cancer statistics review, 1975–2008*. Bethesda, MD: National Cancer Institute.

Hu, C. M., & Zhang, L. (2009). Therapeutic nanoparticles to combat cancer drug resistance. *Current Drug Metabolism, 10*, 836–841.

Huang, Y. F., Shangguan, D., Liu, H., Phillips, J. A., Zhang, X., Chen, Y., et al. (2009). Molecular assembly of an aptamer-drug conjugate for targeted drug delivery to tumor cells. *Chembiochem, 10*, 862–868.

Hughes, G. A. (2005). Nanostructure-mediated drug delivery. *Nanomedicine, 1*, 22–30.

Hu-Lieskovan, S., Heidel, J. D., Bartlett, D. W., Davis, M. E., & Triche, T. J. (2005). Sequence-specific knockdown of EWS-FLI1 by targeted, nonviral delivery of small interfering RNA inhibits tumor growth in a murine model of metastatic Ewing's sarcoma. *Cancer Research, 65*, 8984–8992.

Husseini, G. A., & Pitt, W. G. (2008). Micelles and nanoparticles for ultrasonic drug and gene delivery. *Advanced Drug Delivery Reviews, 60,* 1137–1152.

Iinuma, H., Maruyama, K., Okinaga, K., Sasaki, K., Sekine, T., Ishida, O., et al. (2002). Intracellular targeting therapy of cisplatin-encapsulated transferrin-polyethylene glycol liposome on peritoneal dissemination of gastric cancer. *International Journal of Cancer, 99,* 130–137.

Iyer, A. K., Khaled, G., Fang, J., & Maeda, H. (2006). Exploiting the enhanced permeability and retention effect for tumor targeting. *Drug Discovery Today, 11,* 812–818.

Jain, R. K., Safabakhsh, N., Sckell, A., Chen, Y., Jiang, P., Benjamin, L., et al. (1998). Endothelial cell death, angiogenesis, and microvascular function after castration in an androgen-dependent tumor: Role of vascular endothelial growth factor. *Proceedings of the National Academy of Sciences of the United States of America, 95,* 10820–10825.

Jeong, B., Bae, Y. H., & Kim, S. W. (2000). Drug release from biodegradable injectable thermosensitive hydrogel of PEG-PLGA-PEG triblock copolymers. *Journal of Controlled Release, 63,* 155–163.

Jin, L., Hope, K. J., Zhai, Q., Smadja-Joffe, F., & Dick, J. E. (2006). Targeting of CD44 eradicates human acute myeloid leukemic stem cells. *Nature Medicine, 12,* 1167–1174.

Jokerst, J. V., Lobovkina, T., Zare, R. N., & Gambhir, S. S. (2011). Nanoparticle PEGylation for imaging and therapy. *Nanomedicine (London, England), 6,* 715–728.

Kabanov, A. V., Batrakova, E. V., & Alakhov, V. Y. (2002). Pluronic block copolymers as novel polymer therapeutics for drug and gene delivery. *Journal of Controlled Release, 82,* 189–212.

Kaczorowska, M. A., & Cooper, H. J. (2008). Electron capture dissociation, electron detachment dissociation, and collision-induced dissociation of polyamidoamine (PAMAM) dendrimer ions with amino, amidoethanol, and sodium carboxylate surface groups. *Journal of the American Society for Mass Spectrometry, 19,* 1312–1319.

Kakar, S. S., Jin, H., Hong, B., Eaton, J. W., Kang, K. A., Kang, K. A., et al. (2008). *LHRH receptor targeted therapy for breast cancer oxygen transport to tissue XXIX* (Vol. 614). New York: Springer, 285–296.

Kam, N. W., & Dai, H. (2005). Carbon nanotubes as intracellular protein transporters: Generality and biological functionality. *Journal of the American Chemical Society, 127,* 6021–6026.

Kamada, H., Tsutsumi, Y., Yoshioka, Y., Yamamoto, Y., Kodaira, H., Tsunoda, S., et al. (2004). Design of a pH-sensitive polymeric carrier for drug release and its application in cancer therapy. *Clinical Cancer Research, 10,* 2545–2550.

Kaminskas, L. M., Boyd, B. J., & Porter, C. J. (2011). Dendrimer pharmacokinetics: The effect of size, structure and surface characteristics on ADME properties. *Nanomedicine (London, England), 6,* 1063–1084.

Kaneko, Y., Nakamura, S., Sakai, K., Kikuchi, A., Aoyagi, T., Sakurai, Y., et al. (1999). Synthesis and swelling-deswelling kinetics of poly(N-isopropylacrylamide) hydrogels grafted with LCST modulated polymers. *Journal of Biomaterials Science. Polymer Edition, 10,* 1079–1091.

Kaptain, S., Tan, L. K., & Chen, B. (2001). Her-2/neu and breast cancer. *Diagnostic Molecular Pathology, 10,* 139–152.

Kedar, U., Phutane, P., Shidhaye, S., & Kadam, V. (2010). Advances in polymeric micelles for drug delivery and tumor targeting. *Nanomedicine, 6,* 714–729.

Khajavinia, A., Varshosaz, J., & Dehkordi, A. J. (2012). Targeting etoposide to acute myelogenous leukaemia cells using nanostructured lipid carriers coated with transferrin. *Nanotechnology, 23,* 405101.

Kim, S., Kim, J.-H., Jeon, O., Kwon, I. C., & Park, K. (2009). Engineered polymers for advanced drug delivery. *European Journal of Pharmaceutics and Biopharmaceutics, 71,* 420–430.

Kim, S., Shi, Y., Kim, J. Y., Park, K., & Cheng, J.-X. (2010). Overcoming the barriers in micellar drug delivery: Loading efficiency, *in vivo* stability, and micelle-cell interaction. *Expert Opinion on Drug Delivery, 7,* 49–62.

Kitchens, D. L., Snyder, E. Y., & Gottlieb, D. I. (1994). FGF and EGF are mitogens for immortalized neural progenitors. *Journal of Neurobiology, 25*, 797–807.

Klein, S., & Levitzki, A. (2007). Targeted cancer therapy: Promise and reality. In George F. Vande Woude & George Klein (Eds.), *Advances in Cancer Research*, Vol. 97, (pp. 295–319). New York: Academic Press.

Kobayashi, T. (2011). Cancer hyperthermia using magnetic nanoparticles. *Biotechnology Journal, 6*, 1342–1347.

Kobayashi, T., Ishida, T., Okada, Y., Ise, S., Harashima, H., & Kiwada, H. (2007). Effect of transferrin receptor-targeted liposomal doxorubicin in P-glycoprotein-mediated drug resistant tumor cells. *International Journal of Pharmaceutics, 329*, 94–102.

Kolhatkar, R. B., Kitchens, K. M., Swaan, P. W., & Ghandehari, H. (2007). Surface acetylation of polyamidoamine (PAMAM) dendrimers decreases cytotoxicity while maintaining membrane permeability. *Bioconjugate Chemistry, 18*, 2054–2060.

Kong, G., Braun, R. D., & Dewhirst, M. W. (2001). Characterization of the effect of hyperthermia on nanoparticle extravasation from tumor vasculature. *Cancer Research, 61*, 3027–3032.

Koo, Y.-E. L., Reddy, G. R., Bhojani, M., Schneider, R., Philbert, M. A., Rehemtulla, A., et al. (2006). Brain cancer diagnosis and therapy with nanoplatforms. *Advanced Drug Delivery Reviews, 58*, 1556–1577.

Kos, J., Obermajer, N., Doljak, B., Kocbek, P., & Kristl, J. (2009). Inactivation of harmful tumour-associated proteolysis by nanoparticulate system. *International Journal of Pharmaceutics, 381*, 106–112.

Kremer, C., Breier, G., Risau, W., & Plate, K. H. (1997). Up-regulation of flk-1/vascular endothelial growth factor receptor 2 by its ligand in a cerebral slice culture system. *Cancer Research, 57*, 3852–3859.

Kreuter, J. (2001). Nanoparticulate systems for brain delivery of drugs. *Advanced Drug Delivery Reviews, 47*, 65–81.

Kukowska-Latallo, J. F., Candido, K. A., Cao, Z., Nigavekar, S. S., Majoros, I. J., Thomas, T. P., et al. (2005). Nanoparticle targeting of anticancer drug improves therapeutic response in animal model of human epithelial cancer. *Cancer Research, 65*, 5317–5324.

Kullberg, E. B., Nestor, M., & Gedda, L. (2003). Tumor-cell targeted epidermal growth factor liposomes loaded with boronated acridine: Uptake and processing. *Pharmaceutical Research, 20*, 229–236.

Kumari, A., Yadav, S. K., & Yadav, S. C. (2009). Biodegradable polymeric nanoparticles based drug delivery systems. *Colloids and Surfaces. B, Biointerfaces, 75*, 1–18.

Kwon, G., Suwa, S., Yokoyama, M., Okano, T., Sakurai, Y., & Kataoka, K. (1994). Enhanced tumor accumulation and prolonged circulation times of micelle-forming poly (ethylene oxide-aspartate) block copolymer-adriamycin conjugates. *Journal of Controlled Release, 29*, 17–23.

Labieniec, M., Ulicna, O., Vancova, O., Kucharska, J., Gabryelak, T., & Watala, C. (2009). Effect of poly(amido)amine (PAMAM) G4 dendrimer on heart and liver mitochondria in an animal model of diabetes. *Cell Biology International, 34*, 89–97.

Lacroix, M. (2008). Persistent use of "false" cell lines. *International Journal of Cancer, 122*, 1–4.

Laske, D. W., Youle, R. J., & Oldfield, E. H. (1997). Tumor regression with regional distribution of the targeted toxin TF-CRM107 in patients with malignant brain tumors. *Nature Medicine, 3*, 1362–1368.

Laskin, J. J., & Sandler, A. B. (2004). Epidermal growth factor receptor: A promising target in solid tumours. *Cancer Treatment Reviews, 30*, 1–17.

Lee, R. J., & Low, P. S. (1994). Delivery of liposomes into cultured KB cells via folate receptor-mediated endocytosis. *Journal of Biological Chemistry, 269*, 3198–3204.

Lee, C. C., MacKay, J. A., Frechet, J. M., & Szoka, F. C. (2005). Designing dendrimers for biological applications. *Nature Biotechnology, 23*, 1517–1526.

Lee, R. J., Wang, S., & Low, P. S. (1996). Measurement of endosome pH following folate receptor-mediated endocytosis. *Biochimica et Biophysica Acta, 1312*, 237–242.

Lee, P., Zhang, R., Li, V., Liu, X., Sun, R. W., Che, C. M., et al. (2012). Enhancement of anticancer efficacy using modified lipophilic nanoparticle drug encapsulation. *International Journal of Nanomedicine, 7*, 731–737.

Leemans, C. R., Braakhuis, B. J. M., & Brakenhoff, R. H. (2011). The molecular biology of head and neck cancer. *Nature Reviews. Cancer, 11*, 9–22.

Leslie, E. M., Deeley, R. G., & Cole, S. P. (2005). Multidrug resistance proteins: Role of P-glycoprotein, MRP1, MRP2, and BCRP (ABCG2) in tissue defense. *Toxicology and Applied Pharmacology, 204*, 216–237.

Li, X., Zhou, H., Yang, L., Du, G., Pai-Panandiker, A. S., Huang, X., et al. (2011). Enhancement of cell recognition *in vitro* by dual-ligand cancer targeting gold nanoparticles. *Biomaterials, 32*, 2540–2545.

Liang, F., & Chen, B. (2010). A review on biomedical applications of single-walled carbon nanotubes. *Current Medicinal Chemistry, 17*, 10–24.

Liechty, W. B., & Peppas, N. A. (2012). Expert opinion: Responsive polymer nanoparticles in cancer therapy. *European Journal of Pharmaceutics and Biopharmaceutics, 80*, 241–246.

Lin, S. Y., Makino, K., Xia, W., Matin, A., Wen, Y., Kwong, K. Y., et al. (2001). Nuclear localization of EGF receptor and its potential new role as a transcription factor. *Nature Cell Biology, 3*, 802–808.

Liu, Z., Chen, K., Davis, C., Sherlock, S., Cao, Q., Chen, X., et al. (2008). Drug delivery with carbon nanotubes for *in vivo* cancer treatment. *Cancer Research, 68*, 6652–6660.

Liu, D.-Z., Hsieh, J.-H., Fan, X.-C., Yang, J.-D., & Chung, T.-W. (2007). Synthesis, characterization and drug delivery behaviors of new PCP polymeric micelles. *Carbohydrate Polymers, 68*, 544–554.

Liu, Y., Miyoshi, H., & Nakamura, M. (2007). Nanomedicine for drug delivery and imaging: A promising avenue for cancer therapy and diagnosis using targeted functional nanoparticles. *International Journal of Cancer, 120*, 2527–2537.

Liu, D., Mori, A., & Huang, L. (1992). Role of liposome size and RES blockade in controlling biodistribution and tumor uptake of GM1-containing liposomes. *Biochimica et Biophysica Acta, 1104*, 95–101.

Liu, Z., Sun, X., Nakayama-Ratchford, N., & Dai, H. (2007). Supramolecular chemistry on water-soluble carbon nanotubes for drug loading and delivery. *ACS Nano, 1*, 50–56.

Loh, X. J., Yee, B. J. H., & Chia, F. S. (2012). Sustained delivery of paclitaxel using thermogelling poly(PEG/PPG/PCL urethane)s for enhanced toxicity against cancer cells. *Journal of Biomedical Materials Research. Part A, 100*, 2686–2694.

Lorusso, V., Manzione, L., & Silvestris, N. (2007). Role of liposomal anthracyclines in breast cancer. *Annals of Oncology, 18*, vi70–vi73.

Lovett, K. M., Liang, B. A., & Mackey, T. K. (2012). Risks of online direct-to-consumer tumor markers for cancer screening. *Journal of Clinical Oncology, 30*, 1411–1414.

Low, P. S., & Antony, A. C. (2004). Folate receptor-targeted drugs for cancer and inflammatory diseases. *Advanced Drug Delivery Reviews, 56*, 1055–1058.

Low, P. S., & Kularatne, S. A. (2009). Folate-targeted therapeutic and imaging agents for cancer. *Current Opinion in Chemical Biology, 13*, 256–262.

Lu, J., Jackson, J. K., Gleave, M. E., & Burt, H. M. (2008). The preparation and characterization of anti-VEGFR2 conjugated, paclitaxel-loaded PLLA or PLGA microspheres for the systemic targeting of human prostate tumors. *Cancer Chemotherapy and Pharmacology, 61*, 997–1005.

Lu, Y., & Low, P. S. (2002). Folate-mediated delivery of macromolecular anticancer therapeutic agents. *Advanced Drug Delivery Reviews, 54*, 675–693.

Lurje, G., & Lenz, H. J. (2009). EGFR signaling and drug discovery. *Oncology, 77*, 400–410.

Maeda, H. (2001). The enhanced permeability and retention (EPR) effect in tumor vasculature: The key role of tumor-selective macromolecular drug targeting. *Advances in Enzyme Regulation, 41*, 189–207.

Maeda, H., Bharate, G. Y., & Daruwalla, J. (2009). Polymeric drugs for efficient tumor-targeted drug delivery based on EPR-effect. *European Journal of Pharmaceutics and Biopharmaceutics, 71*, 409–419.

Maeda, H., Sawa, T., & Konno, T. (2001). Mechanism of tumor-targeted delivery of macromolecular drugs, including the EPR effect in solid tumor and clinical overview of the prototype polymeric drug SMANCS. *Journal of Controlled Release, 74*, 47–61.

Malam, Y., Loizidou, M., & Seifalian, A. M. (2009). Liposomes and nanoparticles: Nanosized vehicles for drug delivery in cancer. *Trends in Pharmacological Sciences, 30*, 592–599.

Malik, A. K., & Gerber, H.-P. (2003). Targeting VEGF ligands and receptors in cancer. *Targets, 2*, 48–57.

Malik, N., Wiwattanapatapee, R., Klopsch, R., Lorenz, K., Frey, H., Weener, J. W., et al. (2000). Dendrimers: Relationship between structure and biocompatibility in vitro, and preliminary studies on the biodistribution of 125I-labelled polyamidoamine dendrimers in vivo. *Journal of Controlled Release, 65*, 133–148.

Masson, O., Prebois, C., Derocq, D., Meulle, A., Dray, C., Daviaud, D., et al. (2011). Cathepsin-D, a key protease in breast cancer, is up-regulated in obese mouse and human adipose tissue, and controls adipogenesis. *PLoS One, 6*, e16452.

Matsumura, Y., & Maeda, H. (1986). A new concept for macromolecular therapeutics in cancer chemotherapy: Mechanism of tumoritropic accumulation of proteins and the antitumor agent smancs. *Cancer Research, 46*, 6387–6392.

Medina, S. H., & El-Sayed, M. E. H. (2009). Dendrimers as carriers for delivery of chemotherapeutic agents. *Chemical Reviews, 109*, 3141–3157.

Meng, S., Su, B., Li, W., Ding, Y., Tang, L., Zhou, W., et al. (2010). Enhanced antitumor effect of novel dual-targeted paclitaxel liposomes. *Nanotechnology, 21*, 415103.

Merdan, T., Kopecek, J., & Kissel, T. (2002). Prospects for cationic polymers in gene and oligonucleotide therapy against cancer. *Advanced Drug Delivery Reviews, 54*, 715–758.

Milhem, O. M., Myles, C., McKeown, N. B., Attwood, D., & D'Emanuele, A. (2000). Polyamidoamine Starburst dendrimers as solubility enhancers. *International Journal of Pharmaceutics, 197*, 239–241.

Moghimi, S. M., Hunter, A. C., & Murray, J. C. (2005). Nanomedicine: Current status and future prospects. *The FASEB Journal, 19*, 311–330.

Mohanty, C., Das, M., Kanwar, J. R., & Sahoo, S. K. (2011). Receptor mediated tumor targeting: An emerging approach for cancer therapy. *Current Drug Delivery, 8*, 45–58.

Momekova, D., Rangelov, S., Yanev, S., Nikolova, E., Konstantinov, S., Romberg, B., et al. (2007). Long-circulating, pH-sensitive liposomes sterically stabilized by copolymers bearing short blocks of lipid-mimetic units. *European Journal of Pharmaceutical Sciences, 32*, 308–317.

Montet, X., Funovics, M., Montet-Abou, K., Weissleder, R., & Josephson, L. (2006). Multivalent effects of RGD peptides obtained by nanoparticle display. *Journal of Medicinal Chemistry, 49*, 6087–6093.

Murphy, E. A., Majeti, B. K., Barnes, L. A., Makale, M., Weis, S. M., Lutu-Fuga, K., et al. (2008). Nanoparticle-mediated drug delivery to tumor vasculature suppresses metastasis. *Proceedings of the National Academy of Sciences of the United States of America, 105*, 9343–9348.

Nakamura, K., Abu Lila, A. S., Matsunaga, M., Doi, Y., Ishida, T., & Kiwada, H. (2011). A double-modulation strategy in cancer treatment with a chemotherapeutic agent and siRNA. *Molecular Therapy, 19*, 2040–2047.

Nakayama, M., Okano, T., Miyazaki, T., Kohori, F., Sakai, K., & Yokoyama, M. (2006). Molecular design of biodegradable polymeric micelles for temperature-responsive drug release. *Journal of Controlled Release, 115*, 46–56.

Napier, M. P., Sharma, S. K., Springer, C. J., Bagshawe, K. D., Green, A. J., Martin, J., et al. (2000). Antibody-directed enzyme prodrug therapy: Efficacy and mechanism of action in colorectal carcinoma. *Clinical Cancer Research*, *6*, 765–772.

Nigavekar, S. S., Sung, L. Y., Llanes, M., El-Jawahri, A., Lawrence, T. S., Becker, C. W., et al. (2004). 3H dendrimer nanoparticle organ/tumor distribution. *Pharmaceutical Research*, *21*, 476–483.

Nimjee, S. M., Rusconi, C. P., Harrington, R. A., & Sullenger, B. A. (2005). The potential of aptamers as anticoagulants. *Trends in Cardiovascular Medicine*, *15*, 41–45.

Norman, J., Badie-Dezfooly, B., Nord, E. P., Kurtz, I., Schlosser, J., Chaudhari, A., et al. (1987). EGF-induced mitogenesis in proximal tubular cells: Potentiation by angiotensin II. *American Journal of Physiology. Renal Physiology*, *253*, F299–F309.

Oerlemans, C., Bult, W., Bos, M., Storm, G., Nijsen, J., & Hennink, W. (2010). Polymeric micelles in anticancer therapy: Targeting, imaging and triggered release. *Pharmaceutical Research*, *27*, 2569–2589.

Ohlsson, B., Rehfeld, J. F., & Axelson, J. (1999). CCK stimulates growth of both the pancreas and the liver. *International Journal of Surgical Investigation*, *1*, 47–54.

Onyuksel, H., Mohanty, P. S., & Rubinstein, I. (2009). VIP-grafted sterically stabilized phospholipid nanomicellar 17-allylamino-17-demethoxy geldanamycin: A novel targeted nanomedicine for breast cancer. *International Journal of Pharmaceutics*, *365*, 157–161.

Owens, D. E., III., & Peppas, N. A. (2006). Opsonization, biodistribution, and pharmacokinetics of polymeric nanoparticles. *International Journal of Pharmaceutics*, *307*, 93–102.

Pantarotto, D., Briand, J. P., Prato, M., & Bianco, A. (2004). Translocation of bioactive peptides across cell membranes by carbon nanotubes. *Chemical Communications (Cambridge, England)*, *1*, 16–17.

Panyam, J., & Labhasetwar, V. (2003). Biodegradable nanoparticles for drug and gene delivery to cells and tissue. *Advanced Drug Delivery Reviews*, *55*, 329–347.

Pardridge, W. M. (2002). Drug and gene delivery to the brain: The vascular route. *Neuron*, *36*, 555–558.

Parhi, P., Mohanty, C., & Sahoo, S. K. (2012). Nanotechnology-based combinational drug delivery: An emerging approach for cancer therapy. *Drug Discovery Today*, *17*, 1044–1052.

Park, J. W., Hong, K., Kirpotin, D. B., Colbern, G., Shalaby, R., Baselga, J., et al. (2002). Anti-HER2 immunoliposomes: Enhanced efficacy attributable to targeted delivery. *Clinical Cancer Research*, *8*, 1172–1181.

Park, J. W., Kirpotin, D. B., Hong, K., Shalaby, R., Shao, Y., Nielsen, U. B., et al. (2001). Tumor targeting using anti-her2 immunoliposomes. *Journal of Controlled Release*, *74*, 95–113.

Park, E. K., Lee, S. B., & Lee, Y. M. (2005). Preparation and characterization of methoxy poly(ethylene glycol)/poly(Îμ-caprolactone) amphiphilic block copolymeric nanospheres for tumor-specific folate-mediated targeting of anticancer drugs. *Biomaterials*, *26*, 1053–1061.

Parveen, S., Misra, R., & Sahoo, S. K. (2012). Nanoparticles: A boon to drug delivery, therapeutics, diagnostics and imaging. *Nanomedicine*, *8*, 147–166.

Pastorino, F., Brignole, C., Di Paolo, D., Nico, B., Pezzolo, A., Marimpietri, D., et al. (2006). Targeting liposomal chemotherapy via both tumor cell-specific and tumor vasculature-specific ligands potentiates therapeutic efficacy. *Cancer Research*, *66*, 10073–10082.

Patri, A. K., Majoros, I. J., & Baker, J. R. (2002). Dendritic polymer macromolecular carriers for drug delivery. *Current Opinion in Chemical Biology*, *6*, 466–471.

Peng, C.-L., Tsai, H.-M., Yang, S.-J., Luo, T.-Y., Lin, C.-F., Lin, W.-J., et al. (2011). Development of thermosensitive poly(N-isopropylacrylamide- co -((2-dimethylamino) ethyl methacrylate))-based nanoparticles for controlled drug release. *Nanotechnology*, *22*, 265608.

Peracchia, M. T., Fattala, E., Desmaëleb, D., Besnarda, M., Noëlc, J. P., Gomisc, J. M., et al. (1999). Stealth® PEGylated polycyanoacrylate nanoparticles for intravenous administration and splenic targeting. *Journal of Controlled Release, 60*, 121–128.

Perrault, S. D., Walkey, C., Jennings, T., Fischer, H. C., & Chan, W. C. W. (2009). Mediating tumor targeting efficiency of nanoparticles through design. *Nano Letters, 9*, 1909–1915.

Pestourie, C., Tavitian, B., & Duconge, F. (2005). Aptamers against extracellular targets for *in vivo* applications. *Biochimie, 87*, 921–930.

Podgorski, I., & Sloane, B. F. (2003). Cathepsin B and its role(s) in cancer progression. *Biochemical Society Symposium, 70*, 263–276.

Prakash, S., Malhotra, M., Shao, W., Tomaro-Duchesneau, C., & Abbasi, S. (2011). Polymeric nanohybrids and functionalized carbon nanotubes as drug delivery carriers for cancer therapy. *Advanced Drug Delivery Reviews, 63*, 1340–1351.

Pulkkinen, M., Pikkarainen, J., Wirth, T., Tarvainen, T., Haapa-aho, V., Korhonen, H., et al. (2008). Three-step tumor targeting of paclitaxel using biotinylated PLA-PEG nanoparticles and avidin-biotin technology: Formulation development and *in vitro* anticancer activity. *European Journal of Pharmaceutics and Biopharmaceutics, 70*, 66–74.

Purushotham, S., Chang, P., Rumpel, H., Kee, I., Ng, R., Chow, P., et al. (2009). Thermoresponsive core-shell magnetic nanoparticles for combined modalities of cancer therapy. *Nanotechnology, 20*, 305101.

Radinsky, R., Risin, S., Fan, D., Dong, Z., Bielenberg, D., Bucana, C. D., et al. (1995). Level and function of epidermal growth factor receptor predict the metastatic potential of human colon carcinoma cells. *Clinical Cancer Research, 1*, 19–31.

Rainov, N. G., & Soling, A. (2005). Technology evaluation: TransMID, KS Biomedix/Nycomed/Sosei/PharmaEngine. *Current Opinion in Molecular Therapeutics, 7*, 483–492.

Reubi, J. C. (2003). Peptide receptors as molecular targets for cancer diagnosis and therapy. *Endocrine Reviews, 24*, 389–427.

Roberts, J. C., Bhalgat, M. K., & Zera, R. T. (1996). Preliminary biological evaluation of polyamidoamine (PAMAM) Starburst dendrimers. *Journal of Biomedical Materials Research, 30*, 53–65.

Roda, E., Coccini, T., Acerbi, D., Barni, S., Vaccarone, R., & Manzo, L. (2011). Comparative pulmonary toxicity assessment of pristine and functionalized multi-walled carbon nanotubes intratracheally instilled in rats: Morphohistochemical evaluations. *Histology and Histopathology, 26*, 357–367.

Rubin, P., & Casarett, G. (1966). Microcirculation of tumors. I. Anatomy, function, and necrosis. *Clinical Radiology, 17*, 220–229.

Sahoo, S. K., & Labhasetwar, V. (2003). Nanotech approaches to drug delivery and imaging. *Drug Discovery Today, 8*, 1112–1120.

Sahoo, S. K., & Labhasetwar, V. (2005). Enhanced antiproliferative activity of transferrin-conjugated paclitaxel-loaded nanoparticles is mediated via sustained intracellular drug retention. *Molecular Pharmaceutics, 2*, 373–383.

Sahoo, S. K., Ma, W., & Labhasetwar, V. (2004). Efficacy of transferrin-conjugated paclitaxel-loaded nanoparticles in a murine model of prostate cancer. *International Journal of Cancer, 112*, 335–340.

Saito, Y., Yasunaga, M., Kuroda, J.-I., Koga, Y., & Matsumura, Y. (2007). Antitumour activity of NK012, SN-38-incorporating polymeric micelles, in hypovascular orthotopic pancreatic tumour. *European Journal of Cancer, 46*, 650–658.

Sakai, K., Mori, S., Kawamoto, T., Taniguchi, S., Kobori, O., Morioka, Y., et al. (1986). Expression of epidermal growth factor receptors on normal human gastric epithelia and gastric carcinomas. *Journal of the National Cancer Institute, 77*, 1047–1052.

Salomon, D. S., Brandt, R., Ciardiello, F., & Normanno, N. (1995). Epidermal growth factor-related peptides and their receptors in human malignancies. *Critical Reviews in Oncology/Hematology, 19*, 183–232.

Sampathkumar, S.-G. , & Yarema, K. J. (2007). Nanomaterials for cancer diagnosis. In *Nanotechnologies for the life sciences*: Vol. 7. Weinheim: Wiley-VCH Verlag GmbH & Co. KGaA.

Sandoval, M. A., Sloat, B. R., Lansakara, P. D., Kumar, A., Rodriguez, B. L., Kiguchi, K., et al. (2012). EGFR-targeted stearoyl gemcitabine nanoparticles show enhanced antitumor activity. *Journal of Controlled Release, 157*, 287–296.

Sapra, P., & Allen, T. M. (2003). Ligand-targeted liposomal anticancer drugs. *Progress in Lipid Research, 42*, 439–462.

Saul, J. M., Annapragada, A. V., & Bellamkonda, R. V. (2006). A dual-ligand approach for enhancing targeting selectivity of therapeutic nanocarriers. *Journal of Controlled Release, 114*, 277–287.

Savić, R., Eisenberg, A., & Maysinger, D. (2006). Block copolymer micelles as delivery vehicles of hydrophobic drugs: Micelle–cell interactions. *Journal of Drug Targeting, 14*, 343–355.

Schneider, G. F., Subr, V., Ulbrich, K., & Decher, G. (2009). Multifunctional cytotoxic stealth nanoparticles. A model approach with potential for cancer therapy. *Nano Letters, 9*, 636–642.

Seki, T., Fang, J., & Maeda, H. (2009). Enhanced delivery of macromolecular antitumor drugs to tumors by nitroglycerin application. *Cancer Science, 100*, 2426–2430.

Service, R. F. (2010). Nanotechnology. Nanoparticle Trojan horses gallop from the lab into the clinic. *Science, 330*, 314–315.

Sharma, S. K., Bagshawe, K. D., Reddy, L. H., & Couvreur, P. (2010). *Antibody-directed enzyme prodrug therapy (ADEPT) for cancer macromolecular anticancer therapeutics* (pp. 393–406). New York: Springer.

Sharma, A., & Sharma, U. S. (1997). Liposomes in drug delivery: Progress and limitations. *International Journal of Pharmaceutics, 154*, 123–140.

Shenoy, D., Little, S., Langer, R., & Amiji, M. (2005). Poly(ethylene oxide)-modified Poly($\tilde{I}^2$-amino ester) nanoparticles as a pH-Sensitive system for tumor-targeted delivery of hydrophobic drugs. 1. In vitro evaluations. *Molecular Pharmaceutics, 2*, 357–366.

Shepard, D. M., Ferris, M. C., Olivera, G. H., & Mackie, T. R. (1999). Optimizing the delivery of radiation therapy to cancer patients. *SIAM Review, 41*, 721–744.

Shi Kam, N. W., Jessop, T. C., Wender, P. A., & Dai, H. (2004). Nanotube molecular transporters: Internalization of carbon nanotube-protein conjugates into Mammalian cells. *Journal of the American Chemical Society, 126*, 6850–6851.

Shmeeda, H., Mak, L., Tzemach, D., Astrahan, P., Tarshish, M., & Gabizon, A. (2006). Intracellular uptake and intracavitary targeting of folate-conjugated liposomes in a mouse lymphoma model with up-regulated folate receptors. *Molecular Cancer Therapeutics, 5*, 818–824.

Shuai, X., Merdan, T., Schaper, A. K., Xi, F., & Kissel, T. (2004). Core-cross-linked polymeric micelles as paclitaxel carriers. *Bioconjugate Chemistry, 15*, 441–448.

Shubik, P. (1982). Vascularization of tumors: A review. *Journal of Cancer Research and Clinical Oncology, 103*, 211–226.

Shweiki, D., Itin, A., Soffer, D., & Keshet, E. (1992). Vascular endothelial growth factor induced by hypoxia may mediate hypoxia-initiated angiogenesis. *Nature, 359*, 843–845.

Siegel, R., Naishadham, D., & Jemal, A. (2012). Cancer statistics, 2012. *CA: A Cancer Journal for Clinicians, 62*, 10–29.

Simoes, S., Moreira, J. N., Fonseca, C., Duzgunes, N., & de Lima, M. C. (2004). On the formulation of pH-sensitive liposomes with long circulation times. *Advanced Drug Delivery Reviews, 56*, 947–965.

Singh, M. (1999). Transferrin as a targeting ligand for liposomes and anticancer drugs. *Current Pharmaceutical Design, 5*, 443–451.

Sinha, R., Kim, G. J., Nie, S., & Shin, D. M. (2006). Nanotechnology in cancer therapeutics: Bioconjugated nanoparticles for drug delivery. *Molecular Cancer Therapeutics, 5*, 1909–1917.

Sirsi, S., & Borden, M. (2009). Microbubble compositions, properties and biomedical applications. *Bubble Science Engineering and Technology*, *1*, 3–17.
Smith, J. E., Medley, C. D., Tang, Z., Shangguan, D., Lofton, C., & Tan, W. (2007). Aptamer-conjugated nanoparticles for the collection and detection of multiple cancer cells. *Analytical Chemistry*, *79*, 3075–3082.
Soga, O., van Nostrum, C. F., Fens, M., Rijcken, C. J., Schiffelers, R. M., Storm, G., et al. (2005). Thermosensitive and biodegradable polymeric micelles for paclitaxel delivery. *Journal of Controlled Release*, *103*, 341–353.
Sonavane, G., Tomoda, K., & Makino, K. (2008). Biodistribution of colloidal gold nanoparticles after intravenous administration: Effect of particle size. *Colloids and Surfaces. B, Biointerfaces*, *66*, 274–280.
Sosabowski, J. K., Matzow, T., Foster, J. M., Finucane, C., Ellison, D., Watson, S. A., et al. (2009). Targeting of CCK-2 receptor-expressing tumors using a radiolabeled divalent gastrin peptide. *Journal of Nuclear Medicine*, *50*, 2082–2089.
Springer, C. J., Poon, G. K., Sharma, S. K., & Bagshawe, K. D. (1993). Identification of prodrug, active drug, and metabolites in an ADEPT clinical study. *Cell Biophysics*, *22*, 9–26.
Stella, B., Arpicco, S., Peracchia, M. T., Desmaele, D., Hoebeke, J., Renoir, M., et al. (2000). Design of folic acid-conjugated nanoparticles for drug targeting. *Journal of Pharmaceutical Sciences*, *89*, 1452–1464.
Storm, G., Belliot, S. O., Daemen, T., & Lasic, D. D. (1995). Surface modification of nanoparticles to oppose uptake by the mononuclear phagocyte system. *Advanced Drug Delivery Reviews*, *17*, 31–48.
Straubinger, R. M., Hong, K., Friend, D. S., & Papahadjopoulos, D. (1983). Endocytosis of liposomes and intracellular fate of encapsulated molecules: Encounter with a low pH compartment after internalization in coated vesicles. *Cell*, *32*, 1069–1079.
Stubbs, M., McSheehy, P. M., & Griffiths, J. R. (1999). Causes and consequences of acidic pH in tumors: A magnetic resonance study. *Advances in Enzyme Regulation*, *39*, 13–30.
Sudimack, J., & Lee, R. J. (2000). Targeted drug delivery via the folate receptor. *Advanced Drug Delivery Reviews*, *41*, 147–162.
Sun, L. C., & Coy, D. H. (2011). Somatostatin receptor-targeted anti-cancer therapy. *Current Drug Delivery*, *8*, 2–10.
Sun, B., Ranganathan, B., & Feng, S. S. (2008). Multifunctional poly(D, L-lactide-co-glycolide)/montmorillonite (PLGA/MMT) nanoparticles decorated by Trastuzumab for targeted chemotherapy of breast cancer. *Biomaterials*, *29*, 475–486.
Sutton, D., Nasongkla, N., Blanco, E., & Gao, J. (2007). Functionalized micellar systems for cancer targeted drug delivery. *Pharmaceutical Research*, *24*, 1029–1046.
Svenson, S., & Tomalia, D. A. (2005). Dendrimers in biomedical applications–reflections on the field. *Advanced Drug Delivery Reviews*, *57*, 2106–2129.
Svenson, S., Wolfgang, M., Hwang, J., Ryan, J., & Eliasof, S. (2011). Preclinical to clinical development of the novel camptothecin nanopharmaceutical CRLX101. *Journal of Controlled Release*, *153*, 49–55.
Tagami, T., Ernsting, M. J., & Li, S.-D. (2011). Efficient tumor regression by a single and low dose treatment with a novel and enhanced formulation of thermosensitive liposomal doxorubicin. *Journal of Controlled Release*, *152*, 303–309.
Taheri, A., Dinarvand, R., Atyabi, F., Ahadi, F., Nouri, F. S., Ghahremani, M. H., et al. (2011). Enhanced anti-tumoral activity of methotrexate-human serum albumin conjugated nanoparticles by targeting with luteinizing hormone-releasing hormone (LHRH) peptide. *International Journal of Molecular Sciences*, *12*, 4591–4608.
Tai, W., Mahato, R., & Cheng, K. (2010). The role of HER2 in cancer therapy and targeted drug delivery. *Journal of Controlled Release*, *146*, 264–275.

Tannock, I. F., & Rotin, D. (1989). Acid pH in tumors and its potential for therapeutic exploitation. *Cancer Research, 49,* 4373–4384.

Taratula, O., Garbuzenko, O. B., Kirkpatrick, P., Pandya, I., Savla, R., Pozharov, V. P., et al. (2009). Surface-engineered targeted PPI dendrimer for efficient intracellular and intratumoral siRNA delivery. *Journal of Controlled Release, 140,* 284–293.

Tasis, D., Tagmatarchis, N., Bianco, A., & Prato, M. (2006). Chemistry of carbon nanotubes. *Chemical Reviews, 106,* 1105–1136.

Tejada-Berges, T., Granai, C. O., Gordinier, M., & Gajewski, W. (2002). Caelyx/Doxil for the treatment of metastatic ovarian and breast cancer. *Expert Review of Anticancer Therapy, 2,* 143–150.

Torchilin, V. P. (2005). Recent advances with liposomes as pharmaceutical carriers. *Nature Reviews. Drug Discovery, 4,* 145–160.

Torrisi, M. R., Lotti, L. V., Belleudi, F., Gradini, R., Salcini, A. E., Confalonieri, S., et al. (1999). Eps15 is recruited to the plasma membrane upon epidermal growth factor receptor activation and localizes to components of the endocytic pathway during receptor internalization. *Molecular Biology of the Cell, 10,* 417–434.

Tuerk, C., & Gold, L. (1990). Systematic evolution of ligands by exponential enrichment: RNA ligands to bacteriophage T4 DNA polymerase. *Science, 249,* 505–510.

Ulbrich, K., Etrych, T., Chytil, P., Jelinkova, M., & Rihova, B. (2003). HPMA copolymers with pH-controlled release of doxorubicin: *In vitro* cytotoxicity and *in vivo* antitumor activity. *Journal of Controlled Release, 87,* 33–47.

Ulrich, H., Martins, A. H., & Pesquero, J. B. (2005). RNA and DNA aptamers in cytomics analysis. *Current Protocols in Cytometry,* Chapter 7, Unit 7 28, 7.28.1–7.28.39.

Vasir, J. K., & Labhasetwar, V. (2005). Targeted drug delivery in cancer therapy. *Technology in Cancer Research & Treatment, 4,* 363–374.

Vega-Villa, K. R., Takemoto, J. K., Yanez, J. A., Remsberg, C. M., Forrest, M. L., & Davies, N. M. (2008). Clinical toxicities of nanocarrier systems. *Advanced Drug Delivery Reviews, 60,* 929–938.

Veikkola, T., Karkkainen, M., Claesson-Welsh, L., & Alitalo, K. (2000). Regulation of angiogenesis via vascular endothelial growth factor receptors. *Cancer Research, 60,* 203–212.

Veiseh, O., Gunn, J. W., & Zhang, M. (2010). Design and fabrication of magnetic nanoparticles for targeted drug delivery and imaging. *Advanced Drug Delivery Reviews, 62,* 284–304.

Verbeek, B. S., Adriaansen-Slot, S. S., Vroom, T. M., Beckers, T., & Rijksen, G. (1998). Overexpression of EGFR and c-erbB2 causes enhanced cell migration in human breast cancer cells and NIH3T3 fibroblasts. *FEBS Letters, 425,* 145–150.

Vukelja, S. J., Anthony, S. P., Arseneau, J. C., Berman, B. S., Cunningham, C. C., Nemunaitis, J. J., et al. (2007). Phase 1 study of escalating-dose OncoGel (ReGel/paclitaxel) depot injection, a controlled-release formulation of paclitaxel, for local management of superficial solid tumor lesions. *Anti-Cancer Drugs, 18,* 283–289.

Wang, J., Chen, P., Su, Z.-F., Vallis, K., Sandhu, J., Cameron, R., et al. (2001). Amplified delivery of indium-111 to EGFR-positive human breast cancer cells. *Nuclear Medicine and Biology, 28,* 895–902.

Wang, S., Kim, G., Koo Lee, Y. E., Hah, H. J., Ethirajan, M., Pandey, R. K., et al. (2012). Multifunctional biodegradable polyacrylamide nanocarriers for cancer theranostics—A "see and treat" strategy. *ACS Nano, 6,* 6843–6851.

Watanabe, K., Kaneko, M., & Maitani, Y. (2012). Functional coating of liposomes using a folate- polymer conjugate to target folate receptors. *International Journal of Nanomedicine, 7,* 3679–3688.

Wicki, A., Rochlitz, C., Orleth, A., Ritschard, R., Albrecht, I., Herrmann, R., et al. (2012). Targeting tumor-associated endothelial cells: Anti-VEGFR2 immunoliposomes mediate tumor vessel disruption and inhibit tumor growth. *Clinical Cancer Research, 18,* 454–464.

Wijagkanalan, W., Kawakami, S., & Hashida, M. (2011). Designing dendrimers for drug delivery and imaging: Pharmacokinetic considerations. *Pharmaceutical Research, 28*, 1500–1519.

Wileman, T., Harding, C., & Stahl, P. (1985). Receptor-mediated endocytosis. *Biochemical Journal, 232*, 1–14.

Wilson, C., & Szostak, J. W. (1998). Isolation of a fluorophore-specific DNA aptamer with weak redox activity. *Chemical Biology, 5*, 609–617.

Xu, L., Pirollo, K. F., & Chang, E. H. (1997). Transferrin-liposome-mediated p53 sensitization of squamous cell carcinoma of the head and neck to radiation *in vitro*. *Human Gene Therapy, 8*, 467–475.

Xu, L., Tang, W. H., Huang, C. C., Alexander, W., Xiang, L. M., Pirollo, K. F., et al. (2001). Systemic p53 gene therapy of cancer with immunolipoplexes targeted by anti-transferrin receptor scFv. *Molecular Medicine, 7*, 723–734.

Yang, J., Chen, H., Vlahov, I. R., Cheng, J. X., & Low, P. S. (2007). Characterization of the pH of folate receptor-containing endosomes and the rate of hydrolysis of internalized acid-labile folate-drug conjugates. *Journal of Pharmacology and Experimental Therapeutics, 321*, 462–468.

Ying, X., Wen, H., Lu, W.-L., Du, J., Guo, J., Tian, W., et al. (2010). Dual-targeting daunorubicin liposomes improve the therapeutic efficacy of brain glioma in animals. *Journal of Controlled Release, 141*, 183–192.

Yokoyama, M. (2011). Clinical applications of polymeric micelle carrier systems in chemotherapy and image diagnosis of solid tumors. *Journal of Experimental & Clinical Medicine, 3*, 151–158.

Yonezawa, S., Asai, T., & Oku, N. (2007). Effective tumor regression by anti-neovascular therapy in hypovascular orthotopic pancreatic tumor model. *Journal of Controlled Release, 118*, 303–309.

Yotsumoto, F., Sanui, A., Fukami, T., Shirota, K., Horiuchi, S., Tsujioka, H., et al. (2009). Efficacy of ligand-based targeting for the EGF system in cancer. *Anticancer Research, 29*, 4879–4885.

Young, D., Findlay, H. P., Hahn, S. E., Cechetto, L. M., McConkey, F., inventors, Arius Research Inc., assignee. (2007) Cytotoxicity mediation of cells evidencing surface expression of CD44. United States patent WO 2007098575, July 9.

Yu, B., Tai, H. C., Xue, W., Lee, L. J., & Lee, R. J. (2010). Receptor-targeted nanocarriers for therapeutic delivery to cancer. *Molecular Membrane Biology, 27*, 286–298.

Zentner, G. M., Rathi, R., Shih, C., McRea, J. C., Seo, M. H., Oh, H., et al. (2001). Biodegradable block copolymers for delivery of proteins and water-insoluble drugs. *Journal of Controlled Release, 72*, 203–215.

Zwick, E., Bange, J., & Ullrich, A. (2001). Receptor tyrosine kinase signalling as a target for cancer intervention strategies. *Endocrine-Related Cancer, 8*, 161–173.

CHAPTER TWO

Autophagy: Cancer's Friend or Foe?

Sujit K. Bhutia[*,1], Subhadip Mukhopadhyay[*], Niharika Sinha[*], Durgesh Nandini Das[*], Prashanta Kumar Panda[*], Samir K. Patra[*], Tapas K. Maiti[†], Mahitosh Mandal[‡], Paul Dent[§,¶,‖], Xiang-Yang Wang[¶,‖,#], Swadesh K. Das[#,¶], Devanand Sarkar[¶,‖,#], Paul B. Fisher[¶,‖,#,1]

[*]Department of Life Science, National Institute of Technology Rourkela, Rourkela, Odisha, India
[†]Department of Biotechnology, Indian Institute of Technology Kharagpur, Kharagpur, West Bengal, India
[‡]School of Medical Science and Technology, Indian Institute of Technology Kharagpur, Kharagpur, West Bengal, India
[§]Department of Neurosurgery, Virginia Commonwealth University School of Medicine, Richmond, Virginia, USA
[¶]VCU Institute of Molecular Medicine, Virginia Commonwealth University School of Medicine, Richmond, Virginia, USA
[‖]VCU Massey Cancer Center, Virginia Commonwealth University School of Medicine, Richmond, Virginia, USA
[#]Department of Human and Molecular Genetics, Virginia Commonwealth University School of Medicine, Richmond, Virginia, USA
[1]Corresponding authors: e-mail address: sujitb@nitrkl.ac.in; pbfisher@vcu.edu

Contents

1. Introduction — 62
2. Autophagy and Autophagic Death — 63
   2.1 Phagophore formation and regulation — 65
   2.2 Elongation and multimerization of phagophores — 65
   2.3 Cargo selection — 67
   2.4 Lysosomal fusion — 67
   2.5 Autophagic cell death: An elusive process — 68
3. Autophagy in Tumor Initiation and Development — 70
4. Autophagy in Tumor Progression and Metastasis — 73
5. Autophagy in Tumor Dormancy — 79
6. Autophagy in Cancer Initiating/Stem Cells — 81
7. Autophagy in Cancer Therapy — 82
   7.1 Stimulation of autophagic cell death — 83
   7.2 Inhibition of protective autophagy — 83
8. Conclusions — 86
Acknowledgments — 87
References — 87

Abstract

The functional relevance of autophagy in tumor formation and progression remains controversial. Autophagy can promote tumor suppression during cancer initiation and protect tumors during progression. Autophagy-associated cell death may act as a tumor suppressor, with several autophagy-related genes deleted in cancers. Loss of autophagy induces genomic instability and necrosis with inflammation in mouse tumor models. Conversely, autophagy enhances survival of tumor cells subjected to metabolic stress and may promote metastasis by enhancing tumor cell survival under environmental stress. Unraveling the complex molecular regulation and multiple diverse roles of autophagy is pivotal in guiding development of rational and novel cancer therapies.

1. INTRODUCTION

Stress stimuli, including metabolic stress, activate cellular mechanisms for adaptation that are crucial for cells to either tolerate adverse conditions or to trigger cell suicide mechanisms to eliminate damaged and potentially dangerous cells (Hanahan & Weinberg, 2011). Stress stimulates autophagy, in which double membrane vesicles form and engulf proteins, cytoplasm, protein aggregates, and organelles that are then transported to lysosomes where they are degraded, thereby providing energy (Klionsky & Emr, 2000; Mizushima, Ohsumi, & Yoshimori, 2002). Constitutive, basal autophagy also plays a significant homeostatic function, maintaining protein and organelle quality control and acting simultaneously with the ubiquitin proteasome degradation pathway to prevent the accumulation of polyubiquitinated and aggregated proteins (Klionsky & Emr, 2000). Autophagy-defective mice display signs of energy depletion and reduced amino acid concentrations in plasma and tissues and fail to survive in the neonatal starvation period, providing a clear example of autophagy-mediated maintenance of energy homeostasis (Kuma et al., 2004). Autophagy is also a pathway that is used for the elimination of pathogens (Colombo, 2007) and for the engulfment of apoptotic cells (Qu et al., 2007). Peptides generated from proteins degraded by autophagy can also be used for antigen presentation to T-cells for regulation of immunity and host defense (Crotzer & Blum, 2009; Levine, Mizushima, & Virgin, 2011). The importance of autophagy as a homeostatic and regulatory mechanism is underscored by the association of autophagy defects in the etiology of many diseases, including cancer (Levine & Kroemer, 2008).

Cancer is a multifaceted complex disease characterized by several defining properties, including avoidance of cell death (Hanahan & Weinberg, 2011).

The ability of cancer cells to resist apoptotic cell death is a well-known mechanism that is the key to their survival and aggressiveness. Similarly, the phenomenon of autophagy in cancer has been studied extensively, and it is now firmly established that autophagy can provide both tumor-suppressive and tumor-promoting functions (Høyer-Hansen & Jäättelä, 2008; Maiuri et al., 2009). This review focuses on the tumor-suppressive and tumor-promoting properties of autophagy during different stages of cancer development. It provides insights into how autophagy's tumor-suppressive properties, which are frequently observed at the initial stage of cancer development, are later transformed into tumor-promoting potential during cancer progression.

2. AUTOPHAGY AND AUTOPHAGIC DEATH

Autophagy (from the Greek word "auto," meaning oneself and "phagy," meaning to eat) refers to a process by which cytoplasmic constituents are delivered to the lysosome for bulk degradation (Mizushima & Klionsky, 2007; Mizushima et al., 2002). The term autophagy originated when the Nobel laureate Christian de Duve used it while attending the *Ciba Foundation Symposium on Lysosomes*, which took place in London on February 12–14, 1963. Autophagy is classified into three main types depending on the different pathways in which cargo is delivered to the lysosome or vacuole: chaperone-mediated autophagy, microautophagy, and macroautophagy (Yorimitsu & Klionsky, 2005). In this review, we focus on the most widely investigated autophagic process: macroautophagy (herein, referred to as autophagy) with formation of autophagosomes and autolysosomes. Autophagosomes are double-membrane cytoplasmic vesicles, which engulf various cellular constituents, including cytoplasmic organelles. Autophagosomes fuse to lysosomes to become autolysosomes, where sequestered cellular materials are digested (Mizushima et al., 2002). The molecular basis of autophagy has been extensively studied, mainly in yeasts, through investigation of autophagy-defective mutants to identify the responsible genes (designated as AuTophaGy; *atg*), and presently 35 *atg* genes have been discovered in yeast (Nakatogawa, Suzuki, Kamada, & Ohsumi, 2009). The basic mechanism of autophagy is well conserved during evolution as varied organisms, including plants, flies, yeast, and mammals, all of which contain a related group of *atg* genes, in spite of the fact that there are some differences between yeast and man (Klionsky, 2007).

The fundamental components of the autophagic process (Fig. 2.1) include phagophore formation, elongation and multimerization of

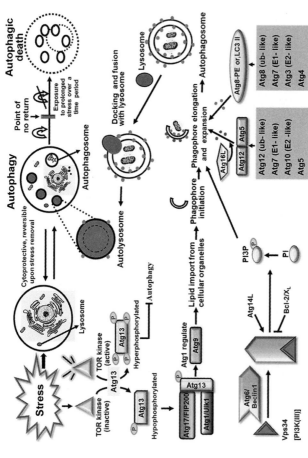

Figure 2.1 Molecular events in the autophagy pathway. A stress response, such as nutrient withdrawal, causes cells to initiate autophagy. The stress sensor TOR kinase remains inactivated in low-nutrient condition and maintains hypophosphorylated Atg13. Atg1/Ulk1 interacts with Atg13 and Atg17 and regulates transmembrane protein Atg9 involved in lipid import from cellular organelles to act as a "phagophore" formation initiator. Next, Vps34/Beclin1 converts PI to PI3P followed by Atg5–Atg12 conjugation and interaction with Atg16L resulting in multimerization at the phagophore and formation of nascent curvature; coupled to these changes, LC3 processing helps elongation and expansion. Random or selective cargoes are targeted for degradation, followed by formation of a complete double-membrane ring called an "autophagosome." Lysosomes dock and fuse with the autophagosome, forming an "autolysosome" where degraded cargoes generate amino and fatty acids to be transported back into the cytoplasmic pool. Autophagy acts as a primary response promoting cell viability and serving a cytoprotective role and upon stress removal the cell resumes normal function. However, extreme stress pushes the cell to cross the point of no return and commits it toward autophagic cell death (type II programmed cell death, PCD). (See Page 1 in Color Section at the back of the book.)

phagosomes, cargo selection and lysosomal fusion. These components of autophagy will be discussed below.

2.1. Phagophore formation and regulation

The initial step of phagophore membrane formation in mammals remains elusive and has not been adequately defined, whereas in the yeast system this pathway is well defined. Unlike yeast, in the mammalian system there are no reports of preautophagosomal structures (Klionsky, 2007; Yorimitsu & Klionsky, 2005). Target of rapamycin (TOR) kinase acts as a molecular sensor to various stress responses, including hypoxia, insulin signaling, and energy and nutrient depletion, playing a pivotal role in cellular growth and autophagy control (Kamada et al., 2010). Initial nutrient starvation inactivates TOR kinase, resulting in a hypophosphorylated Atg13 that shows an increased affinity for Atg1 kinase (mammalian homolog of Ulk1) and forms a complex with a scaffold-like protein Atg17 (Fig. 2.1; Mizushima, 2010). Starvation treatment enhances the crosstalk between Atg13, Atg1, and Atg17. Atg13 and Atg17 are both required for appropriate monitoring of the kinase activity of Atg1. In turn, Atg1 regulates the transmembrane protein Atg9. The kinase activity of Atg1 is dispensable; however, it controls the dynamics of Atg9 recruitment to the phagophore in an Atg17-dependent pathway (Sekito, Kawamata, Ichikawa, Suzuki, & Ohsumi, 2009; Simonsen & Tooze, 2009). Atg9 is involved in lipid import from different sources like the endoplasmic reticulum (ER), endosomes, mitochondria, golgi bodies, and nuclear envelope and also helps in the assembly of the intact phagophore membrane (Axe et al., 2008; Simonsen & Tooze, 2009; Yorimitsu & Klionsky, 2005).

2.2. Elongation and multimerization of phagophores

The phagophore is elongated when Class III PI3 kinases, for example, Vps34 (vesicular protein sorting), bind to Beclin1 (mammalian homolog of yeast Atg6) increasing its catalytic activity to produce PI3P (phosphatidyl inositol-3-phosphate) (Simonsen & Tooze, 2009). PI3P acts as an important localization signal and may facilitate fusion at the final step of autophagosome formation (Axe et al., 2008). Vps34–Beclin1 interaction is upregulated by proteins like Ambra (activating molecule in Beclin1-regulated autophagy protein 1), UVRAG (ultraviolet radiation resistance-associated gene), and Bif 1 (Bax-interacting factor 1). In contrast, this interaction is downregulated by Bcl-2, Bcl-X_L, and Rubicon (RUN domain and cysteine-rich domain-containing Beclin1-interacting protein) (Funderbur, Wang, & Yue, 2010).

Two ubiquitin-like conjugation systems are part of the vesicle elongation process. In one experimental system, Atg12 binds Atg7 (E1 ubiquitin-like activating enzyme) in an ATP-dependent manner. Next, Atg12 non-covalently binds Atg10 (E2-like ubiquitin carrier) linking Atg12–Atg5; Atg16 dimers conjugate with this complex via C-terminal coiled-coil domain to facilitate the creation of the expanding phagophore. The Atg5–Atg12–Atg16 complex induces curvature formation in the growing phagophore. However, this trimeric structure dissociates when the phagophore progresses into a double-membrane ring called an "autophagosome" (Geng & Klionsky, 2008). The second ubiquitin-like system, which plays a pivotal role in autophagosome formation, helps in the processing of microtubule-associated light chain 3 (LC3) (mammalian homolog of Atg8). Cysteine proteinase Atg4 (also known as autophagin) cleaves LC3 to produce LC3BI that binds E1-like Atg7 in an energy-expending pathway resulting in activation of LC3BI. Atg3, an E2-like carrier, interacts with activated LC3BI and promotes lipidation, giving rise to LC3BI–phosphatidylethanolamine (PE) conjugate or LC3BII (Kabeya et al., 2000).

The sequential order of mammalian autophagosome biogenesis begins with activation of an Ulk1/2 (UNC-51-like kinase 1/2) complex, which associates with the initiating phagophore membrane. The Vps34 complex is recruited to the phagophore and phosphorylates phosphoinositides (PIs), leading to the production of PI3P. WIPI1/WIPI2 (WD repeat protein-interacting with PIs) and DFCP1 (double FYVE domain-containing protein), two PI3P effectors, contribute to this nascent elongating membrane. The Atg12–Atg5 complex serves an E3-type enzyme function and acts as a supporting framework to which Atg8s arrive at the phagophore (Hanada et al., 2007; Tanida, Ueno, & Kominami, 2004). Atg16L joins the Atg12–Atg5 complex and helps to govern at which site of the membrane the downstream conjugation of LC3 occurs (Fujita et al., 2008). Atg8 and GATE-16 (Golgi-associated ATPase enhancer) are recruited and conjugated to PE on the phagophore membrane, which starts elongating mediated through the action of LC3–PE. GATE-16 acts downstream of LC3 in a step coupled to the disassociation of the Atg12–Atg5 with Atg16L. Subsequently, a mature autophagosome bearing a double-membrane structure is formed (Weidberg, Shvets, & Elazar, 2011).

A recent report highlights the role of Atg14L, a subunit of PI3kinase involved in localization to the ER via four N-terminal cysteines, which accumulate in omegasomes (PI3P-rich Ω-shaped structure formed at the periphery of ER) upon autophagy induction in the process of autophagosome biogenesis (Matsunaga et al., 2010). Atg14L is able to target the PI3-kinase complex to the ER, enabling PI3P generation for omegasome and autophagosome formation. Although the role of specific kinases in autophagosome is well documented and emphasized, the role of antagonistic partners in this process, that is, phosphatases, is equally important (Vergne & Deretic, 2010). The phosphatidylinositol 3-phosphate (PI3P) phosphatase *Jumpy* (MTMR14) associates with isolated membranes during early autophagosome biogenesis that is guided by Atg16, which helps in the subsequent development and localization of autophagic organelles (Noda, Matsunaga, Taguchi-Atarashi, & Yoshimori, 2010; Vergne et al., 2009). *Jumpy* coordinates the recruitment of Atg factors in an orderly manner by interacting with PI3P through WIPI-1 (Atg18) thereby affecting the distribution of Atg9 and LC3, the factors responsible for controlling growth of the autophagic membrane.

2.3. Cargo selection

In general, autophagy has been considered a random process as it appears to engulf cytoplasm indiscriminately. Electron micrographs often show autophagosomes with wide-ranging contents comprising mitochondria, ER, and golgi membranes. However, there is accumulating evidence that the growing phagophore membrane can interact selectively with protein aggregates and organelles. LC3B-II provides the role of "receptor" at the phagophore and interacts with "adaptor" molecules on the target including protein aggregates, damaged mitochondria, thereby helping in promotion of their selective uptake and degradation (Weidberg et al., 2011; Yorimitsu & Klionsky, 2005). The best-characterized molecule in this process is the multiadaptor molecule p62/SQSTM1, which binds Atg8/LC3 and promotes degradation of polyubiquitinated protein aggregates (Ichimura & Komatsu, 2010). Similarly, Atg32 has been identified in yeast as a protein that promotes selective uptake of mitochondria, a process known as mitophagy (Okamoto, Kondo-Okamoto, & Ohsumi, 2009).

2.4. Lysosomal fusion

Autophagosomes dock with lysosomes and fuse to give rise to structures known as "autolysosomes," where the acidic lysosomal components digest all cargos. Migrating bidirectionally along microtubules, the autophagosomes have a natural propensity toward the lysosome-enriched microtubule organizing center, supervised by the function of dynein motor proteins (Kimura, Noda, & Yoshimori, 2008; Ravikumar et al., 2005; Williams et al., 2008). Small

GTPases, like Rabs (Rab7), ESCRT (endosomal sorting complex required for transport), SNARE (soluble N-ethylmaleimide-sensitive factor activating protein receptor), and class C Vps proteins play key roles in the coordinated vesicular docking and fusion with target components (Atlashkin et al., 2003; Gutierrez, Munafo, Beron, & Colombo, 2004; Jager et al., 2004; Lee, Beigneux, Ahmad, Young, & Gao, 2007; Zerial & McBride, 2001). Accumulation of Rab proteins along specific intracellular niches triggers the last fusion step mediated by SNARE complexes (Martens & McMahon, 2008). It is worthwhile highlighting important proteins like ESCRT whose mutation or loss of function culminates in inhibition of autophagosome maturation. Although the role of UVRAG is most recognized as a Beclin1-interacting protein, it also has an independent function in the final maturation step by engaging the fusion machinery on autophagosomes. UVRAG is also known to engage the class C Vps proteins thereby activating Rab7, which helps to promote fusion with late endosomes and lysosomes. Another Beclin1-interacting protein Rubicon also modulates autophagosomal maturity. Rubicon remains part of a complex containing varied proteins, like UVRAG, hVps34, and hVps15, and is known to suppress autophagosomal maturation (Matsunaga et al., 2009; Zhong et al., 2009). Further research will help to clarify all of the key-interacting players involved in the crucial autophagy pathway.

Shifting the focus from the autophagosome to the lysosome, inhibiting the lysosomal H^+ATPase by chemicals like bafilomycin A1, nocodazole, or vinblastine, will prevent the fusion of autophagosomes with endosomes/lysosomes (Fass, Shvets, Degani, Hirschberg, & Elazar, 2006; Köchl, Hu, Chan, & Tooze, 2006). A recent scientific report suggests an alternative autophagic pathway in mouse cells lacking Atg5 or Atg7 when treated with stress inducers, like etoposide. The key autophagic proteins operational in this Atg5/Atg7-independent route include Ulk1 and Beclin1 and proceed in a Rab9-dependent manner (Nishida et al., 2009).

Lysosomal permeases and transporters export essential products like fatty acids and amino acids back into the cytosolic pool. This replenishing phenomenon plays a survival role for starving cells, contributing to what is called "protective autophagy" (Mizushima et al., 2002; Yorimitsu & Klionsky, 2005).

2.5. Autophagic cell death: An elusive process

The current concept of programmed cell death involves three areas including apoptosis, autophagic cell death (ACD) and necroptosis. The autophagic pathway initially functions as an adaptive response to stress; however, when the cell continues to face extreme stress over a protracted period of time it reaches a point of no return and becomes committed to undergo cell death. This type of cell death that is associated with autophagosomes and is dependent on autophagic proteins is called "Autophagic cell death." As indicated, extended autophagy beyond the optimal survival limit culminates in ACD (Fig. 2.1). ACD has gained immense attention among scientists since the 1990s, which has progressed rapidly due to the discovery of the *atg* genes, establishing a caspase-independent "type II programmed cell death" (Kroemer & Levine, 2008). The Nomenclature Committee on Cell Death (NCCD, 2005) classified ACD as cell death through autophagy, referring to this process as cell death with autophagy (Kroemer et al., 2005). Later in 2008, Kroemer and Levine characterized ACD morphologically (by transmission electron microscopy) as a cell death process occurring in the absence of chromatin condensation but characterized by large-scale sequestration of cytoplasmic components into autophagosomes, imparting a characteristic vacuolated appearance to the cell.

ACD is mainly a morphological phenomenon, and there currently is no conclusive evidence that a specific mechanism of autophagic death exists (Tsujimoto & Shimizu, 2005; Yorimitsu & Klionsky, 2005). It is difficult to define the pathophysiological role of ACD; hence, the attempt to hypothesize the reason for this mode of cell death remains elusive. Pinpointing the exact function of autophagy in programmed cell death is not only challenging but also equally complicated due to the simultaneous occurrence of caspase-dependent apoptosis, often occurring in the context of autophagy. Kinetically speaking, it can be anticipated that caspase-mediated proteolytic degradation would occur faster than self-digestion by autophagy. Accordingly, the cell would be experiencing a predominant extent of apoptosis in spite of extensive autophagy (Debnath, Baehrecke, & Kroemer, 2005). However, the cell displays a highly multifaceted crosstalk between the apoptotic and ACD mechanism(s); which can promote antagonism, synergism, or mutually independent pathways of induction that are context dependent. It is difficult to conceptualize how on one hand ACD is part of the cell death mechanism, while multiple studies also emphasize a protective role of ACD since autophagic inhibition does not stop cell death but may even promote death. Whether ACD is primarily a mediator of the death mechanism, an innocent bystander, or a double-edged sword in cell survival/death processes remains to be determined. It is also possible that this duality of functions depends on the temporal induction of autophagy and the context in which this pathway is induced or suppressed. It is clear that a complete understanding of the autophagic process as well as its mediators and determinants of expression represents an evolving story that will only become clarified with additional research.

3. AUTOPHAGY IN TUMOR INITIATION AND DEVELOPMENT

Autophagy is believed to play an essential role in tumor initiation and development (Chen & Debnath, 2010; Liang & Jung, 2010). When baseline levels of autophagy fluctuation were compared, the amount of proteolysis or autophagic degradation in cancer cells was less than that of their normal counterparts (Gunn, Clark, Knowles, Hopgood, & Ballard, 1977; Kisen et al., 1993). This differential expression suggests a direct connection between tumorigenesis and decreased levels of autophagy. Intriguingly, many oncogenes and tumor suppressor genes affect autophagic pathways (Maiuri et al., 2009), and the deregulation of the autophagic process contributes to malignant transformation. For example, many tumor suppressor proteins such as p53, phosphatase and tensin homolog (PTEN), death-associated protein kinase (DAPK), tuberous sclerosis 1 (TSC1), and TSC2 that provide constitutive input signals to activate autophagy are mutated in multiple cancers.

PTEN, a dual lipid/protein phosphatase, dephosphorylates PIP3 to PIP2, preventing inhibition of autophagy by the PI3K/Akt/mTOR pathway. In human tumors, PTEN is mutated, resulting in activation of Akt that suppresses autophagy (Maehama, 2007; Yin & Shen, 2008) and also participates in increased protein translation, cell growth, and cell proliferation, which is a contributor to tumorigenesis (Hafner et al., 2007; Horn et al., 2008; LoPiccolo, Blumenthal, Bernstein, & Dennis, 2008; Vivanco & Sawyers, 2002). Additionally, p53 plays a divergent role in the regulation of autophagy. Within the nucleus, p53 can act as an autophagy-inducing transcription factor through AMPK and TSC1/TSC2 dependent activation (Tasdemir et al., 2008). In contrast, cytoplasmic p53 exerts an autophagy-inhibitory function, and its degradation is actually required for the induction of autophagy. Although the relationship between autophagy and p53 is complicated, it is clear that p53 mutation(s) cause alterations in p53-mediated autophagy that leads to cancer development. Additionally, another target of p53 that is present in nucleus, DRAM (damage-regulated autophagy modulator), has been shown to positively regulate autophagy. DRAM is essential for p53-mediated autophagy and apoptosis in response to DNA-damaging agents, and the overexpression of DRAM is sufficient to activate autophagy without affecting apoptosis (Crighton et al., 2006).

The fact that DRAM is also deleted in multiple types of cancer underscores its importance and highlights the possibility that autophagy might play a fundamentally important role in cancer.

The death-associated protein kinase (DAPK), a cytoskeleton-associated calmodulin-regulated serine/threonine protein kinase, has been shown to possess multiple tumor- and metastasis-suppressor functions (Bialik & Kimchi, 2006; Eisenberg-Lerner & Kimchi, 2009). It is often decreased or lost in many human cancers, and ectopic expression is associated with p53-mediated apoptosis and decreased cell migration and invasion. Moreover, DAPK expression has been shown to suppress formation of metastatic foci in Lewis lung carcinoma in mice (Inbal, Bialik, Sabanay, Shani, & Kimchi, 2002). Recently, DAPK and DAPK-related protein kinase-1 (DRP-1) have been found to activate autophagy in MCF-7 and HeLa cells. Expression of these genes triggered membrane blebbing and ACD and conversely inhibition of DAPKs resulted in decreased autophagy (Moretti, Yang, Kim, & Lu, 2007). Metabolic stress results in a decline in ATP:ADP ratios and induction of the tumor suppressor *LKB-I* (*STK-I*), which is serine threonine kinase that is upregulated in the LKB-I/AMPK/mTOR pathway. It phosphorylates the α subunit at Thr172 (Shaw et al., 2004) and activates AMPK. AMPK either activates TSC2 and the regulatory-associated protein raptor by promoting phosphorylation of TSC2 or directly phosphorylating raptor, leading to inhibition of mTOR and autophagy induction (Corradetti, Inoki, Bardeesy, DePinho, & Guan, 2004). Moreover, the LKB-I/AMPK/mTOR axis activates p27, a cyclin-dependent kinase inhibitor, inducing cell cycle arrest for energy conservation (Liang et al., 2007).

Conversely, oncogenes including *Akt*, mTOR, Bcl-2, and FLICE-like inhibitory protein (FLIP) inhibit autophagic processes indicating that elevated autophagy signaling may contribute to tumor suppression (Lee et al., 2009; Morselli et al., 2009). In most cancers, the PI3K–Akt axis undergoes mutation in either upstream or downstream regulators (Shaw & Cantley, 2006). Thus, downstream activation of mTOR favors cell growth stimulation and inhibition of autophagy. Constitutive activation of Akt inhibits autophagy *in vitro* and *in vivo* in Bax and Bak double mutants. Monoallelic knockout of *beclin1* or biallelic knockout of *atg5* in mice promotes genomic alterations due to activation of Akt (Karantza-Wadsworth et al., 2007; Mathew, Karantza-Wadsworth, & White, 2007; Mathew, Kongara, et al., 2007). The BH3 receptor domain of Bcl-2 and the multidomain antiapoptotic proteins bind with the amphipathic BH3 helix of Beclin1 (Oberstein, Jeffrey, & Shi, 2007), promoting its sequestration and blocking its interaction with Vps34. Overexpressed Bcl-2 along with deletion in one allele of beclin1 accelerates tumor growth *in vivo* (Degenhardt et al., 2006). Notably, the Bcl2 gene is overexpressed in a majority of cancers (Levine, Sinha, & Kroemer, 2008; Pattingre & Levine, 2006), and knockdown or gene silencing of Bcl-2 through antisense oligonucleotides or siRNA heteroduplexes, respectively, in MCF-7 cells results in induction of autophagy (Akar et al., 2008).

Identification of *beclin1* as a haploinsufficient tumor suppressor frequently containing deletion of one allele in a large proportion of human breast, ovarian, and prostate cancers provided initial evidence for a potential direct link between autophagy and cancer. Notably, overexpression of Beclin1 in MCF-7 breast carcinoma cells induced autophagy and restricted proliferation and clonigenicity *in vitro*, and it also inhibited tumorigenesis in nude mice (Liang et al., 1999). The contribution of the allelic deletion of *Beclin1* to carcinogenesis is supported by the observation that $beclin1+/-$ mice have an increased incidence of lung cancer, hepatocellular carcinoma (HCC), and lymphoma (Yue, Jin, Yang, Levine, & Heintz, 2003; Qu et al., 2003). Furthermore, allelic loss of *beclin1* in HCC causes accumulation of p62/SQSTM (Mathew et al., 2009). ER chaperones and damaged mitochondria result in an elevated production of ROS (reactive oxygen species). Generation of ROS promotes a cascade of events, including increased oxidative stress, DNA damage, and chromosomal instability, which ultimately lead to inhibition of the NF-κB pathway and development of HCC. Consequently, tumor-suppressive or -supportive properties of Beclin1-interacting molecules including UVRAG, Bif-1, and Rubicon have been documented (Funderbur et al., 2010). UVRAG activates Beclin1 and induces autophagosome formation. Ectopic expression of UVRAG suppresses the proliferation and tumorigenicity of HCT116 tumor cells and sensitizes these cells to undergo self-directed autophagy even without starvation treatment (Liang et al., 2006). Monoallelic loss of *UVRAG* is observed in various colon cancer cells and tissues. Nonsense mutations of *UVRAG* are observed in colon and gastric cancers resulting in inactivation of autophagy (Ionov, Nowak, Perucho, Markowitz, & Cowell, 2004; Kim et al., 2008). Moreover, Bif 1 (also known as Endophilin B1) interacts with Beclin1 through UVRAG (Takahashi et al., 2007). Downregulation of *Bif 1* is frequently found in different types of cancer, and loss of *Bif 1* leads to suppression of autophagy by decreasing Vps34 kinase activity, which promotes colon adenocarcinoma formation (Coppola et al., 2008). Moreover, $Bif\ 1^{-/-}$ mice are cancer prone (Takahashi et al., 2007). By contrast, Rubicon, a newly identified Beclin1-interacting protein, reduces Vps34 lipid kinase activity and downregulates autophagy (Matsunaga et al., 2009; Zhong et al., 2009) with aberrant expression in multiple types of cancer. Thus, Beclin1 acts as a nodal point and activates PI3K III, revealing its mandatory role in autophagy and tumorigenesis (Table 2.1).

Additional autophagic genes involved at different stages of autophagy have been identified as contributors to the oncogenic and tumor-suppressive signaling pathways of cancer (Table 2.1). For instance, Atg7 contributes to the maintenance of HSCs (hematopoietic stem cells) (Mortensen et al., 2011). *Atg7*-deficient LSK (Lin$^-$Sca-1$^+$c-Kit$^+$) cells show defects in HSC functions with impairment of production of both lymphoid and myeloid progenitors in lethally irradiated mice. Similarly, suppression of *Atg5* and *Atg16L1* genes leads to tissue injury in intestinal Paneth cells culminating in Crohn's death, a known risk factor for colorectal cancer in humans (Cadwell et al., 2008). Moreover, mice in which *Atg5* and liver-specific *Atg7*$^{-/-}$ have been deleted develop liver cancer and display mitochondrial swelling, p62 accumulation, and oxidative stress and genomic damage responses in isolated hepatocytes (Takamura et al., 2011). Amino acid depletion, which hinders metabolism, occurs during the survival period of neonatal mice deficient in *Atg5*. Moreover, mice with *Atg4C* deficiency have increased tendency for carcinogen-induced tumorigenesis (Marino et al., 2007).

Cellular senescence is a state of stable dynamic cell cycle arrest that limits the proliferation of damaged cells and has been regarded as a tumor suppressor mechanism. A recent study showed that autophagy was activated during senescence, and its activation was correlated with negative feedback in the PI3K–mTOR pathway (Young et al., 2009). A subset of autophagy-related genes is upregulated during senescence. For instance, overexpression of ULK3/Atg3-induced autophagy and senescence contributed to tumor suppression. Furthermore, inhibition of autophagy delayed the senescence phenotype, including senescence-associated secreted factor.

4. AUTOPHAGY IN TUMOR PROGRESSION AND METASTASIS

Tumor metastasis is a complex, multistep process by which tumor cells from a primary site migrate to and colonize at distant organ sites (Fig. 2.2; Das et al., 2012). This process involves multiple, discrete steps including invasion of tumor cells from the primary tumor site, intravasation and survival in

Table 2.1 Autophagy genes in cancer

Autophagy process	Autophagy gene	Cancer	Mechanism	Autophagy in cancer	References
Induction	UKL3	Tumorigenesis	Oncogene-induced cell senescence	Lethal	Young et al. (2009)
Nucleation	Beclin1	Tumorigenesis	Highly mutated in human breast, ovarian, and prostate tumors; haploinsufficient tumor suppressor	Lethal	Liang et al. (1999)
	UVRAG	Tumorigenesis	Mutations detected in human colorectal, breast, and gastric tumors; haploinsufficient tumor suppressor	Lethal	Ionov et al. (2004) and Kim et al. (2008)
	Bif-1	Tumorigenesis	$Bif\text{-}1^{-/-}$ mice are cancer prone; decreased	Lethal	Coppola et al. (2008)
	Ambra1	Tumorigenesis	$Ambra1^{-/-}$ mice have severe neural tube defects	Lethal	Garber (2011)
Elongation	Atg12–Atg5	Tumorigenesis	Atg5 frameshift mutations in gastric cancers	Lethal	Cadwell et al. (2008)
	Atg4C	Tumorigenesis	$Atg4C^{-/-}$ mice develop fibrosarcomas in response to carcinogen treatment	Lethal	Marino et al. (2007)
	Atg7	Tumorigenesis	Defects in hematopoietic stem cell functions, liver-specific $Atg7^{-/-}$ mice develop liver cancer	Lethal	Takamura et al. (2011)
	Atg16	Tumorigenesis	Mutations detected in Crohn's disease	Lethal	Cadwell et al. (2008)

Maturation	Rab7	Tumorigenesis	$Rab7^{-/-}$ aberrant expression in human leukemia	Protective	Liang and Jung (2010)
Cargo selection	p62	Tumorigenesis	ROS accumulation through NF-κB induction leads to tumorigenesis in autophagy-deficient conditions	Lethal	Mathew et al. (2009)
Unknown	Unknown	Metastasis	Resistance to TRAIL, energy metabolism, necrosis, inflammation, HMGB	Lethal/protective (?)	–
Unknown	Unknown	Tumor dormancy	Resistance to TRAIL, Ras homolog member I induction	Protective	–
Unknown	Unknown	Cancer stem cells	Unknown	Lethal/protective (?)	–

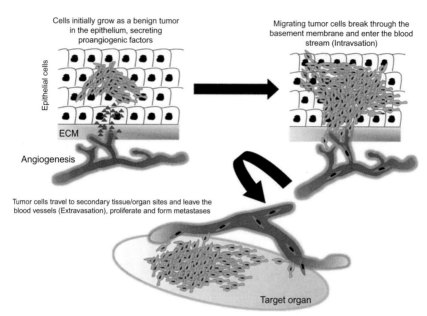

Figure 2.2 Model of the primary events involved in the metastatic cascade. The metastatic process is complex and involves numerous changes in cellular phenotype resulting from both genetic and epigenetic modifications of the cancer genome. The process is initiated by the spread of cancer cells from a primary tumor site to other regions in the body. Cells initiate growth as primary tumors in the epithelium and with genetic and epigenetic modifications, subsets of tumor cells develop metastatic properties allowing them to degrade the basal layer and invade the blood stream (Intravasation). A small percentage of tumor cells escape into blood vessels (Extravasation), survive in the bloodstream, adhere to new target organ sites, and ultimately form secondary tumors (metastases) in distant organ or tissue sites. A key component of both the primary and secondary expansion of the tumor and metastases is the development of a new supply of blood vessels, that is, angiogenesis. *Taken from Das et al. (2012).* (See Page 1 in Color Section at the back of the book.)

the blood stream, extravasation at a distant site, and finally colonization of disseminated tumor cells (DTCs) at distant sites (Bingle, Brown, & Lewis, 2002; Das et al., 2012). In primary tumors, inflammatory cells infiltrate tumor sites in response to necrosis resulting from hypoxia and metabolic stress (Degenhardt et al., 2006; Jin & White, 2007). Protective autophagy, promoted by hypoxia and metabolic stress, inhibits inflammation at primary sites that is required for initiation of metastasis. Interestingly, autophagy reduces necrosis and subsequent macrophage infiltration thereby decreasing primary tumor growth, and genetic inhibition of autophagy can cause cell death, tissue damage, and chronic inflammation (Mathew, Karantza-Wadsworth, et al., 2007;

Mathew, Kongara, et al., 2007). Another important function of autophagy is clearance of cellular debris accumulated as unfolded protein, damaged organelles, and high-cargo receptor p62 in response to metabolic stress during tumor progression. When autophagy is defective, cellular toxic substances are not degraded which cause ROS production followed by DNA damage and chromosomal instability that initiate metastasis (Mathew et al., 2009). Additionally, autophagy activates a proinflammatory immune response by enhancing the release of immunostimulatory molecules including high-mobility group box protein-1 (HMGB-1) from dying tumor cells which mediate antitumor immunity (Fig. 2.3; Thorburn et al., 2008).

Apart from autophagy's antimetastatic properties, it can also have an opposite effect enhancing the metastatic potential of tumor cells. It is well established that death ligand-induced apoptosis, particularly TNF-related apoptosis inducing ligand (TRAIL), plays a critical role in regulating the suppression of metastasis by T-cells and NK cells (Wang, 2008). But recently it was demonstrated that protective autophagy is upregulated in TRAIL-resistant cancer cells, which enhances viability and survival of tumor cells during metastasis (Fig. 2.3; Han et al., 2008; Herrero-Martin et al., 2009).

Another important property of metastatic cancer cells is resistance to anoikis, apoptosis associated with lack of proper extracellular matrix attachment (ECM) (Taddei, Giannoni, Fiaschi, & Chiarugi, 2012). Aberrant activation of growth factor pathways including Ras/MAPK and PI3K/Akt pathways is a common mechanism utilized by cancer cells to evade anoikis. Although autophagy-mediated cell death was initially recognized in association with anoikis, a recent study also supports the protective nature of autophagy in anoikis (Kenific, Thorburn, & Debnath, 2010). Fung, Lock, Gao, Salas, and Debnath (2008) demonstrated autophagy induced by detachment from the substratum or inhibition of the β1-integrin receptor, and inhibition of autophagy by knockdown of *atg* genes enhanced detachment-induced cell death. Anoikis resistance in tumor cells is protective and facilitates survival and expansion of metastatic cells. Although the detailed mechanisms of protective autophagy in anoikis resistance remain largely unknown, a recent study indicates that PERK, an ER kinase, facilitates survival of ECM-detached cells by promoting autophagy and antioxidant activity induced through ROS generation, which may provide fitness to cells without ECM contact during later stages of cancer dissemination and metastasis (Fig. 2.3; Avivar-Valderas et al., 2011).

Alterations in cellular metabolism, an important hallmark of cancer, are employed by neoplastic cells to adapt to specific growth requirements during

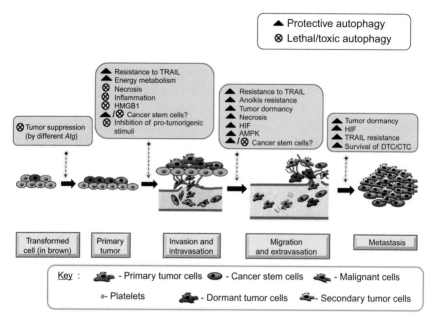

Figure 2.3 Role of autophagy at different stages of cancer. During the initial phase of cancer development, autophagy-related cell death has been regarded as a primary mechanism for tumor suppression. Moreover, autophagy also restricts necrosis and inflammation thus limiting invasion and dissemination of tumor cells from a primary site, resulting in restriction of metastasis at a premature step. Moreover, lethal (toxic) autophagy directly causes the release of immunomodulatory factors such as HMGB-1 from dead tumor cells, which activates immune response and restricts metastasis by inhibiting protumorigenic responses. On the other hand, altered energy metabolism and TRAIL-resistant phenomena of tumor cells are maintained through protective autophagy during tumor progression. Similarly, autophagy may promote metastasis by enhancing tumor cell fitness in response to environmental stresses, such as anoikis during metastatic progression. Protective autophagy is involved in maintaining dormant tumor cells and promoting their survival under stressful conditions. The exact role of autophagy in cancer stem cells is unclear in tumor progression. Finally, tumor cells maintain protective autophagy through activation of HIF-1 (hypoxia-inducible factors) and AMPK (5′-AMP-activated protein kinase) in apoptosis deficient and long-term metabolic stress conditions, in a full-blown cancer. (See Page 2 in Color Section at the back of the book.)

cancer progression (Hanahan & Weinberg, 2011; Mizushima & Klionsky, 2007). Cancer cells specifically consume glucose through anaerobic glycolysis during hypoxic conditions, which is known as the "Warburg effect," with high levels of glycolytic intermediates and lactate as reported in human colon and gastric cancers (Hirayama et al., 2009). In addition, cancer cells also display increased glutamine utilization. Autophagy induced under stressful conditions causes high degradation and recycling of proteins providing substrates for energy and carbon/nitrogen sources for biomass production required for rapidly proliferating cancer cells (Mizushima & Klionsky, 2007). Similarly, intracellular fat storage provides acetyl-CoA for mitochondria to support the TCA cycle through lipophagy to meet the elevated metabolic demands of deregulated tumor cell growth (Rabinowitz & White, 2010). Another important property of autophagy is preservation and maintenance of organelle function, especially mitochondria that are required for cell growth during tumor progression, whereas it prevents tumor growth by reducing tissue damage and necrosis during cancer initiation (Twig et al., 2008). Damaged mitochondria are the major site of ROS production in cells, and mitochondria-selective autophagy, "mitophagy," clears depolarized mitochondria and maintains cellular homeostasis (Wu et al., 2009). Recent studies indicate that Ras-expressing cells have upregulated basal autophagy that is required to maintain the pool of functional mitochondria necessary to support growth of aggressive tumors (Guo et al., 2011).

As tumors grow with defects in apoptosis and following long-term metabolic stress conditions, they may require autophagy to survive in nutrient-limited and low-oxygen conditions, especially in the central area of the tumor, which is often poorly vascularized. Survival through autophagy is a key process enabling long-term tumor cell viability and eventual regrowth and tumor recurrence. Accordingly, induction of autophagy allows cancer cells to survive in low-nutrient and low-oxygen conditions through activation of HIF-1 (hypoxia-inducible factor) and AMPK ($5'$-AMP-activated protein kinase) (Eisenberg-Lerner & Kimchi, 2009; Mathew, Karantza-Wadsworth, et al., 2007; Mathew, Kongara, et al., 2007). For instance, the oncogenic gene astrocyte elevated gene-1 (Emdad et al., 2009; Kang et al., 2005; Su et al., 2002) was associated with protective autophagy through AMPK/mTOR-dependent pathway and inhibition of the protective autophagy by ATG-5 knockdown provided therapeutic benefits (Fig. 2.4; Bhutia, Dash, et al., 2010; Bhutia, Kegelman, et al., 2010). HIF-1 and AMPK are components of a concerted cellular response to maintain energy homeostasis in oxygen- and nutrient-limited tumor microenvironments.

5. AUTOPHAGY IN TUMOR DORMANCY

Tumor dormancy is a protracted stage in tumor progression in which tumors remain occult and asymptomatic for extended periods of time. This state can be present as one of the earliest stages in tumor development, as well

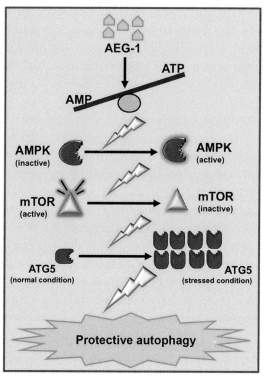

Figure 2.4 Astrocyte elevated gene-1 (AEG-1) and protective autophagy. Model illustrating the possible molecular mechanism of AEG-1-mediated protective autophagy, which promotes escape from apoptosis and resistance to chemotherapy. *Taken from Bhutia, Kegelman, et al. (2010). (See Page 2 in Color Section at the back of the book.)*

as in the in micrometastasis stage, and can occur when minimal residual disease remains after surgical removal or treatment of primary tumors (Almog, 2010). Clinically, the dormant tumor cells are not easily detected representing a major problem in breast cancer, ovarian cancer, and other malignancies. Apart from analysis of autopsies of trauma victims and clinical data accumulating from patients with late recurrence or relapse, recently DTCs and circulating tumor cells (CTCs) in cancer patients provide data relative to the frequency and prevalence of tumor dormancy. Tumor dormancy can result from angiogenesis arrest, a balance between apoptosis and cell proliferation, cell cycle arrest, and immune surveillance (Pantel, Alix-Panabières, & Riethdorf, 2009). Recently, the role of autophagy in tumor dormancy has been recognized (Fig. 2.3). Tumor cells in dormant conditions are not efficiently associated with extracellular matrix and stimulate autophagy for survival and maintenance of dormancy. Impaired β1-integrin signaling, a known inducer of autophagy, has been shown to promote dormancy in MMTV-PyMT breast cancer model (White et al., 2004). In breast cancer metastases to bone, disseminated cells displayed Src-mediated TRAIL resistance and remained dormant in the bone marrow for extended periods of time (Zhang et al., 2009). As mentioned earlier, autophagy can protect cells from TRAIL-induced apoptosis. Based on this consideration, it is speculated that dormancy in the bone marrow could induce protective autophagy and support survival of dormant cells (Kenific et al., 2010). A direct link between autophagy and tumor dormancy was recently demonstrated in ovarian cancer cells. The tumor suppressor aplasia Ras homolog member I (ARHI) induced autophagic death in ovarian tumor cells *in vitro*. However, in xenografted tumors in mice ARHI-induced autophagic death was switched to dormant tumor cell survival in the context of the tumor microenvironment suggesting that autophagy may be a prerequisite for tumor dormancy (Lu et al., 2008). Future studies focused on defining the precise mechanism by which protective autophagy is associated with tumor dormancy is warranted and holds potential for defining new therapeutic strategies for treating cancer.

6. AUTOPHAGY IN CANCER INITIATING/STEM CELLS

The "stem cell hypothesis" embodies the concept of tumor cell heterogeneity, in which only tumor-initiating cells in the heterogeneous tumor population are capable of proliferating and differentiating into new tumor-producing cells. Stem cells form the apex of the tumor hierarchy and retain the capability to replicate and grow into a tumor *in vivo*. It has been argued that cancer initiating or tumor-initiating stem cells (CSCs) do not arise from normal stem cells (Clevers, 2011). They can arise from subpopulation of cancer cells, including cancer initiating stem cells, cancer progenitor cells, or differentiated cancer cells. The phenotype with respect to the self-renewal and differentiating capacity of CSCs depends upon tumor type and is quite predictable (Zhou et al., 2009). Cancer initiating/stem cells have now been identified and isolated from tumors of the hematopoietic system, skin, breast, brain, prostate, colon, head and neck, and pancreas, and reduced CSC numbers can initiate tumor growth in xenograft models (Fu et al., 2009). This subpopulation within the tumor evades therapy, persists, and initiates recurrence thereby enhancing malignant spread of the disease. Several pathways over-expressed in different types of cancers including Wnt/Notch/Hedgehog have been identified and shown to be critical to the self-renewal behavior of CSCs. Moreover, CSCs show resistance to apoptosis; they have high expression of ATP-binding cassette transporters, and display enhanced DNA repair capacity making therapy extremely difficult. Apart from its functions in cancer, autophagy plays a seminal role in maintaining and modulating growth and survival of cancer initiating/stem cells.

Autophagy is downregulated in specific CSCs as compared to the remaining portion of cancer cells. For example, brain CSCs (CD133$^+$) display decreased expression of autophagy-related proteins and are more resistant to temozolomide as compared to putative brain cancer CD133$^-$ nonstem cells (Fu et al., 2009). In contrast, a recent study showed that autophagy in CSCs provides a protective effect to current cancer therapeutics, and inhibition of protective autophagy improves therapeutic response. For example, radiation-induced autophagy in glioma CSCs and the CD133$^+$ cells exhibited a larger degree of autophagy compared with the CD133$^-$ cells. Moreover, the CD133$^+$ cells expressed higher levels of LC3-II, Atg5, and Atg12, and inhibition of autophagy sensitized these cells to γ-radiation (Lomonaco et al., 2009). Similarly, therapy with tyrosine kinase inhibitors is associated with drug resistance in chronic myeloid leukemia (CML) CSCs, and autophagy is one of the mechanisms involved in this protective response. Suppression of autophagy using either pharmacological inhibitors or RNA interference of essential autophagy genes results in nearly complete elimination of phenotypically and functionally defined CML CSCs (Bellodi et al., 2009). However, these contradicting reports need to be reconciled by further experimentation. Understanding the role of autophagy in CSCs holds significant promise for enhancing cancer therapies (Fig. 2.3).

7. AUTOPHAGY IN CANCER THERAPY

Because cancer cells often display defective autophagic capacities, induction of ACD is viewed as a tumor suppressor mechanism. Induction of autophagic death, "type II programmed cell death," could be a useful therapeutic approach for apoptosis-resistant cancer cells and could provide a complementary approach along with apoptosis in promoting cancer cell death. On other hand, autophagy has been shown to provide resistance to therapy-mediated tumor cell death. When tumor cells induce protective autophagy, inhibition of autophagy may provide a way of sensitizing tumor cells to therapy by activating apoptosis. ACD by anticancer drugs may occur depending on cell type and genetic background. Based on the type of treatment, different signaling pathways can be activated in the same cell and produce varied types of autophagy. Understanding whether autophagy will be "protective" or "toxic" is a key area for further development and will define whether it is appropriate to block or promote autophagy in specific cancer contexts (Chen & Karantza, 2011; Kondo, Kanzawa, Sawaya, & Kondo, 2005; White & DiPaola, 2009).

7.1. Stimulation of autophagic cell death

Therapeutic induction of ACD through overstimulation of autophagy remains an important approach for tumor cell elimination. A number of studies have reported that ACD is activated in cancer cells derived from tissues such as breast, colon, prostate, and brain, in response to various anticancer therapies. The consequence of promoting autophagy depends on multiple factors, including extent of induction, duration, and cellular context. Several chemotherapeutic drugs (alkylating agents, actinomycin D, arsenic trioxide), radiation and photodynamic therapy, hormonal therapies (tamoxifen and vitamin D analogs), cytokines (IFN-γ), gene therapies (p53, *mda*-7/IL-24, and p27^{Kip1}), and natural compounds (resveratrol and plant lectins) have been shown to trigger ACD in various cancer cells *in vitro* (Chen & Karantza, 2011). Accumulating evidence indicates that autophagic death contributes to *in vivo* antitumor effects. For instance, a natural BH3-mimetic, small-molecule inhibitor of Bcl2, (−)-gossypol, shows potent antitumor activity in ongoing phase II and III clinical trials for human prostate cancer. The antitumor activity by (−)-gossypol is mediated through induction of both apoptosis and autophagic death (Lian et al., 2011). ACD can occur independently or it can act synergistically or assist apoptotic cell death (Kondo et al., 2005; Maiuri et al., 2009). Interestingly, combining two therapies that trigger autophagy by targeting different pathways increased sensitivity to ACD, an alternative form of programmed cell death to promote synergistic cancer inhibitory effects (White & DiPaola, 2009). But, autophagy appears to serve as a death program primarily when the apoptotic machinery is defective, as observed in most tumors. One major drawback in using autophagy-promoting drugs may involve unwanted paradoxical effects by actually protecting tumors against cell death triggered by simultaneous anticancer therapies or by nutrient deprivation in the tumor environment (Chen & Karantza, 2011).

7.2. Inhibition of protective autophagy

Autophagy, which is decreased in cancer cells as compared to normal cells, can provide a target for enhancing cancer therapy. Although Beclin1 is a haploinsufficient tumor suppressor, deletion of the remaining Beclin1 *in vitro* induces growth arrest in cancer cells (Wirawan et al., 2010). Similarly, elimination of Atg5 induces growth arrest in cancer cells (Yousefi et al., 2006). An *in vivo* study revealed that transplantation of Bcr–Abl-expressing hematopoietic cells depleted of Atg3 to lethally irradiated mice failed to induce leukemia based on ablation of autophagy (Altman et al., 2010). This suggests that a certain level of autophagy is required for tumor growth, and autophagic inhibitors could have relevant anticancer effects even when applied alone. However, it is more likely that autophagy inhibitors will be most effective when used in combination with cytotoxic drugs that activate a protective autophagy to permit cancer cell survival upon treatment. Accordingly, it has been demonstrated that melanoma differentiation-associated gene-7/Interleukin-24 (MDA-7/IL-24), a member of IL-10 gene family, shows nearly ubiquitous antitumor properties *in vitro* and *in vivo* through induction of cancer-specific apoptosis (Dash et al., 2010; Fisher, 2005). A recent report indicates that the apoptosis potential of MDA-7/IL-24 increases by inhibiting protective autophagy with 3-methyladenosine (3-MA) in prostate cancer cells (Bhutia, Dash, et al., 2010; Bhutia, Kegelman, et al., 2010). Similarly, inhibition of autophagy by 3-MA or Atg7 knockdown induced apoptosis in colon cancer cells treated with 5-FU (Li, Hou, Faried, Tsutsumi, & Kuwano, 2010). Inhibition of protective autophagy was shown to sensitize resistant cells to TRAIL-mediated apoptosis in apoptosis-defective leukemic and colon cancer cell lines (Han et al., 2008). Additionally, protective autophagy was accompanied with Ginsenoside F2-induced apoptosis in breast cancer stem cells, and treatment with chloriquine (CQ), an autophagy inhibitor, enhanced Ginsenoside F2-mediated cell death (Mai et al., 2012).

Autophagy also plays an important role in chemoresistance of cancer to some therapeutic agents that typically induce an apoptotic response (Carew, Nawrocki, & Cleveland, 2007; Carew, Nawrocki, Kahue, et al., 2007). Although autophagy has been proposed as a "magic bullet" in fighting apoptosis-resistant cancers (Gozuacik & Kimchi, 2004), a more recent study demonstrated that rapamycin-induced autophagy could protect various tumor cells against apoptosis induced by general apoptotic stimuli (Ravikumar, Berger, Vacher, O'Kane, & Rubinsztein, 2006). Recent reports highlight that treatment of estrogen-receptor-positive breast cancer cells with the anti-estrogen tamoxifen, combined with a histone deacetylase inhibitor, maintains a subpopulation of cells with elevated autophagy that display a remarkable resistance to apoptosis. These apoptosis-resistant cells only become apoptotic after inhibition of autophagy (Thomas, Thurn, Biçaku, Marchion, & Münster, 2011).

The potential to inhibit autophagy and sensitize tumor cells to metabolic stress is another promising approach for cancer therapy. Many current cancer therapies including angiogenesis, growth factor, and receptor inhibitors when combined with autophagy inhibition produced synergistic anticancer effects (Corcelle, Puustinen, & Jäättelä, 2009; Moretti et al., 2007; White & DiPaola, 2009). In addition, autophagy is also involved in removing damaged and potentially dangerous organelles from the cell. Therefore, combining organelle-damaging drugs, such as sigma-2 receptor agonists, with an autophagy inhibitor might be an effective means of anticancer therapy (Ostenfeld et al., 2008). It is likely that ER stress inducers, including thapsigargin and tunicamycin, that trigger cell death in cancer cells will increase cell killing when autophagy is inhibited (Carew, Nawrocki, & Cleveland, 2007; Carew, Nawrocki, Kahue, et al., 2007). Protein turnover by lysosomal degradation through the autophagy pathway is functionally complementary and linked with ubiquitin proteasome protein degradation. Consequently, targeting both proteasome- and autophagy-mediated protein degradations might be an effective antitumor approach for highly metabolically active tumor cells (Ding, Ni, Gao, Hou, et al., 2007; Ding, Ni, Gao, Yoshimori, et al., 2007). The proteasome inhibitor bortezomib has the approval of the US Food and Drug Administration and has demonstrated potent efficiency in treating multiple myeloma (Roccaro et al., 2006).

The metastasis prone state of tumor cells may be particularly susceptible to autophagy inhibition as cells in isolation are expected to be more reliant on autophagy, although this possibility remains to be confirmed. In this regard, chloroquine (CQ), which inhibits lysosome acidification and thereby autophagy, in conjunction with alkylating agents, displayed remarkable efficacy in inhibiting tumor growth in mice as well as in clinical studies (Høyer-Hansen & Jäättelä, 2008; Moretti et al., 2007; White & DiPaola, 2009). Synergy between CQ and the HDAC inhibitor SAHA in killing imatinib refractory chronic myeloid leukemia cells also supports a protective role for autophagy, reinforcing the therapeutic use of autophagy inhibitors in cancer therapy (Carew, Nawrocki, & Cleveland, 2007; Carew, Nawrocki, Kahue, et al., 2007). Similarly, the synergy between CQ and the PI3K–mTOR inhibitor NVP-BEZ235 induced apoptosis in glioma xenografts (Fan et al., 2010). Likewise, CQ enhanced cyclophosphamide-induced tumor cell death in a Myc-induced murine lymphoma similar to that shown by shRNA knockdown of *Atg5*, and it delayed the time-to-tumor recurrence

(Amaravadi et al., 2007). These studies documented that CQ, or its analog hydroxychloroquine, when used as autophagy inhibitors in combination with proapoptotic drugs, increases twofold the median survival of cancer patients (Carew, Nawrocki, & Cleveland, 2007; Carew, Nawrocki, Kahue, et al., 2007; Fimia et al., 2007; Garber, 2011; Savarino, Lucia, Giordano, & Cauda, 2006; Sotelo, Briceño, & López-González, 2006).

One underlying concern is that autophagy inhibitors approved for cancer patients might actually act as promoters of tumor development. However, the tumor-promoting effect of autophagy inhibitors, which depends on necrotic cell lysis that follows the inflammatory response, could prevent this undesirable effect upon cotreatment with immunosuppressive drugs (Chen & Karantza, 2011; Høyer-Hansen & Jäättelä, 2008).

8. CONCLUSIONS

Although the multiple roles of autophagy in cancer require further clarification, it is obvious that autophagy is directly involved in many important physiological processes such as metabolism, response to stress, and cell death pathways in cancer cells. Both tumor suppressor genes and oncogenes are implicated in autophagy regulation, thereby linking autophagy directly to cancer development and progression. Interestingly, autophagy also limits necrosis and inflammation and may restrict the invasion and dissemination of tumor cells from a primary site, thereby inhibiting a critical and early event in metastasis (Fig. 2.3). In contrast, autophagy may paradoxically promote metastasis at later stages by protecting detached and stressed tumor cells as they travel through blood vessel and establish new colonies at distant sites (Fig. 2.3). Accordingly, it is suggested that autophagy might elicit disparate effects in tumors at different stages of progression. Therapy-stimulated accumulation of autophagosomes is by itself not sufficient to draw any conclusions whether autophagy has a lethal or protective function. Accordingly, the role of autophagy in cancer raises a number of intriguing questions. Does autophagy play any direct or indirect role in cancer development and progression? If it does, what is its exact contribution in cancer development and progression? Does autophagy regulate cancer stem cell development, and if it does is its pattern of regulation different from that in normal stem cells? Are there any genetic and cellular physiologic conditions that direct when and how autophagy facilitates cancer cells to survive or causes them to die? Can autophagy be exploited as a means of enhancing cancer therapies? Considering the potential seminal roles of autophagy in both normal and abnormal cellular physiology it is important to unravel its complex regulation. This information will be crucial if one is to exploit autophagy in the future as a potential therapeutic for advanced cancers and potentially other proliferative diseases.

ACKNOWLEDGMENTS

Research support was provided in part by the National Institutes of Health grants R01 CA097318 (P. B. F.), R01 CA127641 (P. B. F.), R01 CA134721 (P. B. F.), R01 CA138540 (D. S.), and P01 CA104177 (P. B. F. and P. D.), and Department of Defense (DOD) Prostate Cancer Research Program (PCRP) Synergistic Idea Development Award W81XWH-10-PCRP-SIDA (P. B. F. and X.-Y. W.), the National Foundation for Cancer Research (P. B. F.), the Samuel Waxman Foundation for Cancer Research (D. S. and P. B. F.), the James S. McDonnell Foundation (D. S.) and a Rapid Grant for Young Investigator (RGYI) award, Department of Biotechnology, Government of India (S. K. B.). D. S. is a Harrison Scholar and a Blick Scholar in the VCU MCC and the VCU SOM. P. B. F. holds the Thelma Newmeyer Corman Chair in Cancer Research in the VCU MCC.

REFERENCES

Akar, U., Chaves-Reyez, A., Barria, M., Tari, A., Sanguino, A., Kondo, Y., et al. (2008). Silencing of Bcl-2 expression by small interfering RNA induces autophagic cell death in MCF-7 breast cancer cells. *Autophagy, 4,* 669–679.
Almog, N. (2010). Molecular mechanisms underlying tumor dormancy. *Cancer Letters, 294,* 139–146.
Altman, B. J., Jacobs, S. R., Mason, E. F., Michalek, R. D., MacIntyre, A. N., Coloff, J. L., et al. (2010). Autophagy is essential to suppress cell stress and to allow BCR-Abl-mediated leukemogenesis. *Oncogene, 30,* 1855–1867.
Amaravadi, R. K., Yu, D., Lum, J. J., Bui, T., Christophorou, M. A., Evan, G. I., et al. (2007). Autophagy inhibition enhances therapy-induced apoptosis in a Myc induced model of lymphoma. *The Journal of Clinical Investigation, 117,* 326–336.
Atlashkin, V., Kreykenbohm, V., Eskelinen, E. L., Wenzel, D., Fayyazi, A., & Fischer von Mollard, G. (2003). Deletion of the SNARE vti1b in mice results in the loss of a single SNARE partner, syntaxin 8. *Molecular and Cellular Biology, 23,* 5198–5207.
Avivar-Valderas, A., Salas, E., Bobrovnikova-Marjon, E., Diehl, J. A., Nagi, C., Debnath, J., et al. (2011). PERK integrates autophagy and oxidative stress responses to promote survival during extracellular matrix detachment. *Molecular and Cellular Biology, 31,* 3616–3629.
Axe, E. L., Walker, S. A., Manifava, M., Chandra, P., Roderick, H. L., Habermann, A., et al. (2008). Autophagosome formation from membrane compartments enriched in phosphatidylinositol 3-phosphate and dynamically connected to the endoplasmic reticulum. *The Journal of Cell Biology, 182,* 685–701.
Bellodi, C., Lidonnici, M. R., Hamilton, A., Helgason, G. V., Soliera, A. R., Ronchetti, M., et al. (2009). Targeting autophagy potentiates tyrosine kinase inhibitor-induced cell death in Philadelphia chromosome-positive cells, including primary CML stem cells. *The Journal of Clinical Investigation, 119,* 1109–1123.
Bhutia, S. K., Dash, R., Das, S. K., Azab, B., Su, Z. Z., Lee, S. G., et al. (2010). Mechanism of autophagy to apoptosis switch triggered in prostate cancer cells by antitumor cytokine

melanoma differentiation-associated gene 7/interleukin-24. *Cancer Research, 70,* 3667–3676.
Bhutia, S. K., Kegelman, T. P., Das, S. K., Azab, B., Su, Z. Z., Lee, S. G., et al. (2010). Astrocyte elevated gene-1 induces protective autophagy. *Proceedings of the National Academy of Sciences of United States of America, 107,* 22243–22248.
Bialik, S., & Kimchi, A. (2006). The death-associated protein kinases: Structure, function and beyond. *Annual Review of Biochemistry, 75,* 189–210.
Bingle, L., Brown, N. J., & Lewis, C. E. (2002). The role of tumour-associated macrophages in tumour progression: Implications for new anticancer therapies. *The Journal of Pathology, 196,* 254–265.
Cadwell, K., Liu, J. Y., Brown, S. L., Miyoshi, H., Loh, J., Lennerz, J. K., et al. (2008). A key role for autophagy and the autophagy gene Atg16L1 in mouse and human intestinal Paneth cells. *Nature, 456,* 259–263.
Carew, J. S., Nawrocki, S. T., & Cleveland, J. L. (2007). Modulating autophagy for therapeutic benefit. *Autophagy, 3,* 464–467.
Carew, J. S., Nawrocki, S. T., Kahue, C. N., Zhang, H., Yang, C., Chung, L., et al. (2007). Targeting autophagy augments the anticancer activity of the histone deacetylase inhibitor SAHA to overcome Bcr-Abl-mediated drug resistance. *Blood, 110,* 313–322.
Chen, N., & Debnath, J. (2010). Autophagy and tumorigenesis. *Federation of European Biochemical Societies Letters, 584,* 1427–1435.
Chen, N., & Karantza, V. (2011). Autophagy as a therapeutic target in cancer. *Cancer Biology & Therapy, 11,* 157–168.
Clevers, H. (2011). The cancer stem cell: Premises, promises and challenges. *Nature Medicine, 17,* 313–319.
Colombo, M. I. (2007). Autophagy: A pathogen driven process. *International Union of Biochemistry and Molecular Biology Life, 59,* 238–242.
Coppola, D., Khalil, F., Eschrich, S. A., Boulware, D., Yeatman, T., & Wang, H. G. (2008). Down-regulation of Bax-interacting factor-1 in colorectal adenocarcinoma. *Cancer, 113,* 2665–2670.
Corcelle, E. A., Puustinen, P., & Jäättelä, M. (2009). Apoptosis and autophagy: Targeting autophagy signalling in cancer cells—'Trick or treats'? *Federation of European Biochemical Societies Journal, 276,* 6084–6096.
Corradetti, M. N., Inoki, K., Bardeesy, N., DePinho, R. A., & Guan, K. L. (2004). Regulation of the TSC pathway by LKB1: Evidence of a molecular link between tuberous sclerosis complex and Peutz–Jeghers syndrome. *Genes & Development, 18,* 1533–1538.
Crighton, D., Wilkinson, S., O'Prey, J., Syed, N., Smith, P., Harrison, P. R., et al. (2006). DRAM, a p53-induced modulator autophagy, is critical for apoptosis. *Cell, 126,* 121–134.
Crotzer, V. L., & Blum, J. S. (2009). Autophagy and its role in MHC-mediated antigen presentation. *Journal of Immunology, 182,* 3335–3341.
Das, S. K., Bhutia, S. K., Kegelman, T. P., Peachy, L., Oyesanya, R. A., Dasgupta, S., et al. (2012). MDA-9/syntenin: A positive gatekeeper of melanoma metastasis. *Frontiers in Bioscience, 17,* 1–15.
Dash, R., Bhutia, S. K., Azab, B., Su, Z. Z., Quinn, B. A., Kegelmen, T. P., et al. (2010). mda-7/IL-24: A unique member of the IL-10 gene family promoting cancer-targeted toxicity. *Cytokine & Growth Factor Reviews, 21,* 381–391.
Debnath, J., Baehrecke, E. H., & Kroemer, G. (2005). Does autophagy contribute to cell death? *Autophagy, 1,* 66–74.
Degenhardt, K., Mathew, R., Beaudoin, B., Bray, K., Anderson, D., Chen, G., et al. (2006). Autophagy promotes tumor cell survival and restricts necrosis, inflammation, and tumorigenesis. *Cancer Cell, 10,* 51–64.

Ding, W. X., Ni, H. M., Gao, W., Hou, Y. F., Melan, M. A., Chen, X., et al. (2007). Differential effects of endoplasmic reticulum stress-induced autophagy on cell survival. *The Journal of Biological Chemistry, 282*, 4702–4710.

Ding, W. X., Ni, H. M., Gao, W., Yoshimori, T., Stolz, D. B., Ron, D., et al. (2007). Linking of autophagy to ubiquitin proteasome system is important for the regulation of endoplasmic reticulum stress and cell viability. *The American Journal of Pathology, 171*, 513–524.

Eisenberg-Lerner, A., & Kimchi, A. (2009). The paradox of autophagy and its implication in cancer etiology and therapy. *Apoptosis, 14*, 376–391.

Emdad, L., Lee, S.-G., Su, Z.-z., Jeong, H. Y., Boukerche, H., Sarkar, D., et al. (2009). Astrocyte elevated gene-1 (AEG-1) functions as an oncogene and regulates angiogenesis. *Proceedings of the National Academy of Sciences of United States of America, 106*, 21300–21305.

Fan, Q. W., Cheng, C., Hackett, C., Feldman, M., Houseman, B. T., Nicolaides, T., et al. (2010). Akt and autophagy cooperate to promote survival of drug resistant glioma. *Science Signal, 3*, ra81.

Fass, E., Shvets, E., Degani, I., Hirschberg, K., & Elazar, Z. (2006). Microtubules support production of starvation-induced autophagosomes but not their targeting and fusion with lysosomes. *The Journal of Biological Chemistry, 281*, 36303–36316.

Fimia, G. M., Stoykova, A., Romagnoli, A., Giunta, L., Bartolomeo, S. D., Nardacci, R., et al. (2007). Ambra1 regulates autophagy and development of the nervous system. *Nature, 447*, 1121–1125.

Fisher, P. B. (2005). Is mda-7/IL-24 a "magic bullet" for cancer? *Cancer Research, 65*, 10128–10138.

Fu, J., Liu, Z. G., Liu, X. M., Chen, F. R., Shi, H. L., Pangjesse, C. S., et al. (2009). Glioblastoma stem cells resistant to temozolomide-induced autophagy. *Chinese Medical Journal (Engl), 122*, 1255–1259.

Fujita, N., Itoh, T., Omori, H., Fukuda, M., Noda, T., & Yoshimori, T. (2008). The Atg16L complex specifies the site of LC3 lipidation for membrane biogenesis in autophagy. *Molecular Biology of the Cell, 19*, 2092–2100.

Funderbur, S. F., Wang, Q. J., & Yue, Z. (2010). The Beclin1–VPS34 complex—At the crossroads of autophagy and beyond. *Trends in Cell Biology, 20*, 355–362.

Fung, C., Lock, R., Gao, S., Salas, E., & Debnath, J. (2008). Induction of autophagy during extracellular matrix detachment promotes cell survival. *Molecular Biology of the Cell, 19*, 797–806.

Garber, K. (2011). Inducing indigestion: Companies embrace autophagy inhibitors. *Journal of the National Cancer Institute, 103*, 708–710.

Geng, J., & Klionsky, D. J. (2008). The Atg8 and Atg12 ubiquitin-like conjugation systems in macroautophagy. *European Molecular Biology Organization Reports, 9*, 859–864.

Gozuacik, D., & Kimchi, A. (2004). Autophagy as a cell death and tumor suppressor mechanism. *Oncogene, 23*, 2891–2906.

Gunn, J. M., Clark, M. G., Knowles, S. E., Hopgood, M. F., & Ballard, F. J. (1977). Reduced rates of proteolysis in transformed cells. *Nature, 266*, 58–60.

Guo, J. Y., Chen, H. Y., Mathew, R., Fan, J., Strohecker, A. M., Karsli-Uzunbas, G., et al. (2011). Activated Ras requires autophagy to maintain oxidative metabolism and tumorigenesis. *Genes & Development, 25*, 460–470.

Gutierrez, M. G., Munafo, D. B., Beron, W., & Colombo, M. I. (2004). Rab7 is required for the normal progression of the autophagic pathway in mammalian cells. *Journal of Cell Science, 117*, 2687–2697.

Hafner, C., Lopez-Knowles, E., Luis, N. M., Toll, A., Baselga, E., Fernandez-Casado, A., et al. (2007). Oncogenic PIK3CA mutations occur in epidermal nevi and seborrheic keratoses with a characteristic mutation pattern. *Proceedings of the National Academy of Sciences of United States of America, 104*, 13450–13454.

Han, J., Hou, W., Goldstein, L. A., Lu, C., Stolz, D. B., Yin, X. M., et al. (2008). Involvement of protective autophagy in TRAIL resistance of apoptosis- defective tumor cells. *The Journal of Biological Chemistry, 283*, 19665–19677.

Hanada, T., Noda, N. N., Satomi, Y., Ichimura, Y., Fujioka, Y., Takao, T., et al. (2007). The Atg12-Atg5 conjugate has a novel E3-like activity for protein lipidation in autophagy. *The Journal of Biological Chemistry, 282*, 37298–37302.

Hanahan, D., & Weinberg, R. A. (2011). Hallmarks of cancer: The next generation. *Cell, 144*, 646–674.

Herrero-Martin, G., Hoyer-Hansen, M., Garcia-Garcia, C., Fumarola, C., Farkas, T., Lopez-Rivas, A., et al. (2009). TAK1 activates AMPK dependent cytoprotective autophagy in TRAIL-treated epithelial cells. *The European Molecular Biology Organization Journal, 28*, 677–685.

Hirayama, A., Kami, K., Sugimoto, M., Sugawara, M., Toki, N., Onozuka, H., et al. (2009). Quantitative metabolome profiling of colon and stomach cancer microenvironment by capillary electrophoresis time-of-flight mass spectrometry. *Cancer Research, 69*, 4918–4925.

Horn, S., Bergholz, U., Jucker, M., McCubrey, J. A., Trumper, L., Stocking, C., et al. (2008). Mutations in the catalytic subunit of class IA PI3K confer leukemogenic potential to hematopoietic cells. *Oncogene, 27*, 4096–4106.

Høyer-Hansen, M., & Jäättelä, M. (2008). Autophagy: An emerging target for cancer therapy. *Autophagy, 4*, 574–580.

Ichimura, Y., & Komatsu, M. (2010). Selective degradation of p62 by autophagy. *Seminars in Immunopathology, 32*, 431–436.

Inbal, B., Bialik, S., Sabanay, I., Shani, G., & Kimchi, A. (2002). DAP kinase and DRP-1 mediate membrane blebbing and the formation of autophagic vesicles during programmed cell death. *The Journal of Cell Biology, 157*, 455–468.

Ionov, Y., Nowak, N., Perucho, M., Markowitz, S., & Cowell, J. K. (2004). Manipulation of nonsense mediated decay identifies gene mutations in colon cancer cells with microsatellite instability. *Oncogene, 23*, 639–645.

Jager, S., Bucci, C., Tanida, I., Ueno, T., Kominami, E., Saftig, P., et al. (2004). Role for Rab7 in maturation of late autophagic vacuoles. *Journal of Cell Science, 117*, 4837–4848.

Jin, S., & White, E. (2007). Role of autophagy in cancer: Management of metabolic stress. *Autophagy, 3*, 28–31.

Kabeya, Y., Mizushima, N., Ueno, T., Yamamoto, A., Kirisako, T., Noda, T., et al. (2000). LC3, a mammalian homologue of yeast Apg8p, is localized in autophagosome membranes after processing. *The European Molecular Biology Organization Journal, 19*, 5720–5728.

Kamada, Y., Yoshino, K., Kondo, C., Kawamata, T., Oshiro, N., Yonezawa, K., et al. (2010). TOR directly controls the Atg1 kinase complex to regulate autophagy. *Molecular and Cellular Biology, 30*, 1049–1058.

Kang, D.-C., Su, Z.-Z., Sarkar, D., Emdad, L., Volsky, D. J., & Fisher, P. B. (2005). Cloning and characterization of HIV-1-inducible astrocyte elevated gene-1, AEG-1. *Gene, 353*, 8–15.

Karantza-Wadsworth, V., Patel, S., Kravchuk, O., Chen, G., Mathew, R., Jin, S., et al. (2007). Autophagy mitigates metabolic stress and genome damage in mammary tumorigenesis. *Genes & Development, 21*, 1621–1635.

Kenific, C. M., Thorburn, A., & Debnath, J. (2010). Autophagy and metastasis: Another double-edged sword. *Current Opinion in Cell Biology, 22*, 241–245.

Kim, M. S., Jeong, E. G., Ahn, C. H., Kim, S. S., Lee, S. H., & Yoo, N. J. (2008). Frameshift mutation of UVRAG, an autophagyrelated gene, in gastric carcinomas with microsatellite instability. *Human Pathology, 39*, 1059–1063.

Kimura, S., Noda, T., & Yoshimori, T. (2008). Dynein-dependent movement of autophagosomes mediates efficient encounters with lysosomes. *Cell Structure and Function, 33*, 109–122.

Kisen, G. O., Tessitore, L., Costelli, P., Gordon, P. B., Schwarze, P. E., Baccino, F. M., et al. (1993). Reduced autophagic activity in primary rat hepatocellular carcinoma and ascites hepatoma cells. *Carcinogenesis, 14,* 2501–2505.

Klionsky, D. J. (2007). Autophagy: From phenomenology to molecular understanding in less than a decade. *Nature Reviews. Molecular Cell Biology, 8,* 931–937.

Klionsky, D. J., & Emr, S. D. (2000). Autophagy as a regulated pathway of cellular degradation. *Science, 290,* 1717–1721.

Köchl, R., Hu, X. W., Chan, E. Y., & Tooze, S. A. (2006). Microtubules facilitate autophagosome formation and fusion of autophagosomes with endosomes. *Traffic, 7,* 129–145.

Kondo, Y., Kanzawa, T., Sawaya, R., & Kondo, S. (2005). The role of autophagy in cancer development and response to therapy. *Nature Reviews. Cancer, 5,* 726–734.

Kroemer, G., El-Deiry, W. S., Goldstein, P., Peter, M. E., Vaux, D., Vandenabeele, P., et al. (2005). Classification of cell death: Recommendations of the Nomenclature Committee on Cell Death. *Cell Death and Differentiation, 12,* 1463–1467.

Kroemer, G., & Levine, B. (2008). Autophagic cell death: The story of a misnomer. *Nature Reviews. Molecular Cell Biology, 9,* 1004–1010.

Kuma, A., Hatano, M., Matsui, M., Yamamoto, A., Nakaya, H., Yoshimori, T., et al. (2004). The role of autophagy during the early neonatal starvation period. *Nature, 432,* 1032–1036.

Lee, J. A., Beigneux, A., Ahmad, S. T., Young, S. G., & Gao, F. B. (2007). ESCRT-III dysfunction causes autophagosome accumulation and neurodegeneration. *Current Biology, 17,* 1561–1567.

Lee, J. S., Li, Q., Lee, J. Y., Lee, S. H., Jeong, J. H., Lee, H. R., et al. (2009). FLIP mediated autophagy regulation in cell death control. *Nature Cell Biology, 11,* 1355–1362.

Levine, B., & Kroemer, G. (2008). Autophagy in the pathogenesis of disease. *Cell, 132,* 27–42.

Levine, B., Mizushima, N., & Virgin, H. W. (2011). Autophagy in immunity and inflammation. *Nature, 469,* 323–335.

Levine, B., Sinha, S., & Kroemer, G. (2008). Bcl-2 family members: Dual regulators of apoptosis and autophagy. *Autophagy, 4,* 600–606.

Li, J., Hou, N., Faried, A., Tsutsumi, S., & Kuwano, H. (2010). Inhibition of autophagy augments 5-fluorouracil chemotherapy in human colon cancer in vitro and in vivo model. *European Journal of Cancer, 46,* 1900–1909.

Lian, J., Wu, X., He, F., Karnak, D., Tang, W., Meng, Y., et al. (2011). A natural BH3 mimetic induces autophagy in apoptosis-resistant prostate cancer via modulating Bcl-2-Beclin1 interaction at endoplasmic reticulum. *Cell Death and Differentiation, 18,* 60–71.

Liang, C., Feng, P., Ku, B., Dotan, I., Canaani, D., Oh, B. H., et al. (2006). Autophagic and tumour suppressor activity of a novel Beclin1-binding protein UVRAG. *Nature Cell Biology, 8,* 688–699.

Liang, X. H., Jackson, S., Seaman, M., Brown, K., Kempkes, B., Hibshoosh, H., et al. (1999). Induction of autophagy and inhibition of tumorigenesis by Beclin1. *Nature, 402,* 672–676.

Liang, C., & Jung, J. U. (2010). Autophagy genes as tumor suppressors. *Current Opinion in Cell Biology, 22,* 226–233.

Liang, J., Shao, S. H., Xu, Z. X., Hennessy, B., Ding, Z., Larrea, M., et al. (2007). The energy sensing LKB1-AMPK pathway regulates p27(kip1) phosphorylation mediating the decision to enter autophagy or apoptosis. *Nature Cell Biology, 9,* 218–224.

Lomonaco, S. L., Finniss, S., Xiang, C., DeCarvalho, A., Umansky, F., Kalkanis, S. N., et al. (2009). The induction of autophagy by γ-radiation contributes to the radioresistance of glioma stem cells. *International Journal of Cancer, 125,* 717–722.

LoPiccolo, J., Blumenthal, G. M., Bernstein, W. B., & Dennis, P. A. (2008). Targeting the PI3K/Akt/mTOR pathway: Effective combinations and clinical considerations. *Drug Resistance Updates, 11*, 32–50.

Lu, Z., Luo, R. Z., Lu, Y., Zhang, X., Yu, Q., Khare, S., et al. (2008). The tumor suppressor gene ARHI regulates autophagy and tumor dormancy in human ovarian cancer cells. *The Journal of Clinical Investigation, 118*, 3917–3929.

Maehama, T. (2007). PTEN: Its deregulation and tumorigenesis. *Biological & Pharmaceutical Bulletin, 30*, 1624–1627.

Mai, T. T., Moon, J., Song, Y., Viet, P. Q., Phuc, P. V., Lee, J. M., et al. (2012). Ginsenoside F2 induces apoptosis accompanied by protective autophagy in breast cancer stem cells. *Cancer Letters, 321*, 144–153.

Maiuri, M. C., Tasdemir, E., Criollo, A., Morselli, E., Vicencio, J. M., Carnuccio, R., et al. (2009). Control of autophagy by oncogenes and tumor suppressor genes. *Cell Death and Differentiation, 16*, 87–93.

Marino, G., Salvador-Montoliu, N., Fueyo, A., Knecht, E., Mizushima, N., & Lopez-Otin, C. (2007). Tissue-specific autophagy alterations and increased tumorigenesis in mice deficient in Atg4C/autophagin-3. *The Journal of Biological Chemistry, 282*, 18573–18583.

Martens, S., & McMahon, H. T. (2008). Mechanism of membrane fusion: Disparate players and common principles. *Nature Review of Molecular and Cellular Biology, 9*, 543–556.

Mathew, R., Karantza-Wadsworth, V., & White, E. (2007). Role of autophagy in cancer. *Nature Review Cancer, 7*, 961–967.

Mathew, R., Karp, C. M., Beaudoin, B., Vuong, N., Chen, G., Chen, H. Y., et al. (2009). Autophagy suppresses tumorigenesis through elimination of p62. *Cell, 137*, 1062–1075.

Mathew, R., Kongara, S., Beaudoin, B., Karp, C. M., Bray, K., Degenhardt, K., et al. (2007). Autophagy suppresses tumor progression by limiting chromosomal instability. *Genes & Development, 21*, 1367–1381.

Matsunaga, K., Morita, E., Saitoh, T., Akira, S., Ktistakis, N. T., Izumi, T., et al. (2010). Autophagy requires endoplasmic reticulum targeting of the PI3-kinase complex via Atg14L. *The Journal of Cell Biology, 190*, 511–521.

Matsunaga, K., Saitoh, T., Tabata, K., Omori, H., Satoh, T., Kurotori, N., et al. (2009). Two Beclin 1-binding proteins, Atg14L and Rubicon, reciprocally regulate autophagy at different stages. *Nature Cell Biology, 11*, 385–396.

Mizushima, N. (2010). The role of the Atg1/ULK1 complex in autophagy regulation. *Current Opinion in Cell Biology, 22*, 132–139.

Mizushima, N., & Klionsky, D. J. (2007). Protein turnover via autophagy: Implications for metabolism. *Annual Review of Nutrition, 27*, 19–40.

Mizushima, N., Ohsumi, Y., & Yoshimori, T. (2002). Autophagosome formation in mammalian cells. *Cell Structure and Function, 27*, 421–429.

Moretti, L., Yang, E. S., Kim, K. W., & Lu, B. (2007). Autophagy signaling in cancer and its potential as novel target to improve anticancer therapy. *Drug Resistance Updates, 10*, 135–143.

Morselli, E., Galluzzi, L., Kepp, O., Vicencio, J. M., Criollo, A., Maiuri, M. C., et al. (2009). Anti- and pro-tumor functions of autophagy. *Biochimica et Biophysica Acta, 1793*, 1524–1532.

Mortensen, M., Soilleux, E. J., Djordjevic, G., Tripp, R., Lutteropp, M., Sadighi-Akha, E., et al. (2011). The autophagy protein Atg7 is essential for hematopoietic stem cell maintenance. *The Journal of Experimental Medicine, 208*, 455–467.

Nakatogawa, H., Suzuki, K., Kamada, Y., & Ohsumi, Y. (2009). Dynamics and diversity in autophagy mechanisms: Lessons from yeast. *Nature Reviews. Molecular Cell Biology, 10*, 458–467.

Nishida, Y., Arakawa, S., Fujitani, K., Yamaguchi, H., Mizuta, T., Kaneseki, T., et al. (2009). Discovery of Atg5/Atg7-independent alternative macroautophagy. *Nature, 461*, 654–658.

Noda, T., Matsunaga, K., Taguchi-Atarashi, N., & Yoshimori, T. (2010). Regulation of membrane biogenesis in autophagy via PI3P dynamics. *Seminars in Cell & Developmental Biology, 21,* 671–676.

Oberstein, A., Jeffrey, P. D., & Shi, Y. (2007). Crystal structure of the BCL-XL-beclin 1 peptide complex: Beclin 1 is a novel BH3-only protein. *The Journal of Biological Chemistry, 282,* 13123–13132.

Okamoto, K., Kondo-Okamoto, N., & Ohsumi, Y. (2009). Mitochondria-anchored receptor Atg32 mediates degradation of mitochondria via selective autophagy. *Developmental Cell, 17,* 87–97.

Ostenfeld, M. S., Høyer-Hansen, M., Bastholm, L., Fehrenbacher, N., Olsen, O. D., Groth-Pedersen, L., et al. (2008). Anti-cancer agent siramesine is a lysosomotropic detergent that induces cytoprotective autophagosome accumulation. *Autophagy, 4,* 487–499.

Pantel, K., Alix-Panabières, C., & Riethdorf, S. (2009). Cancer micrometastases. *Nature Reviews. Clinical Oncology, 6,* 339–351.

Pattingre, S., & Levine, B. (2006). Bcl-2 inhibition of autophagy: A new route to cancer? *Cancer Research, 66,* 2885–2888.

Qu, X., Yu, J., Bhagat, G., Furuya, N., Hibshoosh, H., Troxel, A., et al. (2003). Promotion of tumorigenesis by heterozygous disruption of the beclin1 autophagy gene. *The Journal of Clinical Investigation, 112,* 1809–1820.

Qu, X., Zou, Z., Sun, Q., Luby-Phelps, K., Cheng, P., Hogan, R. N., et al. (2007). Autophagy gene-dependent clearance of apoptotic cells during embryonic development. *Cell, 128,* 931–946.

Rabinowitz, J. D., & White, E. (2010). Autophagy and metabolism. *Science, 330,* 1344–1348.

Ravikumar, B., Acevedo-Arozena, A., Imarisio, S., Berger, Z., Vacher, C., O'Kane, C. J., et al. (2005). Dynein mutations impair autophagic clearance of aggregate-prone proteins. *Nature Genetics, 37,* 771–776.

Ravikumar, B., Berger, Z., Vacher, C., O'Kane, C. J., & Rubinsztein, D. C. (2006). Rapamycin pre-treatment protects against apoptosis. *Human Molecular Genetics, 15,* 1209–1216.

Roccaro, A. M., Hideshima, T., Richardson, P. G., Russo, D., Ribatti, D., & Vacca, A. (2006). Bortezomib as an antitumor agent. *Current Pharmaceutical Biotechnology, 7,* 441–448.

Savarino, A., Lucia, M. B., Giordano, F., & Cauda, R. (2006). Risks and benefits of chloroquine use in anticancer strategies. *The Lancet Oncology, 7,* 792–793.

Sekito, T., Kawamata, T., Ichikawa, R., Suzuki, K., & Ohsumi, Y. (2009). Atg17 recruits Atg9 to organize the pre-autophagosomal structure. *Genes to Cells, 14,* 525–538.

Shaw, R. J., & Cantley, L. C. (2006). Ras, PI(3)K and mTOR signalling controls tumour cell growth. *Nature, 441,* 424–430.

Shaw, R. J., Kosmatka, M., Bardeesy, N., Hurley, R. L., Witters, L. A., DePinho, R. A., et al. (2004). The tumor suppressor LKB1 kinase directly activates AMP-activated kinase and regulates apoptosis in response to energy stress. *Proceedings of the National Academy of Sciences of United States of America, 101,* 3329–3335.

Simonsen, A., & Tooze, S. A. (2009). Coordination of membrane events during autophagy by multiple class III PI3-kinase complexes. *The Journal of Cell Biology, 186,* 773–782.

Sotelo, J., Briceño, E., & López-González, M. A. (2006). Adding chloroquine to conventional treatment for glioblastoma multiforme: A randomized, double blind, placebo-controlled trial. *Annals of Internal Medicine, 144,* 337–343.

Su, Z.-Z., Kang, D.-C., Chen, Y., Pekarskaya, O., Chao, W., Volsky, D. J., et al. (2002). Identification and cloning of human astrocyte genes displaying elevated expression after infection with HIV-1 or exposure to HIV-1 envelope glycoprotein by rapid subtraction hybridization, RaSH. *Oncogene, 21,* 3592–3602.

Taddei, M. L., Giannoni, E., Fiaschi, T., & Chiarugi, P. (2012). Anoikis: An emerging hallmark in health and diseases. *The Journal of Pathology, 226,* 380–393.

Takahashi, Y., Coppola, D., Matsushita, N., Cualing, H. D., Sun, M., Sato, Y., et al. (2007). Bif-1 interacts with Beclin1 through UVRAG and regulates autophagy and tumorigenesis. *Nature Cell Biology*, *9*, 1142–1151.

Takamura, A., Komatsu, M., Hara, T., Sakamoto, A., Kishi, C., Waguri, S., et al. (2011). Autophagy-deficient mice develop multiple liver tumors. *Genes & Development*, *25*, 795–800.

Tanida, I., Ueno, T., & Kominami, E. (2004). LC3 conjugation system in mammalian autophagy. *The International Journal of Biochemistry & Cell Biology*, *36*, 2503–2518.

Tasdemir, E., Chiara Maiuri, M., Morselli, E., Criollo, A., D'Amelio, M., & Djavaheri-Mergny, M. (2008). A dual role of p53 in the control of autophagy. *Autophagy*, *4*, 810–814.

Thomas, S., Thurn, K. T., Biçaku, E., Marchion, D. C., & Münster, P. N. (2011). Addition of a histone deacetylase inhibitor redirects tamoxifen-treated breast cancer cells into apoptosis, which is opposed by the induction of autophagy. *Breast Cancer Research and Treatment*, *130*, 437–447.

Thorburn, J., Horita, H., Redzic, J., Hansen, K., Frankel, A. E., & Thorburn, A. (2008). Autophagy regulates selective HMGB1 release in tumor cells that are destined to die. *Cell Death and Differentiation*, *16*, 175–183.

Tsujimoto, Y., & Shimizu, S. (2005). Another way to die: Autophagic programmed cell death. *Cell Death and Differentiation*, *12*, 1528–1534.

Twig, G., Elorza, A., Molina, A. J., Mohamed, H., Wikstrom, J. D., Walzer, G., et al. (2008). Fission and selective fusion govern mitochondrial segregation and elimination by autophagy. *The European Molecular Biology Organization Journal*, *27*, 433–446.

Vergne, I., & Deretic, V. (2010). The role of PI3P phosphatases in the regulation of autophagy. *Federation of European Biochemical Societies Letters*, *584*, 1313–1318.

Vergne, I., Roberts, E., Elmaoued, R. A., Tosch, V., Delgado, M. A., Proikas-Cezanne, T., et al. (2009). Control of autophagy initiation by phosphoinositide 3-phosphatase jumpy. *The European Molecular Biology Organization Journal*, *28*, 2244–2258.

Vivanco, I., & Sawyers, C. L. (2002). The phosphatidylinositol 3-Kinase AKT pathway in human cancer. *Nature Reviews Cancer*, *2*, 489–501.

Wang, S. (2008). The promise of cancer therapeutics targeting the TNF-related apoptosis-inducing ligand and TRAIL receptor pathway TRAIL signaling in cancer therapy. *Oncogene*, *27*, 6207–6215.

Weidberg, H., Shvets, E., & Elazar, Z. (2011). Biogenesis and cargo selectivity of autophagosomes. *Annual Review of Biochemistry*, *80*, 125–156.

White, E., & DiPaola, R. S. (2009). The double-edged sword of autophagy modulation in cancer. *Clinical Cancer Research*, *15*, 5308–5316.

White, D. E., Kurpios, N. A., Zuo, D., Hassell, J. A., Blaess, S., Mueller, U., et al. (2004). Targeted disruption of β1-integrin in a transgenic mouse model of human breast cancer reveals an essential role in mammary tumor induction. *Cancer Cell*, *6*, 159–170.

Williams, A., Sarkar, S., Cuddon, P., Ttofi, E. K., Saiki, S., Siddiqi, F. H., et al. (2008). Novel targets for Huntington's disease in an mTOR-independent autophagy pathway. *Nature Chemical Biology*, *4*, 295–305.

Wirawan, E., Vande Walle, L., Kersse, K., Cornelis, S., Claerhout, S., Vanoverberghe, I., et al. (2010). Caspase-mediated cleavage of Beclin-1 inactivates Beclin-1-induced autophagy and enhances apoptosis by promoting the release of proapoptotic factors from mitochondria. *Cell Death & Disease*, *1*, e18.

Wu, J. J., Quijano, C., Chen, E., Liu, H., Cao, L., Fergusson, M. M., et al. (2009). Mitochondrial dysfunction and oxidative stress mediate the physiological impairment induced by the disruption of autophagy. *Aging*, *1*, 425–437.

Yin, Y., & Shen, W. H. (2008). PTEN: A new guardian of the genome. *Oncogene*, *27*, 5443–5453.

Yorimitsu, T., & Klionsky, D. J. (2005). Autophagy: Molecular machinery for self-eating. *Cell Death and Differentiation, 12,* 1542–1552.

Young, A. R., Narita, M., Ferreira, M., Kirschner, K., Sadaie, M., Darot, J. F., et al. (2009). Autophagy mediates the mitotic senescence transition. *Genes & Development, 23,* 798–803.

Yousefi, S., Perozzo, R., Schmid, I., Ziemiecki, A., Schaffner, T., Scapozza, L., et al. (2006). Calpain-mediated cleavage of Atg5 switches autophagy to apoptosis. *Nature Cell Biology, 8,* 1124–1132.

Yue, Z., Jin, S., Yang, C., Levine, A. J., & Heintz, N. (2003). Beclin1, an autophagy gene essential for early embryonic development, is a haploinsufficient tumor suppressor. *Proceedings of the National Academy of Sciences of United States of America, 100,* 15077–15082.

Zerial, M., & McBride, H. (2001). Rab proteins as membrane organizers. *Nature Review of Molecular and Cellular Biology, 2,* 107–117.

Zhang, X. H., Wang, Q., Gerald, W., Hudis, C. A., Norton, L., Smid, M., et al. (2009). Latent bone metastasis in breast cancer tied to Src-dependent survival signals. *Cancer Cell, 16,* 67–78.

Zhong, Y., Wang, Q. J., Li, X., Yan, Y., Backer, J. M., Chait, B. T., et al. (2009). Distinct regulation of autophagic activity by Atg14L and Rubicon associated with Beclin 1-phosphatidylinositol-3-kinase complex. *Nature Cell Biology, 11,* 468–476.

Zhou, B. B., Zhang, H., Damelin, M., Geles, K. G., Grindley, J. C., & Dirks, P. B. (2009). Tumour-initiating cells: Challenges and opportunities for anticancer drug discovery. *Nature Reviews Drug Discovery, 8,* 806–823.

CHAPTER THREE

The Transcription Factor FOXM1 (Forkhead box M1): Proliferation-Specific Expression, Transcription Factor Function, Target Genes, Mouse Models, and Normal Biological Roles

Inken Wierstra[1]
Wißmannstr. 17, D-30173 Hannover, Germany
[1]Corresponding author: e-mail address: iwiwiwi@web.de

Contents

1.	FOXM1 is an Activating Transcription Factor	99
2.	The Four Known FOXM1 Splice Variants	102
	2.1 Four FOXM1 splice variants arise from differential splicing of three facultative exons in the *foxm1* gene	102
	2.2 DNA-binding specificity of the FOXM1 splice variants	103
	2.3 Transcriptional activity of the FOXM1 splice variants	106
	2.4 Comparison of the four FOXM1 splice variants	107
	2.5 Differences between the splice variants FoxM1B and FOXM1c	108
3.	FOXM1 Target Genes	110
	3.1 The target genes of FOXM1	110
	3.2 Unusual properties of the transcription factor FOXM1	160
	3.3 Biological functions of FOXM1 target genes	161
4.	Molecular Mechanisms of FOXM1 for Gene Regulation	164
	4.1 Gene regulation mechanisms of FOXM1	164
	4.2 FOXM1-binding sites	166
5.	The Molecular Function of FOXM1 as a Conventional Transcription Factor	167
	5.1 The conventional transcription factor FOXM1	167
	5.2 Functional domains of the conventional transcription factor FOXM1c	168
	5.3 The DBD	170

The present chapter is Part I of a two-part review on the transcription factor FOXM1. Part II of this FOXM1 review is published in Volume 119 of Advances in Cancer Research: Inken Wierstra, FOXM1 (Forkhead box M1) in tumorigenesis: overexpression in human cancer, implication in tumorigenesis, oncogenic functions, tumor-suppressive properties and target of anti-cancer therapy. Advances in Cancer Research, 2013, Volume 119, in press.

5.4	The TAD	179
5.5	The TRD	185
5.6	The NRD-C	188
5.7	The NRD-N	189
5.8	The three IDs for the TAD	190
5.9	The molecular characterization of FOXM1 as a conventional transcription factor	193
5.10	The autoinhibitory N-terminus of FOXM1	195
5.11	Summary of the function of FOXM1c as a conventional transcription factor	196
6.	The FOXM1 Expression	197
6.1	FOXM1 displays a proliferation-specific expression pattern	197
6.2	The expression of FOXM1 during the life span	199
6.3	The expression of FOXM1 during the cell cycle	204
6.4	The expression of FOXM1 upon entry into the cell cycle	207
6.5	Cessation of FOXM1 expression upon cell cycle exit	212
6.6	Antagonistic regulation of the FOXM1 expression by proliferation versus antiproliferation signals	218
6.7	The expression of FOXM1 in response to DNA damage	220
6.8	Control of the FOXM1 protein stability	222
6.9	Control of the *foxm1* promoter by transcription factors	229
6.10	Regulation of the FOXM1 expression by miRNAs	237
6.11	Selected pathways that regulate the expression of FOXM1	238
6.12	The expression of FOXM1 in human disease	254
6.13	Additional regulators of the FOXM1 expression	257
7.	FOXM1 Mouse Models	266
7.1	FOXM1 knockout mice	266
7.2	FOXM1 transgenic mouse models	282
7.3	Xenopus	283
8.	Biological Functions of FOXM1	283
8.1	Cell proliferation and cell cycle progression	283
8.2	Embryonic development	294
8.3	Interference with contact inhibition	296
8.4	Cellular senescence	297
8.5	Adult tissue repair after injury	304
8.6	Liver repopulation by transplanted hepatocytes in a model of chronic liver injury	308
8.7	Insulin-producing β-cells in pancreatic islets	310
8.8	Maintenance of the proliferative capacity of cells	312
8.9	Homologous recombination (HR) repair of DNA damage	312
8.10	Maintenance of genomic stability	316
8.11	Maintenance of stem cell pluripotency	318
8.12	Stem cell self-renewal	321
8.13	Balance between progenitor cell renewal and the commitment to differentiation	322
8.14	Cellular differentiation	323
8.15	Apoptosis and cellular survival	327

8.16	Autophagy	333
8.17	Global DNA hypomethylation and focal DNA hypermethylation	333
8.18	Heat shock response	336
8.19	Allergen-induced lung inflammation	337
8.20	Radiation-induced pulmonary fibrosis	340
References		343

Abstract

FOXM1 (Forkhead box M1) is a typical proliferation-associated transcription factor, which stimulates cell proliferation and exhibits a proliferation-specific expression pattern. Accordingly, both the expression and the transcriptional activity of FOXM1 are increased by proliferation signals, but decreased by antiproliferation signals, including the positive and negative regulation by protooncoproteins or tumor suppressors, respectively. FOXM1 stimulates cell cycle progression by promoting the entry into S-phase and M-phase. Moreover, FOXM1 is required for proper execution of mitosis. Accordingly, FOXM1 regulates the expression of genes, whose products control G1/S-transition, S-phase progression, G2/M-transition, and M-phase progression. Additionally, FOXM1 target genes encode proteins with functions in the execution of DNA replication and mitosis. FOXM1 is a transcriptional activator with a forkhead domain as DNA binding domain and with a very strong acidic transactivation domain. However, wild-type FOXM1 is (almost) inactive because the transactivation domain is repressed by three inhibitory domains. Inactive FOXM1 can be converted into a very potent transactivator by activating signals, which release the transactivation domain from its inhibition by the inhibitory domains. FOXM1 is essential for embryonic development and the *foxm1* knockout is embryonically lethal. In adults, FOXM1 is important for tissue repair after injury. FOXM1 prevents premature senescence and interferes with contact inhibition. FOXM1 plays a role for maintenance of stem cell pluripotency and for self-renewal capacity of stem cells. The functions of FOXM1 in prevention of polyploidy and aneuploidy and in homologous recombination repair of DNA-double-strand breaks suggest an importance of FOXM1 for the maintenance of genomic stability and chromosomal integrity.

1. FOXM1 IS AN ACTIVATING TRANSCRIPTION FACTOR

FOXM1 (**Fo**rkhead b**ox M1**) is a typical proliferation–associated transcription factor (Wierstra & Alves, 2007c) that is also intimately involved in oncogenesis (Costa, 2005; Costa, Kalinichenko, Holterman, & Wang, 2003; Costa, Kalinichenko, Major, & Raychaudhuri, 2005; Kalin, Ustiyan, & Kalinichenko, 2011; Koo, Muir, & Lam, 2011; Laoukili, Stahl, & Medema, 2007; Raychaudhuri & Park, 2011; Wierstra & Alves, 2007c). FOXM1 is an activating transcription factor (Fig. 3.1) with a very strong

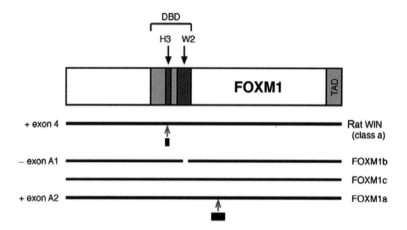

Figure 3.1 FOXM1 splice variants. FOXM1 (Forkhead box M1) is depicted as a white rectangle and its functional domains as light gray boxes. Dark gray boxes mark the recognition helix H3 and the wing W2 of the DBD (DNA-binding domain). The aa (amino acid) sequences of the four different FOXM1 splice variants are shown as thick black lines. Rat WIN (class a transcript) is the only splice variant with exon 4 (12 aa) in helix H3 of the DBD. FOXM1b is the only splice variant without exon A1 (15 aa) in wing W2 of the DBD. FOXM1a is the only splice variant with exon A2 (38 aa). The table indicates the (consensus) DNA-binding site and the transcriptional activity of the FOXM1 splice variants. FOXM1b and FOXM1c transactivate whereas FOXM1a does not. The transcriptional activity of rat WIN (class a transcript) has not been analyzed so far (?). TACTCAATCT was selected as DNA-binding site for rat WIN (class a transcript) in a PCR (polymerase chain reaction)-based binding site selection procedure, that is, SELEX (systematic evolution of ligands by exponential enrichment). A-T/C-AAA-T/C-AA was deduced as consensus sequence for FOXM1-binding sites from DNA-binding studies with FOXM1a, FOXM1b, and FOXM1c. Identical nucleotides in these two DNA-binding sites are marked (bold). DBD, DNA-binding domain; TAD, transactivation domain; H3, recognition helix H3; W2, wing W2.

acidic TAD (transactivation domain) (Wierstra, 2013a; Wierstra & Alves, 2006a) and a forkhead domain as DBD (DNA-binding domain) (Korver, Roose, & Clevers, 1997; Littler et al., 2010; Wierstra, 2011a; Wierstra & Alves, 2006a, 2006d; Yao, Sha, Lu, & Wong, 1997; Ye et al., 1997).

FOXM1 is a member of the FOX (**Fo**rkhead b**ox**) transcription factor family, which is characterized by a conserved DBD, the forkhead domain or forkhead box (Benayoun, Caburet, & Veitia, 2012; Brennan, 1993; Carlsson & Mahlapuu, 2002; Clark, Halay, Lai, & Burley, 1993; Gajiwala & Burley, 2000; Hromas & Costa, 1995; Kaufmann & Knöchel, 1996; Lai, Clark, Burley, & Darnell, 1993; Lalmansingh, Karmakar, Jin, & Nagaich, 2012; Obsil & Obsilova, 2008, 2011; Pohl & Knöchel, 2005; Weigel & Jäckle, 1990; Wijchers, Burbach, & Smidt, 2006). More than 100 known FOX transcription factors or forkhead proteins (http://www.biol ogy.pomona.edu/fox.html) (Hannenhalli & Kaestner, 2009; Jackson, Carpenter, Nebert, & Vasiliou, 2010; Kästner et al., 2000; Katoh & Katoh, 2004; Tuteja & Kaestner, 2007a, 2007b) belong to the larger ensemble of winged-helix proteins, which constitute a subgroup of the HTH (helix–turn–helix) transcription factor class (Gajiwala & Burley, 2000; Teichmann, Dumay-Odelot, & Fribourg, 2012; Wolberger & Campbell, 2000). A winged-helix domain is a DBD that adopts a HTH fold flanked by two loops called wings (Brennan, 1993; Clark et al., 1993; Gajiwala & Burley, 2000).

The canonical forkhead domain adopts a compact α/β-structure consisting of three α-helices (H1, H2, H3), three β-strands (S1, S2, S3), and two loops or wings (W1, W2) arranged in the order H1–S1–H2–turn–H3–S2–W1–S3–W2 (**structures:** Boura, Rezabkova, Brynda, Obsilova, & Obsil, 2010; Brent, Anand, & Marmorstein, 2008; Chu et al., 2011; Clark et al., 1993; Jin, Marsden, Chen, & Liao, 1999; Littler et al., 2010; Liu et al., 2002; Marsden, Chen, Jin, & Liao, 1997; Marsden, Jin, & Liao, 1998; Sheng, Rance, & Liao, 2002; Stroud et al., 2006; Tsai et al., 2006, 2007; van Dongen, Cederberg, Carlsson, Enerbäck, & Wikström, 2000; Wang, Marshall, et al., 2008; Weigelt, Climent, Dahlman-Wright, & Wikström, 2001; **reviewed in:** Benayoun et al., 2012; Brennan, 1993; Carlsson & Mahlapuu, 2002; Clark et al., 1993; Gajiwala & Burley, 2000; Kaufmann & Knöchel, 1996; Lai et al., 1993; Lalmansingh et al., 2012; Obsil & Obsilova, 2008, 2011; Wijchers et al., 2006). The helices H2 and H3 are connected by a turn so that they form a variant of the HTH motif. In the three-dimensional structure, the loops W1 and W2 flank helix H3 like the wings of a butterfly, inspiring the name winged-helix motif (Brennan, 1993; Clark et al., 1993; Gajiwala & Burley, 2000). Strand S1 interacts with strands S2 and S3 to form a three-stranded, twisted, antiparallel β-sheet. Variations on the typical order H1–S1–H2–turn–H3–S2–W1–S3–W2 include additional α-helices and truncated wings.

FOXM1 has been described in human (*Homo sapiens*), house mouse (*Mus musculus*), Norway rat (*Rattus norvegicus*), common chimpanzee (*Pan troglodytes*), bonobo (*Pan paniscus*), gorilla (*Gorilla gorilla*), Sumatran orangutan (*Pongo abelili*), olive baboon (*Papio anubis*), black-capped squirrel monkey (*Saimiri boliviensis*), rhesus monkey (*Macacus mulattus*, *Macacus rhesus*), common marmoset (*Callithrix jacchus*), domestic dog (*Canis lupus familiaris*), domestic cat (*Felis catus*), wild boar (*Sus scrofa*), domestic cattle (*Bos taurus*), sheep (*Ovis aries*), domestic horse (*Equus caballus*), African elephant (*Loxodonta africana*), Northern greater galago (*Otolemur garnettii*), Tasmanian devil (*sarcophilus harrisii*), giant panda (*Ailuropoda melanoleuca*), domestic chicken (*Gallus gallus*), African clawed frog (*Xenopus laevis*), and zebrafish (*Danio rerio*) (http://www.ncbi.nlm.nih.gov/sites/entrez?db=gene&cmd =search&term=foxm1). In contrast, no FOXM1 orthologs were identified in fruit fly (*Drosophila melanogaster*), malaria mosquito (*Anopheles gambiae*), and nematode worm (*Caenorhabditis elegans*) (Mazet, Yu, Liberles, Holland, & Shimeld, 2003).

The transcription factor FOXM1 is also known as MPP2 (MPM2-reactive phosphoprotein 2, M-phase phosphoprotein 2), HFH-11 (HNF-3/fkh (hepatocyte nuclear factor-3/forkhead) homolog-11), FKHL16 (fkh, *Drosophila*, homolog-like 16), WIN (winged helix from INS-1 cells), and Trident due to its independent cloning by several groups (Korver, Roose, & Clevers, 1997; Korver, Roose, Heinen, et al., 1997; Lüscher-Firzlaff et al., 1999; Matsumoto-Taniura, Pirollet, Monroe, Gerace, & Westendorf, 1996; Westendorf, Rao, & Gerace, 1994; Yao et al., 1997; Ye et al., 1997).

2. THE FOUR KNOWN FOXM1 SPLICE VARIANTS

2.1. Four FOXM1 splice variants arise from differential splicing of three facultative exons in the *foxm1* gene

Four different splice variants of FOXM1 have been described in mammals (human, mouse, rat) (Laoukili et al., 2007; Wierstra & Alves, 2007c), namely, FoxM1A, FoxM1B, FOXM1c, and rat WIN (class a transcript) (Fig. 3.1; Korver, Roose, & Clevers, 1997; Korver, Roose, Heinen, et al., 1997; Lüscher-Firzlaff et al., 1999; Yao et al., 1997; Ye et al., 1997).

Human FOXM1 is expressed in three distinct splice variants (Fig. 3.1), which arise from the same gene through differential splicing of the two facultative exons A1 and A2 (Korver, Roose, Heinen, et al., 1997; Lüscher-Firzlaff et al., 1999; Yao et al., 1997; Ye et al., 1997). In addition, rats express a fourth FOXM1 splice variant called rat WIN (class a transcript) (Fig. 3.1), which is characterized by the additional exon A4 (Yao et al., 1997).

FoxM1A, which contains exon A1 and exon A2, is the only splice variant with exon A2 (Fig. 3.1). FoxM1B, which lacks exon A1 and exon A2, is the only splice variant without exon A1 (Fig. 3.1). FOXM1c contains exon A1, but lacks exon A2 (Fig. 3.1). Rat WIN (class a transcript), which contains exon 4 and exon A1, but lacks exon A2, is the only splice variant with exon 4 (Fig. 3.1).

According to the nomenclature for FOX transcription factors, they are divided into alphabetical subclasses and their names contain all uppercase letters for human (FOXM1), only the first letter capitalized for mouse (Foxm1) and the first and subclass letters capitalized for all other chordate proteins (FoxM1; Kästner et al., 2000). Thus, the correct spelling is human FOXM1c, mouse Foxm1c, and rat FoxM1c.

However, the incorrect usage of FoxM1 for the human and mouse proteins is often found in the literature. In order to facilitate the differentiation of the mostly analyzed splice variants FoxM1B and FOXM1c, this review will use the names F**ox**M1**B** and F**OX**M1**c** irrespective of species. Therefore, please note that the names can deviate from the nomenclature (Kästner et al., 2000).

Human FOXM1c has been cloned with either 762 or 763 aa (amino acids), the latter of which contains an additional alanine after cysteine 167.

Because of the four FOXM1 splice variants and the cloning of human FOXM1c with 762 and 763 aa the numbering of aa varies in the FOXM1 literature. Throughout this review, the aa numbering refers to human FOXM1c with 762 aa.

2.2. DNA-binding specificity of the FOXM1 splice variants

2.2.1 The importance of exon A4 and exon A1 in the forkhead domain for the DNA-binding specificity of FOXM1

FOXM1 possesses three alternative exons (Fig. 3.1). Two of them are positioned in the forkhead domain, namely, exon A4 in recognition helix H3 and exon A1 in wing W2 (Fig. 3.1), so that they may affect the DNA-binding specificity of FOXM1. In contrast, an influence of the third exon

A2 on the DNA-binding specificity is unlikely because of its positioning outside the forkhead DBD of FOXM1 (Fig. 3.1). The 12-aa long exon A4 (CW**H**QAY**HK**LGPQ) in recognition helix H3 possesses three basic aa, namely, one lysine and two histidines. The 15-aa long exon A1 (PLDPGSPQLPEHLES) in wing W2 contains no basic aa.

The three human or murine splice variants FoxM1A, FoxM1B, and FOXM1c, which either possess or lack exon A1 in wing W2 of the forkhead domain (Fig. 3.1), display the same DNA-binding specificity (Korver, Roose, & Clevers, 1997; Wierstra & Alves, 2006a; Ye et al., 1997) and bind to DNA-binding sites with the consensus sequence 5′-A-C/T-AAA-C/T-AA-3′ (Hedge, Sanders, Rodriguez, & Balasubramanian, 2011; Korver, Roose, & Clevers, 1997; Littler et al., 2010; Wierstra, 2011a; Wierstra & Alves, 2006a, 2006c, 2007b; Ye et al., 1997). In contrast, rat WIN (class a transcript), which is the sole splice variant with exon A4 in recognition helix H3 of the forkhead domain (Fig. 3.1), exhibits a completely different DNA-binding specificity (Wierstra & Alves, 2006a; Yao et al., 1997). Consequently, the presence of exon A4 in recognition helix H3 changes the DNA-binding specificity of the forkhead domain, whereas the presence or absence of exon A1 in wing W2 does not (Fig. 3.1; Wierstra & Alves, 2006a).

This difference is explained by the distinct roles of the recognition helix H3 and the wing W2 in the binding of the forkhead domain to DNA:

First, according to the general mode of DNA binding by forkhead domains (see below) (Boura et al., 2010; Brent et al., 2008; Clark et al., 1993; Gajiwala et al., 2000; Jin et al., 1999; Sheng et al., 2002; Stroud et al., 2006; Tsai et al., 2006, 2007), the insertion of exon 4 into recognition helix H3 alters the DNA-binding specificity (Fig. 3.1) because the recognition helix H3 makes all base-specific DNA contacts of the forkhead DBD (Gajiwala & Burley, 2000; Wierstra & Alves, 2006a). In contrast, the addition or removal of exon A1 in wing W2 does not alter the DNA-binding specificity (Fig. 3.1) because the wing W2 does not contribute base-specific DNA contacts (Gajiwala & Burley, 2000; Wierstra & Alves, 2006a).

Second, according to the special DNA-binding features of the forkhead DBD of FOXM1c (see below) (Littler et al., 2010), the insertion of exon 4 into recognition helix H3 alters the DNA-binding specificity (Fig. 3.1; Wierstra & Alves, 2006a) because the recognition helix H3 of FOXM1c makes all base contacts of the FOXM1c–DBD (Littler et al., 2010). In contrast, the addition or removal of exon A1 in wing W2 does not alter the

DNA-binding specificity (Fig. 3.1; Wierstra & Alves, 2006a) because the wing W2 of FOXM1c does not contribute any DNA contacts (Littler et al., 2010).

2.2.2 The general mode of DNA binding by forkhead domains

The structures of several forkhead DBDs have been resolved (Boura et al., 2010; Brent et al., 2008; Chu et al., 2011; Clark et al., 1993; Jin et al., 1999; Littler et al., 2010; Liu et al., 2002; Marsden et al., 1997, 1998; Sheng et al., 2002; Stroud et al., 2006; Tsai et al., 2006, 2007; van Dongen et al., 2000; Wang, Marshall, et al., 2008; Weigelt et al., 2001). The forkhead domain binds to DNA in the major groove (Boura et al., 2010; Brent et al., 2008; Clark et al., 1993; Jin et al., 1999; Littler et al., 2010; Stroud et al., 2006; Tsai et al., 2006, 2007). It bends the DNA toward the major groove (Boura et al., 2010, 2007; Brent et al., 2008; Clark et al., 1993; Pierrou, Hellqvist, Samuelsson, Enerbäck, & Carlsson, 1994; Tsai et al., 2006, 2007), and can bind to DNA prebent toward the major groove (Bravieri, Shiyanova, Chen, Overdier, & Liao, 1997). Forkhead proteins generally bind to DNA as monomers (Boura et al., 2010; Brent et al., 2008; Clark et al., 1993; Jin et al., 1999; Littler et al., 2010; Sheng et al., 2002; Tsai et al., 2006), with the exceptions of a FOXO3a homodimer (Tsai et al., 2007) and the homo- or heterodimers formed by the FOXP subclass, that is, by FOXP1, FOXP2, FOXP3, and FOXP4 (Bandukwala et al., 2011; Chu et al., 2011; Koh, Sundrud, & Rao, 2009; Li, Weidenfeld, & Morrisey, 2004; Stroud et al., 2006).

The principal DNA contact surface of the forkhead domain is provided by the highly conserved recognition helix H3, which lies in the major groove of the DNA (Boura et al., 2010; Brent et al., 2008; Clark et al., 1993; Littler et al., 2010; Stroud et al., 2006; Tsai et al., 2006, 2007). In general, forkhead DBDs make the majority of their DNA contacts with recognition helix H3 and wing W2 (Boura et al., 2010; Brent et al., 2008; Clark et al., 1993; Littler et al., 2010; Stroud et al., 2006; Tsai et al., 2006, 2007). The recognition helix H3 makes most base contacts and all base-specific DNA contacts (Gajiwala & Burley, 2000). Also wing W2, and in one case wing W1, can make base contacts, but they do not contribute base-specific DNA contacts (Gajiwala & Burley, 2000). Thus, forkhead DBDs make all their base-specific DNA contacts with the recognition helix H3 (Gajiwala & Burley, 2000), whereas their few base contacts outside recognition helix H3 are not base-specific because they are mediated by a water molecule or target in the minor groove either the O2 atom of a pyrimidine or the N3 atom of a purine. For example, R210 in wing W2 of FOXA3 contacts the O2 atom of a thymine (Clark et al., 1993) and K73 in wing W1 of FOXK1a contacts the N3 atoms of two adenines (Tsai et al., 2006).

2.2.3 The special DNA-binding features of the forkhead DBD of FOXM1c

For human FOXM1c, the X-ray crystal structure of its forkhead DBD has been resolved at 2.2 Å (Littler et al., 2010), but the structures of the three other FOXM1 splice variants are unknown. The forkhead domain of FOXM1c deviates from other forkhead DBDs because, in the case of FOXM1c, the recognition helix H3 makes all base contacts, whereas the wings W2 and W1 do not make any DNA contact at all (Littler et al., 2010). Therefore, the predominant importance of the recognition helix H3 is even more pronounced for the forkhead DBD of FOXM1c (Littler et al., 2010).

Actually, the FOXM1c wing W1 and wing W2 regions adopt unique conformations that are not involved in DNA binding (Littler et al., 2010). Instead, both the wing W1 loop and the wing W2 loop of FOXM1c diverge away from the DNA (Littler et al., 2010). Wing W1 is only short and wing W2 forms an extended structure distal from the DNA unique to FOXM1c (Littler et al., 2010).

2.3. Transcriptional activity of the FOXM1 splice variants

Both FoxM1B (Balli et al., 2012, 2013; Balli, Zhang, Snyder, Kalinichenko, & Kalin, 2011; Bhat, Halasi, & Gartel, 2009a, 2009b; Chen, Chien, et al., 2009; Chen, Müller, et al., 2013; Dai et al., 2007; Fu et al., 2008; Gemenetzidis et al., 2009; Huang et al., 2012; Kalin et al., 2008; Kalinichenko et al., 2004; Kim et al., 2005; Li, Peng, et al., 2013; Li et al., 2009; Ma, Tong, et al., 2005; Major, Lepe, & Costa, 2004; Malin et al., 2007; Park et al., 2012; Park, Kim, et al., 2008; Park, Wang, et al., 2008; Petrovic, Costa, Lau, Raychaudhuri, & Tyner, 2008; Radhakrishnan et al., 2006; Ren et al., 2013, 2010; Sengupta, Kalinichenko, & Yutzey, 2013; Tan et al., 2010; Tan, Raychaudhuri, & Costa, 2007; Tan, Yoshida, Hughes, & Costa, 2006; Ustiyan et al., 2012; Wang, Chen, et al., 2008; Wang et al., 2005; Wang, Kiyokawa, Dennewitz, & Costa, 2002; Wang, Lin, et al., 2011; Wang, Meliton, et al., 2009; Wang, Park, et al., 2011; Wang, Meliton, et al., 2008; Wang, Quail, et al., 2001; Wang, Zhang, Snyder, et al., 2010; Xie et al., 2010; Ye et al., 1997; Zhang et al., 2011, 2008; Zhou, Liu, et al., 2010; Zhou, Wang, et al., 2010) and FOXM1c (Alvarez-Fernandez, Halim,

Aprelia, Mohammed, & Medema, 2011; Alvarez-Fernandez, Halim, et al., 2010; Anders et al., 2011; Karadedou et al., 2012; Korver, Roose, & Clevers, 1997; Laoukili, Alvarez, et al., 2008; Laoukili, Alvarez-Fernandez, Stahl, & Medema, 2008; Laoukili et al., 2005; Leung et al., 2001; Li et al., 2008; Littler et al., 2010; Lüscher-Firzlaff et al., 1999; Ma, Tong, Leung, & Yao, 2010; Ma, Tong, et al., 2005; Madureira et al., 2006; Sullivan et al., 2012; Wierstra, 2011a, 2011b, 2013a; Wierstra & Alves, 2006a, 2006b, 2006c, 2006d, 2007a, 2007b, 2008) are transactivators so that the transcriptional activity of FOXM1 is unaffected by exon A1 (Fig. 3.1). In contrast, FoxM1A is transcriptionally inactive (Fig. 3.1) because the addition of exon A2 abolishes any transcriptional activity of FOXM1 (Ye et al., 1997). Since the transcriptional activity of rat WIN (class a transcript) has never been analyzed the effect of exon 4 on the transcriptional activity of FOXM1 is unknown (Fig. 3.1).

2.4. Comparison of the four FOXM1 splice variants

FoxM1B and FOXM1c are functionally similar because they both are transcriptional activators (Alvarez-Fernandez et al., 2011; Alvarez-Fernandez, Halim, et al., 2010; Anders et al., 2011; Balli et al., 2012, 2013, 2011; Bhat et al., 2009a, 2009b, Chen, Dominguez-Brauer, et al., 2009; Chen, Müller, et al., 2013; Dai et al., 2007; Fu et al., 2008; Gemenetzidis et al., 2009; Grant et al., 2012; Huang et al., 2012; Kalin et al., 2008; Kalinichenko et al., 2004; Karadedou et al., 2012; Kim et al., 2005; Korver, Roose, & Clevers, 1997; Laoukili, Alvarez, et al., 2008; Laoukili, Alvarez-Fernandez, et al., 2008; Laoukili et al., 2005; Leung et al., 2001; Li, Peng, et al., 2013; Li et al., 2008, 2009; Littler et al., 2010; Lüscher-Firzlaff et al., 1999; Ma, Tong, et al., 2005; Ma et al., 2010; Madureira et al., 2006; Major et al., 2004; Malin et al., 2007; Monteiro et al., 2012; Park et al., 2012; Park, Kim, et al., 2008; Park, Wang, et al., 2008; Petrovic et al., 2008; Radhakrishnan et al., 2006; Ren et al., 2013, 2010; Sengupta et al., 2013; Sullivan et al., 2012; Tan et al., 2010, 2007, 2006; Ustiyan et al., 2012; Wan et al., 2012; Wang, Chen, et al., 2008; Wang et al., 2005; Wang, Kiyokawa, et al., 2002; Wang, Meliton, et al., 2009, 2008; Wang, Park, et al., 2011; Wang, Quail, et al., 2001; Wang, Snyder, et al., 2012; Wang, Zhang, Snyder, et al., 2010; Wierstra, 2011a, 2011b, 2013a; Wierstra & Alves, 2006a, 2006b, 2006c, 2006d, 2007a, 2007b, 2008; Xia, Huang, et al., 2012; Xie et al., 2010; Ye et al., 1997; Zhang et al., 2011; Zhang, Wu, et al., 2012; Zhang et al., 2008; Zhou,

Liu, et al., 2010; Zhou, Wang, et al., 2010) with the same DNA-binding specificity (Fig. 3.1; Korver, Roose, & Clevers, 1997; Wierstra & Alves, 2006a; Ye et al., 1997). In contrast, the two other splice variants deviate from FoxM1B and FOXM1c considerably (Fig. 3.1) because FoxM1A is no transactivator (Ye et al., 1997) and rat WIN (class a transcript) exhibits a different DNA-binding specificity (Wierstra & Alves, 2006a; Yao et al., 1997).

Both FoxM1B and FOXM1c are biologically active, and they play similar biological roles (Laoukili et al., 2007; Wierstra & Alves, 2007c). Accordingly, their abundant expression was observed in various tissues and cell lines (Bellelli et al., 2012; Chaudhary, Mosher, Kim, & Skinner, 2000; Dibb et al., 2012; Gemenetzidis et al., 2010; Huynh et al., 2010; Liu, Gampert, Nething, & Steinacker, 2006; Lok et al., 2011; Ma, Tong, et al., 2005; Nakamura, Hirano, et al., 2010; Newick et al., 2012; Park, Kim, Park, Kim, & Lim, 2009; Teh et al., 2002; Yao et al., 1997; Yokomine et al., 2009; Zhang et al., 2011).

Whether FoxM1A and rat WIN (class a transcript) are biologically relevant *in vivo* is unclear. Therefore, in general, only FoxM1B and FOXM1c are considered so that the general term FOXM1 means FoxM1B and FOXM1c.

2.5. Differences between the splice variants FoxM1B and FOXM1c

Several differences between the splice variants FoxM1B and FOXM1c have been described:

First, FoxM1B displays a higher DNA-binding affinity than FOXM1c (Hedge et al., 2011). In EMSAs (electrophoretic mobility shift assays) with purified GST (glutathione-S-transferase)-FOXM1 fusion proteins: FoxM1B ($K_D = 0.2$ µM) is bound to the consensus FOXM1-binding site 5'-AAACAAACAAA-3' with a higher affinity than FOXM1c ($K_D = 0.4$ µM; Hedge et al., 2011). Of course, these different DNA-binding affinities of the splice variants FoxM1B and FOXM1c may be attributed to the absence or presence of exon A1 in wing W2 of their forkhead DBDs, respectively (Fig. 3.1).

Second, ca (constitutively active) MEK1 (MAPK/ERK (mitogen-activated protein kinase/extracellular signal-regulated kinase) kinase 1) increases the transactivation of the *cyclin B1* promoter by FOXM1c but not that by FoxM1B (Ma, Tong, et al., 2005; Ma et al., 2010). This discrepancy may be explained by the finding that the activating effect of ca MEK1 on the FOXM1c-mediated transactivation of the *cyclin B1* promoter depends on two putative ERK1/2 phosphorylation sites (S_{330}, S_{703}) in FOXM1c (Ma, Tong, et al., 2005) and that one site (S_{330}) is missing in FoxM1B because it is located in exon A1.

Third, FoxM1B, which directly binds to β-catenin, strongly activates the β-catenin/TCF-4 (T-cell factor-4) reporter construct TOP-Flash that is driven by TCF-4-binding sites (Zhang et al., 2011). In contrast, TOP-Flash is only weakly activated by FOXM1c, which interacts with β-catenin only very weakly or not at all (Zhang et al., 2011). Since β-catenin binds to the DBD of FoxM1B (Zhang et al., 2011) the clear deviation of the FOXM1c–DBD, which includes exon A1, from the FoxM1B–DBD, which lacks exon A1 (Fig. 3.1), may explain these two differences between FOXM1c and FoxM1B.

Fourth, the interaction of FoxM1B with Cdk2 (cyclin-dependent kinase 2) depends on one single leucine within a putative cyclin/Cdk-binding LXL-motif of FoxM1B (Major et al., 2004), whereas the activation of FOXM1c by both cyclin E/Cdk2 and cyclin A/Cdk2 requires neither this leucine nor this LXL-motif nor the surrounding aa sequence (Wierstra & Alves, 2008). The essential leucine is not located in exon A1.

Fifth, the interaction of FoxM1B with p27 (CDKN1B (Cdk inhibitor 1B), KIP1 (kinase inhibitor protein 1)) depends on the same single leucine within the same LXL-motif of FoxM1B (Kalinichenko et al., 2004), whereas the repression of FOXM1c by p27 requires neither this leucine nor this LXL-motif nor the surrounding aa sequence (Wierstra & Alves, 2008). Again, the essential leucine is not located in exon A1.

Sixth, in A2780cp and OVCA433 ovarian cancer cells, FOXM1c is more effective than FoxM1B in stimulating cell proliferation whereas, conversely, FoxM1B is more effective than FOXM1c in promoting cell migration and invasion (Lok et al., 2011).

Seventh, the expression patterns of FoxM1B and FOXM1c may differ during the cell cycle (Gemenetzidis et al., 2010). When normal telomerase-immortalized N/TERT keratinocytes and metastatic HeLa cervical carcinoma cells were synchronized at G1/S-phase by double-thymidine block and then released from this cell cycle block, a dramatic increase in the *foxm1b* mRNA level but only a modest increase in the *foxm1c* mRNA level were observed, suggesting that only the FoxM1B expression strongly oscillates in a cell cycle-dependent manner whereas the FOXM1c expression may remain rather constant throughout the cell cycle (Gemenetzidis et al., 2010).

3. FOXM1 TARGET GENES

3.1. The target genes of FOXM1

3.1.1 Direct and possibly indirect FOXM1 target genes, which are regulated by FOXM1 at the transcript level

FOXM1 regulates the expression of over 220 genes (Fig. 3.2). It upregulates the mRNA expression of 199 protein-coding genes and downregulates the mRNA expression of 21 protein-coding genes (Fig. 3.2; Table 3.1). Additionally, FOXM1 upregulates the expression of one miRNA and downregulates the expression of five miRNAs (Fig. 3.2; Table 3.1).

Eighty genes are direct FOXM1 target genes, that is, FOXM1 was shown to associate with their promoters in ChIP (chromatin immunoprecipitation) assays, EMSAs, or DNAPs (DNA precipitation assays, DAPAs (DNA affinity pull-down assays)) (Table 3.1). Among them are the two exceptions *KIS* (*kinase interacting stathmin*) and *catalase*, where the FOXM1-binding sites are not located in their promoters but instead in intron 6 of the *KIS* gene (Petrovic et al., 2008) and in intron 1 of the *catalase* gene (Park, Carr, et al., 2009). The true direct binding of purified FOXM1 to the promoter DNA *in vitro* was only demonstrated for three genes (Table 3.1), namely, for *ERα* (*estrogen receptor α*) in fluorescence anisotropy assays (Littler et al., 2010) as well as for *c-myc* (*MYC*) (Wierstra & Alves, 2006d), and *E-cadherin* (*epithelial cadherin*, *CDH1* (*cadherin 1, type 1*)) (Wierstra, 2011a) in EMSAs. Therefore, it remains possible that FOXM1 associates with the promoters of some of the other 77 direct FOXM1 target genes only indirectly through protein–protein interaction with DNA-binding transcription factors.

In fact, recent evidence indicates an indirect association of FOXM1 with WREs/TBEs (Wnt (wingless-type)-responsive elements/TCF-binding elements) via β-catenin/TCF-4 complexes (Bowman & Nusse, 2011; Gong & Huang, 2012; Zhang et al., 2011), with CHR (cell cycle genes homology region) elements via the MuvB (synthetic multivulva class B) complex (Chen, Müller, et al., 2013), and with Myb-binding sites via B-Myb (MYBL2 (MYB-like 2)) (Down et al., 2012).

In accordance with the splice variants FoxM1B (Balli et al., 2012, 2013, 2011; Bhat et al., 2009a, 2009b, Chen, Dominguez-Brauer, et al., 2009; Chen, Müller, et al., 2013; Dai et al., 2007; Fu et al., 2008; Gemenetzidis et al., 2009; Huang et al., 2012; Kalin et al., 2008; Kalinichenko et al., 2004; Kim et al., 2005; Li, Peng, et al., 2013; Li et al., 2009; Ma, Tong,

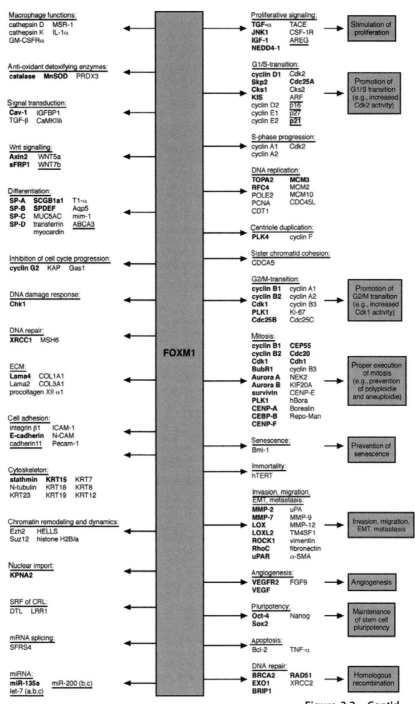

Figure 3.2—Cont'd

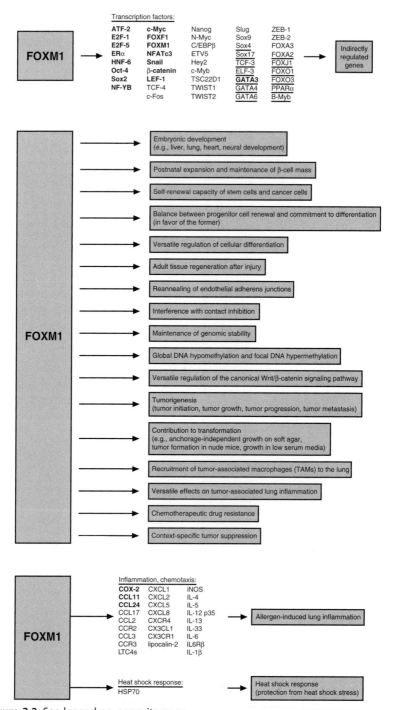

Figure 3.2 See legend on opposite page.

Figure 3.2 FOXM1 target genes and biological functions of FOXM1. Listed are direct FOXM1 target genes (bold) and possibly indirect FOXM1 target genes (normal). It is indicated whether FOXM1 upregulates (not underlined) or downregulates (underlined) the expression of a target gene. Contradictory findings have been reported for the regulation of the *E-cadherin (epithelial cadherin, CDH1 (cadherin 1, type))* expression by FOXM1: on the one hand, FOXM1 transactivated the *E-cadherin* promoter and increased the E-cadherin protein expression (see figure). On the other hand, FOXM1 decreased the *E-cadherin* mRNA and protein expression (not shown). In addition, the E-cadherin protein level was unaffected by FOXM1. Demonstrated biological functions of FOXM1 are shown in gray boxes. The figure does not show the following findings: (1) FOXM1 is important for the pregnancy-induced expansion of maternal β-cell mass. (2) In accordance with the antioxidant detoxifying enzymes *catalase, MnSOD (manganese superoxide dismutase)*, and *PRDX3 (peroxiredoxin 3)* being FOXM1 target genes, FOXM1 downregulates the intracellular ROS (reactive oxygen species) level. (3) FOXM1 prevents premature oxidative-stress-induced senescence by decreasing the ROS level and the level of active p38. (4) FOXM1 supports formation of the premetastatic niche in an *ARF (alternative reading frame, CDKN2A (Cdk (cyclin-dependent kinase) inhibitor 2A), p14, p19)*-null background. (5) FOXM1 plays a role in radiation-induced and bleomycin-induced pulmonary fibrosis. The FOXM1 target genes are grouped according to their functions. Many FOXM1 target genes have additional functions, which are not shown in the figure. **Abbreviations:** ABCA3, ATP-binding cassette, subfamily A, member 3; ARF, alternative reading frame, CDKN2A, Cdk (cyclin-dependent kinase) inhibitor 2A, p14, p19; ATF-2, activating transcription factor-2; Aqp5, aquaporin 5; AREG, amphiregulin; Aurora B, Aurora B kinase; Axin2, axis inhibitor 2, conductin; Bcl-2, B-cell lymphoma-2; Bmi-1, B lymphoma Mo-MLV (Moloney-murine leukemia virus) insertion region-1, PCGF4, Polycomb group RING (really interesting new gene) finger protein 4; hBora, human Bora; Borealin, Dasra B, CDCA8, cell division cycle-associated protein 8; BRCA2, breast cancer-associated gene 2, FANCD1, Fanconi anemia complementation group D1; BRIP1, BRCA1-interacting protein 1, Bach1, BRCA1-associated C-terminal helicase 1; BubR1, Bub1-related kinase; CaMKIIδ, calcium/calmodulin-dependent protein kinase IIδ; Cav-1, caveolin-1; CCL, chemokine, CC motif, ligand; CCR, chemokine, CC motif, receptor; Cdc, cell division cycle; Cdc20, p55CDC, FZY, Fizzy; CDCA5, cell division cycle-associated protein 5, sororin; CDC45L, CDC45-like, CDC45, cell division cycle 45; Cdh1, Cdc20 homolog 1, FZR1, Fizzy-related protein 1; Cdk, cyclin-dependent kinase; CDT1, chromatin licensing and DNA replication factor 1; C/EBPβ, CCAAT/enhancer-binding protein β; CENP, centromer protein; CEP55, centrosomal protein 55-kDa; Chk, checkpoint kinase; Cks, CDC2-associated protein, Cdk subunit; COL1A1, collagen, type I, α-1; COL3A1, collagen, type III, α-1; COX-2, cyclooxygenase-2; CRL, Cullin-RING E3 ubiquitin ligase; CSF-1R, colony-stimulating factor-1 receptor; CXCL, chemokine, CXC motif, ligand; CXCR, chemokine, CXC motif, receptor; CX3CL, chemokine, CX3C motif, ligand; CX3CR, chemokine, CX3C motif, receptor; cyclin F, CCNF, FBXO1, FBX1, F-box only protein 1; DTL, denticleless, CDT2, DCAF2, Ddb1 (damage-specific DNA-binding protein 1)- and Cul4 (Cullin 4)-associated factor 2; E-cadherin, epithelial cadherin, CDH1, cadherin 1, type 1; ECM, extracellular matrix; ELF-3, E74-like factor-3; EMT, epithelial–mesenchymal transition; ERα, estrogen receptor α; ETV5, Ets variant gene 5, Erm, Ets-related molecule; EXO1, exonuclease 1; Ezh2, enhancer of Zeste homolog 2, KMT6, K-methyltransferase 6; FGF9, fibroblast growth factor 9; c-Fos, FOS; FOX, Forkhead box; Gas1, growth arrest-specific 1; GATA, GATA-binding protein; GM-CSFRα, granulocyte-macrophage colony-stimulating factor receptor α; CSF2RA,

(Continued)

Figure 3.2—Cont'd colony-stimulating factor 2 receptor α; HELLS, helicase, lymphoid-specific, LSH, lymphoid-specific helicase, SMARCA6, SWI/SNF-related, matrix-associated, actin-dependent regulator of chromatin, subfamily A, member 6; Hey2, Hairy/enhancer of split-related with YRPW motif 2; HNF-6, hepatocyte nuclear factor 6; HSP70, heat shock protein 70; ICAM-1, intercellular cell adhesion molecule-1; IGF-1, insulin-like growth factor-1; IGFBP1, IGF-binding protein 1; IL, interleukin; IL6Rβ, IL-6 receptor β; IL-12 p35, IL-12, 35-kDa subunit, IL12A, interleukin 12A; iNOS, inducible nitric oxide synthase, NOS2A, nitric oxide synthase 2A, NOS2; JNK1, c-Jun (JUN) N-terminal kinase 1; KAP, Cdk-associated protein phosphatase, CDKN3, Cdk inhibitor 3; Ki-67, Mki67, proliferation-related Ki-67 antigen; KIF20A, kinesin family member 20A; KIS, kinase interacting stathmin; KPNA2, karyopherin α-2; KRT, cytokeratin; Lama, laminin α; LEF-1, lymphoid enhancer factor-1; LOX, lysyl oxidase; LOXL2, LOX oxidase-like 2; LRR1, leucine-rich repeat protein 1, PPIL5, peptidyl-prolyl isomerase-like 5; LTC4s, leukotriene synthase; MCM, minichromosome maintenance; mim-1, myb-induced myeloid protein-1; MMP, matrix metalloproteinase; MnSOD, manganese superoxide dismutase; MSH6, mutS homolog 6; MSR-1, macrophage scavenger receptor-1; MUC5AC, Mucin 5, subtypes A and C, tracheobronchial; B-Myb, MYBL2, MYB-like 2; c-Myb, MYB; c-Myc, MYC; N-Myc, MYCN, Myc oncogene neuroblastoma-derived; N-CAM, neural cell adhesion molecule; NEDD4-1, neural precursor cell-expressed, developmentally downregulated 4-1; NEK2, NIMA (never in mitosis A)-related kinase 2; NFATc3, nuclear factor of activated T-cell c3; NF-YB, nuclear factor-Y B; Oct-4, octamer-binding transcription factor 4, POU5F1, POU domain, class 5, transcription factor 1; p16, CDKN2A, Cdk inhibitor 2A, INK4A, inhibitor of Cdk4 A; p21, CDKN1A, Cdk inhibitor 1A, CIP1, Cdk-interacting protein 1, WAF1, wild-type p53-activated fragment 1; p27, CDKN1B, Cdk inhibitor 1B, KIP1, kinase inhibitor protein 1; PCNA, proliferating cell nuclear antigen; Pecam-1, platelet endothelial cell adhesion molecule 1; PLK, Polo-like kinase; POLE2, DNA polymerase ε 2; PPARα, peroxisome proliferator-activated receptor α; PRDX3, peroxiredoxin 3; Repo-Man, recruits PP1 (protein phosphatase 1) onto mitotic chromatin at anaphase protein, CDCA2, cell division cycle-associated protein 2; RFC4, replication factor C, subunit 4; RhoC, Ras homolog gene family, member C; ROCK1, Rho-associated coiled-coil-containing protein kinase 1; SCGB1a1, secretoglobin, family 1A, member 1, CCSP, Clara cell secretory protein, CC10, Clara cell-specific 10-KD protein; sFRP1, secreted Frizzled-related protein 1; SFRS4, splicing factor, arginine/serine-rich 4, SRp75, splicing factor, arginine/serine-rich 75-KD; Skp2, S-phase kinase-associated protein 2; Slug, SNAI2, Snail2; Snail, SNAI1, Snail1; α-SMA, α-smooth muscle actin; Sox, SRY (sex-determining region Y)-box, SRY-related HMG (high mobility group)-box; SP, surfactant protein; SPDEF, SAM pointed domain-containing ETS transcription factor, PDEF, prostate-derived ETS transcription factor; SRF, substrate recognition factor; survivin, BIRC5, baculoviral IAP repeat-containing 5; Suz12, suppressor of Zeste 12; T1-α; PDPN, podoplanin; TACE, TNF-α converting enzyme, ADAM-17, a disintegrin and metalloprotease domain-17; TCF, T-cell factor; hTERT, human telomerase reverse transcriptase; TGF, transforming growth factor; TM4SF1, transmembrane 4 superfamily member 1, TAL6, TAAL6, tumor-associated antigen L6; TNF-α, tumor necrosis factor-α; TOP2A, TOPO-2α, DNA topoisomerase 2α; TSC22D1, TSC22 (TGF-β-stimulated clone 22) domain family, member 1, TGF-β-stimulated clone 22 domain 1; TWIST2, Dermo1, Dermis-expressed protein 1; uPA, urokinase-type plasminogen activator; uPAR, uPA receptor; VEGF, vascular endothelial growth factor; VEGFR2, VEGF receptor type II; WNT, wingless-type; XRCC, X-ray cross-completing group; ZEB-1, zinc finger E-box-binding homeobox-1, δEF1, δ-crystallin enhancer-binding factor 1; ZEB-2, zinc finger E-box-binding homeobox-2, SIP1, Smad-interacting protein 1.

Table 3.1 FOXM1 target genes

Gene	FOXM1 binding to promoter		Regulation of endogenous gene expression by FOXM1		Regulation of promoter by FOXM1		References
	Method[a]	Comment[b]	Expression[c]	Manipulation of FOXM1[d]	Manipulation of FOXM1-binding site[e]	Manipulation of FOXM1[d]	
Cyclin D1	C		T, P	↑ OE, P, siRNA, tg, KO	wt	↑ OE	Ye, Holterman, Yoo, Franks, and Costa (1999), Wang, Hung, et al. (2001), Wang et al. (2010), Wang, Banerjee, et al. (2007), Wang, Quail, et al. (2001), Liu, Gampert, et al. (2006), Liu, Hock, Van Beneden, and Li (2013), Tan et al. (2006), Petrovic et al. (2008), Chan et al. (2008), Ustiyan et al. (2009), Xia et al. (2009), Dai et al. (2010), Millour et al. (2010), Ren et al. (2010), Wu et al. (2010), Xie et al. (2010), Bao et al. (2011), Raghavan et al. (2012), Zhang et al. (2011), Xue et al. (2012), Qu et al. (2013)

Continued

Table 3.1 FOXM1 target genes—cont'd

Gene	FOXM1 binding to promoter		Regulation of endogenous gene expression by FOXM1		Regulation of promoter by FOXM1		References
	Method	Comment	Expression	Manipulation of FOXM1	Manipulation of FOXM1-binding site	Manipulation of FOXM1	
Cyclin D1			T	↓ siRNA, KO			Balli et al. (2011), Wang, Snyder, et al. (2012)
Cyclin D2			T	↑ siRNA			Zhao et al. (2006), Wang, Teh, Ji, et al. (2010)
Cyclin E			T, P	↑ OE, siRNA, tg			Wang, Hung, et al. (2001), Kalinichenko et al. (2003), Dai et al. (2010), Davis et al. (2010)
Cyclin E1			T	↑ OE			Davis et al. (2010)
Cyclin E2			T	↑ OE, tg, KO	wt	↑ OE	Wang, Meliton, et al. (2008), Davis et al. (2010), Anders et al. (2011)
Cyclin A			P	↑ OE, ca, siRNA, KO			Laoukili, Alvarez, et al. (2008), Davis et al. (2010), Xue, Chiang, He, Zhao, and Winoto (2010), Faust, Al-Butmeh, Linz, and Dietrich (2012)
Cyclin A1			T	↑ OE			Davis et al. (2010)

Cyclin A2	T, P	↑ OE, ca, siRNA, tg, KO	Wang, Hung, et al. (2001), Wang, Teh, Ji, et al. (2010), Wang, Chen, et al. (2008), Wang et al. (2002), Wang, Quail, et al. (2001), Kalinichenko et al. (2003), Krupczak-Hollis et al. (2004), Kim et al. (2006), Kalin et al. (2006), Zhao et al. (2006), Yoshida, Wang, Yoder, Davidson, and Costa (2007), Alvarez-Fernandez, Halim, et al. (2010), Alvarez-Fernandez, Medema, et al. (2010), Davis et al. (2010), Huang and Zhao (2012)
Cyclin B1	C T, P	↑ OE, dn, ca, siRNA, tg, KO wt, del	↑ OE, del, P, siRNA	Ye et al. (1999), Leung et al. (2001), Wang, Hung, et al. (2001), Wang, Banerjee, et al. (2007), Wang et al. (2005), Wang, Krupczak-Hollis, et al. (2002), Wang, Quail, et al. (2001), Kalinichenko et al.

Continued

Table 3.1 FOXM1 target genes—cont'd

Gene	FOXM1 binding to promoter		Regulation of endogenous gene expression by FOXM1		Regulation of promoter by FOXM1		References
	Method	Comment	Expression	Manipulation of FOXM1	Manipulation of FOXM1-binding site	Manipulation of FOXM1	
							(2003), Krupczak-Hollis et al. (2004), Kim et al. (2006), Laoukili et al. (2005), Laoukili, Alvarez, et al. (2008), Ma, Tong, et al. (2005), Ma et al. (2010), Kalin et al. (2006), Zhao et al. (2006), Schüller et al. (2007), Yoshida et al. (2007), Chan et al. (2008), Fu et al. (2008), Li et al. (2008), Park, Wang, et al. (2008), Ueno, Nakajo, Watanabe, Isoda, and Sagata (2008), Ustiyan et al. (2009), Xia et al. (2009), Gemenetzidis et al. (2009), Alvarez-Fernandez, Halim, et al. (2010), Alvarez-Fernandez, Medema, et al. (2010), Davis et al. (2010),

Cyclin B1		P	↓ siRNA	Nakamura, Hirano, et al. (2010), Wu, Liu, et al. (2010), Xue et al. (2010, 2012), Bolte et al. (2011), Hedge et al. (2011), Liu et al. (2011), Chen, Yuan, Tao, and Xiao (2011a, 2012b); Chen, Müller, et al. (2013), Chen et al. (2012), Bellelli et al. (2012), Down, Millour, Lam, and Watson (2012), Huang and Zhao (2012), Mencalha, Binato, Ferreira, Du Rocher, and Abdelhay (2012), Sadavisam, Duan, and DeCarpio (2012), Xu, Zhang, et al. (2012), Qu et al. (2013)
Cyclin B2	C	T, P	↑ OE, tg, KO	Priller et al. (2011) Wang, Quail, et al. (2001), Wang, Krupczak-Hollis,

Continued

Table 3.1 FOXM1 target genes—cont'd

	FOXM1 binding to promoter		Regulation of endogenous gene expression by FOXM1		Regulation of promoter by FOXM1		
Gene	Method	Comment	Expression	Manipulation of FOXM1	Manipulation of FOXM1-binding site	Manipulation of FOXM1	References
							et al. (2002), Laoukili et al. (2005), Davis et al. (2010), Lefebvre et al. (2010), Chen, Müller, et al. (2013)
Cyclin B3			T	↑ OE, siRNA			Ueno et al. (2008)
Cyclin F			T	↑ OE, siRNA, tg			Wang, Hung, et al. (2001), Wang, Quail, et al. (2001), Kalinichenko et al. (2003), Zhao et al. (2006), Chen, Müller, et al. (2013), Huang and Zhao (2012)
Cdk1	C		T, P	↑ OE, siRNA, tg, KO			Ye et al. (1999), Wang, Krupczak-Hollis, et al. (2002), Kalinichenko et al. (2003), Krupczak-Hollis et al. (2004), Zhao et al. (2006), Calvisi et al. (2009), Davis et al. (2010), Xie et al. (2010), Xue et al.

			(2010), Chen, Müller, et al. (2013), Dai et al. (2013)
Cdk2	T, P	↑ OE, siRNA, KO	Wang, Banerjee, et al. (2007), Ahmad et al. (2010), Davis et al. (2010), Xue et al. (2010, 2012)
Cdc25A[f]	C	↑ OE, siRNA, wt, del, P tg, KO	↑ OE Wang, Hung, et al. (2001), Wang, Banerjee, et al. (2007), Wang et al. (2005), Wang, Kiyokawa, et al. (2002), Davis et al. (2010), Sullivan et al. (2012), Liu et al. (2013)
Cdc25B	C	↑ OE, dn, wt siRNA, tg, KO, CI	↑ OE, del Wang, Hung, et al. (2001), Wang et al. (2005), Wang, Kiyokawa, et al. (2002), Wang, Krupczak-Hollis, et al. (2002), Wang, Quail et al. (2001), Kalinichenko et al. (2004), Krupczak-Hollis et al. (2004), Madureira et al. (2006),

Continued

Table 3.1 FOXM1 target genes—cont'd

Gene	FOXM1 binding to promoter		Regulation of endogenous gene expression by FOXM1		Regulation of promoter by FOXM1		References
	Method	Comment	Expression	Manipulation of FOXM1	Manipulation of FOXM1-binding site	Manipulation of FOXM1	
							Radhakrishnan et al. (2006), Tan et al. (2006, 2007, 2010), Park, Wang, et al. (2008), Ramakrishna et al. (2007), Schüller et al. (2007), Chan et al. (2008), Kalin et al. (2008), Li et al. (2008), Ueno et al. (2008), Ustiyan et al. (2009), Chen et al. (2010), Davis et al. (2010), Nakamura, Hirano, et al. (2010), Xie et al. (2010), Zhou, Wang, et al. (2010), Zhou, Liu, et al. (2010), Bergamaschi, Christensen, and Katzenellenbogen (2011), Bolte et al. (2011), Hedge et al. (2011), Balli et al.

			(2012), Dai et al. (2013), Mencalha et al. (2012), Wang and Gartel (2012)
Cdc25C	T, P	↑ OE, siRNA, tg, KO	Madureira et al. (2006), Zhao et al. (2006), Wang, Meliton, et al. (2008), Davis et al. (2010), Liu et al. (2011), Huang and Zhao (2012)
Cdc20	C	↑ OE, siRNA, tg	Wang, Quail, et al. (2001), Wang, Teh, Ji, et al. (2010), Wang, Krupczak-Hollis, et al. (2002), Davis et al. (2010), Chen, Müller, et al. (2013), Dai et al. (2013)
PLK1	C	↑ OE, dn, ca, P, wt siRNA, KO	↑ OE, del, KO Krupczak-Hollis et al. (2004), Kim et al. (2005), Laoukili et al. (2005), Wang et al. (2005), Wang, Snyder, et al. (2012), Madureira et al. (2006), Gusarova et al. (2007), Fu

Continued

Table 3.1 FOXM1 target genes—cont'd

Gene	FOXM1 binding to promoter		Regulation of endogenous gene expression by FOXM1		Regulation of promoter by FOXM1		References
	Method	Comment	Expression	Manipulation of FOXM1	Manipulation of FOXM1-binding site	Manipulation of FOXM1	
							et al. (2008), Chen, Dominguez-Brauer, et al. (2009), Chen, Müller, et al. (2013), Chen, Yang, et al. (2012), Ustiyan et al. (2009), Alvarez-Fernandez, Halim, et al. (2010), Alvarez-Fernandez, Medema, et al. (2010), Davis et al. (2010), Pellegrino et al. (2010), Bolte et al. (2011, 2012), Anders et al. (2011), Bellelli et al. (2012), Dibb et al. (2012), Down et al. (2012), Faust et al. (2012), Monteiro et al. (2012), Sadavisam et al. (2012)
BUBR1	C		T, P	↑ siRNA	del	↑ siRNA	Lefebvre et al. (2010), Wan et al. (2012)

Aurora A	C	T	↑ siRNA	Lefebvre et al. (2010), Raghavan et al. (2012), Down et al. (2012), Mencalha et al. (2012), Sadavisam et al. (2012)
Aurora B	C	T, P	↑ OE, dn, ca, wt del, P, siRNA, KO	↑ OE, P	Krupczak-Hollis et al. (2004), Kim et al. (2005), Wang et al. (2005), Gusarova et al. (2007), Park, Wang, et al. (2008), Fu et al. (2008), Chen, Dominguez-Brauer, et al. (2009), Davis et al. (2010), Nakamura, Hirano, et al. (2010), Zhou, Wang, et al. (2010), Bergamaschi et al. (2011), Bellelli et al. (2012), Bonet et al. (2012), Down et al. (2012), Wang and Gartel (2012), Xu, Zhang, et al. (2012)
Survivin[g]	C	T, P	↑ OE, P, wt, P siRNA, tg, KO, CI	↑ siRNA	Wang et al. (2005), Wang, Banerjee, et al. (2007), Radhakrishnan et al. (2006), Gusarova et al.

Continued

Table 3.1 FOXM1 target genes—cont'd

Gene	FOXM1 binding to promoter		Regulation of endogenous gene expression by FOXM1		Regulation of promoter by FOXM1		References
	Method	Comment	Expression	Manipulation of FOXM1	Manipulation of FOXM1-binding site	Manipulation of FOXM1	
							(2007), Yoshida et al. (2007), Chen, Dominguez-Brauer, et al. (2009), Chen, Müller, et al. (2013), Dai et al. (2010), Davis et al. (2010), Nakamura, Hirano, et al. (2010), Bergamaschi et al. (2011), Down et al. (2012), Xu, Zhang, et al. (2012)
hBora			T	↑ siRNA			Alvarez-Fernandez, Halim, et al. (2010), Alvarez-Fernandez, Medema, et al. (2010)
CENP-A	C		T, P	↑ OE, P, siRNA, KO			Wang et al. (2005), Wonsey and Follettie (2005), Chen, Dominguez-Brauer, et al. (2009), Chen,

			Müller, et al. (2013), Davis et al. (2010), Zhou, Wang, et al. (2010)
CENP-B	C	T	↑ OE, siRNA, CI	Wang et al. (2005), Radhakrishnan et al. (2006), Zhou, Wang, et al. (2010), Zhou, Liu, et al. (2010), Chen, Müller, et al. (2013)
CENP-E		T	↑ OE	Davis et al. (2010)
CENP-F	C	T, P	↑ OE, siRNA, wt KO	↑ OE, KO	Laoukili et al. (2005), Laoukili, Alvarez, et al. (2008), Davis et al. (2010), Anders et al. (2011), Chen, Müller, et al. (2013)
NEK2		T	↑ OE, siRNA, KO	Laoukili et al. (2005), Wonsey and Follettie (2005), Davis et al. (2010)
KIF20A		T	↑ OE, siRNA	Wonsey and Follettie (2005), Davis et al. (2010)
Stathmin	C	T	↑ OE, tg	Carr, Park, Wang, Kiefer, and Raychaudhuri (2010), Park et al. (2011)
KAP		T	↑ siRNA	Wonsey and Follettie (2005)

Continued

Table 3.1 FOXM1 target genes—cont'd

Gene	FOXM1 binding to promoter		Regulation of endogenous gene expression by FOXM1		Regulation of promoter by FOXM1		References
	Method	Comment	Expression	Manipulation of FOXM1	Manipulation of FOXM1-binding site	Manipulation of FOXM1	
ARF			T	↓ OE, siRNA, KO			Wang et al. (2005), Li et al. (2008)
p21			T, P	↓ OE, siRNA, tg, KO, CI			Wang, Hung, et al. (2001), Wang, Banerjee, et al. (2007), Wang et al. (2005), Wang, Kiyokawa, et al. (2002), Kalinichenko et al. (2003), Kalin et al. (2006), Kim et al. (2006), Radhakrishnan et al. (2006), Ramakrishna et al. (2007), Tan et al. (2007), Chan et al. (2008), Li et al. (2008), Ahmad et al. (2010), Xia et al. (2009), Nakamura, Hirano, et al. (2010), Bolte et al. (2011), Sengupta et al. (2013), Xue et al. (2012), Qu et al. (2013)
p21	C		T, P	↑ OE, tg	wt, del	↑ OE	Tan et al. (2010), Wang, Zhang, Snyder, et al. (2010)

p27	T, P	↓	OE, siRNA, wt KO	Wang, Krupczak-Hollis, et al. (2002), Wang, Banerjee, et al. (2007), Wang et al. (2005), Kalinichenko et al. (2004), Kalin et al. (2006), Liu, Gampert, et al. (2006), Zhao et al. (2006), Chan et al. (2008), Petrovic et al. (2008), Ahmad et al. (2010), Zeng et al. (2009), Nakamura, Hirano, et al. (2010), Wu, Liu, et al. (2010), Chen et al. (2011), Wang and Gartel (2012), Sengupta et al. (2013), Xue et al. (2012), Zhang, Zeng, et al. (2012), Qu et al. (2013)
		↓	siRNA	
p53	P	↓	OE, siRNA	Li et al. (2008)
p53	P	↑	siRNA	Chetty, Bhoopathi, Rao, and Lakka (2009)

Continued

Table 3.1 FOXM1 target genes—cont'd

Gene	FOXM1 binding to promoter		Regulation of endogenous gene expression by FOXM1			Regulation of promoter by FOXM1		References
	Method	Comment	Expression	Manipulation of FOXM1		Manipulation of FOXM1-binding site	Manipulation of FOXM1	
Bmi-1[h]			T, P	↑ OE, siRNA, KO		wt, del, P	↑ OE	Li et al. (2008), Wang, Park, et al. (2011)
Skp2	C		T, P	↑ OE, dn, del, siRNA, KO		wt	↑ OE	Wang et al. (2005), Liu, Gampert, et al. (2006), Park, Wang, et al. (2008), Park, Kim, Park, Kim, and Lim (2009), Calvisi et al. (2009), Zeng et al. (2009), Davis et al. (2010), Nakamura, Hirano, et al. (2010), Xie et al. (2010), Zhou, Wang, et al. (2010), Zhou, Liu, et al. (2010), Anders et al. (2011), Bellelli et al. (2012), Chen, Müller, et al. (2013), Mencalha et al. (2012), Xu, Zhang, et al. (2012), Zhang, Zeng, et al. (2012)

Cks1	C	T, P	↑ OE, siRNA, KO	Wang et al. (2005), Calvisi et al. (2009), Davis et al. (2010)
Cks2		T	↑ OE	Davis et al. (2010)
KIS[i,j]	C	T, P	↑ OE, siRNA, P, art KO	Petrovic et al. (2008), Nakamura, Hirano, et al. (2010)
hTERT		T	↑ siRNA	Zeng et al. (2009), Zhang, Zeng, et al. (2012)
MCM3	C	T	↑ siRNA	Lefebvre et al. (2010)
Gas1		T, P	↑ OE, KO	Laoukili et al. (2005)
BRCA2	C	T, P	↑ OE, ca, wt siRNA, KO	Tan et al. (2007, 2010), Kwok et al. (2010), Zhang, Wu, et al. (2012)
XRCC1	C	T, P	↑ OE, ca, wt siRNA, KO	Tan et al. (2007, 2010), Chetty et al. (2009), Kwok et al. (2010)
BTG2		P	↓ OE, siRNA	Park, Kim, Park, Kim, and Lim (2009b)
Chk1	E, C S, C	T, P	↑ siRNA, tg wt, del	Chetty et al. (2009), Tan et al. (2010)
Chk2		P	↑ siRNA	Chetty et al. (2009), Zhang, Wu, et al. (2012)

Continued

Table 3.1 FOXM1 target genes—cont'd

Gene	FOXM1 binding to promoter		Regulation of endogenous gene expression by FOXM1		Regulation of promoter by FOXM1		References
	Method	Comment	Expression	Manipulation of FOXM1	Manipulation of FOXM1-binding site	Manipulation of FOXM1	
Histone H2B/a					wt, H	↑ OE, del	Wierstra and Alves (2006d)
Hsp70					wt, del, H	↑ OE, del	Wierstra and Alves (2006d)
c-Myc[k]	E, C	IV, S, C	T, P	↑ OE, P, siRNA, tg, KO	wt, del, P, H	↑ OE, dn, del	Wierstra and Alves (2006d, 2007a, 2007b, 2008), Li et al. (2008), Ustiyan et al. (2009), Zeng et al. (2009), Ren et al. (2010), Wang, Zhang, Snyder, et al. (2010), Green et al. (2011), Hedge et al. (2011), Zhang et al. (2011), Zhang, Zeng, et al. (2012)
c-Myc			T	↓ siRNA, KO			Balli et al. (2011), Wang, Snyder, et al. (2012)
c-Myc[l]	C	P					Zhang et al. (2011)

N-Myc		T	↑ KO		Bolte et al. (2011)
c-Fos				wt, H	↑ OE, del	Wierstra and Alves (2006d)
ERα	C, D, F IV, C, P	T, P	↑ OE, siRNA	wt, del, P	↑ OE	Madureira et al. (2006), Millour et al. (2010), Littler et al. (2010), Hedge et al. (2011)
ERα		T	↓ OE, KO			Carr et al. (2012)
E2F-1	C	T	↑ siRNA			Wang, Banerjee, et al. (2007), Ahmad et al. (2010), Lefebvre et al. (2010)
E2F-5	C	T	↑ siRNA			Lefebvre et al. (2010)
NF-YB	C	T	↑ siRNA			Lefebvre et al. (2010)
TCF-4		T, P	↑ OE, tg, KO			Yoshida et al. (2007), Zhang et al. (2011)
TCF-4		T	↓ siRNA, KO			Wang, Snyder, et al. (2012)
ERM		T	↑ OE, KO			Laoukili et al. (2005)
NFATc3	C	T	↑ siRNA, KO			Ramakrishna et al. (2007), Bolte et al. (2011)

Continued

Table 3.1 FOXM1 target genes—cont'd

Gene	FOXM1 binding to promoter		Regulation of endogenous gene expression by FOXM1		Regulation of promoter by FOXM1		References
	Method	Comment	Expression	Manipulation of FOXM1	Manipulation of FOXM1-binding site	Manipulation of FOXM1	
C/EBPβ			T	↑ tg			Ye et al. (1999)
HNF-6	C				wt, del	↑ OE	Tan et al. (2006)
FoxF1	C		T	↑ ca, siRNA, KO	wt	↑ OE	Kim et al. (2005), Balli et al. (2011, 2013)
FOXM1	C		T	↑ OE	wt	↑ OE	Halasi and Gartel (2009), Wang, Teh, Ji, et al. (2010), Xie et al. (2010), Down et al. (2012)
PPARα			T	↓ siRNA, KO			Wang, Meliton, et al. (2009)
ATF-2	C		T, P	↑ siRNA			Wang, Chen, et al. (2008)
JNK1	C		T, P	↑ siRNA, KO	wt	↑ OE	Wang, Chen, et al. (2008), Wang, Snyder, et al. (2012), Ustiyan et al. (2009), Balli et al. (2013)
TOP2A	C		T, P	↑ siRNA, KO	wt	↑ OE	Wang, Meliton, et al. (2009)

	E, C	S, C	T, P	↑ OE, siRNA	wt, P	↑ OE, siRNA	
MMP-2							Dai et al. (2007), Wang, Banerjee, et al. (2007), Wang, Chen, et al. (2008), Ahmad et al. (2010), Chen, Chien, et al. (2009), Chen et al. (2011, 2013), Wu, Liu, et al. (2010), Ahmad et al. (2011), Uddin et al. (2011, 2012), He et al. (2012), Lynch et al. (2012), Xue et al. (2012)
MMP-9			T, P	↑ OE, siRNA			Wang, Banerjee, et al. (2007), Wang, Chen, et al. (2008), Ahmad et al. (2010), Ahmad et al. (2011), Lok et al. (2011), Uddin et al. (2011, 2012), He et al. (2012), Xue et al. (2012)
MMP-9			T	↓ KO			Bolte et al. (2012)
MMP-12			T	↑ tg, KO			Wang, Meliton, et al. (2008), Balli et al. (2012)
uPA			T, P	↑ siRNA			Ahmad et al. (2010), Wu, Liu, et al. (2010), Bellelli et al. (2012)

Continued

Table 3.1 FOXM1 target genes—cont'd

Gene	FOXM1 binding to promoter		Regulation of endogenous gene expression by FOXM1		Regulation of promoter by FOXM1		References
	Method	Comment	Expression	Manipulation of FOXM1	Manipulation of FOXM1-binding site	Manipulation of FOXM1	
uPAR	C		↑ T, P	OE, siRNA	wt	↑ OE, siRNA	Wang, Banerjee, et al. (2007), Ahmad et al. (2010), Lok et al. (2011), Li, Peng, et al. (2012)
TGF-α	C				wt	↑ OE	Tan et al. (2006)
DUSP1			↓ P	OE, siRNA			Calvisi et al. (2009)
PTEN			↓ P	OE, siRNA			Dai et al. (2010)
NEDD4-1	C		↑ T, P	OE, siRNA			Dai et al. (2010), Kwak et al. (2012)
VEGFR2	C		↑ T	siRNA, KO	wt	↑ OE	Kim et al. (2005), Balli et al. (2011)
VEGF	E, C, D	S, C, P	↑ T, P	OE, siRNA	wt, del, P	↑ OE, del, P, siRNA	Zhang et al. (2008), Calvisi et al. (2009), Li et al. (2009), Gemenetzidis et al. (2009), Bao et al. (2011), Karadedou et al. (2012), Chen et al. (2011), Lynch et al. (2012), Xue et al. (2012)

EPO		P	↑ OE, siRNA	Calvisi et al. (2009)
TM4SF1		T	↑ OE, KO	Laoukili et al. (2005)
Pecam-1		T	↑ KO	Kim et al. (2005)
Lama2		T	↑ KO	Kim et al. (2005)
Lama4	E S, C	T	↑ KO wt ↑ OE	Kim et al. (2005)
Procollagen type XII α1		T	↑ KO	Kim et al. (2005)
Integrin β1		T	↑ KO	Kim et al. (2005)
E-cadherin[m]	E, C IV, C	P	↑ OE wt ↑ OE, del	Ye et al. (1997), Zhou, Wang, et al. (2010), Zhou, Liu, et al. (2010), Bao et al. (2011), Wierstra (2011a)
E-cadherin		T, P	↓ OE,ca, siRNA, tg	Park et al. (2011), Li, Wang, et al. (2012), Balli et al. (2013)
N-CAM		T	↑ siRNA	Ueno et al. (2008)
N-tubulin		T	↑ siRNA	Ueno et al. (2008)
Tubulin β III		P	↓ siRNA	Wang, Lin, et al. (2011)
Catalase[n]	C	T, P	↑ OE, siRNA	Park et al. (2009)
MnSOD	C	T, P	↑ OE, siRNA	Park, Carr, et al. (2009)
PRDX3		T, P	↑ siRNA	Park, Carr, et al. (2009)

Continued

Table 3.1 FOXM1 target genes—cont'd

Gene	FOXM1 binding to promoter		Regulation of endogenous gene expression by FOXM1		Regulation of promoter by FOXM1		References
	Method	Comment	Expression	Manipulation of FOXM1	Manipulation of FOXM1-binding site	Manipulation of FOXM1	
Cox-2	C		T	↑ siRNA, tg, KO	wt	↑ OE	Wang, Meliton, et al. (2008), Balli et al. (2012), Ren et al. (2013)
Lipocalin-2			T	↑ tg			Wang, Meliton, et al. (2008)
MSR-1			T	↑ tg			Wang, Meliton, et al. (2008)
Cathepsin D			T	↑ tg			Wang, Meliton, et al. (2008)
Cathepsin K			T	↑ tg			Wang, Meliton, et al. (2008)
CCL3			T	↑ tg, KO			Wang, Meliton, et al. (2008), Balli et al. (2012)
CXCL1			T	↑ tg			Wang, Meliton, et al. (2008)
CXCL5°			T	↑ ca, siRNA, tg, KO	wt	↑ OE	Wang, Meliton, et al. (2008), Wang, Teh, Ji, et al. (2010), Balli et al. (2013)

CXCL8		T	↑ siRNA		Wang, Teh, Ji, et al. (2010)
SP-A	C	T	↑ KO	wt ↑ OE	Kalin et al. (2008)
SP-B	C	T	↑ KO	wt ↑ OE	Kalin et al. (2008)
SP-B		T	↓ KO		Ustiyan et al. (2012)
SP-C	C	T	↑ KO		Kalin et al. (2008)
SP-C		T	↓ KO		Ustiyan et al. (2012)
SP-D	C	T	↑ KO		Kalin et al. (2008)
Aqp5		T	↑ KO		Kalin et al. (2008), Liu et al. (2011)
T1-α		T	↑ KO		Kalin et al. (2008)
CEP55	C	T	↑ OE		Gemenetzidis et al. (2009)
HELLS[p]		T	↑ OE		Gemenetzidis et al. (2009)
TACE		T	↑ KO		Kim et al. (2005)
Mim-1				wt ↑ OE, del	Wierstra and Alves (2007b)
Transferrin				wt ↑ dn, as	Chaudhary et al. (2000)

Continued

Table 3.1 FOXM1 target genes—cont'd

Gene	FOXM1 binding to promoter		Regulation of endogenous gene expression by FOXM1		Regulation of promoter by FOXM1		References
	Method	Comment	Expression	Manipulation of FOXM1	Manipulation of FOXM1-binding site	Manipulation of FOXM1	
Oct4	C		T, P	↑ OE, siRNA	wt, del, P	↑ OE, siRNA	Xie et al. (2010), Wang, Park, et al. (2011)
Sox2[q]	C		T, P	↑ OE, P, siRNA, KO	wt, P, H	↑ OE	Xie et al. (2010), Wang, Park, et al. (2011), Zhang et al. (2011), Ustiyan et al. (2012)
Sox2			T	↓ siRNA, tg			Wang, Zhang, Snyder, et al. (2010), Wang, Snyder, et al. (2012)
Nanog			T, P	↑ OE, siRNA			Xie et al. (2010), Wang, Park, et al. (2011)
GATA4			T	↓ siRNA			Xie et al. (2010)
SSEA-1			P	↑ OE, siRNA			Xie et al. (2010), Zhang et al. (2011)
β-Catenin	C		T, P	↑ siRNA, KO			Mirza et al. (2010), Liu et al. (2011)

PCNA	T, P	↑	OE, siRNA		Davis et al. (2010), Raghavan et al. (2012)
Borealin	T	↑	OE, siRNA		Davis et al. (2010), Bergamaschi et al. (2011)
Ki-67	T	↑	OE		Davis et al. (2010)
Repo-Man	T	↑	OE		Davis et al. (2010)
Sox9	T	↑	tg		Wang, Zhang, Snyder, et al. (2010)
Sox4	T	↓	tg, KO		Wang, Zhang, Snyder, et al. (2010), Wang, Snyder, et al. (2012)
Sox17	T	↓	siRNA, tg		Wang, Zhang, Snyder, et al. (2010), Wang, Snyder, et al. (2012)
FoxJ1	T	↓	tg		Wang, Zhang, Snyder, et al. (2010)
CCR2	T	↑	KO	wt, del ↑ OE	Ren et al. (2010, 2013)
Vimentin	T, P	↑	OE, ca, siRNA, tg		Gemenetzidis et al. (2010), Park et al. (2011), Bao et al. (2011), Li, Wang, et al. (2012), Balli et al. (2013)

Continued

Table 3.1 FOXM1 target genes—cont'd

Gene	FOXM1 binding to promoter		Regulation of endogenous gene expression by FOXM1		Regulation of promoter by FOXM1		References
	Method	Comment	Expression	Manipulation of FOXM1	Manipulation of FOXM1-binding site	Manipulation of FOXM1	
Vimentin			T	↓ KO			Bolte et al. (2012)
KRT7			T	↑ OE			Gemenetzidis et al. (2010)
KRT8			T	↑ OE			Gemenetzidis et al. (2010)
KRT12			T	↑ OE			Gemenetzidis et al. (2010)
KRT15		C	T	↑ OE			Gemenetzidis et al. (2010), Bose et al. (2012)
KRT18			T	↑ OE			Gemenetzidis et al. (2010)
KRT18			T	↓ OE, KO			Carr et al. (2012)
KRT19			T	↑ OE			Gemenetzidis et al. (2010)
KRT23			T	↑ OE			Gemenetzidis et al. (2010)
α-SMA			T, P	↑ OE, ca, siRNA, tg, KO			Park et al. (2011), Balli et al. (2013)

α-SMA		T	↓ KO		Bolte et al. (2012)
Snail	C	T, P	↑ OE, ca, siRNA, tg, KO	wt, P	↑ OE	Park et al. (2011), He et al. (2012), Balli et al. (2013)
Slug		T, P	↑ OE, del		Bao et al. (2011), Balli et al. (2013)
LOX	C	T	↑ OE, tg		Park et al. (2011)
LOXL2	C	T	↑ OE, tg		Park et al. (2011)
sFRP1	C	T	↑ KO		Balli et al. (2011)
RelA (p65)		P	↑ OE		Bao et al. (2011)
Hes-1		P	↑ OE		Bao et al. (2011)
ZEB-1		T, P	↑ OE, ca, siRNA, KO		Bao et al. (2011), Balli et al. (2013)
ZEB-2		T, P	↑ OE, ca, siRNA, KO		Bao et al. (2011), Balli et al. (2013)
CD44		P	↑ OE		Bao et al. (2011)

Continued

Table 3.1 FOXM1 target genes—cont'd

Gene	FOXM1 binding to promoter		Regulation of endogenous gene expression by FOXM1		Regulation of promoter by FOXM1		References
	Method	Comment	Expression	Manipulation of FOXM1	Manipulation of FOXM1-binding site	Manipulation of FOXM1	
EpCAM			P ↑	OE			Bao et al. (2011)
let-7a			T ↓	OE			Bao et al. (2011)
let-7b			T ↓	OE			Bao et al. (2011)
let-7c			T ↓	OE			Bao et al. (2011)
miR-200b			T ↓	OE			Bao et al. (2011)
miR-200c			T ↓	OE			Bao et al. (2011)
Ezh2			T ↑	siRNA			Wang, Park, et al. (2011)
Suz12			T ↑	siRNA			Wang, Park, et al. (2011)
miR-135a		C	T ↑	OE, siRNA			Liu, Guo, Shi, et al. (2012)
NF-M			P ↓	siRNA			Wang, Park, et al. (2011)
Nestin			P ↑	OE, P, siRNA, KO			Wang, Park, et al. (2011), Zhang et al. (2011)
CaMKIIδ			T ↑	KO			Bolte et al. (2011)
Hey2			T ↑	KO			Bolte et al. (2011, 2012)

Myocardin		T	↑ KO		Bolte et al. (2011)
IL-1β		T	↓ KO		Bolte et al. (2011)
IL-1β		T	↑ ca, KO		Balli et al. (2013)
Axin2	C	T, P	↑ OE, P, siRNA, KO	wt	Zhang et al. (2011), Wang, Snyder, et al. (2012)
LEF-1ʳ	C	P	↑ OE, P, siRNA, KO	wt, P	Zhang et al. (2011)
CD133		P	↑ OE, P, siRNA		Zhang et al. (2011)
Musashi-1		P	↑ OE, P, siRNA		Zhang et al. (2011)
GFAP		P	↑ OE, P, siRNA		Zhang et al. (2011)
DTL		T	↑ KO	wt	↑ OE	Anders et al. (2011)
MSH6		T	↑ KO	wt	↑ OE	Anders et al. (2011)
LRR1		T	↑ KO	wt	↑ OE	Anders et al. (2011)
CDCA5				wt	↑ OE	Anders et al. (2011)
XRCC2		T	↑ KO	wt	↑ OE	Anders et al. (2011)
IGFBP1				wt	↑ OE	Anders et al. (2011)

Continued

Table 3.1 FOXM1 target genes—cont'd

Gene	FOXM1 binding to promoter		Regulation of endogenous gene expression by FOXM1		Regulation of promoter by FOXM1		References
	Method	Comment	Expression	Manipulation of FOXM1	Manipulation of FOXM1-binding site	Manipulation of FOXM1	
Cyclin G2	C				wt	↑ OE	Anders et al. (2011), Chen, Müller, et al. (2013)
TSC22D1					wt	↑ OE	Anders et al. (2011)
CDT1			T	↑ KO			Anders et al. (2011)
SFRS4			T	↑ KO			Anders et al. (2011)
MCM2			T	↑ KO			Anders et al. (2011)
MCM10			T	↑ KO			Anders et al. (2011)
c-Myb			T	↑ KO			Anders et al. (2011)
CXCL2			T	↑ OE, KO			Balli et al. (2012), Huang and Zhao (2012)
INOS			T	↑ KO			Balli et al. (2012)
IL-6			T	↑ OE, KO			Balli et al. (2012), Huang and Zhao (2012)
IL6Rβ			T	↑ KO			Balli et al. (2012)
CCL11	C		T, P	↑ KO			Balli et al. (2012), Ren et al. (2013)

CX3CL1		T	↑ KO	Balli et al. (2012), Ren et al. (2013)
CX3CR1		T	↑ KO	Balli et al. (2012), Ren et al. (2013)
CSF-1R		T	↑ KO	Balli et al. (2012)
CXCR4		T	↑ KO	Balli et al. (2012)
Cav-1	C	T, P	↑ OE, siRNA wt	Huang et al. (2012)
RACGAP1	C			Sadavisam et al. (2012)
B-Myb		T, P	↓ siRNA	Sadavisam et al. (2012)
BCL-2		T, P	↑ siRNA	Halasi and Gartel (2012)
p16^s		T, P	↓ OE	Teh et al. (2012)
POLE2		T	↑ siRNA	Park et al. (2012)
CDC45L		T	↑ siRNA	Park et al. (2012)
RFC4	C	T	↑ siRNA wt	Park et al. (2012)
PLK4	C	T	↑ siRNA wt	Park et al. (2012)
EXO1	C	T	↑ siRNA wt	Park et al. (2012)

Continued

Table 3.1 FOXM1 target genes—cont'd

Gene	FOXM1 binding to promoter		Regulation of endogenous gene expression by FOXM1		Regulation of promoter by FOXM1		References
	Method	Comment	Expression	Manipulation of FOXM1	Manipulation of FOXM1-binding site	Manipulation of FOXM1	
MMP-7	C		T, P	↑ OE, siRNA	wt, P	↑ OE	Xia, Huang, et al. (2012)
RhoC	C		T, P	↑ OE, siRNA	wt, P	↑ OE	Xia, Huang, et al. (2012)
ROCK1	C		T, P	↑ OE, siRNA	wt, P	↑ OE	Xia, Huang, et al. (2012)
p107			P	↑ KO			Xue et al. (2010)
Menin			P	↓ KO			Zhang et al. (2010)
Wnt5a			T	↑ KO			Wang, Snyder, et al. (2012)
FGF-9			T	↑ KO			Wang, Snyder, et al. (2012)
GATA6			T	↓ KO			Wang, Snyder, et al. (2012)
Wnt7b			T	↓ KO			Wang, Snyder, et al. (2012)
TCF-3			T	↓ KO			Wang, Snyder, et al. (2012)
AREG			T	↓ OE, KO			Carr et al. (2012)

Cadherin11		T	↓ OE, KO		Carr et al. (2012)
GATA3	C	T, P	↓ OE, KO		Carr et al. (2012)
GATA3^t		T	↑ OE		Carr et al. (2012)
α-Casein		P	↑ KO		Carr et al. (2012)
β-Casein		P	↑ KO		Carr et al. (2012)
SCGB1A1	C	T	↑ KO	wt	↑ OE	Ustiyan et al. (2012)
ABCA3		T	↓ KO		Ustiyan et al. (2012)
IL-13		T	↓ KO		Ustiyan et al. (2012)
IL-13		T, P	↑ KO		Ren et al. (2013)
FOXA3		T	↓ KO		Ustiyan et al. (2012)
FOXA3		T	↑ KO		Ren et al. (2013)
FOXA2		T	↓ KO		Ren et al. (2013)
ELF-3		T	↓ KO		Ustiyan et al. (2012)
RAD51	C	T, P	↑ siRNA	wt, del	↑ siRNA	Zhang, Wu, et al. (2012)
Cdh1	C	T	↑ siRNA		Chen, Müller, et al. (2013)
KPNA2	C	T	↑ siRNA		Chen, Müller, et al. (2013)

Continued

Table 3.1 FOXM1 target genes—cont'd

	FOXM1 binding to promoter		Regulation of endogenous gene expression by FOXM1		Regulation of promoter by FOXM1		
Gene	Method	Comment	Expression	Manipulation of FOXM1	Manipulation of FOXM1-binding site	Manipulation of FOXM1	References
BUB3	C						Chen, Müller, et al. (2013)
ETV4	C						Chen, Müller, et al. (2013)
PRC1	C						Chen, Müller, et al. (2013)
PTMS	C						Chen, Müller, et al. (2013)
UBE2C	C						Chen, Müller, et al. (2013)
UBE2S	C						Chen, Müller, et al. (2013)
TACO1	C						Chen, Müller, et al. (2013)
BRIP1	C		T, P	↑ OE, del, siRNA, KO	wt, del, P	↑ del	Monteiro et al. (2012)
IGF-1	C		T	↑ KO	wt, P	↑ OE	Sengupta et al. (2013)
FOXO1			T	↓ KO			Sengupta et al. (2013)

FOXO3	T	↓ KO		Sengupta et al. (2013)
Fibronectin	T	↓ KO		Bolte et al. (2012)
Fibronectin	T, P	↑ ca, siRNA		Balli et al. (2013)
IL-12 p35	T	↑ KO		Ren et al. (2013)
IL-12 p40	P	↑ KO		Ren et al. (2013)
IL-4	T, P	↑ KO		Ren et al. (2013)
IL-5	T, P	↑ KO		Ren et al. (2013)
IL-1α	T	↑ KO		Ren et al. (2013)
IL-33	T	↑ KO		Ren et al. (2013)
LTC4s	T	↑ KO		Ren et al. (2013)
MUC5AC	T	↑ KO		Ren et al. (2013)
CCL2	T, P	↑ ca, KO wt	↑ OE	Ren et al. (2013), Balli et al. (2013)
CCL24 C	T	↑ KO		Ren et al. (2013)
CCR3	T	↑ KO		Ren et al. (2013)
CCL17	T	↑ KO		Ren et al. (2013)

Continued

GM-CSFRα		T	↑ KO		Ren et al. (2013)
SPDEF	C	T	↑ KO	↑ OE	Ren et al. (2013)
ICAM-1		T	↑ OE	wt	Huang and Zhao (2012)
TNF-α		T	↑ OE		Huang and Zhao (2012)
Integrin β3	C				Malin et al. (2007)
CDC6		P	↑ siRNA		Liu et al. (2013)
COL1A1		T	↑ ca, KO		Balli et al. (2013)
COL3A1		T	↑ ca, KO		Balli et al. (2013)
TWIST1		T	↑ KO		Balli et al. (2013)
TWIST2		T	↑ ca, siRNA, KO		Balli et al. (2013)
TGF-β		T	↑ KO		Balli et al. (2013)
PTTG1	C				Lefebvre et al. (2010)
FANCI	C				Lefebvre et al. (2010)
XBP-1	C				Hedge et al. (2011)

GREB1	C		Hedge et al. (2011)
Cdx-2	E	IV, C	Ye et al. (1997)
Fabpi	E	IV, C	Ye et al. (1997)
MHC II	P	↑ KO	Ren et al. (2013)
CD86	P	↑ KO	Ren et al. (2013)

ABCA3, ATP-binding cassette, subfamily A, member 3; ARF, alternative reading frame, CDKN2A, Cdk (cyclin-dependent kinase) inhibitor 2A, p14, p19; ATF-2, activating transcription factor-2; Aqp5, aquaporin 5; AREG, amphiregulin; Aurora B, Aurora B kinase; Axin2, axis inhibitor 2, conductin; Bcl-2, B-cell lymphoma-2; Bmi-1, B lymphoma Mo-MLV (Moloney-murine leukemia virus) insertion region–1, PCGF4, Polycomb group RING (really interesting new gene) finger protein 4; hBora, human Bora; Borealin, Dasra B, CDCA8, cell division cycle-associated protein 8; BRCA2, breast cancer-associated gene 2, FANCD1, Fanconi anemia complementation group D1; BRIP1, BRCA1-interacting protein 1, Bach1, BRCA1-associated C-terminal helicase 1; BTG2, B-cell translocation gene 2, TIS21, TPA (12-O-tetradecanoylphorbol-13-acetate)-inducible sequence 21, APRO1, antiproliferative 1; BUB3, budding uninhibited by benzimidazoles 3; BubR1, Bub1-related kinase; CaMKIIδ, calcium/calmodulin-dependent protein kinase IIδ; Cav-1, caveolin-1; CCL, chemokine, CC motif, ligand; CCR, chemokine, CC motif, receptor; CD133, CD133 antigen, PROM1, prominin 1; Cdc, cell division cycle; Cdc20, p55CDC, FZY, Fizzy; CDCA5, cell division cycle-associated protein 5, sororin; CDC45L, CDC45-like, CDC45, cell division cycle 45; Cdh1, Cdc20 homolog 1, FZR1, Fizzy-related protein 1; Cdk, cyclin-dependent kinase; CDT1, chromatin licensing and DNA replication factor 1; Cdx-2, caudal type homeobox-2; C/EBPβ, CCAAT/enhancer-binding protein β; CENP, centromer protein; CEP55, centrosomal protein 55-kDa; Chk, checkpoint kinase; Cks, CDC2-associated protein, Cdk subunit; COL1A1, collagen, type I, α–1; COL3A1, collagen, type III, α–1; COX-2, cyclooxygenase-2; CSF-1R, colony-stimulating factor-1 receptor; CXCL, chemokine, CXC motif, ligand; CXCR, chemokine, CXC motif, receptor; CX3CL, chemokine, CX3C motif, ligand; CX3CR, chemokine, CX3C motif, receptor; cyclin F, CCNF, FBXO1, FBX1, F-box only protein 1; DTL, denticleless, CDT2, DCAF2, Ddb1 (damage-specific DNA-binding protein 1)– and Cul4 (Cullin 4)-associated factor 2; DUSP1, dual-specificity phosphatase 1, MKP-1, MAPK (mitogen-activated protein kinase) phosphatase–1; E-cadherin, epithelial cadherin, CDH1, cadherin 1, type 1; ELF-3, E74-like factor–3; EpCAM, epithelial cell adhesion molecule; EPO, erythropoietin; ERα, estrogen receptor α; ETV5, Ets variant gene 5, Erm, Ets-related molecule; EXO1, exonuclease 1; Ezh2, enhancer of Zeste homolog 2, KMT6, K-methyltransferase 6; Fabpi, fatty acid-binding protein intestinal; FANCI, Fanconi anemia complementation group I; FGF9, fibroblast growth factor 9; c-Fos, FOS, FOX, Forkhead box; Gas1, growth arrest-specific 1; CSF2RA, colony-stimulating factor 2 receptor α; GFAP, glial fibrillary acidic protein; GM-CSFRα, granulocyte-macrophage colony-stimulating factor receptor α; GREB1, growth regulation by estrogen in breast cancer 1; HELLS, helicase, lymphoid-specific, LSH, lymphoid-specific helicase, SMARCA6, SWI/SNF-related, matrix-associated, actin-dependent regulator of chromatin, subfamily A, member 6; Hes-1, Hairy/enhancer of split-1; Hey2, Hairy/enhancer of split-related with YRPW motif 2; HNF-6, hepatocyte nuclear factor 6; HSP70, heat shock protein 70; ICAM-1, intercellular cell adhesion molecule–1; IGF-1, insulin-like growth factor–1;1GFBP1, IGF-binding protein 1; IL, interleukin; IL6Rβ, IL–6 receptor β; IL-12 p35, IL-12, 35-kDa subunit, IL12A, interleukin 12A; IL-12 p40, IL-12, 40-kDa subunit, IL12B, interleukin 12B; iNOS, inducible nitric oxide synthase, NOS2A, nitric oxide synthase 2A, NOS2; JNK1, c-Jun (JUN) N-terminal kinase 1; KAP, Cdk-associated protein phosphatase, CDKN3, Cdk inhibitor 3; Ki-67, Mki67, proliferation-related

Ki-67 antigen; KIF20A, kinesin family member 20A; KIS, kinase interacting stathmin; KPNA2, karyopherin α-2; KRT, cytokeratin; Lama, laminin α; LEF-1, lymphoid enhancer factor-1; LOX, lysyl oxidase; LOXL2, LOX-like 2; LRR1, leucine-rich repeat protein 1; PPIL5, peptidyl-prolyl isomerase-like 5; LTC4s, leukotriene synthase; MCM, minichromosome maintenance; Menin, MEN1, multiple endocrine neoplasia 1; MHC II, major histocompatibility complex class II; mim−1, myb-induced myeloid protein−1; MMP, matrix metalloproteinase; MnSOD, manganese superoxide dismutase; MSH6, mutS homolog 6; MSR-1, macrophage scavenger receptor−1; MUC5AC, Mucin 5, subtypes A and C, tracheobronchial; B-Myb, MYBL2, MYB-like 2; c-Myc, MYB; c-Myc, MYC; N-Myc, MYCN, Myc oncogene neuroblastoma-derived; N-CAM, neural cell adhesion molecule; NEDD4-1, neural precursor cell-expressed, developmentally downregulated 4-1; NEK2, NIMA (never in mitosis A)-related kinase 2; NFATc3, nuclear factor of activated T-cell c3; NF-M, neurofilament protein, medium polypeptide; NF-YB, nuclear factor-Y B; Oct-4, octamer-binding transcription factor 4; POU5F1, POU domain, class 5, transcription factor 1; p16, CDKN2A, Cdk inhibitor 2A, INK4A, inhibitor of Cdk4 A; p21, CDKN1A, Cdk inhibitor 1A, CIP1, Cdk-interacting protein 1, WAF1, wild-type p53-activated fragment 1; p27, CDKN1B, Cdk inhibitor 1B, KIP1, kinase inhibitor protein 1; p53, TP53, tumor protein p53; p107, RBL1, retinoblastoma-like 1; PCNA, proliferating cell nuclear antigen; Pecam−1, platelet endothelial cell adhesion molecule 1; PLK, Polo-like kinase; POLE2, DNA polymerase ε 2; PPARα, peroxisome proliferator-activated receptor α; PRC1, protein regulating cytokinesis 1; PRDX3, peroxiredoxin 3; PTEN, phosphatase and tensin homolog deleted on chromosome ten; PTMS, parathymosin; PTTG1, pituitary tumor-transforming gene 1, securin; RACGAP1, Rac GTPase-activating protein 1; RelA, RELA, p65; Repo-Man, recruits PP1 (protein phosphatase 1) onto mitotic chromatin at anaphase protein, CDCA2, cell division cycle-associated protein 2; RFC4, replication factor C, subunit 4; RhoC, Ras homolog gene family, member C; ROCK1, Rho-associated coiled-coil-containing protein kinase 1; SCGB1a1, secretoglobin, family 1A, member 1, CCSP, Clara cell secretory protein, CC10, Clara cell-specific 10-KD protein; sFRP1, secreted Frizzled-related protein 1; SFRS4, splicing factor, arginine/serine-rich 4, SRp75, splicing factor, arginine/serine-rich 75-KD; Skp2, S-phase kinase-associated protein 2; Slug, SNAI2, Snail2; Snail, SNAI1, Snail1; α-SMA, α-smooth muscle actin; Sox, SRY (sex-determining region Y)-box, SRY-related HMG (high mobility group)-box; SP, surfactant protein; SPDEF, SAM pointed domain-containing ETS transcription factor, PDEF, prostate-derived ETS transcription factor; SSEA-1, stage-specific embryonic antigen-1; survivin, BIRC5, baculoviral IAP repeat-containing 5; Suz12, suppressor of Zeste 12; T1-α, PDPN, podoplanin; TACE, TNF-α converting enzyme, ADAM-17, a disintegrin and metalloprotease domain-17; TACO1, translational activator of mitochondrially encoded cytochrome c oxidase subunit 1; TCF, T-cell factor; hTERT, human telomerase reverse transcriptase; TGF, transforming growth factor; TM4SF1, transmembrane 4 superfamily member 1, TAL6, TAAL6, tumor-associated antigen L6; TNF-α, tumor necrosis factor-α; TOP2A, TOPO-2α, DNA topoisomerase 2α; TSC22D1, TSC22 (TGF-β-stimulated clone 22) domain family, member 1, TGF-β-stimulated clone 22 domain 1; TWIST2, Dermo1, Dermis-expressed protein 1; UBE2C, ubiquitin-conjugating enzyme E2C; uPA, urokinase-type plasminogen activator; uPAR, uPA receptor; VEGF, vascular endothelial growth factor; VEGFR2, VEGF receptor type II; WNT, wingless-type; XBP-1, X box-binding protein-1; XRCC, X-ray cross-completing group; ZEB-1, zinc finger E-box-binding homeobox-1, δEF1, δ-crystallin enhancer-binding factor 1; ZEB-2, zinc finger E-box-binding homeobox-2, SIP1, Smad-interacting protein 1.

Method: E = EMSA = electrophoretic mobility shift assay; C = ChIP = chromatin immunoprecipitation assay; P = FOXM1-binding site was point-mutated. affinity pull-down assay; F = fluorescence anisotropy assay.

[a]**Comment:** IV = *in vitro* = purified FOXM1; S = supershift experiments; C = competition experiments; P = FOXM1-binding site was point-mutated.

[b]**Expression:** T = transcript = mRNA = endogenous mRNA level affected; P = protein = endogenous protein level affected.

[c]**Manipulation of FOXM1:** OE = overexpression of wild-type FOXM1; dn = dominant-negative form of FOXM1; ca = constitutively active form of FOXM1; del = analyzed with deletion mutant of FOXM1; P = analyzed with point mutant of FOXM1; siRNA = siRNA-mediated or shRNA-mediated knockdown of FOXM1; as = addition of antisense oligonucleotide to FOXM1; tg = *foxm1* transgenic cells/mice; KO = *foxm1* knockout cells/mice; CI = chemical inhibitor of FOXM1 (siomycin A).

a Manipulation of FOXM1-binding site: wt = "wild-type" promoter; del = analyzed with deletion mutant of promoter; P = FOXM1-binding site was point-mutated; H = FOXM1-binding site in context of heterologous core promoter; art = analyzed with an artificial construct.

b The point-mutated versions of the *Cdc25A* promoter had point mutations either in three putative FOXM1-binding sites or in two known E2F-binding sites.

g The point-mutated version of the *survivin* promoter had point mutations in the four B-Myb (MYBL2 (MYB-like 2))-binding sites (but not in the two FOXM1-binding sites).

h *Bmi-1* is probably an indirect FOXM1 target gene because the transactivation of the *Bmi-1* promoter by FOXM1c depends on the presence of an E-box in the *Bmi-1* promoter (Li et al., 2008), which is bound by c-Myc (Guney, Wu, & Sedivy, 2006; Guo, Datta, Band, & Dimri, 2007; Hydbring et al., 2009), and because shRNA-mediated knockdown of c-Myc attenuated the stimulation of Bmi-1 protein expression by FOXM1c (Li et al., 2008). Since *Bmi-1* is a direct c-Myc target gene (Guney et al., 2006; Guo et al., 2007; Hydbring et al., 2009) and since *c-myc* is a direct FOXM1c target gene (Hedge et al., 2011; Wierstra & Alves, 2006d; Zhang et al., 2011), FOXM1 may indirectly activate the *Bmi-1* transcription through c-Myc (Li et al., 2008).

i The 1.1 kb (kilobases) proximal human *KIS* promoter is not FOXM1-responsive, but FOXM1 activated an artificial construct, in which the 23 kb putative FOXM1-binding site from intron 6 of the *KIS* genes was inserted upstream of the 1.1 kb proximal *KIS* promoter (Petrovic et al., 2008).

j Putative FOXM1-binding site is positioned in intron 6.

k In EMSAs (electrophoretic mobility shift assays), the purified DBD (DNA-binding domain) of FOXM1c bound directly to the TATA-box of the *c-myc* P1 promoter and to the TATA-box of the *c-myc* P2 promoter (Wierstra & Alves, 2006d).

l The point-mutated transfected *c-myc* promoter had point mutations in TBE1 and TBE2, two known TCF-4/LEF-1 (T-cell factor-4/lymphoid enhancer factor-1)-binding sites (but not in the *c-myc* P1 or P2 TATA-boxes).

m In the presence of genistein, FOXM1 had a positive effect on the E-cadherin protein expression because the E-cadherin protein level was higher in cells stably overexpressing FOXM1 than in parental cells (Bao et al., 2011). However, in the absence of genistein, the E-cadherin protein expression was unaffected by FOXM1 because the E-cadherin protein level was similar in FOXM1-overexpressing and parental cells (Bao et al., 2011).

n Putative FOXM1-binding site is positioned in intron 1.

o *Cxcl5* seems to be an indirect FOXM1 target gene because FOXM1 transactivates the *cxcl5* promoter (Balli et al., 2013; Wang, Teh, Ji, et al., 2010) and increases the *cxcl5* mRNA expression (Balli et al., 2013; Wang, Meliton, et al., 2008; Wang, Teh, Ji, et al., 2010), but an *in vivo* occupancy of the *cxcl5* promoter by FOXM1 was not detected (Wang, Teh, Ji, et al., 2010).

p *HELLS* is probably an indirect FOXM1 target gene because FOXM1 increased the *HELLS* mRNA level but did not occupy the *HELLS* promoter *in vivo* (Gemenetzidis et al., 2009).

q FOXM1 activated the 5.7 kb mouse *sox2* promoter (Ustiyan et al., 2012). Additionally, FOXM1 activated an artificial reporter construct, in which a far upstream region (−15178/−14836) of the *sox2* promoter containing the putative FOXM1-binding site (−15008/−14991) was inserted into the vector pGL3 (Wang, Lin, et al., 2011).

r The point-mutated version of the *LEF-1* promoter had point mutations in the three TBEs/WREs (TCF (T-cell factor)-binding elements/Wnt (wingless-type)-responsive elements) (but not in any putative FOXM1-binding site).

s *p16* is probably an indirect FOXM1 target gene because the *p16* promoter was not *in vivo* occupied by FOXM1 (Gemenetzidis et al., 2009), which decreased the *p16* mRNA and protein expression (Teh et al., 2012).

t FOXM1 overexpression increased the *GATA3* mRNA level only in RB-deficient cells, namely, in the RB (retinoblastoma protein, RB1 (retinoblastoma 1), p105) mutant breast cancer cell line MDA-MB-468 and in MCF-7 breast cancer cells depleted of RB by shRNA (Carr et al., 2012).

et al., 2005; Major et al., 2004; Malin et al., 2007; Park et al., 2012; Park, Kim, et al., 2008; Park, Wang, et al., 2008; Petrovic et al., 2008; Radhakrishnan et al., 2006; Ren et al., 2013, 2010; Sengupta et al., 2013; Tan et al., 2010, 2007, 2006; Ustiyan et al., 2012; Wang, Chen, et al., 2008; Wang et al., 2005; 2009; Wang, Kiyokawa, et al., 2002; Wang, Meliton, et al., 2009, 2008; Wang, Park, et al., 2011; Wang, Quail, et al., 2001; Wang, Snyder, et al., 2012; Wang, Zhang, Snyder, et al., 2010; Xie et al., 2010; Ye et al., 1997; Zhang et al., 2011, 2008; Zhou, Liu, et al., 2010; Zhou, Wang, et al., 2010) and FOXM1c (Alvarez-Fernandez et al., 2011; Alvarez-Fernandez, Halim, et al., 2010; Anders et al., 2011; Karadedou et al., 2012; Korver, Roose, & Clevers, 1997; Laoukili, Alvarez, et al., 2008; Laoukili, Alvarez-Fernandez, et al., 2008; Laoukili et al., 2005; Leung et al., 2001; Li et al., 2008; Littler et al., 2010; Lüscher-Firzlaff et al., 1999; Ma, Tong, et al., 2005; Ma et al., 2010; Madureira et al., 2006; Sullivan et al., 2012; Wierstra, 2011a, 2011b, 2013a; Wierstra & Alves, 2006a, 2006b, 2006c, 2006d, 2007a, 2007b, 2008) being transcriptional activators, FOXM1 transactivates the promoters of 66 genes (Table 3.1). Additionally, FOXM1 activated an artificial *KIS* promoter construct, in which the FOXM1-binding site from intron 6 of the *KIS* gene was inserted upstream of the non-FOXM1 responsive 1.1-kb proximal *KIS* promoter (Table 3.1; Petrovic et al., 2008).

Unexpectedly, the activity of the *p27* promoter was 1.4- to 1.6-fold increased by siRNA-mediated depletion of FOXM1 suggesting that FOXM1 slightly represses the *p27* promoter (Table 3.1; Zeng et al., 2009). Yet, *p27* is no direct FOXM1 target gene (Table 3.1) so that FOXM1 may indirectly repress the transcription of *p27* (Zeng et al., 2009) by activating the expression of a known transcriptional repressor of the *p27* promoter (e.g., c-Myc).

Similarly, FOXM1 may indirectly downregulate the transcript level of 18 other protein-coding genes and 5 miRNAs (Fig. 3.2) by activating the expression of transcription factors, which repress their transcription, because binding of FOXM1 to the promoters of these 18 protein-coding genes and 5 miRNAs has not been shown so far (Table 3.1). Alternatively, FOXM1 may downregulate the mRNA levels of these 18 protein-coding genes without an (indirect) effect on their promoters but instead by decreasing the stability of their mRNAs, for example, by activating the expression of an appropriate miRNA.

In contrast, FOXM1 *in vivo* occupied the *GATA3* (*GATA-binding protein 3*) (Carr et al., 2012) and *p21* (*CDKN1A* (*Cdk inhibitor 1A*), *CIP1* (*Cdk-interacting protein 1*), *WAF1* (*wild-type p53* (*TP53, tumor protein p53*)-*activated fragment*)) (Tan et al., 2010) promoters so that the FOXM1-mediated downregulation of the *GATA3* and *p21* mRNA levels (Fig. 3.2; Table 3.1) has other explanations:

FOXM1 represses the *GATA3* transcription (Table 3.1) through recruitment of the corepressors RB (retinoblastoma protein, RB1 (retinoblastoma 1), p105) and DNMT3b (DNA methyltransferase 3b) to the *GATA3* promoter and through DNMT3b-mediated, RB-dependent DNA methylation of the *GATA3* promoter (Carr et al., 2012).

Surprisingly, the murine *p21* promoter was activated by FoxM1B in Hepa 1–6 mouse hepatoma cells (Tan et al., 2010). Moreover, FoxM1B was surprisingly reported to upregulate the *p21* mRNA level in MEFs (mouse embryonic fibroblasts) (Tan et al., 2010) and mouse lungs (Wang, Zhang, Snyder, et al., 2010) as well as to upregulate the p21 protein level in MEFs and U2OS cells (Table 3.1; Tan et al., 2010). In clear contrast, sixteen other studies reported a downregulation of the *p21* mRNA and protein expression by FOXM1 in diverse cell types and tissues (Table 3.1; Ahmad et al., 2010; Bolte et al., 2011; Chan et al., 2008; Kalin et al., 2006; Kalinichenko et al., 2003; Li et al., 2008; Nakamura, Hirano, et al., 2010; Qu et al., 2013; Ramakrishna et al., 2007; Sengupta et al., 2013; Tan et al., 2007; Wang, Banerjee, Kong, Li and Sarkar, 2007; Wang, Hung, & Costa, 2001; Wang, Kiyokawa, et al., 2002; Wang et al., 2005; Xia et al., 2009; Xue et al., 2009). These contradictory findings suggest that FOXM1 generally represses *p21* transcription indirectly via other transcription factors, whereas it rarely activates *p21* transcription through direct binding to the *p21* promoter in a few exceptional biological settings.

FOXM1 was found to control the expression of additional genes in microarrays (Guerra et al., 2013; Kim et al., 2005; Laoukili et al., 2005; Lefebvre et al., 2010; Mencalha et al., 2012; Park et al., 2012; Ustiyan et al., 2009; Wang et al., 2003; Wang, Hung, et al., 2001; Wang, Meliton, et al., 2008; Wonsey & Follettie, 2005; Ye et al., 1999; Zhang, Wu, et al., 2012), which are not included in Fig. 3.2 or Table 3.1.

3.1.2 Probably indirect FOXM1 target genes

Four target genes of FOXM1 are probably only indirect FOXM1 target genes (Fig. 3.2; Table 3.1):

First, *HELLS* (*helicase, lymphoid-specific*); *LSH* (*lymphoid-specific helicase*); *SMARCA6* (*SWI/SNF* (*switching defective/sucrose non fermenting*)-*related, matrix-associated, actin-dependent regulator of chromatin, subfamily A, member 6*)) is probably an indirect FOXM1 target gene because FOXM1 increased the *HELLS* mRNA level but did not occupy the *HELLS* promoter *in vivo* (Gemenetzidis et al., 2009).

Second, *Cxcl5* (*chemokine, CXC motif, ligand 5*) seems to be an indirect FOXM1 target gene because FOXM1 transactivated the *cxcl5* promoter (Balli et al., 2013; Wang, Teh, Ji, et al., 2010) and increased the *cxcl5* mRNA expression (Balli et al., 2013; Wang, Meliton, et al., 2008; Wang, Teh, Ji, et al., 2010), but an *in vivo* occupancy of the *cxcl5* promoter by FOXM1 was not detected (Wang, Teh, Ji, et al., 2010).

Third, *p16* (*CDKN2A* (*Cdk inhibitor 2A*), *INK4A* (*inhibitor of Cdk4*)) is probably an indirect FOXM1 target gene because the *p16* promoter was not *in vivo* occupied by FOXM1 (Gemenetzidis et al., 2009), which decreased the *p16* mRNA and protein expression (Teh et al., 2012).

Fourth, *Bmi-1* (*B lymphoma Mo-MLV* (*Moloney-murine leukemia virus*) *insertion region-1*, *PCGF4* (*Polycomb group RING* (*really interesting new gene*) *finger protein 4*)) is probably an indirect FOXM1 target gene because the transactivation of the *Bmi-1* promoter by FOXM1c depended on the presence of an E-box (i.e., a c-Myc/Max (Myc-associated factor X)-binding site) in the *Bmi-1* promoter (Li et al., 2008), which was bound by c-Myc (Guney et al., 2006; Guo et al., 2007; Hydbring et al., 2009), and because shRNA-mediated depletion of c-Myc attenuated the stimulation of Bmi-1 protein expression by FOXM1c (Li et al., 2008). Since *c-myc* is a direct FOXM1c target gene (Hedge et al., 2011; Wierstra & Alves, 2006d; Zhang et al., 2011) and since *Bmi-1* is a direct c-Myc target gene (Guney et al., 2006; Guo et al., 2007; Hydbring et al., 2009) FOXM1 may indirectly activate the *Bmi-1* transcription through c-Myc (Li et al., 2008).

3.1.3 Proteins, which are regulated by FOXM1 at the protein level

In addition to regulating the transcript levels of more than 220 FOXM1 target genes (Fig. 3.2), FOXM1 affects the protein expression of several genes, for which an effect of FOXM1 on their mRNA expression has not been reported so far (Table 3.1). On the one hand, FOXM1 increased the protein levels of CDC6 (cell division cycle 6) (Liu et al., 2013), SSEA-1 (stage-specific embryonic antigen-1) (Xie et al., 2010; Zhang et al., 2011), Nestin (Wang, Park, et al., 2011; Zhang et al., 2011), EPO (erythropoietin) (Calvisi et al., 2009), IL-12 p40 (interleukin-12 40-kDa subunit, IL12B (interleukin 12B)), CD86 (CD86 antigen), MHC II (major histocompatibility complex class II) (Ren et al., 2013), p107 (RBL1 (retinoblastoma-like 1)) (Xue et al., 2010), Chk2 (checkpoint kinase 2) (Chetty et al., 2009; Zhang, Wu, et al., 2012), CD133, Musashi-1 (Zhang et al., 2011), RelA (p65), Hes-1 (Hairy/enhancer of split-1), EpCAM (epithelial cell adhesion molecule), and CD44 (Table 3.1) (Bao et al., 2011). On the other hand, FOXM1 decreased the protein levels of BTG2 (B-cell translocation gene 2, TIS21 (TPA (12-O-tetradecanoylphorbol-13-acetate)-inducible sequence 21), APRO1 (antiproliferative 1)) (Park, Kim, Park, Kim, & Lim, 2009), Menin (MEN1 (multiple endocrine neoplasia 1)) (Zhang et al., 2010), DUSP1 (dual-specificity phosphatase 1, MKP-1 (MAPK phosphatase-1)) (Calvisi et al., 2009), PTEN (phosphatase and tensin homolog deleted on chromosome ten) (Dai et al., 2010), GFAP (glial fibrillary acidic protein) (Zhang et al., 2011), α-casein, β-casein (Carr et al., 2012), NF-M (neurofilament protein, medium polypeptide), and tubulin β III (Table 3.1; Wang, Park, et al., 2011). Surprisingly, the p53 protein level was either upregulated (Chetty et al., 2009) or downregulated (Li et al., 2008) by FOXM1 (Table 3.1).

3.1.4 Genes, which are bound by FOXM1

ChIP assays revealed an *in vivo* occupancy of 13 genes by FOXM1 (Table 3.1), namely, of *pttg1* (*pituitary tumor-transforming gene, securin*), *racgap1* (*RAC GTPase-activating protein 1*), *fanci* (*Fanconi anemia complementation group I*), *xbp-1* (*X box-binding protein-1*), *greb1* (*growth regulation by estrogen in breast cancer 1*), *integrin β3*, *bub3* (*budding uninhibited by benzimidazoles 3*), *etv4* (*ETS variant gene 4*), *prc1* (*protein regulating cytokinesis 1*), *ptms* (*parathymosin*), *taco1* (*translational activator of mitochondrially encoded cytochrome c oxidase subunit 1*), *ube2c* (*ubiquitin-conjugating enzyme E2C*), *and ube2s* (*ubiquitin-conjugating enzyme E2S*) (Chen, Müller, et al., 2013; Hedge et al., 2011; Lefebvre et al., 2010; Malin et al., 2007; Sadavisam et al., 2012).

Additionally, oligonucleotides derived from the murine *cdx-2* (*caudal type homeobox-2*) and rat *fabpi* (*fatty acid-binding protein intestinal*) genes were bound by the purified FOXM1–DBD in EMSAs *in vitro* (Table 3.1; Ye et al., 1997).

Furthermore, genome-wide ChIP-Seq (ChIP followed by sequencing) assays with an α-FOXM1 antibody showed that FOXM1 *in vivo* occupies

270 binding regions in human U2OS osteosarcoma cells accumulated in late G2- and M-phase (Chen, Müller, et al., 2013), which are not included in Table 3.1.

3.2. Unusual properties of the transcription factor FOXM1

Two findings for FOXM1 are rather surprising for a transcription factor, namely, first, its low DNA-binding affinity and its low DNA-binding selectivity (Littler et al., 2010) as well as, second, the low number of FOXM1-binding regions in the human genome in late G2- or M-phase and the scarcity of the consensus sequence for FOXM1-binding sites in these regions (Chen, Müller, et al., 2013).

First, the FOXM1c–DBD exhibits an anomalous low DNA-binding affinity because it bound to a FOXM1 consensus binding site with an apparent dissociation constant (K_D) of 7000 nM (Littler et al., 2010). Thus, the FOXM1c–DBD recognizes its consensus site only with micromolar affinity (Littler et al., 2010). This is at least an order of magnitude lower than the affinity reported for the FOXO3a–DBD ($K_D = 300$ nM) and four orders of magnitude lower than that reported for the FOXD3–DBD ($K_D = 0.3$ nM; Littler et al., 2010) so that the FOXM1c–DBD displays a lower DNA-binding affinity than other forkhead domains (Littler et al., 2010). In addition, the FOXM1c–DBD exhibits a low DNA-binding selectivity because it showed only a three- to fourfold higher affinity for its consensus sequence over random DNA (Littler et al., 2010). Hence, the FOXM1c–DBD binds to DNA not only weakly, but also with a relatively low selectivity (Littler et al., 2010).

A comparison of the splice variants FoxM1B and FOXM1c in EMSAs with purified full-length GST-FOXM1 fusion proteins revealed that FoxM1B displays a higher DNA-binding affinity than FOXM1c because FoxM1B ($K_D = 0.2$ μM) bound to a FOXM1 consensus binding site with a higher affinity than FOXM1c ($K_D = 0.4$ μM; Hedge et al., 2011).

Second, genome-wide ChIP-seq assays with an α-FOXM1 antibody in human U2OS osteosarcoma cells accumulated in late G2- and M-phase showed that FOXM1 *in vivo* occupies only 270 binding regions (Chen, Müller, et al., 2013). The majority of these FOXM1-binding regions is located in close proximity to promoters, with 74% either in the 5′-UTR (untranslated region) or within 1 kb (kilobase) upstream of the transcription start site (Chen, Müller, et al., 2013). Thus, in comparison to other transcription factors, the number of FOXM1-binding sites in the human genome is extremely low, at least in late G2- and M-phase (Chen, Müller, et al., 2013). Analysis of these 270 FOXM1-binding regions revealed that 58% are located within 200 bp of the transcription start site, that the consensus sequence for FOXM1-binding sites is not overrepresented in the FOXM1-binding regions, and that those motifs present are not localized at the summit of the FOXM1-binding peak (Chen, Müller, et al., 2013). Hence, the number of FOXM1-binding regions in the human genome is low in late G2- and M-phase, and these regions are surprisingly not enriched with the consensus sequence for FOXM1-binding sites (Chen, Müller, et al., 2013).

Together, its low DNA-binding affinity and selectivity (Littler et al., 2010), its relatively few DNA-binding sites in late G2- or M-phase, and their scarcity of FOXM1 consensus motifs (Chen, Müller, et al., 2013) suggest that FOXM1 might not only function as a classical transcription factor, which binds directly to the DNA in order to transactivate target gene promoters, but that it might also dispose of DNA-binding-independent gene regulation mechanisms. In fact, FOXM1 interacts with β-catenin/TCF-4 (Zhang et al., 2011), the MuvB complex (namely, with the subunits LIN9 (cell lineage-abnormal 9), LIN37, and LIN52) (Chen, Müller, et al., 2013; Sadavisam et al., 2012), and B-Myb (Chen, Müller, et al., 2013) and controls the expression of their target genes without direct binding to DNA (Chen, Müller, et al., 2013; Down et al., 2012; Zhang et al., 2011).

One obvious question is whether FOXM1 *in vivo* occupies more than 270 DNA-binding sites in other cell cycle phases outside late G2- and M-phase and if yes, whether it binds to FOXM1 consensus binding sites in these genes.

The 270 FOXM1-binding regions, which were *in vivo* occupied by FOXM1 in late G2- and M-phase in ChIP-Seq assays with U2OS cells (Chen, Müller, et al., 2013), are not included in Table 3.1.

3.3. Biological functions of FOXM1 target genes

As expected, the genes associated with the 270 FOXM1-binding regions, which were identified in U2OS cells accumulated in late G2- and M-phase, encode proteins with functions in mitotic events and in regulation of mitosis (Chen, Müller, et al., 2013). Accordingly, their expression in HeLa cells peaked in late G2- and M-phases following release from a double-thymidine block that synchronized the cells at G1/S-phase (Chen, Müller, et al., 2013).

In general, FOXM1 regulates the expression of many cell cycle-related genes (Fig. 3.2). The FOXM1 target genes encode proteins, which control the G1/S-transition, the G2/M-transition, and the progression through S-phase and M-phase (Fig. 3.2). Moreover, the protein products of FOXM1 target genes have important functions in DNA replication and execution of mitosis (Fig. 3.2). Others play roles for centriole duplication or sister chromatid cohesion (Fig. 3.2). Furthermore, some FOXM1 target genes encode components of proliferative signal transduction pathways (Fig. 3.2). In accordance with this transcriptome, FOXM1 stimulates cell proliferation and cell cycle progression by promoting the entry into both S-phase and M-phase (Ackermann Misfeldt, Costa, & Gannon, 2008; Ahmad et al., 2010, 2011; Anders et al., 2011; Bao et al., 2011; Barsotti & Prives, 2009; Bergamaschi et al., 2011; Bhat, Jagadeeswaran, Halasi, & Gartel, 2011; Bolte et al., 2011; Brezillon et al., 2007; Calvisi et al., 2009; Carr et al., 2010; Chan et al., 2008; Chen, Müller, et al., 2013; Chen et al., 2010, 2011; Chu et al., 2012; Dai et al., 2013; Davis et al., 2010; Faust et al., 2012; Fu et al., 2008; Gusarova et al., 2007; Ho et al., 2012; Huang & Zhao, 2012; Kalin et al., 2006; Kalinichenko et al., 2003, 2004; Kim et al., 2005, 2006; Krupczak-Hollis, Wang, Dennewitz, & Costa, 2003; Krupczak-Hollis et al., 2004; Laoukili, Alvarez, et al., 2008; Laoukili et al., 2005; Lefebvre et al., 2010; Leung et al., 2001; Li et al., 2011; Liu, Gampert, et al., 2006; Liu et al., 2011; Lok et al., 2011; Mencalha et al., 2012; Millour et al., 2010, 2011; Monteiro et al., 2012; Nakamura, Hirano, et al., 2010; Ning, Li, Xiang, Liu, & Cao, 2012; Park, Carr, et al., 2009; Park, Costa, Lau, Tyner, & Raychaudhuri, 2008; Park, Wang, et al., 2008; Park et al., 2012; Pellegrino et al., 2010; Raghavan et al., 2012; Ramakrishna et al., 2007; Schüller et al., 2007; Sengupta et al., 2013; Uddin et al., 2011; Ueno et al., 2008; Ustiyan et al., 2009, 2012; Wang, Banerjee, et al. 2007; Wang, Chen, et al., 2008; Wang, Hung, et al., 2001; Wang, Kiyokawa, et al., 2002; Wang, Krupczak-Hollis, et al., 2002; Wang, Meliton, et al., 2008; Wang, Park, et al., 2011; Wang, Quail, et al., 2001; Wang, Teh, Ji, et al., 2010; Wang et al., 2005, 2010; Wierstra & Alves, 2006d; Wonsey & Follettie, 2005; Wu, Liu, et al., 2010; Xia, Huang, et al., 2012; Xia et al., 2009; Xie et al., 2010; Xue et al., 2010, 2012; Ye et al., 1999; Yoshida et al., 2007; Zeng et al., 2009; Zhang, Ackermann, et al., 2006; Zhang, Wu, et al., 2012,

2009; Zhang et al., 2011; Zhao et al., 2006). Additionally, FOXM1 is required for proper execution of mitosis (Chan et al., 2008; Fu et al., 2008; Gusarova et al., 2007; Laoukili et al., 2005; Priller et al., 2011; Ramakrishna et al., 2007; Schüller et al., 2007; Wang et al., 2005; Wonsey & Follettie, 2005; Yoshida et al., 2007).

In accordance with its roles in angiogenesis (Kim et al., 2005; Li et al., 2009; Lynch et al., 2012; Wang, Banerjee, et al., 2007; Xue et al., 2012; Zhang et al., 2008), migration, invasion (Ahmad et al., 2010; Balli et al., 2012; Bao et al., 2011; Behren et al., 2010; Bellelli et al., 2012; Chen, Chien, et al., 2009; Chen et al., 2013; Chu et al., 2012; Dai et al., 2007; He et al., 2012; Li et al., 2011; Li, Peng, et al., 2013; Lok et al., 2011; Lynch et al., 2012; Mizuno et al., 2012; Park et al., 2011; Park, Wang, et al., 2008; Uddin et al., 2011, 2012; Wang, Banerjee, et al., 2007; Wang, Chen, et al., 2008; Wang, Teh, Ji, et al., 2010; Wang, Wen, et al., 2012; Wu, Liu, et al., 2010; Xia, Huang, et al., 2012; Xue et al., 2012), EMT (epithelial–mesenchymal transition) (Balli et al., 2013; Bao et al., 2011; Li, Wang, et al., 2012; Park et al., 2011), metastasis (Li, Peng, et al., 2013; Li et al., 2009; Park et al., 2011; Xia, Huang, et al., 2012), prevention of senescence (Anders et al., 2011; Li et al., 2008; Park, Carr, et al., 2009; Qu et al., 2013; Rovillain et al., 2011; Wang et al., 2005; Zeng et al., 2009; Zhang, Ackermann, et al., 2006), heat shock response (Dai et al., 2013), and homologous recombination repair (Monteiro et al., 2012; Park et al., 2012) FOXM1 regulates genes, the products of which have pivotal functions in these processes (Fig. 3.2). Also, the three key pluripotency transcription factors Oct-4 (octamer-binding transcription factor 4, POU5F1 (POU (Pit-Oct-Unc) domain, class 5, transcription factor 1)), Sox2 (SRY (sex-determining region Y)-box 2, SRY-related HMG (high mobility group)-box 2), and Nanog are encoded by FOXM1 target genes, in accordance with FOXM1's role in maintenance of stem cell pluripotency (Fig. 3.2; Xie et al., 2010). Numerous protein products of FOXM1 target genes play roles in inflammation, chemotaxis, and macrophage functions, in accordance with the functions of FOXM1 in allergen-induced lung inflammation (Ren et al., 2013) and TAM (tumor-associated macrophages) recruitment to the lung (Fig. 3.2; Balli et al., 2012).

Additional FOXM1 target genes encode proteins with functions in apoptosis, immortality, DNA repair, DNA damage response, inhibition of cell cycle progression, signal transduction, chromatin remodeling and dynamics, nuclear import, mRNA splicing, and the Wnt signaling pathway (Fig. 3.2). Among the products of FOXM1 target genes are cell adhesion proteins as well as components of the cytoskeleton and the ECM (extracellular matrix) (Fig. 3.2). Several FOXM1 target genes encode proteins with specific functions in differentiated cell types (Fig. 3.2). Also antioxidant detoxifying enzymes, subunits of E3 ubiquitin ligases, and miRNAs are encoded by FOXM1 target genes (Fig. 3.2).

Notably, the FOXM1 transcriptome includes numerous protooncogenes, tumor-suppressor genes, candidate tumor-suppressor genes, and (candidate) metastasis suppressor genes (see Part II of this two-part review, that is see Wierstra, 2013b) in accordance with the implication of FOXM1 in tumorigenesis (Costa, Kalinichenko, Major, et al., 2005; Gong & Huang, 2012; Kalin et al., 2011; Koo et al., 2011; Laoukili et al., 2007; Myatt & Lam, 2007; Raychaudhuri & Park, 2011; Wierstra & Alves, 2007c).

Among the FOXM1 target genes are 43 transcription factors (Fig. 3.2) so that FOXM1 could indirectly regulate a large number of genes through these 43 transcription factors. In particular, the transactivator FOXM1 has the opportunity for indirect downregulation of genes if it activates the expression of those transcription factors, which are transrepressors.

4. MOLECULAR MECHANISMS OF FOXM1 FOR GENE REGULATION

4.1. Gene regulation mechanisms of FOXM1

FOXM1 is an activating transcription factor with a forkhead domain as DBD (Korver, Roose, & Clevers, 1997; Littler et al., 2010; Wierstra, 2011a; Wierstra & Alves, 2006a, 2006d, Yao et al., 1997; Ye et al., 1997) and with a very strong, acidic TAD (Wierstra, 2013a; Wierstra & Alves, 2006a). Since FOXM1 is a strong transactivator (Wierstra, 2013a; Wierstra & Alves, 2006a) it can transactivate the promoters of target genes as a classical (conventional) transcription factor by binding with its DBD to a FOXM1-binding site upstream (or downstream) of the core promoter and by activating transcription through its TAD (Alvarez-Fernandez et al., 2011; Alvarez-Fernandez, Halim, et al., 2010; Alvarez-Fernandez, Medema, & Lindqvist, 2010; Anders et al., 2011; Bhat et al., 2009a, 2009b, Chen, Dominguez-Brauer, et al., 2009; Fu et al., 2008; Gemenetzidis et al., 2009; Grant et al., 2012; Kalinichenko et al., 2004; Korver, Roose, & Clevers, 1997; Laoukili, Alvarez, et al., 2008; Laoukili, Alvarez-Fernandez, et al., 2008; Laoukili et al., 2005; Littler et al., 2010; Lüscher-Firzlaff et al., 1999; Madureira et al., 2006; Major et al., 2004; Park et al., 2012; Park, Wang, et al., 2008; Radhakrishnan et al., 2006; Sullivan et al., 2012; Tan et al., 2006; Wang, Snyder, et al., 2012; Wang, Zhang, Snyder, et al., 2010; Wierstra, 2011a, 2011b, 2013a; Wierstra & Alves, 2006a, 2006b, 2006c, 2007a, 2007b, 2007c, 2008; Ye et al., 1997; Zhang et al., 2011; Zhou, Liu, et al., 2010).

In addition, FOXM1 disposes of several other mechanisms for the control of target gene promoters:

First, FOXM1c can transactivate via certain TATA-boxes so that it can transactivate some isolated core promoters by directly binding with its DBD to their TATA-boxes (Wierstra & Alves, 2006d, 2007a, 2007b, 2007c, 2008). This represents a completely new transactivation mechanism (Wierstra & Alves, 2006d, 2007b, 2007c), which has never been described before.

Second, FOXM1 can bring about DNA hypermethylation at target gene promoters, which results in the repression of transcription (Carr et al., 2012; Teh et al., 2012). This chromatin remodeling-based mechanism offers the transactivator FOXM1 the possibility to repress the transcription of some target genes (Carr et al., 2012; Teh et al., 2012). FOXM1 can accomplish DNA hypermethylation at a promoter not only by binding to the target gene (*GATA3*) (Carr et al., 2012; Teh et al., 2012) and recruiting corepressors, but also without its own association with the target gene (*p16*) (Gemenetzidis et al., 2009; Teh et al., 2012).

Third, FOXM1 can activate the transcription of some target genes in a DNA-binding independent manner through protein–protein interactions with DNA-binding transcription factors (Chen, Müller, et al., 2013; Down et al., 2012; Zhang et al., 2011). Consequently, these genes are shared target genes of FOXM1 and its interaction partner, and the FOXM1-mediated activation of their promoters depends on the DNA-binding site for the interacting transcription factor but not on a FOXM1-binding site (Chen, Müller, et al., 2013; Down et al., 2012; Zhang et al., 2011). Examples for this DNA-binding-independent mechanism are the interactions of FOXM1 with β-catenin (Zhang et al., 2011), with B-Myb (Chen, Müller, et al., 2013), and with the MuvB complex (namely, with the subunits LIN9, LIN37, and LIN52) (Chen, Müller, et al., 2013; Sadavisam et al., 2012), which enable FOXM1 to activate target genes via WREs/TBEs through β-catenin/TCF4 (Bowman & Nusse, 2011; Gong & Huang, 2012; Zhang et al., 2011), via Myb-binding sites through B-Myb (Down et al., 2012), and via CHR elements through the MuvB complex (Chen,

Müller, et al., 2013). Interestingly, the interaction domain of FOXM1 for β-catenin (Zhang et al., 2011), B-Myb, and LIN9 (Chen, Müller, et al., 2013) is the forkhead DBD of FOXM1 so that the binding of the FOXM1–DBD to other transcription factors substitutes for its direct binding to DNA. (The MuvB complex binds to CHR elements through its subunit LIN54; Müller & Engeland, 2009; Schmit, Cremer, & Gaubatz, 2009.)

4.2. FOXM1-binding sites

The diversity of gene regulation mechanisms of FOXM1 demands a careful analysis where and how FOXM1 associates with the promoter of a target gene. In particular, the true direct binding of FOXM1c to the promoter DNA *in vitro* was only demonstrated for three FOXM1 target genes, namely, for *ERα* in fluorescence anisotropy assays (Littler et al., 2010) as well as for *c-myc* (Wierstra & Alves, 2006a), and *E-cadherin* (Wierstra, 2011a) in EMSAs.

In the *ERα* (Littler et al., 2010; Madureira et al., 2006) and *E-cadherin* (Wierstra, 2011a) promoters, FOXM1c binds to (a) FOXM1 consensus binding site(s) upstream of the core promoter so that it transactivates the *ERα* (Madureira et al., 2006) and *E-cadherin* (Wierstra, 2011a) promoters as a conventional (classical) transcription factor. In contrast, in the human *c-myc* promoter, FOXM1c binds to the TATA-boxes of the P1 and P2 promoters (Wierstra & Alves, 2006d) so that it transactivates the isolated for *c-myc* P1 and P2 core promoters directly via their TATA-boxes through a completely new, unprecedented transactivation mechanism (Wierstra & Alves, 2006d, 2007a, 2007b, 2007c, 2008).

For most FOXM1 target genes only ChIP assays were performed (Table 3.1), which leaves open not only the exact position of the FOXM1-binding site, but also whether FOXM1 binds to the promoter DNA directly or whether FOXM1 associates with the target gene promoter only indirectly via its protein–protein interaction with a DNA-binding transcription factor.

For a few FOXM1 target genes EMSAs (*Chk1*, *Lama4* (*laminin α 4*), *MMP-2* (*matrix metalloproteinase-2*), *VEGF* (*vascular endothelial growth factor*)), or DNAPs (*VEGF*) with cell lysates were carried out (Table 3.1), which do not preclude that the association of FOXM1 with the oligonucleotide is mediated by another transcription factor. This issue should be considered because not all oligonucleotides were good matches to the consensus sequence for FOXM1-binding sites and because several other FOX transcription factors display the same DNA-binding specificity as FOXM1 or at least very similar specificities. Nevertheless, the FOXM1-binding sites in the *Chk1* (Tan et al., 2010), *Lama4* (Kim et al., 2005), *MMP-2* (Dai et al., 2007), and *VEGF* (Karadedou et al., 2012; Zhang et al., 2008) promoters are located outside the core promoter so that FOXM1 would transactivate these four promoters as a conventional transcription factor.

5. THE MOLECULAR FUNCTION OF FOXM1 AS A CONVENTIONAL TRANSCRIPTION FACTOR

5.1. The conventional transcription factor FOXM1

FOXM1 functions as a conventional transcription factor if it binds to a conventional FOXM1-binding site upstream (or downstream) of a non-FOXM1-responsive core promoter and transactivates the promoter through its very strong acidic TAD (Alvarez-Fernandez et al., 2011; Alvarez-Fernandez, Halim, et al., 2010; Alvarez-Fernandez, Medema, et al., 2010; Anders et al., 2011; Bhat et al., 2009a, 2009b, Chen, Dominguez-Brauer, et al., 2009; Fu et al., 2008; Gemenetzidis et al., 2009; Grant et al., 2012; Kalinichenko et al., 2004; Korver, Roose, & Clevers, 1997; Laoukili, Alvarez, et al., 2008; Laoukili, Alvarez-Fernandez, et al., 2008; Laoukili et al., 2005; Littler et al., 2010; Lüscher-Firzlaff et al., 1999; Madureira et al., 2006; Major et al., 2004; Park et al., 2012; Park, Wang, et al., 2008; Radhakrishnan et al., 2006; Sullivan et al., 2012; Tan et al., 2006; Wang, Snyder, et al., 2012; Wang, Zhang, Snyder, et al., 2010; Wierstra, 2011a, 2011b, 2013a; Wierstra & Alves, 2006a, 2006b, 2006c, 2007a, 2007b, 2007c, 2008; Ye et al., 1997; Zhang et al., 2011; Zhou, Liu, et al., 2010).

The consensus sequence for conventional FOXM1-binding sites is 5′-A-C/T-AAA-C/T-AA-3′ (Wierstra & Alves, 2006a). Such conventional FOXM1-bindings sites are bound *in vitro* by the purified splice variants FoxM1A (Ye et al., 1997), FoxM1B (Hedge et al., 2011; Ye et al., 1997), and FOXM1c (Hedge et al., 2011; Korver, Roose, & Clevers, 1997; Littler et al., 2010; Wierstra, 2011a; Wierstra & Alves, 2006a, 2006c, 2007b).

FOXM1 transactivates the natural *ERα* (Madureira et al., 2006) and *E-cadherin* (Wierstra, 2011a) promoters as a conventional transcription factor. Moreover, the function of FOXM1 as conventional transcription factor was analyzed with artificial reporter constructs, which contained several copies of a conventional FOXM1-binding site upstream of a non-FOXM1-responsive core promoter. Both the splice variants FoxM1B (Bhat et al., 2009a, 2009b; Chen, Dominguez-Brauer, et al., 2009; Fu et al., 2008;

Gemenetzidis et al., 2009; Kalinichenko et al., 2004; Major et al., 2004; Park et al., 2012; Park, Wang, et al., 2008; Radhakrishnan et al., 2006; Tan et al., 2006; Wang, Snyder, et al., 2012; Wang, Zhang, Snyder, et al., 2010; Ye et al., 1997; Zhang et al., 2011; Zhou, Liu, et al., 2010) and FOXM1c (Alvarez-Fernandez et al., 2011; Alvarez-Fernandez, Halim, et al., 2010; Alvarez-Fernandez, Medema, et al., 2010; Anders et al., 2011; Korver, Roose, & Clevers, 1997; Laoukili, Alvarez, et al., 2008; Laoukili, Alvarez-Fernandez, et al., 2008; Laoukili et al., 2005; Littler et al., 2010; Lüscher-Firzlaff et al., 1999; Sullivan et al., 2012; Wierstra, 2011b, 2013a; Wierstra & Alves, 2006a, 2006b, 2006c, 2007a, 2007b, 2008) work as conventional transcription factors in transactivation of these reporter constructs, whereas the transcriptionally inactive third splice variant FoxM1A failed to transactivate such a reporter construct (Fig. 3.1; Ye et al., 1997). FOXM1c transactivated also the murine *E-cadherin* promoter (Wierstra, 2011a) and the human *ERα* promoter A, but not the *ERα* promoter B (Madureira et al., 2006), as a conventional transcription factor.

5.2. Functional domains of the conventional transcription factor FOXM1c

The conventional transcription factor FOXM1c possesses five functional domains (Fig. 3.3; Wierstra & Alves, 2007c). They coincide with those FOXM1 segments, which show a very high sequence conservation among mammals ($\geq$90% sequence identity), verifying their biological importance (Wierstra & Alves, 2006a).

The five functional domains are (Fig. 3.3; Wierstra & Alves, 2007c):
- the DBD, that is, the forkhead domain or forkhead box, in the middle of FOXM1c (Korver, Roose, & Clevers, 1997; Littler et al., 2010; Wierstra, 2011a; Wierstra & Alves, 2006a)
- the very strong acidic TAD at the outermost C-terminus of FOXM1c (Wierstra, 2013a; Wierstra & Alves, 2006a)
- the strong TRD (transrepression domain) at the center of FOXM1c (Wierstra, 2013a; Wierstra & Alves, 2006a, 2007b)
- the NRD-C (negative-regulatory domain-C) at the center of FOXM1c (Wierstra, 2013a; Wierstra & Alves, 2006a, 2006b, 2006c, 2007b, 2008)
- the NRD-N (negative-regulatory domain-N) at the outermost N-terminus of FOXM1c (Wierstra, 2011a, 2011b, 2013a; Wierstra & Alves, 2006a, 2006b, 2006c, 2008)

The TRD and the NRD-C comprise the central domain of FOXM1c (Fig. 3.3).

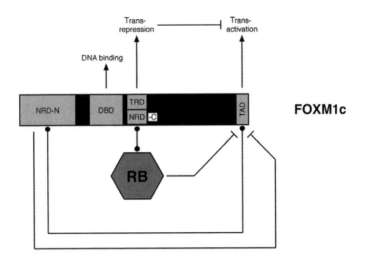

Figure 3.3 The five functional domains of the conventional transcription factor FOXM1c. FOXM1c is depicted as a black rectangle and its functional domains as light gray boxes. The tumor suppressor RB (retinoblastoma protein, RB1 (retinoblastoma 1), p105) is shown as a dark gray hexaeder. Direct protein–protein interactions are depicted as dot-ended lines. The conventional transcription factor FOXM1c possesses five functional domains, namely, the forkhead DBD, the acidic TAD (transactivation domain), the RB-independent TRD (transrepression domain), the autoinhibitory NRD-N (negative-regulatory domain-N), and the RB-recruiting NRD-C. The DBD binds to a conventional FOXM1c-binding site upstream (or downstream) of a non-FOXM1 c-responsive core promoter. The TAD activates transcription. The TRD, the NRD-N, and the NRD-C function as IDs (inhibitory domains) for the TAD: The TRD transrepresses actively against the transactivation by the TAD. The NRD-N binds directly to the TAD and thereby inhibits the TAD completely. The NRD-C recruits RB as corepressor, which in turn represses the TAD of FOXM1c indirectly, that is, without interacting with the FOXM1c-TAD. Thus, RB binds directly to the NRD-C of FOXM1c and thereby represses indirectly the TAD of FOXM1c. DBD, DNA-binding domain; TAD, transactivation domain; TRD, transrepression domain; NRD, negative-regulatory domain; RB, retinoblastoma protein; RB1, retinoblastoma 1, p105. *Reprinted from Biochemical and Biophysical Research Communications, volume 431, Inken Wierstra, Cyclin D1/Cdk4 increases the transcriptional activity of FOXM1c without phosphorylating FOXM1c, pages 753–759 (Supplementary Fig. S-1), 2013, with permission from Elsevier.*

The DBD binds to DNA (Korver, Roose, & Clevers, 1997; Littler et al., 2010; Wierstra, 2011a; Wierstra & Alves, 2006a) and the very potent TAD activates transcription (Fig. 3.3; Wierstra, 2013a; Wierstra & Alves, 2006a) so that FOXM1c is principally a very strong transactivator (Wierstra & Alves, 2006a). However, full-length FOXM1c is inactive because three functional domains work as IDs (inhibitory domains) for the TAD (Fig. 3.3), which keep in check the very high transactivation potential of the TAD (Wierstra & Alves, 2006a, 2006b). These three IDs are the TRD (Wierstra, 2013a; Wierstra & Alves, 2006a, 2007b), the NRD-C (Wierstra, 2013a; Wierstra & Alves, 2006b, 2008), and the NRD-N (Fig. 3.3; Wierstra, 2013a; Wierstra & Alves, 2006a, 2006b, 2008). FOXM1c can be converted from an inactive form into a very potent transactivator if the TAD is released from its inhibition by the three IDs (Wierstra, 2011a, 2011b, 2013a; Wierstra & Alves, 2006a, 2006b, 2006c, 2007b, 2008). This release is accomplished by activating signals *in vivo* (Wierstra, 2011b, 2013a; Wierstra & Alves, 2006b, 2006c, 2008), and it can be mimicked by deletion of the IDs *in vitro* (Wierstra, 2011a, 2011b, 2013a; Wierstra & Alves, 2006a, 2006b, 2006c, 2007b, 2008).

Since the central domain functions not only as TRD (Wierstra, 2013a; Wierstra & Alves, 2006a), but also as NRD-C (Wierstra, 2013a; Wierstra & Alves, 2006b), it represents a dual ID for the TAD (Fig. 3.3; Wierstra & Alves, 2007c).

Both the central domain and the N-terminus function as IDs for the TAD, but they act independently from each other (Wierstra & Alves, 2006a) and inhibit the TAD through different mechanisms (see below) (Wierstra, 2013a; Wierstra & Alves, 2006a, 2006b, 2007b, 2007c, 2008).

Moreover, the N-terminus inhibits the TAD completely whereas the central domain represses the TAD only partially so that the N-terminus works as a dominant on/off switch for FOXM1c (Wierstra & Alves, 2006a). According to this hierarchy of the two IDs for the TAD, inactive wild-type FOXM1c can be activated in two steps (Wierstra & Alves, 2006a): deletion of the N-terminus results in a strong transactivator and additional deletion of the central domain creates a very strong transactivator (Wierstra & Alves, 2006a).

5.3. The DBD

5.3.1 The DBD of FOXM1c contains the additional exon A1 in wing W2 of the forkhead domain

The splice variants FoxM1B and FOXM1c differ only in their forkhead DBDs because the 15-aa-long exon A1 in wing W2 of the DBD is present in FOXM1c but absent from FoxM1B (Fig. 3.1). Actually, FOXM1c is the only member of the FOX transcription factor family with an additional insertion (exon A1) in wing W2 of the forkhead DBD (see sequence comparisons in: Carlsson & Mahlapuu, 2002; Hannenhalli & Kaestner, 2009; Katoh &

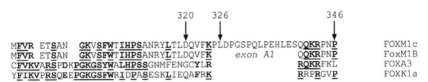

Figure 3.4 Alignment of the forkhead DBDs of FOXM1c, FoxM1B, FOXA3, and FOXK1a. Identical amino acids (bold and underlined) and conservative mutations (bold; D/E, K/R, S/T, N/Q, F/W/Y, I/L/M/V) are marked. Shown are aa 233–346 of FOXM1c (with 762 aa). The forkhead domain of FOXM1c includes aa 235–346, but the 2.2 Å crystal structure of the FOXM1c–DBD was only obtained for aa 231–320 or aa 234–326 because neither of the two FOXM1c molecules displayed electron density for residues C-terminal to aa 320 or 326, indicating a degree of disorder in this region (Littler et al., 2010). The DBDs of the splice variants FOXM1c and FoxM1B differ in the presence or absence of the 15-aa-long exon A1 (PLDPGSPQLPEHLES) in wing W2 at the C-terminus of the forkhead domain, respectively. Please note that the structure of the FoxM1B–DBD has not been resolved.

Katoh, 2004; Kaufmann & Knöchel, 1996; Obsil & Obsilova; 2008; Pohl & Knöchel, 2005; Wijchers et al., 2006), whereas FoxM1B shares the normal wing W2 structure with the remainder of the FOX family (Fig. 3.4).

The structure of the FOXM1c–DBD has been resolved by Littler et al. (2010), but the structure of the FoxM1B–DBD is unknown. Please note that the aa numbering in Littler et al. (2010), who analyzed human FOXM1c with 763 aa, deviates by 1 aa from that in this review, which refers to human FOXM1c with 762 aa.

The canonical structure of a forkhead DBD is H1–S1–H2–turn–H3–S2–loop (wing W1)–S3–loop (wing W2) (Benayoun et al., 2012; Brennan, 1993; Carlsson & Mahlapuu, 2002; Clark et al., 1993; Gajiwala & Burley, 2000; Kaufmann & Knöchel, 1996; Lai et al., 1993; Lalmansingh et al., 2012; Obsil & Obsilova, 2008, 2011; Wijchers et al., 2006). Instead of the turn, the DBD of FOXM1c possesses an additional helix H4 between H2 and H3 so that the FOXM1c–DBD has the structure H1–S1–H2–H4–H3–S2–W1–S3–W2 (Fig. 3.5A; Littler et al., 2010).

The DBD of FOXM1c includes aa 235–346 (Fig. 3.4). Although the FOXM1c fragment aa 221–359 was purified by Littler et al. (2010) their 2.2 Å X-ray crystal structure of the FOXM1c–DBD was only resolved

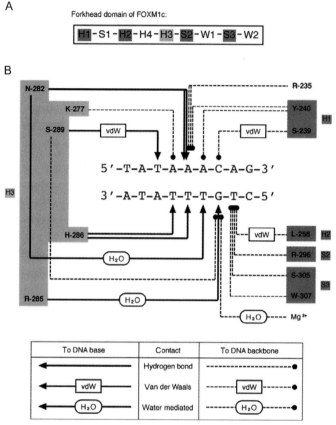

Figure 3.5 DNA contacts of the DNA-binding domain (DBD) of FOXM1c. The figure summarizes the results of Littler et al. (2010), who obtained a 2.2 Å X-ray crystal structure of the FOXM1c–DBD bound to DNA. (A) FOXM1c possesses a forkhead domain as DBD, which consists of three β-strands (S), four α-helices (H), and two wings (W). Those structural elements of the FOXM1c–DBD, which are involved in DNA contacts, are shown in gray. The recognition helix H3 (light gray) lies in the major groove and makes all base contacts. (B) Depicted are those amino acids of the FOXM1c–DBD, which contact the DNA, and it is indicated, to which structural elements they belong. The recognition helix H3 (to the left) is shown in light gray and the other structural elements (to the right) in dark gray. R-235 is positioned N-terminal to helix H1. Littler et al. (2010) did not specify exactly, which residue belongs to which structural element, so that this information was deduced from indirect hints in their paper and from the structures of other forkhead domains (not shown). The aa numbering is for FOXM1c with 762 aa and differs from that in Littler et al. (2010), who analyzed FOXM1c with 763 aa. All contacts to DNA bases are depicted as thick black lines with arrows and all contacts to the DNA backbone as dashed dot-ended lines. Van der Waals contacts (vdW) and water-mediated contacts (H_2O) are indicated. The other DNA contacts are hydrogen bonds. H, α-helix; S, β-strand; W, wing.

for aa 231–320 or 234–326 (Fig. 3.4) because of the lack of electron density for residues C-terminal to aa 320 or 326 in both FOXM1c–DBD molecules (Littler et al., 2010). Therefore, the exon A1 (aa 325–339), which is unique to FOXM1c, as well as the C-terminal end (aa 340–346) of wing W2, which is common to all forkhead DBDs, are both missing in the 2.2 Å X-ray crystal structure of the FOXM1c–DBD (Fig. 3.4).

Since FOXM1c represents the sole exception with an additional insertion (exon A1) in wing W2 of the forkhead domain (Fig. 3.4) it would be desirable to have the structures of both the complete FOXM1c–DBD and the FoxM1B–DBD in order to compare them among each other and with the structures of other forkhead DBDs.

5.3.2 The X-ray crystal structure of the FOXM1c–DBD (Littler et al., 2010)

Littler et al. (2010) resolved the 2.2 Å X-ray crystal structure of the forkhead DBD of human FOXM1c (Figs. 3.5 and 3.6). In this structure, the FOXM1c–DBD bound to the DNA sequence 5'-T$\underline{A}$TAAACAG-3' (Fig. 3.5B; Littler et al., 2010), which matches the consensus sequence for conventional FOXM1-binding sites 5'-$\underline{A}$-C/$\underline{T}$-$\underline{AAA}$-C/$\underline{T}$-$\underline{AA}$-3' (Wierstra & Alves, 2006a). The FOXM1c–DBD contacted the DNA bases of five nucleotides (TAAAC) and the DNA backbone of five nucleotides (AAACA; Fig. 3.5B; Littler et al., 2010).

Since the 19-bp DNA duplex used for crystallization contains two FOXM1-binding sites in a palindromic orientation (atTGTTTA-TAAACAgcccg) this forward-to-back tandem repeat is bound by two FOXM1c molecules, which are nearly identical in structure and make similar DNA contacts (Littler et al., 2010). These two FOXM1c–DBDs show few interprotein contacts and none that could be described as being specific to FOXM1c so that there is no obvious protein–protein interface between symmetry-related subunits, consistent with the model of the isolated FOXM1c–DBD being monomeric (Littler et al., 2010). Most likely, this palindromic intermolecule interface with minimal contacts is promoted artificially through the requirements of crystallization (Littler et al., 2010). Thus, FOXM1c binds to DNA as a monomer with the weakly interacting "dimer" resulting from the choice of a DNA duplex with two palindromic FOXM1-binding sites (Littler et al., 2010).

Neither FOXM1c molecule (aa 221–359) displays electron density for the first 10 N-terminal residues or the 25–30 C-terminal ones, indicating a degree of disorder in these regions (Littler et al., 2010). Therefore, the

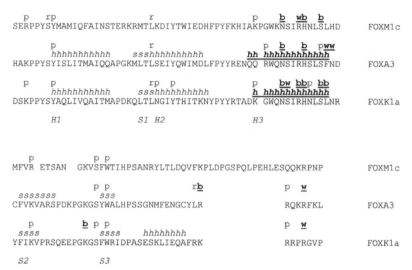

Figure 3.6 DNA contacts of the forkhead DBDs of FOXM1c, FOXA3, and FOXK1a as well as secondary structures of the forkhead DBDs of FOXA3 and FOXK1a. Secondary structures (italics) are indicated above the aa sequences: s, β-strand, h, α-helix. Since Littler et al. (2010) did not specify the exact positions of the structural elements in the FOXM1c–DBD they are not indicated for the FOXM1c–DBD. The three α-helices (H1, H2, H3) and three β-strands (S1, S2, S3) (italics) of a canonical forkhead DBD with the order H1–S1–H2–turn–H3–S2–wing W1–S3–wing W2 are numbered below the aa sequences. Instead of wing W2, FOXK1a has an additional helix H4 after S3. Instead of the turn, FOXM1c has an additional helix H4 between H2 and H3. DNA contacts are indicated above the aa sequences: b, base contact; w, water-mediated base contact; p, phosphate contact; r, ribose contact. The base contacts and the recognition helix H3, which lies in the major groove of the DNA, are marked bold and underlined. Please note that some residues participate in more than one DNA contact but only the most specific one is indicated. If a residue contacts a DNA base and the DNA backbone, the DNA backbone contact is left out. If a residue makes a direct and a water-mediated base contact, the water-mediated base contact is left out. Shown are aa 233–346 of FOXM1c (with 762 aa). The forkhead domain of FOXM1c includes aa 235–346, but the 2.2 Å crystal structure of the FOXM1c–DBD was only obtained for aa 231–320 or aa 234–326 because neither of the two FOXM1c molecules displayed electron density for residues C-terminal to aa 320 or 326, indicating a degree of disorder in this region (Littler et al., 2010). The DNA sequences, which were bound by the forkhead DBDs in the X-ray crystal structures, are listed in Fig. 3.7. **References for the DBD structures:** FOXM1c: Littler et al. (2010) (X-ray crystallography at 2.2 Å). FOXA3: Clark et al. (1993) (X-ray crystallography at 2.5 Å). FOXK1a: Liu et al. (2002) (NMR spectroscopy), Tsai et al. (2006) (X-ray crystallography at 2.4 Å).

```
FOXM1c    ATTGTTTATAAACAGCCCG
FOX3A         GACTAAGTCAACC
FOXK1a      TGTTGTAAACAATACA
```

Figure 3.7 DNA sequences, which were bound by the forkhead DBDs in the X-ray crystal structures. Underlined are those nucleotides, which are contacted by the FOXM1c–DBD in the X-ray crystal structure. The FOXM1c–DBD makes DNA base contacts to TAAAC and DNA backbone contacts to AAACA. Nucleotides, which are identical with the DNA sequence bound by the FOXM1c–DBD, are marked bold.

crystal structure was obtained for aa 231–320 or aa 234–326 of FOXM1c (Fig. 3.4; Littler et al., 2010).

The FOXM1c–DBD adopts a forkhead winged-helix fold (Littler et al., 2010). This is constructed from three α-helices and three β-strands with the topology αβααββ (Fig. 3.5A; Littler et al., 2010). The α-helices make up the center of the fold and pack tightly into a squat-cylindrical volume that sits perpendicularly alongside the DNA, an arrangement that allows recognition helix H3 to insert into the major groove where it can make sequence-specific DNA contacts (Figs. 3.5B and 3.6; Littler et al., 2010). The mixed three-stranded β-sheet lies at one end of this helical bundle with strands S2 and S3 running parallel to the DNA helix, contacting the phosphate backbone at several points (Figs. 3.5B and 3.6; Littler et al., 2010). An additional small 3_{10}-helix (helix H4) lies within the loop connecting helices H2 and H3 so that the order of structural elements in the FOXM1c–DBD is H1–S1–H2–H4–H3–S2–W1–S3–W2 (Fig. 3.5A; Littler et al., 2010).

The FOXM1c–DBD binds to the major groove of the DNA (Littler et al., 2010). FOXM1c makes both sequence-specific DNA contacts to the DNA bases and sequence-independent DNA contacts to the DNA backbone (Figs. 3.5B and 3.6; Littler et al., 2010). The latter are in the majority (Figs. 3.5B and 3.6; Littler et al., 2010).

Direct contact with the DNA bases ($T_1A_2A_3A_4C_5A_6$) is made through protein residues from within the recognition helix H3 (Fig. 3.5B; Littler et al., 2010). The major contributors to specificity are three invariantly conserved residues: N-282, R-285, and H-286 (Fig. 3.5B; Littler et al., 2010). Specificity for the A_3 base derives from two hydrogen bonds with the side chain of N-282, and the A_2 specificity arises from a hydrogen bond between its complementary base and H-286 (Fig. 3.5B; Littler et al., 2010). Indirect water-mediated interactions with the complementary bases possibly yield a preference for the A_4 position (via N-282) and the C_5 position (via R-285; Fig. 3.5B; Littler et al., 2010). A van der Waals contact between S-289 and the T_1 base could promote selectivity at this position (Fig. 3.5B; Littler et al., 2010). No contact between the A_6 base and the protein was observed (Fig. 3.5B; Littler et al., 2010).

Like other forkhead family members, FOXM1c binds to DNA in the major groove and makes all base-specific DNA contacts with recognition helix H3 (Figs. 3.5B and 3.6), which lies in the major groove of the DNA (Littler et al., 2010). In fact, FOXM1c makes all base contacts with recognition helix H3 whereas neither wing W2 nor wing W1 make any base contact (Figs. 3.5B and 3.6; Littler et al., 2010). Actually, FOXM1c does not at all use wing W2 or wing W1 for DNA contacts (Figs. 3.5B and 3.6) because the wing W1 loop does not contact the DNA and because the wing W2 loop is entirely absent (Littler et al., 2010).

In contrast to other forkhead domains, the two wings W1 and W2 of the forkhead winged-helix fold of FOXM1c make minimal contacts with the DNA (Littler et al., 2010).

Wing W1 of FOXM1c is short, namely 6-residues long (Littler et al., 2010). In the asymmetric unit, the two FOXM1c molecules have their wing W1 loops in slightly different conformations but, in both subunits, this loop diverges away from the DNA and does not contact the minor groove (Littler et al., 2010). Consequently, wing W1 in FOXM1c is unable to directly affect protein–DNA interactions (Littler et al., 2010). (Wing W1 of the FOXM1c molecule A crosses the unligated phosphate backbone and is therefore less likely to represent a physiologically relevant state (Littler et al., 2010)).

Wing W2 of FOXM1c adopts an unusual ordered conformation across the back of the molecule (Littler et al., 2010). No 4th helix C-terminal to strand S3 is seen, although a helical turn does encompass residues 310–313 (Littler et al., 2010). The C-terminal residues of the FOXM1c–DBD adopt a unique structure that consists of a sharp turn leading into an extended loop packed perpendicularly across H1 (Littler et al., 2010). This loop diverges away from the DNA so is not merely a different conformation of the wing W2 loop (Littler et al., 2010). Thus, the structure of the C-terminus of the FOXM1c–DBD differs significantly from that of other forkhead domains because the C-terminal residues of the FOXM1c–DBD form an extended structure distal from the DNA unique to FOXM1c (Littler et al., 2010).

Despite crystallizing a FOXM1c construct that extends to residue 359, neither FOXM1c molecule displays electron density for the 25–30

C-terminal residues limiting the crystal structure of the FOXM1c–DBD to 320 or 326 aa at the C-terminus (Littler et al., 2010). Any additional C-terminal residues that could make up a DNA-interacting wing appear to remain distant from the DNA and disordered (Littler et al., 2010). This could imply two possibilities (Littler et al., 2010). Either FOXM1 does not have a 2nd wing-like loop or this wing does not bind to the minor groove but instead the major groove of the DNA, in which case the palindromic DNA sequence may perturb its binding (Littler et al., 2010).

In summary, the FOXM1c wing W1 and wing W2 regions adopt unique conformations that are not involved in DNA binding (Littler et al., 2010). Unlike other forkhead domains, the FOXM1c–DBD displays a relatively short wing W1 loop, which does not interact with the DNA, and the FOXM1c–DBD structure shows no evidence for the presence of the wing W2 loop, indicating that this DNA-binding element is missing (Littler et al., 2010). As a consequence, the FOXM1c–DBD lacks contacts between the DNA and the wing W1 and wing W2 loop regions (Littler et al., 2010).

The FOXM1c–DBD is special because its wing W2 is not involved in DNA binding (Littler et al., 2010), whereas forkhead domains generally make the majority of their DNA contacts with recognition helix H3 and wing W2 (Fig. 3.6; Boura et al., 2010; Brent et al., 2008; Clark et al., 1993; Stroud et al., 2006; Tsai et al., 2006, 2007). However, as in all other forkhead domains (Boura et al., 2010; Brent et al., 2008; Clark et al., 1993; Stroud et al., 2006; Tsai et al., 2006, 2007), the recognition helix H3 of FOXM1c represents its principal DNA contact surface (Fig. 3.6; Littler et al., 2010).

5.3.3 Differences between the DBD of FOXM1c and other forkhead DBDs

5.3.3.1 General differences between forkhead domains

Despite their highly conserved recognition helix H3 (Fig. 3.4) different forkhead domains bind to different DNA sequences, which is generally attributed to three less-conserved regions (namely, the region in front of H3 and the wings W1 and W2) as well as to differences in the electrostatic potential of the surface of the forkhead domain (Benayoun et al., 2012; Boura et al., 2010; Chu et al., 2011; Jin et al., 1999; Liu et al., 2002; Marsden et al., 1997, 1998; Obsil & Obsilova, 2011; Overdier, Porcella, & Costa, 1994; Pierrou et al., 1994; Sheng et al., 2002; Tsai et al., 2006, 2007; van Dongen et al., 2000; Weigelt et al., 2001). These differences are thought to result in a different positioning of the recognition helix H3 in the major groove so that either other residues are involved in base contacts or the same residues contact other bases (Boura et al., 2010; Carlsson & Mahlapuu, 2002; Chu et al., 2011; Gajiwala & Burley, 2000; Hromas & Costa, 1995; Kaufmann & Knöchel, 1996; Obsil & Obsilova, 2008, 2011; Pohl & Knöchel, 2005; Wijchers et al., 2006).

Also the aa sequence of the DBD of FOXM1c differs from those of other forkhead DBDs in each of the three less-conserved regions, that is, the region in front of H3 and the wings W1 and W2 (see sequence comparisons in: Carlsson and Mahlapuu (2002), Kaufmann and Knöchel (1996), Katoh and Katoh (2004), Pohl and Knöchel (2005), Wijchers et al. (2006), Obsil and Obsilova (2008), Hannenhalli and Kaestner (2009)). Additionally, the distribution of positively and negatively charged residues in the FOXM1c–DBD deviates from that in other forkhead DBDs (data not shown).

5.3.3.2 The wings W1 and W2 of FOXM1c adopt unique conformations distal from the DNA, and they are not involved in DNA binding

The differences between the FOXM1c–DBD and other forkhead domains in the aa sequence of wing W1 and wing W2 (Fig. 3.4) manifest in unique conformations of the wing W1 and wing W2 regions of FOXM1c (Littler et al., 2010). As a result, the FOXM1c–DBD lacks contacts between the DNA and the wing W1 and wing W2 loop regions (Figs. 3.5B and 3.6) because they both diverge away from the DNA (Littler et al., 2010). This lack of DNA contacts of the FOXM1c wings W1 and W2 contrasts with the DNA contacts made by the wings W1 and W2 of other forkhead DBDs (Fig. 3.6; Littler et al., 2010).

Wing W1 of FOXM1c is relatively short, namely, only 6-residues long, and this wing W1 loop diverges away from the DNA so that it is unable to contact the DNA (Littler et al., 2010).

The wing W2 loop is entirely absent in FOXM1c as its wing W2 adopts an unusual ordered conformation across the back of the molecule, which consists of a sharp turn leading into an extended loop distal from the DNA (Littler et al., 2010). Since this unique structure diverges away from the DNA wing W2 of FOXM1c is not involved in DNA binding (Littler et al., 2010).

Thus, in contrast to other forkhead domains, neither wing W1 nor wing W2 of the FOXM1c–DBD makes DNA contacts (Fig. 3.6) because they adopt unique conformations distal from the DNA (Littler et al., 2010).

Prototypic forkhead DBDs make the majority of their DNA contacts with recognition helix H3 and wing W2 (Fig. 3.6; Boura et al., 2010; Brent et al., 2008; Clark et al., 1993; Littler et al., 2010; Stroud et al., 2006; Tsai et al., 2006, 2007). In doing so, the recognition helix H3 makes most base contacts and all base-specific DNA contacts, whereas wing W2 does not contribute base-specific DNA contacts (Gajiwala & Burley, 2000). Hence, the principal DNA contact surface of the forkhead domain is provided by the recognition helix H3 (Boura et al., 2010; Brent et al., 2008; Clark et al., 1993; Littler et al., 2010; Stroud et al., 2006; Tsai et al., 2006, 2007).

In contrast, the exceptional DBD of FOXM1c makes all base contacts and thus all base-specific DNA contacts with recognition helix H3 (Fig. 3.5B) whereas not a single DNA contact is made by wing W2 (Fig. 3.6) so that the predominant importance of the recognition helix H3 for DNA binding is even more pronounced in the case of the FOXM1c–DBD (Littler et al., 2010).

5.3.4 Reflections on wing W2 of the FOXM1c–DBD

Relating to the uncommon wing W2 of FOXM1c, Littler et al. (2010) mention the possibility that the chosen DNA duplex with two palindromic FOXM1-binding sites and the resulting binding of two adjacent FOXM1c molecules might give rise to an artificial conformation of wing W2 in their crystal structure of the FOXM1c–DBD:

Wing W2 forms an extended structure distal from the DNA unique to FOXM1c, which consists of a sharp turn leading into an extended loop packed perpendicularly across H1 (Littler et al., 2010). Since wing W2 of FOXM1c adopts an unusual ordered conformation, which extends away from the DNA and which does not resemble the wing W2 loops in other forkhead domains, Littler et al. (2010) concluded that a wing W2 loop is missing in the FOXM1c–DBD. However, they remark also that their 2.2 Å crystal structure of the FOXM1c–DBD was obtained for a DNA duplex with two palindromic FOXM1-binding sites, which may perturb the binding of wing W2 to the major groove of the DNA so that the observed absence of a wing W2 loop might artificially arise from the requirements of crystallization (Littler et al., 2010).

5.4. The TAD

5.4.1 The acidic TAD of FOXM1c

FOXM1c possesses a very strong TAD at its outermost C-terminal end, which was mapped to aa 721–762 (Fig. 3.8A; Wierstra, 2013a; Wierstra &

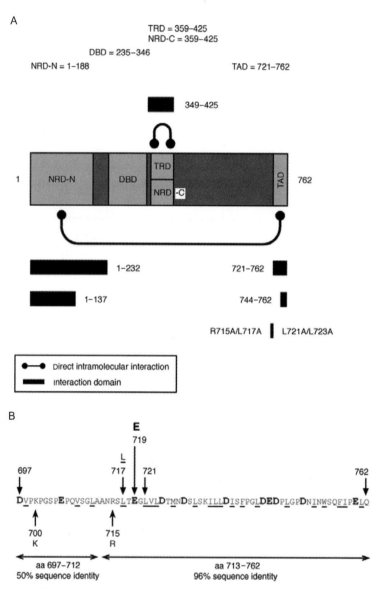

Figure 3.8 The transcription factor FOXM1. (A) Intramolecular interactions of FOXM1. FOXM1 is depicted as a dark gray rectangle and its functional domains as light gray boxes. Numbers indicate the aa of FOXM1. The aa numbering refers to human FOXM1c with 762 aa. The two direct intramolecular interactions within FOXM1 are shown as dot-ended lines between the participating FOXM1 domains. First, the central domain (aa 349–425) interacts directly with itself. Second, the N-terminus (aa 1–232) binds directly to the TAD (aa 721–762). The respective interaction domains are depicted as thick black lines. The interaction between the N-terminus and the TAD was also

Alves, 2006a). In contrast, the FOXM1c fragment aa 588–696 lacks any transactivation potential (Wierstra, 2013a; Wierstra & Alves, 2006a). The TAD of FOXM1c is a typical acidic TAD with a preponderance of acidic residues and interspersed large hydrophobic residues (Fig. 3.8B; Wierstra & Alves, 2006a).

Since the basic residue R-715 does definitely not belong to the acidic TAD of FOXM1c while the acidic residue E-719 and the large hydrophobic residue L-717 may be part of its acidic TAD (Fig. 3.8B) the N-terminal end of the FOXM1c-TAD may be situated at aa 717 or 716 (Wierstra & Alves, 2006a). This conclusion is strongly supported by the position of the FOXM1 segments with high sequence conservation among mammals because the FOXM1 segment aa 713–762 is very highly conserved (96% sequence identity) whereas the N-terminal adjacent FOXM1 segment aa 697–712 is not (50% sequence identity; Fig. 3.8B; Wierstra & Alves, 2006a). In addition to its low sequence conservation, the FOXM1c segment aa 697–712 contains the basic residue K-700 but no closely packed accumulation of acidic and large hydrophobic residues (Fig. 3.8B), which both argues against an N-terminal extension of the acidic FOXM1c-TAD beyond aa 717 or 716.

---

observed with the C-terminal part (aa 744–762) of the TAD or with the N-terminal part (aa 1–137) of the N-terminus. Combined point mutations in an RXL-motif (R715A/L717A) and a neighboring LXL-motif (L721A/L723A) reduced the binding of an N-terminal FOXM1 fragment to full-length FOXM1, suggesting that aa 715–723, which are located in or adjacent to the TAD, participate in the interaction with the N-terminus. (B) The acidic TAD of FOXM1. Shown is the aa sequence of human FOXM1c (762 aa). The TAD of FOXM1c was mapped to aa 721–762. The FOXM1c-TAD represents a typical acidic TAD. Acidic TADs are characterized by the preponderance of acidic aa (bold and enlarged) and interspersed large hydrophobic aa (underlined). The acidic aa E-719 may belong to the FOXM1-TAD because it is only four residues apart from the next acidic aa (D-724) within the FOXM1-TAD (aa 721–762). Also the large hydrophobic aa L-717 may belong to the FOXM1-TAD because it is only three residues apart from the next large hydrophobic aa (L-721) within the FOXM1-TAD. In contrast, the basic aa R-715 is definitely not part of the acidic TAD of FOXM1. Therefore, the N-terminal end of the FOXM1-TAD may be at aa 716 or 717 because E-719 and L-717 may belong to the acidic TAD of FOXM1, whereas R-715 does definitely not. This position of the N-terminal end of the FOXM1-TAD is in accordance with the given fact that the FOXM1 segment aa 713–762 shows a high sequence conservation among mammals (96% sequence identity), whereas the FOXM1 segment aa 697–712 does not (50% sequence identity). The acidic aa E-705 and D-697 are very unlikely to belong to the FOXM1-TAD because they are separated from E-719 by the intervening basic aa R-715 and/or K-700, which are definitely not part of the acidic TAD of FOXM1, and because they are 13 or 21 residues apart from E-719, respectively. DBD, DNA-binding domain; TAD, transactivation domain; TRD, transrepression domain; NRD, negative-regulatory domain; aa, amino acid.

In summary, the TAD of FOXM1c is located at aa 721–762 (Fig. 3.8A; Wierstra, 2013a; Wierstra & Alves, 2006a), and it may N-terminally slightly extend until aa 717 or 716 (Wierstra & Alves, 2006a).

Accordingly, each deletion, which removes the C-terminal TAD (plus additional parts of FOXM1), abolished the transcriptional activity of FOXM1c (Laoukili et al., 2005; Ma, Tong, et al., 2005; Ma et al., 2010; Wierstra, 2013a; Wierstra & Alves, 2006a, 2008) and FoxM1B (Fu et al., 2008; Kalinichenko et al., 2004; Wang, Kiyokawa, et al., 2002; Ye et al., 1997).

5.4.2 Molecular targets of the FOXM1c-TAD
5.4.2.1 The TAD of FOXM1c does not interact with TBP, TFIIB, and TFIIEα

The acidic TAD of FOXM1c is very potent (Wierstra, 2013a; Wierstra & Alves, 2006a), but its targets are completely unknown. Surprisingly, the FOXM1c-TAD binds neither to TBP (TATA-binding protein) nor to TFIIB (transcription factor IIB) (Wierstra & Alves, 2006d) although both are typical targets of acidic TADs (Choy et al., 1993; Tjian & Maniatis, 1994; Triezenberg, 1995; Stargell & Struhl, 1996; Struhl, 1995, 1996). Moreover, the FOXM1c-TAD does not interact with TFIIEα (Wierstra & Alves, 2006d).

5.4.2.2 The TAD of FOXM1c recruits neither P/CAF nor p300/CBP as coactivators

The coactivators and KATs (K-acetyltransferases) P/CAF (p300/CBP-associated factor, KAT2B) and p300 (KAT3B)/CBP (CREB (cAMP response element-binding protein)-binding protein, KAT3A) seem to be not involved in the transactivation by the TAD of FOXM1c:

P/CAF inhibited the transactivation of the reporter construct p(MBS)$_3$-mintk-luc by FOXM1c (Fig. 3.9A), demonstrating that the TAD of FOXM1c does not recruit the coactivator and KAT P/CAF in order to transactivate (Wierstra & Alves, 2008). (p(MBS)$_3$-mintk-luc contains three copies of a FOXM1c-binding site upstream of the minimal *TK* (*thymidine kinase*) promoter of HSV (herpes simplex virus) (Lüscher-Firzlaff et al., 1999)).

p300 and CBP did not increase, but rather slightly decreased, the transactivation of p(MBS)$_3$-mintk-luc by FOXM1c (Fig. 3.9B and C), which indicates that the FOXM1c-TAD does not recruit the coactivator and KAT p300/CBP for transactivation (Wierstra & Alves, 2008).

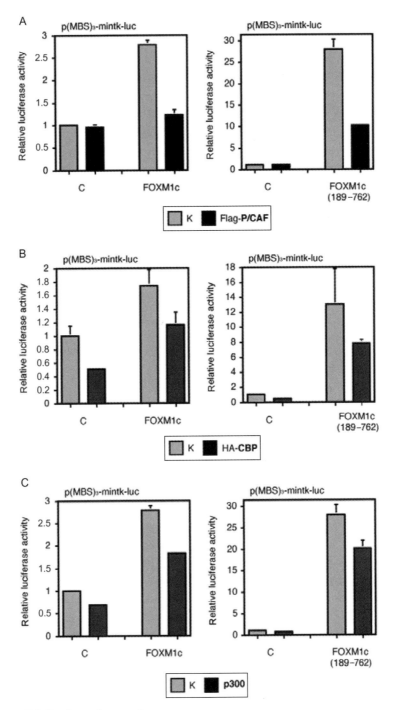

Figure 3.9 See legend on next page.

Accordingly, also the transactivation of the reporter construct 6x-FoxM1B-TATA-Luc by the splice variant FoxM1B was almost unaffected by CBP because CBP had only a 1.4-fold positive effect on the FoxM1B-mediated transactivation (and only a 1.8-fold positive effect in the presence of Cdc25B) (Major et al., 2004). (6x-FoxM1B-TATA-Luc contains six copies of a FoxM1B-binding site adjacent to the TATA-box sequence from the CMV (cytomegalovirus) promoter (Major et al., 2004)).

The conclusion that FOXM1 does not recruit CBP as a coactivator is underscored by the finding in U2OS cells that siRNA-mediated depletion of FOXM1 barely diminished the occupancy of the *JNK1* (*c-Jun N-terminal kinase 1*) promoter with CBP although it obliterated the association of Pol II (RNA polymerase II) with the direct FOXM1 target gene *JNK1* (Wang, Chen, et al., 2008).

In summary, P/CAF, p300, and CBP do not work as coactivators for FOXM1c (Fig. 3.9; Wierstra & Alves, 2008) so that neither P/CAF nor p300/CBP are targets of the TAD of FOXM1c. Accordingly, CBP seems also not to act as a coactivator for the splice variant FoxM1B (Major et al., 2004).

Nonetheless, FOXM1c interacts possibly indirectly with P/CAF (Wierstra & Alves, 2008) and FoxM1B interacts possibly indirectly with CBP (Chen, Dominguez-Brauer, et al., 2009; Major et al., 2004). However, P/CAF binds to the N-terminal aa 1–477 of FOXM1c (Wierstra & Alves, 2008), which are located far outside the C-terminal TAD (aa 721–762). Similarly, the binding of CBP to FoxM1B depends on T-596 (Major et al., 2004) and S-251 (Chen, Dominguez-Brauer, et al., 2009) of FoxM1B (corresponding to T-610 and S-250 of FOXM1c), which are positioned outside the TAD (aa 721–762). Hence, if P/CAF and p300/CBP should turn out to be direct interaction partners of FOXM1 their failure to work

---

Figure 3.9 The FOXM1c-mediated transactivation is (A) inhibited by P/CAF and (B and C) not increased but rather slightly decreased by p300/CBP. RK-13 cells were transiently transfected with p(MBS)$_3$-mintk-luc and with pFOXM1c, pFOXM1c(189–762) or the empty vector (control C). Either the expression plasmid for (A) Flag-P/CAF or (B) HA (hemagglutinin)-CBP or (C) p300 or the empty vector (control K) were cotransfected. The relative luciferase activity of p(MBS)$_3$-mintk-luc in the combination of control C with control K was set to 1. The left and right panels show parts of the same experiment. CBP, CREB-binding protein, KAT3A, K-acetyltransferase 3A; CREB, cAMP (cyclic adenosine monophosphate) response element-binding protein; p300, KAT3B, K-acetyltransferase 3B; P/CAF, p300/CBP-associated factor, KAT2B, K-acetyltransferase 2B.

as coactivators for the TAD of FOXM1 (Major et al., 2004; Wierstra & Alves, 2008) might be attributed to the circumstance that the FOXM1-TAD is apparently not their target (Chen, Dominguez-Brauer, et al., 2009; Major et al., 2004; Wierstra & Alves, 2008).

An unusual transient transfection experiment was performed by Sullivan et al. (2012): their reporter construct contained only the upstream region (−1962 to −549) of the *Cdc25A* promoter, but no core promoter, that is, neither the *Cdc25A* core promoter nor a heterologous core promoter, so that it is unclear how and where transcription is initiated (Sullivan et al., 2012). This reporter construct was fourfold activated by FOXM1c, which *in vivo* occupied the upstream region of the *Cdc25A* promoter, but it is unknown how FOXM1c could activate transcription despite the absence of a core promoter (Sullivan et al., 2012).

The FOXM1c-mediated activation of this reporter construct was unaffected by P/CAF and CBP whereas it was barely enhanced (approx. 1.7-fold) by p300 (Sullivan et al., 2012). Although the significance of this unusual transient transfection experiment is questionable it might indicate that P/CAF and CBP do not act as coactivators for FOXM1c (Sullivan et al., 2012). In addition, the weak effect of p300 (Sullivan et al., 2012) might argue against a coactivator role of p300 for FOXM1c.

5.5. The TRD

5.5.1 The TRD of FOXM1c

FOXM1c possesses a strong TRD, which was mapped to aa 359–425 (Fig. 3.8A; Wierstra, 2013a; Wierstra & Alves, 2006a). The high potency of the FOXM1c-TRD is demonstrated by its ability to transrepress even against transactivation by the very strong TAD of VP16 (virion protein 16), a viral transactivator of HSV (Wierstra & Alves, 2006a). The TRD of FOXM1c functions independently of RB because the FOXM1c-TRD can efficiently transrepress in RB-deficient human SAOS-2 osteosarcoma cells (Wierstra, 2013a; Wierstra & Alves, 2006a).

The TRD of FOXM1c acts as an ID for the TAD of FOXM1c because the transrepression by the FOXM1c-TRD counteracts the transactivation by the FOXM1c-TAD (Fig. 3.3; Wierstra, 2013a; Wierstra & Alves, 2006a).

FOXM1c proteins, which encompass the C-terminal TAD as well as the central TRD, are strong transactivators, demonstrating that the transactivation potential of the FOXM1c-TAD is higher than the transrepression potential of the FOXM1c-TRD (Wierstra, 2013a; Wierstra & Alves, 2006a). Therefore, FOXM1c is a transactivator (Wierstra, 2013a; Wierstra & Alves, 2006a).

Accordingly, FOXM1c (Alvarez-Fernandez, Halim, et al., 2010; Alvarez-Fernandez et al., 2011; Anders et al., 2011; Karadedou et al., 2012; Korver, Roose, & Clevers, 1997; Laoukili, Alvarez, et al., 2008; Laoukili, Alvarez-Fernandez, et al., 2008; Laoukili et al., 2005; Leung et al., 2001; Li et al., 2008; Littler et al., 2010; Lüscher-Firzlaff et al., 1999; Ma, Tong, et al., 2005; Ma et al., 2010; Madureira et al., 2006; Sullivan et al., 2012; Wierstra, 2011a, 2011b, 2013a; Wierstra & Alves, 2006a, 2006b, 2006c, 2006d, 2007a, 2007b, 2008) and the splice variant FoxM1B (Balli et al., 2012, 2013, 2011; Bhat et al., 2009a, 2009b, Chen, Dominguez-Brauer, et al., 2009; Chen, Müller, et al., 2013; Dai et al., 2007; Fu et al., 2008; Gemenetzidis et al., 2009; Huang et al., 2012; Kalin et al., 2008; Kalinichenko et al., 2004; Kim et al., 2005; Li, Peng, et al., 2013; Li et al., 2009; Ma, Tong, et al., 2005; Major et al., 2004; Malin et al., 2007; Park et al., 2012; Park, Kim, et al., 2008; Park, Wang, et al., 2008; Petrovic et al., 2008; Radhakrishnan et al., 2006; Ren et al., 2013, 2010; Sengupta et al., 2013; Tan et al., 2010, 2007, 2006; Ustiyan et al., 2012; Wang, Chen, et al., 2008; Wang et al., 2005; Wang, Kiyokawa, et al., 2002; Wang, Meliton, et al., 2009, 2008; Wang, Park, et al., 2011; Wang, Quail, et al., 2001; Wang, Snyder, et al., 2012; Wang, Zhang, Snyder, et al., 2010; Xie et al., 2010; Ye et al., 1997; Zhang et al., 2011, 2008; Zhou, Liu, et al., 2010; Zhou, Wang, et al., 2010) were always found to be transcriptional activators.

5.5.2 Molecular targets of the FOXM1c-TRD
5.5.2.1 The TRD of FOXM1c probably targets the basal transcription complex
The TRD of FOXM1c does probably not recruit corepressors or chromatin remodeling enzymes but instead interferes directly with the basal transcription complex (Wierstra & Alves, 2007b).

These two transrepression mechanisms can be discriminated through the effect of the TRD *in trans* on the TRD *in cis* (Wierstra & Alves, 2007b): an excess of the isolated FOXM1c-TRD (TRD *in trans*) was added to a transactivating FOXM1c protein, which contained both the TAD and the TRD (TRD *in cis*) (Wierstra & Alves, 2007b). If the TRD of FOXM1c would recruit a corepressor or a chromatin remodeling enzyme for transrepression, the TRD *in trans* could titrate this corepressor or chromatin remodeling enzyme away from the TRD *in cis* so that the addition of the TRD *in trans* would decrease the transrepression by the TRD *in cis*, which would result in an increased transactivation by the FOXM1c protein with the TRD *in cis* because the transactivation through its TAD would no longer be counteracted by the transrepression through its TRD (Wierstra & Alves, 2007b). However, the addition of the isolated FOXM1c-TRD *in trans* (aa 359–425) reduced the transactivation by FOXM1c proteins with the TRD *in cis*, demonstrating that the TRD of FOXM1c recruits neither corepressors nor chromatin remodeling enzymes for transrepression (Wierstra & Alves, 2007b). Instead, the FOXM1c-TRD seems to interfere directly with the basal transcription complex (Wierstra & Alves, 2007b).

Accordingly, the central domain of FOXM1c interacts with five components of the basal transcription apparatus, namely, directly with TBP, TFIIB, and TFIIAα/β as well as possibly indirectly with TFIIAγ and TAF1 (TBP-associated factor 1, KAT4) (Wierstra & Alves, 2006d). Consequently, the central TRD of FOXM1c may directly interfere with the basal transcription complex through its interactions with these five components of the basal transcription machinery (TBP, TFIIB, TFIIAα/β, TFIIAγ, TAF1) (Wierstra & Alves, 2007b). In contrast, the central domain of FOXM1c does not interact with TFIIEα (Wierstra & Alves, 2006d) so that the central TRD of FOXM1c cannot target TFIIEα for transrepression (Wierstra & Alves, 2007b).

5.5.2.2 The direct interaction of the TRD with itself is involved in the transrepression by the FOXM1c-TRD

The FOXM1c-TRD *in trans* (aa 359–425) reduced the transactivation by those FOXM1c proteins, which contained the TRD *in cis*, whereas it left unaffected a FOXM1c deletion mutant, which lacked the TRD *in cis* (Wierstra & Alves, 2007b). Thus, the repression of FOXM1c proteins by the FOXM1c-TRD *in trans* depends on the presence of the FOXM1c-TRD *in cis*, which suggests that the TRD *in trans* reinforces the transrepression by the TRD *in cis* (Wierstra & Alves, 2007b). Actually, the central domain of FOXM1c interacts directly with itself (Fig. 3.8A) so that the FOXM1c-TRD *in trans* can bind to the FOXM1c-TRD *in cis* (Wierstra & Alves, 2007b). Since this binding of the TRD *in trans* to the TRD *in cis* results in increased transrepression by the FOXM1c-TRD *in cis*, which manifests in a decreased transcriptional activity of the FOXM1c protein with the TRD *in cis*, the direct interaction of the central domain with itself (Fig. 3.8A) seems to be involved in the transrepression by the TRD of FOXM1c (Wierstra & Alves, 2007b). Hence, the transrepression by the FOXM1c-TRD appears to include its direct intramolecular interaction with itself (Fig. 3.8A; Wierstra & Alves, 2007b).

5.5.2.3 Summary
In summary, the central RB-independent TRD of FOXM1c (Wierstra, 2013a; Wierstra & Alves, 2006a) could transrepress by targeting TBP, TFIIB, and TFIIα/β (Wierstra & Alves, 2007b) because the central domain of FOXM1c binds directly to these three components of the basal transcription complex (TBP, TFIIB, TFIIα/β) (Wierstra & Alves, 2007b). The TRD of FOXM1c functions as an ID for the TAD of FOXM1c because the transrepression by the FOXM1c-TRD counteracts the transactivation by the FOXM1c-TAD (Fig. 3.3; Wierstra, 2013a; Wierstra & Alves, 2006a).

5.6. The NRD-C
The tumor suppressor RB binds directly to the central domain of FOXM1c (Wierstra, 2013a; Wierstra & Alves, 2006b) but does not interact with the C-terminal TAD of FOXM1c (Fig. 3.3; Wierstra & Alves, 2006b). Nonetheless, RB represses the transactivation by the FOXM1c-TAD (Fig. 3.3; Wierstra, 2013a; Wierstra & Alves, 2006b, 2008) whereas it does not affect the DNA binding by the FOXM1c–DBD (Wierstra & Alves, 2006b). RB binds directly to the central domain of FOXM1c (Fig. 3.3; Wierstra, 2013a; Wierstra & Alves, 2006b) and thereby represses indirectly the transactivation by the C-terminal TAD of FOXM1c (Fig. 3.3; Wierstra, 2013a; Wierstra & Alves, 2006b, 2008). Since the central domain of FOXM1c recruits the corepressor RB, which in turn indirectly represses the FOXM1c-TAD, the central domain functions as a RB-recruiting negative-regulatory domain (NRD) for the TAD and therefore is named NRD-C (Fig. 3.3; Wierstra & Alves, 2006b). The NRD-C functions as an ID for the TAD of FOXM1c through recruitment of the corepressor RB (Fig. 3.3; Wierstra, 2013a; Wierstra & Alves, 2006b, 2008).

In addition, the central domain of FOXM1c is a RB-independent TRD (Wierstra, 2013a; Wierstra & Alves, 2006a) so that the central domain represents a dual ID for the FOXM1c-TAD because it functions as both RB-independent TRD and RB-recruiting NRD-C (Fig. 3.3). Whether the central domain of FOXM1c can simultaneously work as TRD and NRD-C or whether its functions as TRD and NRD-C are mutually exclusive is presently unknown.

5.7. The NRD-N

The N-terminus of FOXM1c is a dominant ID for the TAD of FOXM1c because it inhibits the TAD completely (Fig. 3.3; Wierstra & Alves, 2006a). Consequently, all FOXM1c proteins, which contain the autoinhibitory N-terminus, are inactive (Wierstra, 2011a, 2011b; Wierstra & Alves, 2006a, 2006b, 2006c, 2008).

This inhibition of the FOXM1c-TAD by the N-terminus is caused by a direct interaction between the N-terminus and the TAD of FOXM1c (Fig. 3.8A; Wierstra & Alves, 2006a), which blocks the TAD completely (Fig. 3.3; Wierstra, 2013a; Wierstra & Alves, 2006a, 2006b, 2008). Probably, the binding of the N-terminus to the TAD prevents those interactions of the FOXM1c-TAD with the basal transcription machinery, coactivators, or/and chromatin remodeling enzymes, which are required for transactivation, so that the N-terminus blocks the TAD by masking the interaction domains for the targets of the FOXM1c-TAD. Thus, the autoinhibitory N-terminus of FOXM1c inhibits the FOXM1c-TAD completely (Fig. 3.3; Wierstra, 2013a; Wierstra & Alves, 2006a, 2006b, 2008) through its direct interaction with the TAD (Fig. 3.8A; Wierstra & Alves, 2006a).

Therefore, the N-terminus works as a central on/off switch for FOXM1c so that FOXM1c remains inactive unless its N-terminus is disabled (Wierstra & Alves, 2006a). This disabling of the N-terminus can be accomplished either by activating signals *in vivo* (Wierstra, 2011b, 2013a; Wierstra & Alves, 2006b, 2006c, 2008) or through deletion of the N-terminus *in vitro* (Wierstra, 2011a, 2011b; Wierstra & Alves, 2006a, 2006b, 2006c, 2008).

The autoinhibitory N-terminus of FOXM1c is named NRD-N (Figs. 3.3 and 3.8A; Wierstra & Alves, 2006a). The NRD-N functions as an ID for the TAD of FOXM1c because the NRD-N blocks the TAD completely (Fig. 3.3) by directly binding to the TAD (Fig. 3.8A; Wierstra & Alves, 2006a). Due to the resulting inactivity of wild-type FOXM1c (Wierstra & Alves, 2006a) FOXM1c deletion mutants without NRD-N are often used in experiments because inactive full-length FOXM1c can be activated through deletion of the NRD-N (Wierstra, 2011a, 2011b; Wierstra & Alves, 2006a, 2006b, 2006c, 2008).

Since the NRD-N inhibits the FOXM1c-TAD through its direct interaction with the TAD (Fig. 3.8A; Wierstra & Alves, 2006a) this inhibition takes place not only within one FOXM1c molecule (Fig. 3.3), but also if the NRD-N and the FOXM1c-TAD are distributed over two separate proteins

(Wierstra, 2013a; Wierstra & Alves, 2006a, 2006b, 2006d, 2008): an excess of the isolated NRD-N (NRD-N *in trans*) inhibits the transactivation by a Gal-FOXM1c fusion protein, which consists of the TAD (aa 721–762) of FOXM1c (FOXM1c-TAD *in cis*) and the heterologous DBD of yeast GAL4 (Wierstra & Alves, 2006a). Hence, the NRD-N *in trans* inhibits the FOXM1c-TAD *in cis* (Wierstra, 2013a; Wierstra & Alves, 2006a, 2006b, 2006d, 2008). This experiment demonstrates that the NRD-N blocks the TAD (Fig. 3.3) through the direct interaction between the NRD-N and the TAD of FOXM1c (Fig. 3.8A) and that this inhibition of the TAD by the NRD-N does not require other domains of FOXM1c (Wierstra & Alves, 2006a).

Accordingly, the mapping of the participating interaction domains within FOXM1c showed that the N-terminus (aa 1–232) binds directly to the TAD (aa 721–762; Wierstra & Alves, 2006a) and even to the C-terminal half (aa 744–762) of the TAD (Fig. 3.8A). In contrast, the N-terminus (aa 1–232) does not bind to aa 195–596 of FOXM1c so that the NRD-N interacts neither with the DBD nor with the central domain of FOXM1c (Fig. 3.8A; Wierstra & Alves, 2006a).

The NRD-N does not possess any transrepression potential so that it is no TRD (Wierstra & Alves, 2006a).

The NRD-N inhibits the TAD of FOXM1c independent of RB because the FOXM1c-TAD is efficiently repressed by the isolated NRD-N *in trans* in RB-deficient SAOS-2 cells (Wierstra & Alves, 2006a). Accordingly, all FOXM1c proteins with the NRD-N *in cis* are inactive in SAOS-2 cells because their TAD is inhibited by the NRD-N independent of RB (Wierstra & Alves, 2006a).

In summary, the NRD-N binds directly to the TAD of FOXM1c (Fig. 3.8A; Wierstra & Alves, 2006a) and their interaction blocks the TAD so that the NRD-N functions as an ID for the TAD (Fig. 3.3; Wierstra, 2013a; Wierstra & Alves, 2006a, 2006b, 2006d, 2008). Since the NRD-N inhibits the TAD completely (Wierstra & Alves, 2006a) wild-type FOXM1c is inactive (Wierstra, 2011a, 2011b; Wierstra & Alves, 2006a, 2006b, 2006c, 2006d, 2007b, 2008).

5.8. The three IDs for the TAD
5.8.1 Three IDs inhibit the very strong TAD of FOXM1c
FOXM1c possesses three IDs for the TAD (Wierstra & Alves, 2007c), namely, the central TRD (Wierstra, 2013a; Wierstra & Alves, 2006a, 2007b), the central NRD-C (Wierstra, 2013a; Wierstra & Alves, 2006b, 2008), and the

N-terminal NRD-N (Fig. 3.3; Wierstra, 2013a; Wierstra & Alves, 2006a, 2006b, 2008). Since the central domain of FOXM1c functions as both RB-independent TRD (Wierstra, 2013a; Wierstra & Alves, 2006a) and RB-recruiting NRD-C (Wierstra, 2013a; Wierstra & Alves, 2006b) it represents a dual ID for the TAD (Fig. 3.3; Wierstra & Alves, 2007c).

The N-terminus and the central domain of FOXM1c act as IDs for the TAD (Fig. 3.3) independently from each other (Wierstra & Alves, 2006a) and through different mechanisms (Wierstra, 2013a; Wierstra & Alves, 2006a, 2006b, 2007b, 2007c, 2008): first, the N-terminus inhibits the TAD completely, but the central domain represses the TAD only partially (Wierstra & Alves, 2006a). Second, the N-terminus binds directly to the TAD (Wierstra & Alves, 2006a) whereas the central domain does not interact with the TAD (Fig. 3.8A; Wierstra & Alves, 2007b). Third, in contrast to the strong central TRD (Wierstra, 2013a; Wierstra & Alves, 2006a), the NRD-N is no TRD because it lacks any transrepression potential (Wierstra & Alves, 2006a). Fourth, the NRD-C binds directly to RB (Fig. 3.3; Wierstra, 2013a; Wierstra & Alves, 2006b) and recruits RB as a corepressor (Fig. 3.3; Wierstra & Alves, 2006b) whereas both the TRD (Wierstra, 2013a; Wierstra & Alves, 2006a) and the NRD-N (Wierstra & Alves, 2006a) work independently of RB.

FOXM1c shows two intramolecular interactions (Fig. 3.8A; Wierstra & Alves, 2006a, 2007b). The N-terminus binds directly to the TAD (Wierstra & Alves, 2006a) and the central domain interacts directly with itself (Fig. 3.8A; Wierstra & Alves, 2007b). The direct interaction of the central TRD with itself is involved in the transrepression by the TRD (Wierstra & Alves, 2007b). The direct binding of the NRD-N to the TAD causes the complete inhibition of the TAD by the NRD-N (Fig. 3.3; Wierstra, 2013a; Wierstra & Alves, 2006a, 2006b, 2006d, 2008).

Because of its very strong acidic TAD, FOXM1c is principally a very potent transactivator (Wierstra, 2013a; Wierstra & Alves, 2006a). However, wild-type FOXM1c is inactive because the TAD is repressed by three IDs, that is, the NRD-N, the TRD, and the NRD-C (Fig. 3.3; Wierstra & Alves, 2006a, 2006b). In order to uncover the very high transactivation potential of the TAD, the TAD must be released from its inhibition by the three IDs (Wierstra, 2011a, 2011b, 2013a; Wierstra & Alves, 2006a, 2006b, 2006c, 2007b, 2008). Therefore, activating signals, which disable the IDs for the TAD, can convert FOXM1c from an inactive form into a very strong transactivator (Wierstra, 2011b, 2013a; Wierstra & Alves, 2006b, 2006c, 2008).

5.8.2 The biological principle of the tight control of the FOXM1c-TAD by three IDs

The object behind the tight control of the FOXM1c-TAD by three IDs (Fig. 3.3) is to prevent the inappropriate activity of this very potent TAD when the expression of FOXM1 target genes is not desired or even harmful (Wierstra & Alves, 2006a, 2006b, 2007c). Since FOXM1 stimulates cell proliferation and promotes cell cycle progression at the G1/S- and G2/M-transitions (Ackermann Misfeldt et al., 2008; Ahmad et al., 2010, 2011; Anders et al., 2011; Bao et al., 2011; Barsotti & Prives, 2009; Bergamaschi et al., 2011; Bhat et al., 2011; Bolte et al., 2011; Brezillon et al., 2007; Calvisi et al., 2009; Carr et al., 2010; Chan et al., 2008; Chen, Müller, et al., 2013; Chen et al., 2010, 2011; Chu et al., 2012; Dai et al., 2013; Davis et al., 2010; Faust et al., 2012; Fu et al., 2008; Gusarova et al., 2007; Ho et al., 2012; Huang & Zhao, 2012; Kalin et al., 2006; Kalinichenko et al., 2003, 2004; Kim et al., 2005, 2006; Krupczak-Hollis et al., 2003, 2004; Laoukili, Alvarez, et al., 2008; Laoukili et al., 2005; Lefebvre et al., 2010; Leung et al., 2001; Li et al., 2011; Liu, Gampert, et al., 2006; Liu et al., 2011; Lok et al., 2011; Mencalha et al., 2012; Millour et al., 2010, 2011; Monteiro et al., 2012; Nakamura, Hirano, et al., 2010; Ning et al., 2012; Park, Carr, et al., 2009; Park, Costa, et al., 2008; Park, Wang, et al., 2008; Park et al., 2012; Pellegrino et al., 2010; Raghavan et al., 2012; Ramakrishna et al., 2007; Schüller et al., 2007; Sengupta et al., 2013; Uddin et al., 2011; Ueno et al., 2008; Ustiyan et al., 2009, 2012; Wang, Ahmad, Banerjee, et al., 2010; Wang, Banerjee, et al. 2007; Wang, Chen, et al., 2008; Wang, Hung, et al., 2001; Wang, Kiyokawa, et al., 2002; Wang, Krupczak-Hollis, et al., 2002; Wang, Meliton, et al., 2008; Wang, Park, et al., 2011; Wang, Quail, et al., 2001; Wang, Teh, Ji, et al., 2010; Wang et al., 2005; Wierstra & Alves, 2006d; Wonsey & Follettie, 2005; Wu, Liu, et al., 2010; Xia, Mo, et al., 2012; Xia et al., 2009; Xie et al., 2010; Xue et al., 2010, 2012; Ye et al., 1999; Yoshida et al., 2007; Zeng et al., 2009; Zhang, Ackermann, et al., 2006; Zhang, Wu, et al., 2012; Zhang et al., 2011, 2009; Zhao et al., 2006) its aberrant activity could lead to tumorigenesis (see Part II of this two-part review, that is see Wierstra, 2013b) (Costa, Kalinichenko, Major, et al., 2005; Costa, Kalinichenko, Tan, & Wang, 2005; Kalin et al., 2011; Koo et al., 2011; Laoukili et al., 2007; Myatt & Lam, 2007; Raychaudhuri & Park, 2011; Wierstra & Alves, 2007c). Consequently, the very strong FOXM1c-TAD must be kept in check by three IDs (Fig. 3.3; Wierstra & Alves, 2006a, 2006b, 2007c). This concept of

FOXM1c regulation is strongly supported by the given fact that the NRD-C recruits the tumor suppressor RB (Fig. 3.3; Wierstra, 2013a; Wierstra & Alves, 2006b), which in turn indirectly represses the TAD of FOXM1c (Fig. 3.3; Wierstra, 2013a; Wierstra & Alves, 2006b, 2008).

Obviously, when the expression of FOXM1 target genes is needed, appropriate signals must be capable of activating the inactive wild-type FOXM1c (Wierstra & Alves, 2006b, 2006c, 2007c). Accordingly, several activating signals can release the FOXM1c-TAD from its inhibition by the NRD-N and from its repression by RB bound to the NRD-C so that these activating signals are able to transform inactive wild-type FOXM1c into a very potent transactivator (Wierstra, 2011b, 2013a; Wierstra & Alves, 2006b, 2006c, 2008). These activating signals are the kinases cyclin D1/Cdk4, cyclin E/Cdk2, cyclin A/Cdk2, cyclin A/Cdk1, c-Src, PKA (protein kinase A, cAMP (cyclic adenosine monophosphate)-dependent protein kinase), and CK2 (protein kinase CK2, CKII (casein kinase II)) (Wierstra, 2011b, 2013a; Wierstra & Alves, 2006b, 2006c, 2008). Yet, a signal, which can overcome the transrepression by the TRD of FOXM1c, has so far not been described.

5.9. The molecular characterization of FOXM1 as a conventional transcription factor

The conventional transcription factor FOXM1c possesses five functional domains (Wierstra & Alves, 2007c), namely, the forkhead DBD, the acidic TAD, the autoinhibitory NRD-N, the RB-recruiting NRD-C, and the RB-independent TRD (Fig. 3.3). These functional domains were characterized by Wierstra and Alves (2006a, 2006b, 2006c, 2007b, 2008) and Wierstra (2011a, 2011b, 2013a).

Gal-FOXM1 fusion proteins bind to DNA via their heterologous yeast GAL4–DBD independent of the FOXM1–DBD so that they represent only the transactivation potential of FOXM1 but not the DNA-binding activity of FOXM1. Moreover, Gal-FOXM1 fusion proteins are localized to the nucleus through their heterologous NLS (nuclear localization signal) of yeast GAL4 (Lohr, Venkov, & Zlatanova, 1995; Nelson & Silver, 1986; Silver, Brent, & Ptashne, 1986; Silver, Keegan, & Ptashne, 1984) independent of the genuine NLSs of FOXM1. Therefore, Gal-FOXM1 fusion proteins allow the discrimination of the transactivation potential of FOXM1 from its DNA-binding activity and its nuclear translocation. This discrimination is mandatory for the definition of the TAD and the TRD as well as for the identification of the IDs for the TAD, that is, the NRD-N and the NRD-C. Since only Wierstra and Alves (2006a, 2006b, 2006c, 2008) and Wierstra (2011b, 2013a) analyzed Gal-FOXM1 fusion proteins—whereas FOXM1 was never fused to a heterologous DBD in other studies—only Wierstra and Alves (2006a, 2006b, 2006c, 2008) and Wierstra (2011b, 2013a) defined the TAD, the TRD, the NRD-N, and the NRD-C of FOXM1.

In vitro DNA-binding studies with purified FOXM1 and conventional FOXM1-binding sites were performed by several groups, namely, for FoxM1A (Ye et al., 1997), FoxM1B (Hedge et al., 2011; Ye et al., 1997), FOXM1c (Hedge et al., 2011; Korver, Roose, & Clevers, 1997; Littler et al., 2010; Wierstra, 2011a; Wierstra & Alves, 2006a, 2006c, 2007b), and rat WIN (class a transcript) (Yao et al., 1997). These studies defined the forkhead domain as DBD of FOXM1 (Korver, Roose, & Clevers, 1997; Littler et al., 2010; Wierstra, 2011a; Wierstra & Alves, 2006a; Yao et al., 1997; Ye et al., 1997).

The conventional transcription factor FOXM1c engages in two direct intramolecular interactions (Fig. 3.8A), which were first described by Wierstra and Alves (2006a, 2006b, 2007b, 2008): the N-terminus binds directly to the TAD (Fig. 3.8A; Wierstra & Alves, 2006a) and the central domain interacts directly with itself (Fig. 3.8A; Wierstra & Alves, 2007b).

After the publications of Wierstra & Alves (2006a, 2006b, 2006c, 2006d, 2007c, 2008), two other groups repeated a few of their experiments with the autoinhibitory N-terminus of FOXM1 (Laoukili, Alvarez, et al., 2008; Park, Wang, et al., 2008), confirming the previous findings of Wierstra & Alves (2006a, 2006b, 2006c, 2006d, 2007c, 2008).

These two repetition studies revealed that aa 1–138 in the N-terminus of FoxM1B (corresponding to aa 1–137 of FOXM1c) can interact with the C-terminus of FoxM1B (Fig. 3.8A; Park, Wang, et al., 2008) and that the interaction of the FOXM1c N-terminus with the remainder of FOXM1c involves R-716, L-718, L-722, and L-724 (corresponding to R-715, L-717, L-721, and L-723 of FOXM1c with 762 aa) in the C-terminus of FOXM1c (Laoukili, Alvarez, et al., 2008), which are either part of or located immediately adjacent to the FOXM1c-TAD (aa 721–762; Fig. 3.8A). Hence, the findings of Laoukili, Alvarez et al. (2008a) and Park, Wang, et al. (2008a) verified the previous result of Wierstra and Alves

(2006a) that the N-terminus (aa 1–232) of FOXM1c binds directly to the TAD (aa 721–762) of FOXM1c (Fig. 3.8A).

5.10. The autoinhibitory N-terminus of FOXM1

The NRD-N serves as a dominant on/off switch for FOXM1c because it inhibits the TAD completely (Fig. 3.3) by directly binding to the TAD (Fig. 3.8A; Wierstra & Alves, 2006a). Consequently, wild-type FOXM1c is inactive, but deletion of the NRD-N is sufficient to activate the inactive full-length FOXM1c (Wierstra, 2011a, 2011b; Wierstra & Alves, 2006a, 2006b, 2006c, 2006d, 2007b, 2008). Conversely, addition of an excess of the NRD-N *in trans* represses the transactivation by the FOXM1c-TAD *in cis* (Wierstra, 2013a; Wierstra & Alves, 2006a, 2006b, 2006d, 2008).

The central importance of the NRD-N as an ID for the TAD of FOXM1c (Fig. 3.3; Wierstra & Alves, 2006a) was also documented *in vivo*:

The deletion of the N-terminus of FoxM1B increased its positive effects on the *Skp2* (*S-phase kinase-associated protein 2*) mRNA expression in murine NIH3T3 fibroblasts and on the Aurora B (Aurora B kinase) protein expression in human U2OS osteosarcoma cells (Park, Wang, et al., 2008).

Conversely, overexpression of the isolated FoxM1B N-terminus decreased the mRNA levels of the direct FOXM1 target genes *Skp2*, *Cdc25B*, and *Aurora B* in NIH3T3 fibroblasts (Park, Wang, et al., 2008).

The deletion of the N-terminus of FoxM1B enhanced its positive effects on normal cell proliferation, anchorage-independent growth on soft agar, growth in low serum media, and invasion ability of NIH3T3 fibroblasts (Park, Wang, et al., 2008).

The overexpression of a FOXM1 mutant with a deletion of the N-terminus prevented the 4-OHT (4-hydroxytamoxifen)-induced G1-phase cell cycle arrest in human MCF-7 breast carcinoma cells whereas overexpression of full-length FOXM1 failed to do so (Millour et al., 2010).

The overexpression of a FOXM1c mutant with a deletion of the N-terminus readily bypassed senescence in HMFs (human mammary fibroblasts) whereas the overexpression of full-length FOXM1c had no effect (Rovillain et al., 2011). These HMFs had first been immortalized by hTERT (human telomerase reverse transcriptase) and SV40 (simian virus 40) large T antigen and were then induced to senesce by inactivation of SV40 large T (Rovillain et al., 2011).

IFN-β (interferon-β) + mezerein induce the terminal differentiation of human HO-1 metastatic melanoma cells (Fisher, Prignoli, Hermo, Weinstein, & Pestka, 1985).

IFN-β + mezerein treatment decreased the fraction of HO-1 cells in S-phase, and increased that in G1-phase (Huynh et al., 2010). Overexpression of a FOXM1 mutant without N-terminus prevented this IFN-β/mezerein-induced decrease in the percentage of S-phase cells whereas overexpression of wild-type FOXM1 had no effect (Huynh et al., 2010). However, neither FOXM1 protein affected the cell cycle distribution of untreated HO-1 cells (Huynh et al., 2010).

IFN-β + mezerein treatment caused apoptosis of HO-1 cells (Huynh et al., 2010). Overexpression of the FOXM1 mutant without N-terminus, but not wild-type FOXM1, slightly alleviated this IFN-β/mezerein-induced increase in the number of apoptotic cells (Huynh et al., 2010). Yet, the rate of apoptosis in untreated HO-1 cells was unaffected by both FOXM1 proteins (Huynh et al., 2010).

The above-mentioned observations that FOXM1 develops some biological effects only if its N-terminus is deleted entail two difficulties for the interpretation of experiments with FOXM1. On the one hand, the failure of wild-type FOXM1 to exert a certain biological effect does not imply a general inability of FOXM1 to bring about this effect because it may still give rise to this effect if the inhibition of FOXM1 by its own autoinhibitory N-terminus is somehow overcome. On the other hand, the finding that a FOXM1 deletion mutant without N-terminus exerts a certain biological effect does not imply that wild-type FOXM1 will also bring about this effect because it remains unproven whether the cell manages to overcome the inhibition of FOXM1 by its own autoinhibitory N-terminus.

5.11. Summary of the function of FOXM1c as a conventional transcription factor

As a conventional transcription factor, FOXM1c binds with its forkhead DBD to a conventional FOXM1c-binding site upstream (or downstream) of a non-FOXM1c-responsive core promoter (Fig. 3.3; Korver, Roose, & Clevers, 1997; Littler et al., 2010; Wierstra, 2011a; Wierstra & Alves,

2006a). FOXM1c activates the transcription of target genes through its very strong acidic TAD (Fig. 3.3; Wierstra, 2013a; Wierstra & Alves, 2006a).

Yet, the prerequisite for the transactivation of a promoter by FOXM1c is that the very potent TAD is released from its inhibition by the three IDs (Wierstra & Alves, 2006a, 2006b), which otherwise repress the TAD (Fig. 3.3) so that wild-type FOXM1c remains inactive (Wierstra, 2011a, 2011b, 2013a; Wierstra & Alves, 2006a, 2006b, 2006c, 2007b, 2008) until activating signals switch of the three IDs (Wierstra, 2011b, 2013a; Wierstra & Alves, 2006b, 2006c, 2008). The three IDs for the TAD are the autoinhibitory NRD-N (Wierstra, 2013a; Wierstra & Alves, 2006a, 2006b, 2008), the RB-recruiting NRD-C (Wierstra, 2013a; Wierstra & Alves, 2006b, 2008), and the RB-independent TRD (Wierstra, 2013a; Wierstra & Alves, 2006a, 2007b; Fig. 3.3). This tight control of the FOXM1c-TAD by three IDs versus activating signals allows an ingenious fine-tuning of the conventional transcription factor FOXM1c (Wierstra & Alves, 2006a, 2006b, 2006c, 2007c).

Several cellular signals are known to regulate FOXM1c by targeting the IDs for the TAD (Wierstra, 2011b, 2013a; Wierstra & Alves, 2006b, 2006c, 2008). In accordance with the role of FOXM1 in stimulation of cell proliferation (see below) and with its implication in tumorigenesis (see Part II of this two-part review, that is see Wierstra, 2013b) (Costa, 2005; Costa et al., 2003; Costa, Kalinichenko, Major, et al., 2005; Costa, Kalinichenko, Tan, et al., 2005; Gong & Huang, 2012; Kalin et al., 2011; Koo et al., 2011; Laoukili et al., 2007; Mackey, Singh, & Darlington, 2003; Myatt & Lam, 2007; Raychaudhuri & Park, 2011; Wierstra & Alves, 2007c), these signals include proliferation signals (cyclin E, cyclin A, Cdk2, Cdk1, CK2) and antiproliferation signals (p21, p27) as well as protooncoproteins (cyclin D1, Cdk4, c-Src) and tumor suppressors (RB, p16) (Wierstra, 2011b, 2013a; Wierstra & Alves, 2006b, 2006c, 2007c, 2008).

6. THE FOXM1 EXPRESSION

6.1. FOXM1 displays a proliferation-specific expression pattern

FOXM1 is a typical proliferation-associated transcription factor (Wierstra & Alves, 2007c). Accordingly, FOXM1 exhibits a strictly proliferation-specific expression pattern so that it is expressed in all proliferating cells but not in resting cells (Costa et al., 2003; Costa, Kalinichenko, & Lim, 2001; Kalin et al., 2011; Koo et al., 2011; Laoukili et al., 2007; Murakami, Aiba, Nakanishi, &

Murakami-Tonami, 2010; Raychaudhuri & Park, 2011; Wierstra & Alves, 2007c).

FOXM1 is ubiquitously expressed in the developing embryo but its expression is restricted to a few proliferating cell types and self-renewing tissues in adults (Bolte et al., 2011; Chaudhary et al., 2000; Kalin et al., 2011, 2008; Kalinichenko et al., 2003; Korver, Roose, & Clevers, 1997; Korver, Roose, Wilson, & Clevers, 1997; Pohl, Rössner, & Knöchel, 2004; Schüller et al., 2007; Sengupta et al., 2013; Ueno et al., 2008; Ustiyan et al., 2009; Wang, Zhang, Snyder, et al., 2010; Yao et al., 1997; Ye et al., 1997; Yokomine et al., 2009; Zhang, Ackermann, et al., 2006). FOXM1 is generally not expressed in resting cells, like quiescent cells (Balli et al., 2012; Bolte et al., 2011; Chaudhary et al., 2000; Coller, Sang, & Roberts, 2006; Down et al., 2012; Faust et al., 2012; Kalin et al., 2008; Kalinichenko et al., 2003; Kalinichenko, Lim, Shin, & Costa, 2001; Kalinina et al., 2003; Korver, Roose, & Clevers, 1997; Küppers et al., 2010; Laoukili et al., 2005; Leung et al., 2001; Li et al., 2008; Matsuo et al., 2003; Mirza et al., 2010; Newick et al., 2012; Park, Carr, et al., 2009; Park, Costa, Lau, Tyner, & Raychaudhuri, 2008; Petrovic et al., 2008; Ren et al., 2013, 2010; Sengupta et al., 2013; Tan et al., 2006; Wang, Chen, et al., 2008; Wang, Hung, et al., 2001; Wang, Zhang, Snyder, et al., 2010; Xue et al., 2010; Yao et al., 1997; Ye et al., 1997) and senescent cells (Hardy et al., 2005; Rovillain et al., 2011) and terminally differentiated cells (Ahn, Lee, et al., 2004; Bose et al., 2012; Carr et al., 2012; Gemenetzidis et al., 2010; Huynh et al., 2009, 2010; Korver, Roose, Wilson, et al., 1997; Prots, Skapenko, Lipsky, & Schulze-Koops, 2011; Takahashi et al., 2005; Wang, Park, et al., 2011; Wierstra & Alves, 2006d; Xie et al., 2010; Ye et al., 1997). Yet, its expression can be reinduced in quiescent cells upon their reentry into the cell cycle in response to various proliferation signals (Blanco-Bose et al., 2008; Bräuning et al., 2011; Chang et al., 2004; Cicatiello et al., 2004; Down et al., 2012; Karadedou, 2006; Korver, Roose, & Clevers, 1997; Korver, Roose, Heinen, et al., 1997; Korver, Roose, Wilson, et al., 1997; Krupczak-Hollis et al., 2003; Laoukili et al., 2005; Ledda-Columbano et al., 2004; Leung et al., 2001; Li et al., 2008; Masumoto et al., 2007; Millour et al., 2010; Newick et al., 2012; Park, Carr, et al., 2009; Petrovic et al., 2008; Takizawa et al., 2011; Wang, Chen, et al., 2008; Ye et al., 1997; Zhang et al., 2010) or in the course of adult tissue regeneration after injury (Ackermann Misfeldt et al., 2008; Chen et al., 2010; Gieling et al., 2010; Huang & Zhao, 2012; Kalinichenko et al., 2003, 2001; Krupczak-Hollis et al., 2003; Kurinna et al., 2013; Lehmann et al., 2012; Liu et al., 2011; Matsuo et al., 2003; Meng et al., 2011, 2010; Mukhopadhyay et al., 2011; Ren et al., 2010; Tan et al., 2006; Wang et al., 2003; Wang, Hung, et al., 2001; Wang, Kiyokawa, et al., 2002; Wang, Krupczak-Hollis, et al., 2002; Wang, Quail, et al., 2001; Weymann et al., 2009; Ye et al., 1999, 1997; Zhang, Li, Yu, Forman, & Huang, 2012; Zhang et al., 2012; Zhao et al., 2006).

6.2. The expression of FOXM1 during the life span

In the developing embryo, FOXM1 is ubiquitously expressed whereas, in adults, the expression of FOXM1 is limited to self-renewing tissues and a few proliferating cell types (Bolte et al., 2011; Chaudhary et al., 2000; Kalin et al., 2011, 2008; Kalinichenko et al., 2003; Korver, Roose, & Clevers, 1997; Korver, Roose, Wilson, et al., 1997; Pohl et al., 2004; Schüller et al., 2007; Sengupta et al., 2013; Ueno et al., 2008; Ustiyan et al., 2009; Wang, Zhang, Snyder, et al., 2010; Yao et al., 1997; Ye et al., 1997; Yokomine et al., 2009; Zhang, Ackermann, et al., 2006). Accordingly, the *foxm1* mRNA expression decreases during prenatal lung maturation (from E17.5 to PN0) (Xu, Wang, et al., 2012). In adult mammals, the expression of FOXM1 is high in colon, small intestine, testis, and thymus but low in lung, kidney, spleen, and ovary, so that tissues with a large percentage of proliferating cells (colon, small intestine, testis, thymus) display a high FOXM1 expression whereas tissues with a small percentage of proliferating cells (lung, kidney, spleen, ovary) exhibit a low FOXM1 expression (Chaudhary et al., 2000; Kalin et al., 2008; Kalinichenko et al., 2003; Korver, Roose, & Clevers, 1997; Yao et al., 1997; Ye et al., 1997; Yokomine et al., 2009). Hence, the FOXM1 expression levels correlate with the size of the proliferating cell population in the individual tissue.

The *foxm1* mRNA expression decreases during human aging (Fig. 3.10; Ly et al., 2000). The analysis of transcripts in actively dividing, early-passage fibroblasts, which were derived from young, middle-age, or old-age human donors, revealed a strong downregulation of *foxm1* mRNA expression during aging because the *foxm1* mRNA level was 3.3-fold and 9-fold lower in fibroblasts from middle-age or old-age humans, respectively, than in fibroblasts from young humans (Ly et al., 2000).

In accordance with this age-related decline of the *foxm1* mRNA expression during human aging (Ly et al., 2000), the *foxm1* mRNA level in the heart was significantly decreased (6.5-fold) in old mice (16 month) compared to young adult mice (Bolte et al., 2012).

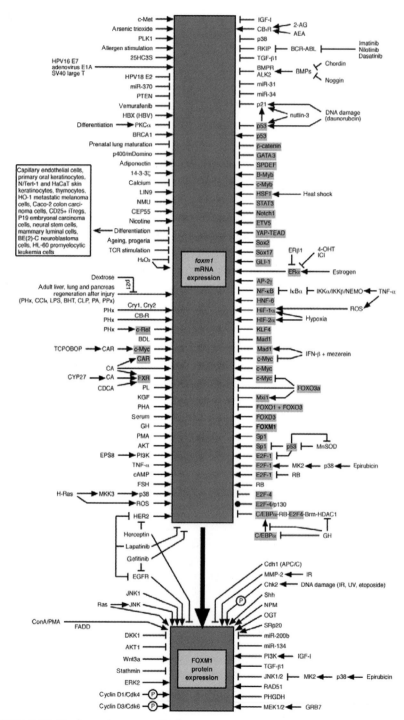

Figure 3.10 See legend on opposite page.

Figure 3.10 Control of the FOXM1 expression. Shown are the factors, which regulate the expression of *foxm1* at the mRNA or/and protein level. Transcription factors are marked in gray and their effects on the *foxm1* expression are summarized in Table 3.2. The top left box lists those differentiation systems, in which the *foxm1* expression was downregulated during differentiation. Phosphorylation events are depicted as an encircled P (i.e., a white circle with a P). CAR (constitutive androstane receptor) and FXR (farnesoid X receptor) are nuclear receptors. Cholic acid (CA) is a bile acid. Nuclear receptor (CAR, FXR)-dependent bile acid (CA) signaling is required for normal liver regeneration after PHx (partial hepatectomy). The xenobiotic and nongenotoxic carcinogen TCPOBOP (1,4-bis[2-(3,5-dichloropyridyloxy)]benzene), a halogenated hydrocarbon, is a synthetic ligand for CAR. **References:** Korver, Roose, and Clevers (1997) (serum), Korver, Roose, Heinen, et al. (1997) (serum), Korver, Roose, Wilson, et al. (1997) (PHA, PMA, differentiation of thymocytes), Ye et al. (1997) (KGF, H_2O_2, PHx, differentiation of Caco-2 colon carcinoma cells toward intestinal enterocytes), Ye et al. (1999) (PHx), Chaudhary et al. (2000) (FSH, cAMP), Ly, Lockhart, Lerner, and Schultz (2000) (aging, progeria), Kalinichenko et al. (2001) (BHT), Leung et al. (2001) (serum), Matsushima-Nishiu et al. (2001) (PTEN), Wang, Hung, et al. (2001) (CCl_4), Wang, Quail, et al. (2001) (PHx), Teh et al. (2002) (GLI-1), Wang, Krupczak-Hollis, et al. (2002) (PHx), Wang, Kiyokawa, et al. (2002) (PHx), Fernandez et al. (2003) (c-Myc), Kalinichenko et al. (2003) (BHT), Krupczak-Hollis et al. (2003) (PHx, GH), Matsuo et al. (2003) (PHx, Cry1, Cry2), Ahn, Urist, et al. (2004) (differentiation of neural stem cells), Cicatiello et al. (2004) (estrogen), Huang, Li, Winoto, and Robey (2004) (TCR stimulation), Ledda-Columbano et al. (2004) (TCPOBOP), Thierry et al. (2004) (HPV18 E2), Bae et al. (2005) (BRCA1), Laoukili et al. (2005) (serum, FOXO3a), Takahashi et al. (2005) (FGF-2 (fibroblast growth factor-2)-induced differentiation of capillary endothelial cells), Untergasser et al. (2005) (TGF-β1), Bhonde et al. (2006) (p53), Huang et al. (2006) (CA, CAR, FXR), Karadedou (2006) (estrogen, ICI), Spurgers et al. (2006) (p53), Takahashi et al. (2006) (NMU), Tan et al. (2006) (PHx, HNF-6), Wierstra and Alves (2006d) (TPA (12-O-tetradecanoylphorbol-13-acetate)-induced differentiation of human promyelocytic HL-60 leukemia cells toward macrophages), Zhao et al. (2006) (TNF-α, H_2O_2, LPS), Chang et al. (2007b) (miR-34), Delpuech et al. (2007) (c-Myc, Mxi1, FOXO3a), He, He, Lim, et al. (2007), (miR-34), Keller et al. (2007) (c-Myc), Markey et al. (2007) (RB), Masumoto et al. (2007) (GH), Osborn, Sohn, and Winoto (2007) (ConA/PMA, FADD), Schüller et al. (2007) (Shh), Wang, Shi, et al. (2007) (E2F-1, p130, C/EBPα, E2F-4, RB, Brm, GH), Woodfield et al. (2007) (AP-2γ), Ackermann Misfeldt et al. (2008) (PPx), Blanco-Bose et al. (2008) (TCPOBOP, c-Myc), Halasi and Gartel (2009) (FOXM1), Hallstrom et al. (2008) (E2F-1), Laoukili, Alvarez-Fernandez, et al. (2008) (Cdh1, APC/C), Park, Costa, Lau, Tyner, and Raychaudhuri (2008c) (Cdh1, APC/C), Ueno et al. (2008) (BMPs, BMPR, ALK2, Noggin, Chordin), Wang, Chen, et al. (2008) (serum, JNK1), Wang, Salisbury, et al. (2008) (HDAC1, C/EBPα, E2F-4, RB, Brm, GH), Knight, Notaridou, and Watson (2009) (LIN9), Zou et al. (2008) (FOXO3a), Adam et al. (2009) (p38), Barsotti and Prives (2009) (p53, p21, nutlin-3, daunorubicin, E2F-1, pocket proteins), Calvisi et al. (2009) (ERK2, GLI-1), Chen, Chien, et al. (2009) (CEP55), Chetty et al. (2009) (MMP-2, IR), Chua, Yip, and Bay (2009) (H_2O_2), Francis et al. (2009) (HER2, lapatinib), Gemenetzidis et al. (2009) (nicotine), Huynh et al. (2009) (differentiation of HO-1 human metastatic melanoma cells induced by IFN-β (interferon-β) + mezerein), Lange et al. (2009) (Sox17), McGovern et al. (2009) (gefitinib, EGFR, FOXO3a), Pandit et al. (2009) (p53), Park, Carr, et al. (2009) (serum, H_2O_2, H-Ras, ROS, JNK, AKT1), Penzo et al. (2009)

(Continued)

Figure 3.10—Cont'd (TNF-α, IKKα, IKKβ, NEMO, IκBα, NF-κB), Weymann et al. (2009) (PHx, dextrose, p21), Xia et al. (2009) (HIF-1α, hypoxia), Behren et al. (2010) (H-Ras, MKK3, p38), Berger, Rome, Vega, Ciancia, and Vidal (2010) (adiponectin), Caldwell et al. (2010) (OGT), Carr et al. (2010) (Herceptin), Chen et al. (2010) (PHx, CCl_4, FXR, CDCA, C/EBPα), Christoffersen et al. (2010) (miR-34), Fujii, Ueda, Nagata, and Funaga (2010) (p400/mDomino), Gemenetzidis et al. (2010) (differentiation of primary oral keratinocytes), Gieling et al. (2010) (PHx, c-Rel), Huynh et al. (2010) (IFN-β + mezerein, c-Myc, Mad1, PKCα, differentiation of HO-1 human metastatic melanoma cells induced by IFN-β + mezerein), Jia, Li, McCoy, Deng, and Zheng (2010) (SRp20), Jin et al. (2010) (C/EBPα), Lefebvre et al. (2010) (c-Myb), Lorvellec et al. (2010) (B-Myb), Meng et al. (2010) (FXR, CCl_4), Millour et al. (2010) (ERα, estrogen, 4-OHT, ICI), Pellegrino et al. (2010) (Ha-Ras), Petrovic et al. (2010) (Sp1), Ren et al. (2010) (CCl_4), Sanchez-Calderon et al. (2010) (IGF-I), Takemura et al. (2010) (RKIP, BCR-ABL, imatinib, nilotinib, dasatinib), Tan et al. (2010) (p53, IR, etoposide), Wang, Teh, Ji, et al. (2010) (EPS8, PI3K, FOXM1, nicotine), Wang, Li, Ahmad, et al. (2010) (AKT, PTEN, Notch1), Wu, Liu, et al. (2010) (p53), Xie et al. (2010) (FOXM1, RA (retinoic acid)-induced neural differentiation of mouse P19 embryonal carcinoma cells), Zhang et al. (2010) (PL), Anders et al. (2011) (cyclin D1/Cdk4, cyclin D3/Cdk6, Cdh1), Bao et al. (2011) (miR-200b), Bergamaschi et al. (2011) (14-3-3τ), Bhat et al. (2011) (NPM), Bräuning et al. (2011) (TCPOBOP, β-catenin), Horimoto et al. (2011) (ERα, ERβ1), Kim et al. (2011) (H_2O_2), Liu et al. (2011) (PA), Lok et al. (2011) (p53), Meng et al. (2011) (PHx, FXR, CA, CYP27), Millour et al. (2011) (E2F-1, RB, p53, p21), Mitra et al. (2011) (stathmin), Mukhopadhyay et al. (2011) (PHx, CB_1R, AEA, 2-AG), Plank et al. (2011) (FOXD3), Prots et al. (2011) (IL-4 (interleukin-4)-induced differentiation of CD25 + iTregs (induced regulatory T-cells)), Rovillain et al. (2011) (E2F, p53, p21 HPV16 E7, adenovirus E1A, SV40 large T antigen), Takizawa et al. (2011) (CAR, TCPOBOP), Tompkins et al. (2011) (Sox2), Wang, Park, et al. (2011) (RA-induced differentiation of BE(2)-C neuroblastoma cells toward neuronal lineage), Wang, Lin, et al. (2011) (calcium), Yang et al. (2011) (BDL), Zhang et al. (2011) (Wnt3a, DKK1), Aburto, Magarinos, Leon, Varela-Nieto, and Sanchez-Calderon (2012) (IGF-I, PI3K), Balli et al. (2012) (BHT (FOXM1 in macrophages)), Bellelli et al. (2012) (E2F-4, p21, p53), Bonet et al. (2012) (vemurafenib), Bose et al. (2012) (PKC, differentiation of HaCaT keratinocytes (spontaneously immortalized human skin keratinocytes) induced by suspension culture in methylcellulose (anchorage deprivation), differentiation of N/Tert-1 keratinocytes (hTERT- and p16 loss-immortalized normal human skin keratinocytes) induced by methylcellulose suspension culture, PMA (phorbol-12-myristate-13-acetate)-induced differentiation of N/Tert-1 keratinocytes, differentiation of N/Tert-1 keratinocytes induced by high cell density), Carr et al. (2012) (GATA3, RB, differentiation of mammary luminal cells), Chan et al. (2012) (GRB7, MEK1/2), Chen, Müller, et al. (2012) (LIN9), Dai et al. (2013) (heat shock, HSF1), Dean, McClendon, Stengel, and Knudsen (2012) (RB), Demirci et al. (2012) (c-Met), de Olano et al. (2012) (E2F-1, H_2O_2, p38, MK2, JNK1/2, epirubicin), Dibb et al. (2012) (PLK1), Down et al. (2012) (FOXM1, B-Myb, p130, serum), Halasi and Gartel (2012) (IR), He et al. (2013) (FOXO3a), Ho et al. (2012) (N-Ras, AKT), Huang and Zhao (2012) (CLP), Karadedou et al. (2012) (FOXO3a, lapatinib), Lehmann et al. (2012) (PHx), Liu, Guo, Li, et al. (2012) (PHGDH), Llaurado et al. (2012) (ETV5), Li, Jia, et al. (2012) (KLF4), Li, Wang, et al. (2012) (TGF-β1, miR-134), Lin et al. (2013) (miR-31), Lynch et al. (2012) (OGT), Mencalha et al. (2012) (STAT3, imatinib), Mizuno et al. (2012) (YAP, TEAD),

(Continued)

Figure 3.10—Cont'd Newick et al. (2012) (serum), Raghavan et al. (2012) (HIF-1α, HIF-2α, hypoxia), Ren et al. (2012) (SPDEF, allergen stimulation), Sadavisam et al. (2012) (B-Myb), Sengupta et al. (2013) (FOXO1 + FOXO3), Stoyanova et al. (2012) (p21), Xia, Huang, et al. (2012) (HBX (HBV)), Xia, Mo, et al. (2012) (TNF-α, HIF-1α, ROS), Xu, Zhang, et al. (2012) (gefitinib), Xu, Wang, et al. (2012) (prenatal lung maturation), Zhang, Li, et al. (2012) (FXR, CDCA, LPS), Zhang, Wang, et al. (2012) (FXR, PHx, CCl$_4$), Zhang, Bai, et al. (2012) (25HC3S). Zhang, Zeng, et al. (2012) (miR-370), Zhang, Wu, et al. (2012) (RAD51), Chen et al. (2013) (MnSOD, p53, E2F-1, Sp1), Kurinna et al. (2013) (p53, PHx), Liu et al. (2013) (arsenic trioxide), Qu et al. (2013) (p53). **Abbreviations:** P, phosphorylation; ALK2, activin receptor-like kinase 2, BMP7 (bone morphogenetic protein 7) receptor; AEA, arachidonoyl ethanolamide, anandamide; 2-AG, 2-arachidonoylglycerol; AP-2γ, activator protein-2γ; APC/C, anaphase-promoting complex/cyclosome; BDL, bile duct ligation in liver; BHT, BHT (butylated hydroxytoluene) lung injury; BCR-ABL, breakpoint cluster region-Abelson; BMP, bone morphogenetic protein; BMPR, BMP receptor; BRCA1, breast cancer-associated gene 1; Brm, Brahma, SMARCA4 (SWI/SNF-related, matrix-associated, actin-dependent regulator of chromatin, subfamily A, member 4); CA, cholic acid; cAMP, cyclic adenosine monophosphate; CAR, constitutive androstane receptor; CB$_1$R, cannabinoid type 1 receptor; CCl$_4$, CCl$_4$ (carbon tetrachloride) liver injury; CDCA, chenodeoxycholic acid; Cdh1, Cdc20 homolog 1; Cdk, cyclin-dependent kinase; C/EBPα, CCAAT/enhancer-binding protein α; CEP55, centrosomal protein 55-kDa; Chk2, checkpoint kinase 2; CLP, CLP (cecal ligation and puncture)-induced lung vascular injury; ConA, concanavalin A; Cry, Cryptochrome; CYP27, sterol 27-hydroxylase; dasatinib, Sprycel, BMS354825; DKK1, Dickkopf 1; EGFR, EGF (epidermal growth factor) receptor, ErβB1; EPS8, EGFR pathway substrate 8; ER, estrogen receptor; ERK2, extracellular signal-regulated kinase 2; ETV5, ETS variant 5, ERM; FOX, Forkhead box; FADD, Fas-associated death domain; FSH, follicle-stimulating hormone; FXR, farnesoid X receptor; gefitinib, Iressa, ZD1839; GH, growth hormone; GLI-1, glioma transcription factor-1; GRB7, growth factor receptor-bound protein 7; HBV, hepatitis B virus; HBX, HBV X protein; 25HC3S, 5-cholesten-3b, 25-diol-3-sulfate; HDAC1, histone deacetylase 1; HER2, human EGF receptor 2, ErβB2, Neu; Herceptin, Trastuzumab; HIF, hypoxia-inducible factor; HNF-6, hepatocyte nuclear factor-6; HPV, human papillomavirus; HSF1, heat shock factor 1; ICI, ICI182780, fulvestrant, Faslodex; IGF-I, insulin-like growth factor-I; IκB, inhibitor of NF-κB; IKK, IκB kinase; imatinib, Gleevec, STI571, imatinib mesylate; IR, ionizing radiation, γ-irradiation; JNK, c-Jun N-terminal kinase; KGF, keratinocyte growth factor; KLF4, Krüppel-like factor 4; lapatinib, Tykerb, GW572016; LIN9, cell lineage-abnormal 9; LPS, LPS (lipopolysaccharide)-induced acute vascular lung injury; Mad1, Max dimerizer 1, MXD1 (Max dimerization protein 1); c-Met, HGFR (HGF (hepatocyte growth factor) receptor); MK2, MAPKAPK2 (MAPK (mitogen-activated protein kinase)-activated protein kinase 2); MKK3, MAPK kinase kinase 3; MMP-2, matrix metalloproteinase-2; MnSOD, manganese superoxide dismutase; Mxi1, Max interactor 1; c-Myb, MYB; B-Myb, MYBL2 (MYB-like 2); c-Myc, MYC; NEMO, NF-κB essential modifier, IKKγ; NF-κB, nuclear factor-κB; nilotinib, Tasigna, AMN107; NMU, neuromedin U; NPM, nucleophosmin, B23; OGT, O-GlcNAc (O-linked b-N-acetylglucosamine) transferase; 4-OHT, 4-hydroxytamoxifen; p21, CDKN1A (Cdk inhibitor 1A); p130, RBL2 (retinoblastoma-like 2); p53, TP53, tumor protein p53; PA, *Pseudomonas aeruginosa*-induced lung alveolar injury; PHA, phytohemagglutinin; PHx, partial hepatectomy; PKCα, protein kinase Cα; PHGDH, phosphoglycerate dehydrogenase; PI3K, phosphatidylinositol-3 kinase; PL, placental lactogen; PLK1, Polo-like kinase 1; PMA,

(Continued)

6.3. The expression of FOXM1 during the cell cycle
6.3.1 The proliferation-specific expression of FOXM1
In all proliferating cells, which were ever analyzed for FOXM1 expression, FOXM1 was found to be abundantly expressed (Costa et al., 2003, 2001; Costa, Kalinichenko, Major, et al., 2005; Kalin et al., 2011; Laoukili et al., 2007; Murakami et al., 2010; Wierstra & Alves, 2007c). Since FOXM1 is expressed in all proliferating cells, but not in resting cells (i.e., quiescent, senescent, and terminally differentiated cells), the FOXM1 expression is restricted to actively dividing cells (Costa et al., 2003, 2001; Kalin et al., 2011; Koo et al., 2011; Laoukili et al., 2007; Murakami et al., 2010; Raychaudhuri & Park, 2011; Wierstra & Alves, 2007c). Thus, FOXM1 displays a strictly proliferation-specific expression pattern (Costa et al., 2003, 2001; Kalin et al., 2011; Koo et al., 2011; Laoukili et al., 2007; Murakami et al., 2010; Raychaudhuri & Park, 2011; Wierstra & Alves, 2007c). Accordingly, the expression of FOXM1 is strongly induced upon cell cycle entry from quiescence, whereas it ceases upon cell cycle exit into quiescence, senescence, or terminal differentiation (see below).

6.3.2 The cell cycle-dependent oscillation of the foxm1 expression in proliferating cells
In proliferating cells, *foxm1* is expressed in a cell cycle-dependent manner. During the entry of quiescent cells into the cell cycle, the expression of *foxm1* is induced in G1-phase (Chaudhary et al., 2000; Down et al., 2012; Korver, Roose, & Clevers, 1997; Leung et al., 2001; Petrovic et al., 2008; Wang, Chen, et al., 2008; Wang, Hung, et al., 2001; Ye et al., 1999, 1997). Thereafter, the *foxm1* expression continues throughout S-, G2-, and M-phase (Alvarez-Fernandez et al., 2011; Chen, Dominguez-Brauer, et al., 2009; Dibb et al., 2012; Down et al., 2012; Fu et al., 2008; Grant et al., 2012; Korver, Roose, Heinen, et al., 1997;

Figure 3.10—Cont'd phorbol-12-myristate-13-acetate; PPx, partial pancreatectomy; PTEN, phosphatase and tensin homolog on chromosome 10; RB, retinoblastoma protein, RB1 (retinoblastoma 1), p105; c-Rel, REL; RKIP, Raf kinase inhibitor protein; ROS, reactive oxygen species; Shh, Sonic hedgehog; Sox, SRY box, SRY-related HMG (high mobility group)-box; Sp1, specificity protein 1; SPDEF, SAM pointed domain-containing ETS transcription factor, PDEF (prostate-derived ETS transcription factor); SRp20, SFRS3 (splicing factor, arginine/serine-rich 3); STAT3, signal transducer and activator of transcription 3; SV40, simian virus 40; TCPOBOP, 1,4-bis[2-(3,5-dichloropyridyloxy)]benzene; TEAD, TEAD1 (TEA domain family member 1); TCR, T-cell receptor; TGF-β1, transforming growth factor-β1; TNF-α, tumor necrosis factor-α; vemurafenib, PLX4032, RG7204, RO5185426; Wnt, wingless-type; YAP, Yes-associated protein.

Laoukili, Alvarez, et al., 2008; Laoukili, Alvarez-Fernandez, et al., 2008; Laoukili et al., 2005; Park, Costa, Lau, Tyner, & Raychaudhuri, 2008; Sadavisam et al., 2012; Sullivan et al., 2012; Ye et al., 1997) until the FOXM1 protein is degraded by the 26S proteasome in late mitosis and early G1-phase after its ubiquitylation by the E3 ubiquitin ligase APC/C^{Cdh1} (anaphase-promoting complex/cyclosome) (Laoukili, Alvarez-Fernandez, et al., 2008; Park, Costa, Lau, Tyner, & Raychaudhuri, 2008).

Thus, the expression of *foxm1* oscillates in a cell cycle-dependent manner.

The *foxm1* mRNA and protein expression reaches its maximum in G2/M-phase (Bar-Joseph et al., 2008; Dibb et al., 2012; Down et al., 2012; Gemenetzidis et al., 2010; Grant et al., 2012; Laoukili, Alvarez, et al., 2008; Laoukili, Alvarez-Fernandez, et al., 2008; Leung et al., 2001; Li et al., 2008; Ma, Tong, et al., 2005; Park, Costa, Lau, Tyner, & Raychaudhuri, 2008; Sadavisam et al., 2012; Sullivan et al., 2012; Wan et al., 2012; Whitfield et al., 2002). In contrast, to this general peak of the *foxm1* mRNA and protein levels in G2/M-phase, some studies reported that their differences between G1/S- and G2/M-cells are rather small (Alvarez-Fernandez et al., 2011; Laoukili, Alvarez, et al., 2008; Laoukili, Alvarez-Fernandez, et al., 2008; Ma, Tong, et al., 2005).

Large-scale gene expression analyses in human cells revealed that *foxm1* is a cell cycle-regulated gene (Whitfield, George, Grant, & Perou, 2006), which is periodically expressed during the cell cycle in synchronized human HeLa cervical carcinoma cells (Whitfield et al., 2002) and synchronized primary human foreskin fibroblasts (Bar-Joseph et al., 2008) with peak expression of FOXM1 in G2/M-phase.

Moreover, in a cross-species analysis combining sequence and expression data in order to identify a core set of cycling cell cycle genes, FOXM1 was grouped into the cell cycle conservation set CCC3 with conservation across three species (human, budding yeast, fission yeast) (Lu et al., 2007).

6.3.3 foxm1 *represents a delayed-early gene*
Serum-starved cells that are in G0-phase or quiescence do (almost) not express *foxm1* mRNA and FOXM1 protein (Down et al., 2012; Korver, Roose, & Clevers, 1997; Laoukili et al., 2005; Leung et al., 2001; Li et al., 2008; Newick et al., 2012; Park, Carr, et al., 2009; Park, Costa, Lau, Tyner, & Raychaudhuri, 2008; Petrovic et al., 2008; Wang, Chen,

et al., 2008). Serum stimulation of such starved cells strongly induces the expression of *foxm1* mRNA and FOXM1 protein (Fig. 3.10; Chang et al., 2004; Down et al., 2012; Korver, Roose, & Clevers, 1997; Laoukili et al., 2005; Leung et al., 2001; Li et al., 2008; Newick et al., 2012; Park, Carr, et al., 2009; Petrovic et al., 2008; Wang, Chen, et al., 2008). Moreover, serum stimulation of starved Rat1 fibroblasts potently activated the *foxm1* promoter (Korver, Roose, Heinen, et al., 1997). The proximal region (−296 to +60) of the TATA-less *foxm1* promoter was sufficient for its response to serum refeeding (Korver, Roose, Heinen, et al., 1997). Accordingly, *foxm1* is one of the genes in the common gene expression signature of the fibroblast serum response (Chang et al., 2004).

Although the requirement of *de novo* protein biosynthesis for the serum-induced transcription of *foxm1* in serum-starved cells has never been examined, *foxm1* is considered a delayed-early gene (Costa et al., 2003; Mackey et al., 2003; Ye et al., 1999, 1997) because of the kinetics of its serum-induced reexpression after serum starvation (Chang et al., 2004; Down et al., 2012; Korver, Roose, & Clevers, 1997; Korver, Roose, Heinen, et al., 1997; Laoukili et al., 2005; Leung et al., 2001; Li et al., 2008; Newick et al., 2012; Park, Carr, et al., 2009; Petrovic et al., 2008; Wang, Chen, et al., 2008). In particular, after serum refeeding of starved immortalized MEFs, the serum-induced *foxm1* mRNA expression paralleled that of the well-known delayed-early gene *cyclin D1* (Petrovic et al., 2008).

6.3.4 The cell cycle-dependent expression of foxm1 in continuously proliferating cells

With regard to the cell cycle-dependent expression of *foxm1* in continuously cycling cells several not mutually exclusive scenarios have been suggested:

First, in continuously proliferating cells, the FOXM1 mRNA and protein expression may be rather constant during the whole cell cycle (Korver, Roose, Heinen, et al., 1997): Korver, Roose, Heinen, et al. (1997) mentioned that the *foxm1* mRNA and protein levels appear to be invariant when exponentially growing cells progress through the cell cycle.

Second, only the expression of the splice variant FoxM1B may oscillate in a cell cycle-dependent manner, whereas the expression of the splice variants FOXM1c and FoxM1A may remain rather constant throughout the cell cycle (Gemenetzidis et al., 2010): when normal telomerase-immortalized N/TERT keratinocytes and metastatic HeLa cervical carcinoma cells were synchronized at G1/S-phase by double-thymidine block and then released from this cell cycle block, a dramatic increase in the *foxm1b* mRNA level but only modest increases in the *foxm1c* and *foxm1a* mRNA levels were observed (Gemenetzidis et al., 2010). Therefore, it seems conceivable that the expression of the splice variants FOXM1c and FoxM1A remains rather constant throughout the cell cycle, whereas the expression of the splice variant FoxM1B strongly oscillates in a cell cycle-dependent manner (Gemenetzidis et al., 2010).

Third, in continuously cycling cells, a basal FOXM1 protein expression throughout the cell cycle may be combined with an additional cell cycle-dependent oscillation of the FOXM1 protein expression above this basal level (Grant et al., 2012): Grant et al. (2012) analyzed the FOXM1 protein expression in U2OS cells after their synchronization at G1/S by a double-thymidine block not only during the first cell cycle after release from this block, but also during the second cell cycle, which more closely resembles continuously proliferating cells. In both successive cell cycle rounds, the FOXM1 protein level peaked in late G2-phase and the FOXM1 protein was degraded in M-phase (Grant et al., 2012). Notably, the FOXM1 protein expression continued during the intervening G1-phase, although at a moderate level, and the FOXM1 protein level re-increased in the following S-phase (Grant et al., 2012). This finding suggests that, in continuously proliferating cells, a basal FOXM1 protein expression is maintained during the whole cell cycle and the FOXM1 protein expression additionally oscillates in a cell cycle-dependent manner above this basal level (Grant et al., 2012).

Together, these three observations (Gemenetzidis et al., 2010; Grant et al., 2012; Korver, Roose, Wilson, et al., 1997) point to the possibility that, in continuously cycling cells, a rather low basal FOXM1 mRNA and protein level is maintained throughout the whole cell cycle and that, additionally, the FOXM1 mRNA and protein expression increases above this basal level in a cell cycle-dependent manner, reaching its maximum in G2/M-phase. In this scenario, the splice variant FOXM1c (and possibly FoxM1A) might contribute the continuous basal *foxm1* expression whereas the splice variant FoxM1B might add the cell cycle-dependent oscillation of the *foxm1* level.

In summary, FOXM1 is expressed in a cell cycle-dependent manner in proliferating cells.

6.4. The expression of FOXM1 upon entry into the cell cycle

Quiescent cells generally do not express *foxm1* (Koo et al., 2011; Laoukili et al., 2007; Murakami et al., 2010; Wierstra & Alves, 2007c). However, in accordance with its strictly proliferation-specific expression pattern, the expression of FOXM1 is reinduced in quiescent cells when they reenter the cell cycle in response to proliferative stimuli (Costa et al., 2003, 2001; Kalin et al., 2011; Koo et al., 2011; Laoukili et al., 2007; Murakami et al., 2010; Raychaudhuri & Park, 2011; Wierstra & Alves, 2007c). This *foxm1* reexpression occurs during the regenerative proliferation for adult tissue repair after injury (see below) (Ackermann Misfeldt et al., 2008; Chen et al., 2010; Gieling et al., 2010; Huang & Zhao, 2012; Kalinichenko et al., 2003, 2001; Krupczak-Hollis et al., 2003; Kurinna et al., 2013; Lehmann et al., 2012; Liu et al., 2011; Matsuo et al., 2003; Meng et al., 2011, 2010; Mukhopadhyay et al., 2011; Ren et al., 2010; Tan et al., 2006; Wang et al., 2003; Wang, Hung, et al., 2001; Wang, Kiyokawa, et al., 2002; Wang, Krupczak-Hollis, et al., 2002; Wang, Quail, et al., 2001; Weymann et al., 2009; Ye et al., 1999, 1997; Zhang, Li, et al., 2012; Zhang, Wang, et al., 2012; Zhao et al., 2006) as well as in response to various proliferation signals, which cause quiescent cells to enter the cell cycle (see below) (Blanco-Bose et al., 2008; Bräuning et al., 2011; Chang et al., 2004; Cicatiello et al., 2004; Down et al., 2012; Karadedou, 2006; Korver, Roose, & Clevers, 1997; Korver, Roose, Heinen, et al., 1997; Korver, Roose, Wilson, et al., 1997; Krupczak-Hollis et al., 2003; Laoukili et al., 2005; Ledda-Columbano et al., 2004; Leung et al., 2001; Li et al., 2008; Masumoto et al., 2007; Millour et al., 2010; Newick et al., 2012; Park, Carr, et al., 2009; Petrovic et al., 2008; Takizawa et al., 2011; Wang, Chen, et al., 2008; Ye et al., 1997; Zhang et al., 2010).

6.4.1 Induction of FOXM1 expression upon cell cycle reentry of quiescent cells

In quiescent cells, the expression of *foxm1* can be reinduced by several proliferation signals that cause the reentry of these cells into the cell cycle (Blanco-Bose et al., 2008; Bräuning et al., 2011; Chang et al., 2004; Cicatiello et al., 2004; Down et al., 2012; Karadedou, 2006; Korver, Roose, & Clevers, 1997; Korver, Roose, Heinen, et al., 1997; Korver, Roose, Wilson, et al., 1997; Krupczak-Hollis et al., 2003; Laoukili et al., 2005; Ledda-Columbano et al., 2004; Leung et al., 2001; Li et al., 2008; Masumoto et al., 2007; Millour et al., 2010; Newick et al., 2012; Park, Carr, et al., 2009; Petrovic et al., 2008; Takizawa et al., 2011; Wang, Chen, et al., 2008; Ye et al., 1997; Zhang et al., 2010). Such proliferation signals, which reactivate the FOXM1 expression in quiescent cells upon their cell cycle reentry, include serum (Chang et al., 2004; Down et al., 2012; Korver, Roose, & Clevers, 1997; Korver, Roose, Heinen, et al.,

1997; Laoukili et al., 2005; Leung et al., 2001; Li et al., 2008; Newick et al., 2012; Park, Carr, et al., 2009; Petrovic et al., 2008; Wang, Chen, et al., 2008), KGF (keratinocyte growth factor) (Ye et al., 1997), GH (growth hormone) (Krupczak-Hollis et al., 2003; Masumoto et al., 2007), TCPOBOP (1,4-bis[2-(3,5-dichloropyridyloxy)]benzene) (Blanco-Bose et al., 2008; Bräuning et al., 2011; Ledda-Columbano et al., 2004; Takizawa et al., 2011), PL (placental lactogen) (Zhang et al., 2010), estrogen (Cicatiello et al., 2004; Karadedou, 2006; Millour et al., 2010), PHA (phytohemagglutinin), and PMA (phorbol-12-myristate-13-acetate) (Fig. 3.10; Korver, Roose, Wilson, et al., 1997).

Accordingly, the FOXM1 expression is also increased by other extracellular or intracellular proliferative stimuli and transmembrane receptors that can lead to cell cycle entry of resting cells, namely, by Wnt3a (Zhang et al., 2011), Shh (Sonic hedgehog) (Schüller et al., 2007), c-Met (HGFR (HGF (hepatocyte growth factor) receptor)) (Demirci et al., 2012), EGFR (EGF (epidermal growth factor) receptor, HER1 (human EGF receptor 1), ErβB1) (McGovern et al., 2009), HER2/Neu (ErβB2) (Francis et al., 2009), H-Ras (Behren et al., 2010; Park, Carr, et al., 2009; Pellegrino et al., 2010), N-Ras (Ho et al., 2012), ERK2 (Calvisi et al., 2009), PLK1 (Polo-like kinase 1) (Dibb et al., 2012), cyclin D1/Cdk4, and cyclin D3/Cdk6 (Fig. 3.10; Anders et al., 2011).

6.4.2 Reinduction of FOXM1 expression for adult tissue repair after injury

Some adult organs (e.g., liver, lung, pancreas) possess a regenerative capacity, which allows tissue repair following injury through regenerative cell proliferation (Poss, 2010; Stoick-Cooper, Moon, & Weidinger, 2007; Zaret & Grompe, 2008).

In accordance with the restriction of *foxm1* expression to proliferating cells (Koo et al., 2011; Laoukili et al., 2007; Murakami et al., 2010; Wierstra & Alves, 2007c), *foxm1* is not expressed in quiescent cells of such uninjured organs (Balli et al., 2012; Chaudhary et al., 2000; Kalin et al., 2008; Kalinichenko et al., 2003, 2001; Kalinina et al., 2003; Matsuo et al., 2003; Ren et al., 2013, 2010; Tan et al., 2006; Wang, Hung, et al., 2001; Wang, Zhang, Snyder, et al., 2010; Yao et al., 1997; Ye et al., 1997). Yet, in these quiescent cells of adult organs (liver, lung, pancreas), the expression of *foxm1* is strongly reinduced in the course of regenerative cell proliferation during tissue repair after injury (Ackermann Misfeldt et al., 2008; Chen et al., 2010; Gieling et al., 2010; Huang & Zhao, 2012;

Kalinichenko et al., 2003, 2001; Krupczak-Hollis et al., 2003; Kurinna et al., 2013; Lehmann et al., 2012; Liu et al., 2011; Matsuo et al., 2003; Meng et al., 2011, 2010; Mukhopadhyay et al., 2011; Ren et al., 2010; Tan et al., 2006; Wang et al., 2003; Wang, Hung, et al., 2001; Wang, Kiyokawa, et al., 2002; Wang, Krupczak-Hollis, et al., 2002; Wang, Quail, et al., 2001; Weymann et al., 2009; Ye et al., 1999, 1997; Zhang, Li, et al., 2012; Zhang, Wang, et al., 2012; Zhao et al., 2006). In general, the *foxm1* expression is limited to the proliferation period and correlates with the proliferation rate and time of the individual cell types. Thus, *foxm1* is transiently reexpressed in quiescent cells of adult liver (Chen et al., 2010; Gieling et al., 2010; Krupczak-Hollis et al., 2003; Kurinna et al., 2013; Lehmann et al., 2012; Matsuo et al., 2003; Meng et al., 2011, 2010; Mukhopadhyay et al., 2011; Ren et al., 2010; Tan et al., 2006; Wang et al., 2003; Wang, Hung, et al., 2001; Wang, Kiyokawa, et al., 2002; Wang, Krupczak-Hollis, et al., 2002; Wang, Quail, et al., 2001; Weymann et al., 2009; Ye et al., 1999, 1997; Zhang, Wang, et al., 2012), lung (Huang & Zhao, 2012; Kalinichenko et al., 2003, 2001; Liu et al., 2011; Zhang, Li, et al., 2012; Zhao et al., 2006), and pancreas (Ackermann Misfeldt et al., 2008), when they reenter the cell cycle for tissue regeneration following injury (Fig. 3.10). This injury-induced reactivation of the *foxm1* expression for adult tissue repair has been studied extensively:

The adult liver has the amazing capacity to restore its original mass after PHx (partial hepatectomy), that is, surgical removal of two-thirds of the liver (Böhm et al., 2010; Costa et al., 2003; Diehl & Rai, 1996; Fausto, 2000, 2004; Fausto, Campbell, & Riehle, 2006; Fausto, Laird, & Webber, 1995; Fausto & Webber, 1993; Koniaris, McKillop, Schwartz, & Zimmers, 2003; Michalopoulos, 2007, 2010; Michalopoulos & DeFrances, 1997; Steer, 1995; Taub, 1996, 2004). During liver regeneration following PHx, approx. 95% of the normally quiescent hepatocytes in the remnant liver reenter the cell cycle, undergo one or two rounds of DNA replication and subsequent cell division, and then return to quiescence. This compensatory hepatocyte proliferation restores the normal liver size.

PHx induces strong FOXM1 mRNA and protein expression in adult mouse livers (Fig. 3.10; Chen et al., 2010; Gieling et al., 2010; Krupczak-Hollis et al., 2003; Kurinna et al., 2013; Lehmann et al., 2012; Matsuo et al., 2003; Meng et al., 2011; Mukhopadhyay et al., 2011; Tan et al., 2006; Wang et al., 2003; Wang, Kiyokawa, et al., 2002; Wang, Krupczak-Hollis, et al., 2002; Wang, Quail, et al., 2001; Weymann et al.,

2009; Ye et al., 1999; Zhang, Wang, et al., 2012), which otherwise do not express *foxm1* (Chaudhary et al., 2000; Kalinichenko et al., 2003; Kalinina et al., 2003; Matsuo et al., 2003; Ren et al., 2010; Tan et al., 2006; Wang, Hung, et al., 2001).

Likewise, PHx also induces strong FOXM1 mRNA and protein expression in adult rat livers (Fig. 3.10), which otherwise do not express *foxm1* (Ye et al., 1997).

Similar to PHx, toxic liver injury by carbon tetrachloride (CCl_4) induces liver regeneration through regenerative cell proliferation, too (Luster et al., 2001; Manibusan, Odin, & Eastmond, 2007; Mehendale, 1995, 2005; Mehendale, Roth, et al., 1994; Mehendale, Thakore, & Rao, 1994; Soni & Mehendale, 1998; Weber, Boll, & Stampfl, 2003; Zimmerman & Lewis, 1995).

Also CCl_4 liver injury causes strong *foxm1* mRNA expression in adult livers (Fig. 3.10; Chen et al., 2010; Meng et al., 2010; Ren et al., 2010; Wang, Hung, et al., 2001; Zhang, Wang, et al., 2012).

In unhurt adult mouse lungs, *foxm1* is not expressed (Balli et al., 2012; Kalin et al., 2008; Kalinichenko et al., 2001; Ren et al., 2013; Wang, Zhang, Snyder, et al., 2010). However, lung injury induces *foxm1* mRNA (and protein) expression during the regenerative cell proliferation for tissue repair (Kalinichenko et al., 2003, 2001; Liu et al., 2011; Zhang, Li, et al., 2012; Zhao et al., 2006). This injury-induced FOXM1 reexpression in adult mouse lungs has been described for BHT (butylated hydroxytoluene) lung injury (Kalinichenko et al., 2001, 2003), for LPS (lipopolysaccaride)-induced acute vascular lung injury (Zhang, Li, et al., 2012; Zhao et al., 2006), for CLP (cecal ligation and puncture)-induced lung vascular injury (Huang & Zhao, 2012), and for *Pseudomonas aeruginosa*-induced lung alveolar injury (Fig. 3.10; Liu et al., 2011).

The uninjured adult murine pancreas shows no expression of *foxm1* (Kalinichenko et al., 2003). Yet, 60% PPx (partial pancreatectomy) induces *foxm1* mRNA expression during the regenerative cell proliferation for tissue repair so that this pancreas injury reactivates the FOXM1 expression in the adult mouse pancreas (Fig. 3.10; Ackermann Misfeldt et al., 2008).

In summary, in the course of adult tissue repair after injury, *foxm1* is reexpressed in otherwise quiescent cells of the adult liver (Chen et al.,

2010; Gieling et al., 2010; Krupczak-Hollis et al., 2003; Kurinna et al., 2013; Lehmann et al., 2012; Matsuo et al., 2003; Meng et al., 2011, 2010; Mukhopadhyay et al., 2011; Ren et al., 2010; Tan et al., 2006; Wang et al., 2003; Wang, Hung, et al., 2001; Wang, Kiyokawa, et al., 2002; Wang, Krupczak-Hollis, et al., 2002; Wang, Quail, et al., 2001; Weymann et al., 2009; Ye et al., 1999, 1997; Zhang, Wang, et al., 2012), lung (Huang & Zhao, 2012; Kalinichenko et al., 2003, 2001; Liu et al., 2011; Zhang, Li, et al., 2012; Zhao et al., 2006), and pancreas (Ackermann Misfeldt et al., 2008) during their regenerative cell proliferation (Fig. 3.10).

In accordance with the general decline of FOXM1 levels during aging (Ly et al., 2000), the PHx-induced *foxm1* expression is reduced in old-aged mice because the livers of old mice (12 or 13–16 month) express significantly less *foxm1* mRNA and protein in response to PHx than the livers of young animals (2 month) (Chen et al., 2010; Wang, Krupczak-Hollis, et al., 2002; Wang, Quail, et al., 2001).

Interestingly, in these old-aged mice, the PHx-induced *foxm1* expression could largely be restored by GH at the mRNA and protein levels (Krupczak-Hollis et al., 2003) as well as by either FXRα2-VP16 or Px20350 at the mRNA level (Fig. 3.10; Chen et al., 2010). The synthetic, highly specific FXR (farnesoid X receptor) ligand Px20350 is a potent FXR agonist. FXRα2-VP16 is constitutively active in the absence of ligands because it is a fusion protein between FXRα2 and the strong TAD of VP16 of HSV (Zhang, Lee, et al., 2006).

6.5. Cessation of FOXM1 expression upon cell cycle exit

In accordance with its strictly proliferation-specific expression pattern (Costa et al., 2003, 2001; Kalin et al., 2011; Koo et al., 2011; Laoukili et al., 2007; Murakami et al., 2010; Raychaudhuri & Park, 2011; Wierstra & Alves, 2007c), the expression of FOXM1 is always down-regulated when cells exit the cell cycle, no matter whether this is a transient (reversible) exit into quiescence or a permanent (irreversible) exit into senescence or terminal differentiation (Koo et al., 2011; Laoukili et al., 2007; Murakami et al., 2010; Wierstra & Alves, 2007c). Consequently, FOXM1 is not expressed in resting cells, i.e. neither in quiescent cells (Balli et al., 2012; Bolte et al., 2011; Chaudhary et al., 2000; Coller et al., 2006; Down et al., 2012; Faust et al., 2012; Kalin et al., 2008; Kalinichenko et al., 2003, 2001; Kalinina et al., 2003; Korver, Roose, & Clevers, 1997;

Küppers et al., 2010; Laoukili et al., 2005; Leung et al., 2001; Li et al., 2008; Matsuo et al., 2003; Mirza et al., 2010; Newick et al., 2012; Park, Carr, et al., 2009; Park, Costa, Lau, Tyner, & Raychaudhuri, 2008; Petrovic et al., 2008; Ren et al., 2013, 2010; Sengupta et al., 2013; Tan et al., 2006; Wang, Chen, et al., 2008; Wang, Hung, et al., 2001; Wang, Zhang, Snyder, et al., 2010; Xue et al., 2010; Yao et al., 1997; Ye et al., 1997) nor in senescent cells (Hardy et al., 2005; Rovillain et al., 2011) nor in terminally differentiated cells (Ahn, Lee, et al., 2004; Bose et al., 2012; Carr et al., 2012; Gemenetzidis et al., 2010; Huynh et al., 2009, 2010; Korver, Roose, Wilson, et al., 1997; Prots et al., 2011; Takahashi et al., 2005; Wang, Lin, et al., 2011; Wierstra & Alves, 2006d; Xie et al., 2010; Ye et al., 1997).

6.5.1 Discontinuance of FOXM1 expression during transient cell cycle exit

6.5.1.1 Downregulation of the FOXM1 expression during quiescence

The FOXM1 expression is generally downregulated upon entrance into quiescence, independent of the trigger (Coller et al., 2006; Faust et al., 2012; Küppers et al., 2010; Mirza et al., 2010; Park, Costa, Lau, Tyner, & Raychaudhuri, 2008), so that *foxm1* is generally not expressed in quiescent cells (Balli et al., 2012; Bolte et al., 2011; Chaudhary et al., 2000; Coller et al., 2006; Down et al., 2012; Faust et al., 2012; Kalin et al., 2008; Kalinichenko et al., 2003, 2001; Kalinina et al., 2003; Korver, Roose, & Clevers, 1997; Küppers et al., 2010; Laoukili et al., 2005; Leung et al., 2001; Li et al., 2008; Matsuo et al., 2003; Mirza et al., 2010; Newick et al., 2012; Park, Carr, et al., 2009; Park, Costa, Lau, Tyner, & Raychaudhuri, 2008; Petrovic et al., 2008; Ren et al., 2013, 2010; Sengupta et al., 2013; Tan et al., 2006; Wang, Chen, et al., 2008; Wang, Hung, et al., 2001; Wang, Zhang, Snyder, et al., 2010; Xue et al., 2010; Yao et al., 1997; Ye et al., 1997).

In fact, the expression of FOXM1 rapidly decreases when cells exit the cell cycle into quiescence induced by either serum depletion (Park, Costa, Lau, Tyner, & Raychaudhuri, 2008c) or mitogen withdrawal (Coller et al., 2006) or loss of adhesion (Coller et al., 2006) or confluence, that is, contact inhibition (Coller et al., 2006; Faust et al., 2012; Küppers et al., 2010; Mirza et al., 2010).

As a result, quiescent cells do not express FOXM1 (Koo et al., 2011; Laoukili et al., 2007; Murakami et al., 2010; Wierstra & Alves, 2007c), which was explicitly demonstrated for:

- serum-deprived (immortalized) MEFs (Laoukili et al., 2005; Park, Carr, et al., 2009; Petrovic et al., 2008; Wang, Chen, et al., 2008),
- serum-starved mouse NIH3T3 fibroblasts (Down et al., 2012; Li et al., 2008; Park, Costa, Lau, Tyner, & Raychaudhuri, 2008),
- serum-deprived rat Rat-1 fibroblasts (Korver, Roose, & Clevers, 1997),
- serum-starved human HeLa cervix cancer cells (Leung et al., 2001),
- serum-deprived human mesothelial LP9 cells (immortalized with hTERT) (Newick et al., 2012),
- fetal human lung diploid fibroblasts subjected to mitogen withdrawal (deprivation of mitogen) (Coller et al., 2006),
- confluent (contact-inhibited) mouse NIH3T3 fibroblasts (Faust et al., 2012; Küppers et al., 2010),
- confluent human HMVEC-L (lung human microvascular endothelial cells) cells (Mirza et al., 2010),
- fetal human lung diploid fibroblasts subjected to contact inhibition (growth at high cell density) (Coller et al., 2006),
- fetal human lung diploid fibroblasts subjected to loss of adhesion (disruption of cell–substratum adhesions) (Coller et al., 2006),
- naive mature T-cells (Xue et al., 2010),
- adult lung (Balli et al., 2012; Kalin et al., 2008; Kalinichenko et al., 2001; Ren et al., 2013; Wang, Zhang, Snyder, et al., 2010),
- adult liver (Chaudhary et al., 2000; Kalinichenko et al., 2003; Kalinina et al., 2003; Matsuo et al., 2003; Ren et al., 2010; Tan et al., 2006; Wang, Hung, et al., 2001; Yao et al., 1997; Ye et al., 1997),
- adult heart (Bolte et al., 2011),
- heart at postnatal day 30 (pd30) (Sengupta et al., 2013),
- adult SMCs (smooth muscle cells) of intestine, bronchi, and blood vessels (Ustiyan et al., 2009).

6.5.1.2 FOXM1 is a member of the fibroblast "quiescence program"

FOXM1 is part of the fibroblast "quiescence program", which represents a set of gene expression changes (116 upregulated and 33 downregulated genes) central to any long-term quiescent state and independent of the specific arrest signal (i.e., mitogen withdrawal, contact inhibition, or loss of adhesion) (Coller et al., 2006). The gene expression pattern defined by the fibroblast "quiescence program" is representative of quiescence *per se* because it includes that set of genes, whose specific expression or non-expression in nondividing fibroblasts is signal-independent and therefore characteristic of multiple, independent quiescent states (Coller et al., 2006). FOXM1 is a member of the fibroblast "quiescence program" because the *foxm1* mRNA level always declines during quiescence of fetal human lung diploid fibroblasts, no matter whether their entry into quiescence is triggered by mitogen withdrawal (deprivation of mitogen), contact inhibition (growth at high cell density), or loss of adhesion (disruption of cell–substratum adhesions) (Coller et al., 2006).

6.5.1.3 Downregulation of the FOXM1 expression upon serum starvation and contact inhibition

The membership of FOXM1 in the fibroblast "quiescence program" demonstrates that the *foxm1* mRNA expression is generally downregulated upon cell cycle exit into quiescence, independent of the trigger (Coller et al., 2006). Accordingly, FOXM1 protein levels decline also when other cell types enter quiescence due to serum depletion (Park, Costa, Lau, Tyner, & Raychaudhuri, 2008c) or contact inhibition (confluence) (Faust et al., 2012; Küppers et al., 2010; Mirza et al., 2010):

First, after seeding at high cell density (100% confluence), the FOXM1 protein expression rapidly decreased in mouse NIH3T3 fibroblasts, showing that contact inhibition causes the cessation of FOXM1 expression (Faust et al., 2012). Actually, due to contact inhibition, the preexisting FOXM1 protein had disappeared 16 h after seeding of NIH3T3 fibroblasts at high cell density (100% confluence), which is 12 h after establishment of cell–cell contacts (because adherence to the culture dish was achieved 4 h after seeding) (Faust et al., 2012). Thus, the FOXM1 expression is rapidly decreased after establishment of cell–cell contacts (Faust et al., 2012). Accordingly, contact inhibition resulted in downregulated FOXM1 protein levels not only in confluent NIH3T3 fibroblasts (Faust et al., 2012; Küppers et al., 2010), but also in confluent mouse C3H10T1/2 fibroblasts (Faust et al., 2012), confluent rat WB-F344 liver oval cells (Faust et al., 2012), confluent human Caco-2 colon epithelial cells (Faust et al., 2012), and confluent HMVEC-L (Mirza et al., 2010).

Second, after serum removal, the FOXM1 protein expression rapidly decreased in mouse NIH3T3 fibroblasts, indicating that serum starvation causes the close of FOXM1 expression (Park, Costa, Lau, Tyner, & Raychaudhuri, 2008c). In fact, the preexisting FOXM1 protein in NIH3T3 fibroblasts had largely disappeared after 24 h serum depletion and was almost undetectable after 48 h (Park, Costa, Lau, Tyner, & Raychaudhuri, 2008c). This dramatic reduction of the FOXM1 protein level was accomplished by the proteasome-dependent destruction of FOXM1 because the serum-starved cells retained much higher amounts of FOXM1 protein following serum removal in the presence of the proteasome inhibitor MG132

(Park, Kim, et al., 2008) and if the APC/C subunit Cdh1 (Cdc20 homolog 1, FZR1 (Fizzy-related protein 1)), which recognizes FOXM1 (Laoukili, Alvarez-Fernandez, et al., 2008; Park, Kim, et al., 2008), was knocked down with siRNA (Park, Kim, et al., 2008). Consequently, serum starvation leads to the ubiquitination of FOXM1 by the E3 ubiquitin ligase APC/C^{Cdh1} and thus to its degradation by the 26S proteasome (Park, Kim, et al., 2008).

6.5.2 Termination of FOXM1 expression during permanent cell cycle exit

6.5.2.1 Terminal differentiation

The expression of FOXM1 is rapidly decreased and finally extinguished during terminal differentiation of various cell types so that *foxm1* is generally not expressed in terminally differentiated cells (Ahn, Lee, et al., 2004; Bose et al., 2012; Carr et al., 2012; Gemenetzidis et al., 2010; Huynh et al., 2009, 2010; Korver, Roose, Wilson, et al., 1997; Prots et al., 2011; Takahashi et al., 2005; Wang, Park, et al., 2011; Wierstra & Alves, 2006d; Xie et al., 2010; Ye et al., 1997). This downregulation of the *foxm1* expression in the course of terminal differentiation was shown for several different cell types (Fig. 3.10), for example, for:

- differentiation of thymocytes (Korver, Roose, Wilson, et al., 1997).
- differentiation of NSCs (neural stem cells) (Ahn, Lee, et al., 2004).
- differentiation of mammary luminal cells (Carr et al., 2012).
- differentiation of primary oral keratinocytes (Gemenetzidis et al., 2010).
- differentiation of Caco-2 colon carcinoma cells toward intestinal enterocytes (Ye et al., 1997).
- TPA-induced differentiation of human promyelocytic HL-60 leukemia cells toward macrophages (Wierstra & Alves, 2006d).
- FGF-2 (fibroblast growth factor-2)-induced differentiation of capillary endothelial cell (Takahashi et al., 2005).
- RA (retinoic acid)-induced neural differentiation of mouse P19 EC (embryonal carcinoma) cells (Xie et al., 2010).
- RA-induced differentiation of BE(2)-C neuroblastoma cells toward neuronal lineage (Wang, Park, et al., 2011).
- differentiation of HO-1 human metastatic melanoma cells induced by IFN-β + mezerein (Huynh et al., 2009, 2010).
- IL-4-induced differentiation of CD25+ iTregs (induced regulatory T-cells) (Prots et al., 2011).

- differentiation of HaCaT keratinocytes (spontaneously immortalized human skin keratinocytes) induced by suspension culture in methylcellulose (i.e., anchorage deprivation) (Bose et al., 2012).
- differentiation of N/Tert-1 keratinocytes (hTERT- and p16 loss-immortalized normal human skin keratinocytes) induced by methylcellulose suspension culture (Bose et al., 2012).
- PMA-induced differentiation of N/Tert-1 keratinocytes (Bose et al., 2012).
- differentiation of N/Tert-1 keratinocytes induced by high cell density (Bose et al., 2012).

6.5.2.2 Cellular senescence

FOXM1 expression ceases when cells undergo cellular senescence so that senescent cells do not express *foxm1* (Hardy et al., 2005; Rovillain et al., 2011). This downregulation of the *foxm1* mRNA expression upon cell cycle exit into senescence has been described in two different scenarios of cellular senescence, replicative senescence (Hardy et al., 2005) as well as senescence upon reactivation of the p16-RB and p53-p21 pathways in conditionally immortalized HMFs (Hardy et al., 2005; Rovillain et al., 2011):

First, the *foxm1* mRNA level decreased during replicative senescence, that is, during the classical serial passaging of primary HMFs to senescence (Hardy et al., 2005).

Second, the *foxm1* mRNA level decreased during senescence upon reactivation of the p16-RB and p53-p21 pathways in conditionally immortalized HMFs (Hardy et al., 2005; Rovillain et al., 2011). In this case, primary HMFs were first immortalized with hTERT and SV40 large T antigen, and then senescence was induced in the resulting conditionally immortalized HMFs by inactivation of SV40 large T (Rovillain et al., 2011), which otherwise inhibits the two key tumor suppressors RB and p53 (Ahuja, Saenz-Robles, & Pipas, 2005; Ali & DeCaprio, 2001; Caracciolo, Reiss, Khalili, De Falco, & Giordano, 2006; Cheng, DeCaprio, Fluck, & Schaffhausen, 2009; DeCaprio, 2009; Felsani, Mileo, & Paggi, 2006; Helt & Galloway, 2003; Lee & Cho, 2002; Levine, 2009; Liu & Marmorstein, 2006; Moens, Van Ghelue, & Johannessen, 2007; O'Shea & Fried, 2005; Pipas, 2009; Pipas & Levine, 2001; Saenz-Robles & Pipas, 2009; Saenz-Robles, Sullivan, & Pipas, 2001; Sullivan & Pipas, 2002; White & Khalili, 2004, 2006).

6.6. Antagonistic regulation of the FOXM1 expression by proliferation versus antiproliferation signals

In general, the expression of FOXM1 is upregulated by proliferation signals but downregulated by antiproliferation and differentiation signals (Fig. 3.10; Koo et al., 2011; Laoukili et al., 2007; Raychaudhuri & Park, 2011; Wang, Ahmad, Li, et al., 2010; Wierstra & Alves, 2007c). This positive regulation of the FOXM1 expression by proliferation signals (e.g., serum, KGF, GH, IGF-I (insulin-like growth factor), estrogen, ERα, PL, PHA, c-Myc, E2F-1, c-Myb (MYB), B-Myb, c-Rel (REL), STAT3 (signal transducer and activator of transcription 3), Notch1, GLI-1 (glioma transcription factor-1), YAP (Yes-associated protein), ETV5, HSF1 (heat shock factor 1), PLK1, AKT, ERK2, H-Ras, N-Ras, EGFR, HER2, GRB7 (growth factor receptor-bound protein 7), cyclin D1/Cdk4, cyclin D3/Cdk6, Wnt3a, c-Met, Shh, HPV16 (human papillomavirus 16) E7 (early ORF (open reading frame) 7), adenovirus E1A (early region 1A), SV40 large T antigen, HBX (HBV (hepatitis B virus) X protein), BCR-ABL (breakpoint cluster region-Abelson)), on the one hand, and this negative regulation of the FOXM1 expression by antiproliferation and differentiation signals (e.g., RB, p53, p21, PTEN, Cdh1, DKK1 (Dickkopf 1), FOXO3a, C/EBPα (CCAAT/enhancer-binding protein α), Mad1 (Max dimerizer 1, MXD1 (Max dimerization protein 1)), Mxi1 (Max interactor 1), KLF4 (Krüppel-like factor 4), GATA3, SPDEF (SAM pointed domain-containing ETS transcription factor, PDEF (prostate-derived ETS transcription factor)), miR-200b, miR-31, miR-34), on the other hand (Fig. 3.10), correlate nicely not only with the strictly proliferation-specific expression pattern of *foxm1*, but also with the function of FOXM1 as a typical proliferation-associated transcription factor.

The antagonistic control of the FOXM1 expression by proliferation versus antiproliferation signals includes the upregulation of FOXM1 by protooncoproteins (c-Myc, c-Myb, c-Rel, STAT3, Notch1, GLI-1, YAP, ETV5, AKT, H-Ras, N-Ras, EGFR, HER2/Neu, cyclin D1/Cdk4, Wnt3a, c-Met), a chimeric oncoprotein (BCR-ABL) and viral oncoproteins (HPV16 E7, E1A, SV40 large T, HBX) as well as the downregulation of FOXM1 by tumor suppressors (RB, p53, PTEN, Cdh1, KLF4, GATA3), candidate tumor suppressors (Mad1, FOXO3a), and tumor suppressor miRNAs (miR-31, miR-34, miR-200b) (Fig. 3.10; Costa, 2005; Costa, Kalinichenko, Major, et al., 2005; Gartel, 2008, 2010; Koo et al., 2011; Laoukili et al., 2007; Murakami et al., 2010; Myatt & Lam, 2007;

Raychaudhuri & Park, 2011; Wang, Ahmad, Li, et al., 2010; Wierstra & Alves, 2007c). Since protooncoproteins and tumor suppressors are frequently activated or inactivated, respectively, in cancer cells (Balmain, Gray, & Ponder, 2003; Benvenuti, Arena, & Bardelli, 2005; Bishop, 1995; Bocchetta & Carbone, 2004; Boulikas, 1995; Futreal et al., 2004; Hanahan & Weinberg, 2000, 2011; Harris & McCormick, 2010; Hesketh, 1997; Lee & Muller, 2010; Massagué, 2004; Polsky & Cordon-Cardo, 2003; Ponder, 2001; Sherr, 2004; Vogelstein & Kinzler, 2004; Weinberg, 2006; Yeang, McCormick, & Levine, 2008) the control of the FOXM1 expression by them in normal cells inevitably predetermines the deregulation of the FOXM1 expression in tumor cells (Wierstra & Alves, 2007c). Accordingly, FOXM1 is really overexpressed in a large variety of human cancers (Banz et al., 2009; Bektas et al., 2008; Bellelli et al., 2012; Calvisi et al., 2009; Chan et al., 2008; Chen, Yang, et al., 2012; Chu et al., 2012; Craig et al., 2013; Dibb et al., 2012; Elgaaen et al., 2012; Francis et al., 2009; Garber et al., 2001; Gemenetzidis et al., 2009; Gialmanidis et al., 2009; Guerra et al., 2013; He et al., 2012; Hodgson et al., 2009; Hui et al., 2012; Iacobuzio-Donahue et al., 2003; Janus et al., 2011; Jiang et al., 2011; Kalin et al., 2006; Kalinina et al., 2003; Kim et al., 2006; Kretschmer, Sterner-Kock, Siedentopf, Schlag, & Kemmner, 2011; Laurendeau et al., 2010; Li, Peng, et al., 2013; Li et al., 2009; Lin, Guo, et al., 2010; Lin, Madan, et al., 2010; Liu, Gampert, Nething and Steinacker, 2006; Llaurado et al., 2012; Lok et al., 2011; Nakamura, Hirano, et al., 2010; Nakamura, Thakore, et al., 2004; Nakamura, Yamashita, et al., 2010; Newick et al., 2012; Obama et al., 2005; Okabe et al., 2001; Park et al., 2012; Pellegrino et al., 2010; Perou et al., 2000; Pignot et al., 2012; Pilarsky, Wenzig, Specht, Saeger, & Grutzmann, 2004; Qu et al., 2013; Rickman et al., 2001; Rodriguez, Li, Yao, Chen, & Fisher, 2005; Romagnoli et al., 2009; Rosty et al., 2005; Salvatore et al., 2007; Santin et al., 2005; Schüller, Kho, Zhao, Ma, & Rowitch, 2006; Sorlie et al., 2001; Sun et al., 2011; Takahashi et al., 2006; Teh et al., 2002; The Cancer Genome Atlas Research Network, 2011; Uddin et al., 2011, 2012; van den Boom et al., 2003; Waseem, Ali, Odell, Fortune, & Teh, 2010; Wonsey & Follettie, 2005; Wu, Lu, et al., 2010; Xia, Huang, et al., 2012; Xia, Wang, et al., 2012; Xue et al., 2012; Yokomine et al., 2009; Yoshida et al., 2007; Zeng et al., 2009; Zhang, Zeng, et al., 2012).

Also other proliferation and antiproliferation signals are often activated or inactivated, respectively, in cancer (Balmain et al., 2003; Benvenuti et al., 2005; Bishop, 1995; Bocchetta & Carbone, 2004; Boulikas, 1995;

Futreal et al., 2004; Hanahan & Weinberg, 2000, 2011; Harris & McCormick, 2010; Hesketh, 1997; Lee & Muller, 2010; Massagué, 2004; Polsky & Cordon-Cardo, 2003; Ponder, 2001; Sherr, 2004; Vogelstein & Kinzler, 2004; Weinberg, 2006; Yeang et al., 2008) so that the normal control of the FOXM1 expression by them provides an additional basis for the deregulated expression of FOXM1 in human tumors (Wierstra & Alves, 2007c).

6.7. The expression of FOXM1 in response to DNA damage

Contradictory findings have been reported regarding the regulation of the FOXM1 expression in response to DNA damage:

DNA damage induced by **doxorubicin** decreased the FOXM1 mRNA and protein levels in human MCF-7 breast cancer cells (Pandit et al., 2009) and it decreased the FOXM1 protein level in human U2OS osteosarcoma cells (Alvarez-Fernandez, Halim, et al., 2010) and human HCT116 colon cancer cells (Halasi & Gartel, 2012).

In contrast, doxorubicin-induced DNA damage increased the FOXM1 protein level in human MIA PaCa-2 pancreatic cancer cells, human MDA-MB-231 breast cancer cells, and human Hep3B liver cancer cells (Halasi & Gartel, 2012).

DNA damage induced by **daunorubicin** reduced the FOXM1 mRNA and protein expression in MCF-7 cells and human HepG2 HCC (hepatocellular carcinoma) cells (Barsotti & Prives, 2009).

In contrast, daunorubicin-induced DNA damage enhanced the FOXM1 protein expression in U2OS cells and HCT116 cells (Barsotti & Prives, 2009).

DNA damage induced by **epirubicin** downregulated the FOXM1 mRNA and protein levels in MCF-7 cells (Millour et al., 2011; Monteiro et al., 2012) and it downregulated the *foxm1* mRNA level in MEFs (Millour et al., 2011).

In contrast, epirubicin-induced DNA damage upregulated the FOXM1 protein level in human MDA-MB-453 breast cancer cells (Millour et al., 2011), U2OS cells (de Olano et al., 2012; Millour et al., 2011), MCF-7 cells, and MEFs (de Olano et al., 2012).

Remarkably, DNA damage induced by epirubicin exerted opposite effects on the FOXM1 protein expression in the same cell type, namely, in MCF-7 cells, where either a positive (de Olano et al., 2012) or negative (Millour et al., 2011; Monteiro et al., 2012) effect of epirubicin-induced DNA damage on the FOXM protein level were reported in three different studies by the same group.

On the one hand, **cisplatin**-induced DNA damage diminished the FOXM1 mRNA and protein expression in MCF-7 cells (Kwok et al., 2010).

Oxaliplatin-induced DNA damage reduced the FOXM1 mRNA and protein expression in human HepG2 and SMMC-7721 HCC cells (Qu et al., 2013).

DNA damage induced by **5-FU (5-fluorouracil)** decreased the FOXM1 protein expression in MCF-7 cells (Monteiro et al., 2012).

On the other hand, DNA damage induced by **IR (ionizing radiation, γ-irradiation)** raised the FOXM1 protein level in human A549 lung adenocarcinoma cells (Chetty et al., 2009), U2OS cells (Tan et al., 2007), MEFs (Tan et al., 2010), MIA PaCa-2 cells, and MDA-MB-231 cells (Halasi & Gartel, 2012).

DNA damage induced by **UV (ultraviolet radiation)** elevated the FOXM1 protein expression in U2OS cells (Tan et al., 2007), and UVB-induced DNA damage elevated the FOXM1 protein expression in hTERT-immortalized keratinocytes (N/TERT) (Teh, Gemenetzidis, Chaplin, Young, & Philpott, 2010).

Etoposide-induced DNA damage augmented the FOXM1 protein level in U2OS cells (Tan et al., 2007, 2010).

In general, the DNA damage-induced changes in the FOXM1 expression depend on both the cell type (Barsotti & Prives, 2009; Halasi & Gartel, 2012; Millour et al., 2011) and the p53 status of the cell (Barsotti & Prives, 2009; Halasi & Gartel, 2012; Millour et al., 2011; Pandit et al., 2009) as well as possibly on the DNA-damaging agent.

Additionally, the effect of the same DNA-damaging drug on the FOXM1 expression in the same cell type can dramatically change or even reverse to the opposite effect when a tumor cell line becomes resistant to this anticancer drug (see Part II of this two-part review, that is see Wierstra, 2013b). Examples include cisplatin-resistant MCF-7 cells (Kwok et al., 2010), 5-FU-resistant MCF-7 cells (Monteiro et al., 2012), and epirubicin-resistant MCF-7 cells (de Olano et al., 2012; Millour et al., 2011; Monteiro et al., 2012).

6.8. Control of the FOXM1 protein stability
6.8.1 Stabilization of FOXM1 after DNA damage through its phosphorylation by Chk2

The checkpoint kinase Chk2, which plays a pivotal role in the DNA damage response (Abraham, 2001; Ahn, Urist, & Prives, 2004; Antoni, Sodha, Collins, & Garrett, 2007; Bartek, Falck, & Lukas, 2001; Bartek & Lukas, 2003, 2007; Bartek, Lukas, & Lukas, 2004; Branzei & Foiani, 2008; Chen & Poon, 2008; Deckbar, Jeggo, & Löbrich, 2011; Finn, Lowndes, & Grenon, 2012; Harper & Elledge, 2007; Harrison & Haber, 2006; Ishikawa, Ishii, & Saito, 2006; Jackson, 2009; Jackson & Bartek, 2009; Kastan, 2008; Kastan & Bartek, 2004; Li & Zou, 2005; Liang, Lin, Brunicardi, Goss, & Li, 2009; Lukas, Lukas, & Bartek, 2004; McGowan, 2002; Nakanishi, Shimada, & Niida, 2006; Nyberg, Michelson, Putnam, & Weinert, 2002; Perona, Moncho-Amor, Machado-Pinilla, Belad-Iniesta, & Sanchez Perez, 2008; Reinhardt & Yaffe, 2009; Rouse & Jackson, 2002; Samuel, Weber, & Funk, 2002; Sancar, Lindsey-Boltz, Ünsal-Kacmaz, & Linn, 2004; Smith, Tho, Xu, & Gillespie, 2010; Stracker, Usui, & Petrini, 2009; Su, 2006; Walworth, 2000; Wang, Ji, et al., 2009; Yang & Zou, 2006; Zhou & Elledge, 2000), phosphorylates FOXM1 at S-375 (corresponding to S-361 of FoxM1B) *in vitro* as well as *in vivo* (Tan et al., 2007). In the absence of experimentally induced DNA damage, overexpression of Chk2 in U2OS cells increased the FOXM1 protein level dependently on the Chk2 phosphorylation site S-375, suggesting a stabilization of the FOXM1 protein through its phosphorylation at S-375 by Chk2 (Fig. 3.10; Tan et al., 2007). This supposition is strongly corroborated by the results for the stabilization of FOXM1 in response to DNA damage induced by IR, UV, and etoposide (see below) (Tan et al., 2007).

As outlined above, the FOXM1 protein expression is elevated by IR-induced DNA damage (Chetty et al., 2009; Halasi & Gartel, 2012; Tan et al., 2010, 2007), UV/UVB-induced DNA damage (Tan et al., 2007; Teh et al., 2010), and etoposide-induced DNA damage (Tan et al., 2007, 2010).

The FOXM1 protein level was considerably upregulated in U2OS cells after induction of DNA damage by either IR or UV or etoposide (Fig. 3.10), whereas the *foxm1* mRNA level remained unchanged, pointing to a DNA damage-induced FOXM1 protein stabilization (Tan et al., 2007). This

DNA damage-induced increase in the FOXM1 protein expression depends on Chk2 because it was abolished by overexpression of dn (dominant-negative) Chk2 and by silencing of Chk2 with siRNA (Tan et al., 2007). Thus, Chk2 mediates the protein stabilization of FOXM1 in response to DNA damage induced by IR, UV, and etoposide (Fig. 3.10; Tan et al., 2007). Accordingly, Chk2 overexpression enhanced the IR-induced rise in the FOXM1 level further (Tan et al., 2007).

In U2OS cells, both IR-induced and etoposide-induced DNA damage caused the phosphorylation of FOXM1 at the Chk2 phosphorylation site S-375, indicating that Chk2 phosphorylates FOXM1 in response to IR and etoposide and that DNA damage induced by IR and etoposide stabilizes FOXM1 by effectuating its Chk2-mediated phosphorylation at S-375 (Tan et al., 2007).

In summary, DNA damage caused by IR, UV, and etoposide induces the stabilization of FOXM1 protein through phosphorylation of FOXM1 by Chk2 (Fig. 3.10; Tan et al., 2007).

6.8.2 Cell cycle-dependent degradation of FOXM1 via the ubiquitin–proteasome pathway

The FOXM1 protein is subject to ubiquitin-dependent proteasomal degradation because it is ubiquitinated by the E3 ubiquitin ligase APC/C^{Cdh1} and subsequently destructed by the 26S proteasome (Fig. 3.10; Laoukili, Alvarez-Fernandez, et al., 2008; Park, Costa, Lau, Tyner, & Raychaudhuri, 2008).

In late mitosis and early G1-phase, FOXM1 is degraded by the 26S proteasome after its ubiquitylation by the E3 ubiquitin ligase APC/C^{Cdh1} (Laoukili, Alvarez-Fernandez, et al., 2008; Park, Costa, Lau, Tyner, & Raychaudhuri, 2008). Thus, FOXM1 is destructed via the ubiquitin–proteasome pathway in a cell cycle-dependent manner (Laoukili, Alvarez-Fernandez, et al., 2008; Park, Costa, Lau, Tyner, & Raychaudhuri, 2008).

FOXM1 interacts with the APC/C subunits Cdh1 (Laoukili, Alvarez-Fernandez, et al., 2008; Park, Costa, Lau, Tyner, & Raychaudhuri, 2008) and Cdc27 (Park, Costa, Lau, Tyner, & Raychaudhuri, 2008c). However, none of these interactions has been demonstrated to be direct. In accordance with the cell cycle-dependent degradation of FOXM in late mitosis and early G1-phase (Laoukili, Alvarez-Fernandez, et al., 2008; Park, Costa, Lau, Tyner, & Raychaudhuri, 2008), the interaction between FOXM1 and Cdc27 is restricted to this cell cycle period but not detected in other cell cycle phases (i.e., neither in the remainder of G1-phase nor in S-phase nor in

G2-phase nor in the remainder of M-phase) (Park, Costa, Lau, Tyner, & Raychaudhuri, 2008c).

The APC/C activator protein Cdh1 recognizes two consecutive D-boxes (destruction boxes) and a KEN-box in the N-terminus of FOXM1 (Laoukili, Alvarez-Fernandez, et al., 2008) so that the proteasome-dependent degradation of FOXM1 depends on Cdh1 as well as on these two D-boxes and/or this KEN-box (Laoukili, Alvarez-Fernandez, et al., 2008; Park, Costa, Lau, Tyner, & Raychaudhuri, 2008). Regarding the importance of the individual Cdh1 recognition motifs, differing results were reported in two studies (Laoukili, Alvarez-Fernandez, et al., 2008; Park, Costa, Lau, Tyner, & Raychaudhuri, 2008). Nevertheless, the combined elimination of the KEN-box and both D-boxes renders FOXM1 resistant to Cdh1-mediated degradation (Laoukili, Alvarez-Fernandez, et al., 2008) and thus creates stable FOXM1 mutants, which are no longer destructed during mitotic exit (Laoukili, Alvarez-Fernandez, et al., 2008; Park, Costa, Lau, Tyner, & Raychaudhuri, 2008). Accordingly, such stable FOXM1 mutants show a severely reduced ubiquitination and a severely reduced interaction with Cdh1 (Laoukili, Alvarez-Fernandez, et al., 2008).

In contrast to Cdh1 (Laoukili, Alvarez-Fernandez, et al., 2008; Park, Costa, Lau, Tyner, & Raychaudhuri, 2008), the second APC/C activator protein Cdc20 (FZY (Fizzy), p55CDC) does not interact with FOXM1 and therefore does not downregulate the FOXM1 protein expression (Park, Costa, Lau, Tyner, & Raychaudhuri, 2008c). Accordingly, in contrast to siRNA against Cdh1, siRNA against Cdc20 inhibits neither the ubiquitination of FOXM1 nor the degradation of FOXM1 in late mitosis and early G1-phase (Park, Costa, Lau, Tyner, & Raychaudhuri, 2008c). Hence, FOXM1 is ubiquitinated only by APC/C^{Cdh1} (Laoukili, Alvarez-Fernandez, et al., 2008; Park, Costa, Lau, Tyner, & Raychaudhuri, 2008) but not by APC/C^{Cdc20} (Park, Costa, Lau, Tyner, & Raychaudhuri, 2008c), so that only Cdh1 (Laoukili, Alvarez-Fernandez, et al., 2008; Park, Costa, Lau, Tyner, & Raychaudhuri, 2008) but not Cdc20 (Park, Costa, Lau, Tyner, & Raychaudhuri, 2008c) triggers the proteasome-dependent degradation of FOXM1.

Also in response to serum depletion, FOXM1 is degraded by the 26S proteasome, and this FOXM1 destruction upon serum starvation depends on Cdh1, too (Park, Costa, Lau, Tyner, & Raychaudhuri, 2008c).

Thus, the E3 ubiquitin ligase APC/C^{Cdh1} induces the proteasomal degradation of FOXM1 through ubiquitylation of FOXM1 (Fig. 3.10) during mitotic exit in each cell cycle round (Laoukili, Alvarez-Fernandez, et al., 2008; Park, Costa, Lau, Tyner, & Raychaudhuri, 2008) and upon transient cell cycle exit into quiescence after serum removal (Park, Costa, Lau, Tyner, & Raychaudhuri, 2008c).

6.8.3 Stabilization of endogenous FOXM1 through its phosphorylation by cyclin D1/Cdk4 and cyclin D3/Cdk6

Cyclin D1/Cdk4 and cyclin D3/Cdk6 phosphorylate FOXM1 and they increase the endogenous FOXM1 protein expression (Fig. 3.10), probably because they stabilize the FOXM1 protein by preventing its proteasomal degradation (Anders et al., 2011).

However, the positive effects of cyclin D/Cdk4,6 on the FOXM1 protein level are limited to endogenous FOXM1 and exogenous FOXM1 expressed from a weak promoter, whereas the protein level of exogenous FOXM1 expressed from a strong promoter is unaffected by cyclin D/Cdk4,6 (Anders et al., 2011). This striking difference was demonstrated in experiments with the Cdk4,6 inhibitor PD0332991, which significantly reduced the ectopic FOXM1 protein expression driven by the weak MMuLV (murine leukemia virus long terminal repeat) promoter but had only a negligible effect on the ectopic FOXM1c protein expression driven by the strong CMV promoter (Anders et al., 2011). This difference is important because cyclin D1/Cdk4 (Anders et al., 2011; Wierstra, 2013a; Wierstra & Alves, 2006b, 2006c, 2008) and cyclin D3/Cdk6 (Anders et al., 2011) enhance also the transcriptional activity of FOXM1c independent of their positive effects on its expression.

In U2OS cells, the FOXM1 protein expression was upregulated by overexpression of either cyclin D1/Cdk4 or cyclin D3/Cdk6 but downregulated by the Cdk4/6 inhibitor PD0332991 (Anders et al., 2011). This PD0332991-induced decrease in the FOXM1 protein level was blocked by the proteasome inhibitor MG123 and alleviated by siRNA-mediated knockdown of the FOXM1-targeting APC/C subunit Cdh1, which indicates that cyclin D/Cdk4,6 prevents the degradation of FOXM1 by the 26S proteasome and suggests that cyclin D/Cdk4,6 partially protects FOXM1 from ubiquitylation by the E3 ubiquitin ligase APC/C^{Cdh1} (Anders et al., 2011). Thus, cyclin D1/Cdk4 and cyclin D3/Cdk6 augment the FOXM1 protein expression (Fig. 3.10) by stabilizing FOXM1 through reduction of its degradation via the ubiquitin–proteasome pathway (Anders et al., 2011).

For this purpose, they phosphorylate FOXM1 at multiple sites because phosphomimetic FOXM1c mutants were partially resistant to PD0332991-induced turnover (Anders et al., 2011).

The cyclin D3/Cdk6 phosphorylation sites in FOXM1c have been mapped: cyclin D3/Cdk6 phosphorylates FOXM1c *in vitro* at S-4, S-450, S-488, S-507, T-509, S-521, T-599, T-610, T-619, T-626, and S-703 as well as *in vivo* at S-4, S-35, T-610, T-619, and T-626 (Anders et al., 2011). The cyclin D1/Cdk4 phosphorylation sites in FOXM1 were not mapped, but cyclin D1/Cdk4 was shown to *in vitro* phosphorylate the same two FOXM1c fragments as cyclin D3/Cdk6, namely, aa 344–536 and aa 528–762 of FOXM1c (Anders et al., 2011).

Anders et al. (2011) reported that three phosphomimetic FOXM1c mutants displayed a reduced susceptibility to PD0332991-induced turnover compared to wild-type FOXM1c. These stabilized FOXM1c mutants harbored replacements of the phosphorylated serine/threonine residues with aspartic acid residues at either two (S4D/S35D mutant) or twelve ("12D" mutant) or the combined fourteen ("14D" mutant) Cdk consensus sites S/T-P (Anders et al., 2011). This result suggests that the cyclin D/Cdk4,6-mediated multisite phosphorylation of FOXM1c interferes with FOXM1c protein degradation and thus converts FOXM1c to a more proteolysis-resistant state (Anders et al., 2011). The 12D and 14D mutants were partially protected from PD0332991-induced degradation, namely to a similar extent, whereas the degradation susceptibility of the S4D/S35D was less reduced (Anders et al., 2011).

The two D-boxes (aa 6–18) and the KEN-box (aa 206–208) of FOXM1c, which are recognized by Cdh1 (Laoukili, Alvarez-Fernandez, et al., 2008), are located in the (outermost) N-terminus of FOXM1c. By contrast, the point-mutated FOXM1c residues in the 12D mutant (between aa 450 and aa 703) are positioned in the C-terminal half of FOXM1c, so that they are unlikely to affect the recognition of FOXM1c by Cdh1. However, the two point-mutated FOXM1c residues in the S4D/S35D mutant (nearly) coincide with the two D-boxes (aa 6–18) so that their phosphorylation by cyclin D/Cdk4,6 could principally render the D-boxes less accessible to Cdh1 (Anders et al., 2011). Yet, this attractive model cannot explain the experimental data because "simulated" phosphorylation at S-4 and S-35 in the phosphomimetic S4D/S35D mutant was not sufficient to protect FOXM1c from PD332991-induced degradation (Anders et al., 2011). Instead, the "simulated" phosphorylation at multiple C-terminal sites in the phosphomimetic 12D mutant was required to efficiently stabilize FOXM1c (Anders et al., 2011).

In summary, cyclin D1/Cdk4 and cyclin D3/Cdk6 stabilize the FOXM1 protein through its phosphorylation (Fig. 3.10) by rendering FOXM1 mostly resistant to proteasomal degradation (Anders et al., 2011).

Accordingly, the Cdk4/6 inhibitor PD0332991 decreased the endogenous FOXM1 protein level independent of RB because the PD0332991-induced FOXM1 degradation was also observed in U2OS cells depleted of RB by siRNA (Anders et al., 2011).

Notably, the endogenous FOXM1 protein expression in U2OS osteosarcoma and SKMEL2 melanoma cells was unaltered by the Cdk2 inhibitor CVT313, whereas it was drastically reduced by the Cdk4/6 inhibitor PD0332991 (Anders et al., 2011) although both cyclin D/Cdk4,6 (Anders et al., 2011) and cyclin A/Cdk2 (Laoukili, Alvarez, et al., 2008) phosphorylate FOXM1c, in part even at the same sites (e.g., at T-599 and T-610). Thus, the Cdk2-mediated phosphorylation of FOXM1 does surprisingly not affect its protein stability, in clear contrast to the Cdk4,6-mediated phosphorylation (Anders et al., 2011).

6.8.4 Stabilization of FOXM1 by Wnt3a

Wnt3a increased the FOXM1 protein level (Fig. 3.10) but not the *foxm1* mRNA level, suggesting that Wnt3a enhances the FOXM1 protein stability (Zhang et al., 2011). In fact, Wnt3a inhibited the degradation of preexisting FOXM1 protein in the presence of the protein synthesis inhibitor cycloheximide, which confirms the stabilization of the FOXM1 protein by Wnt3a (Zhang et al., 2011).

The secreted Wnt inhibitor DKK1 antagonizes Wnt signaling by binding to the Wnt receptor LRP5/6 (low density lipoprotein receptor-related protein 5 or 6) (Clevers & Nusse, 2012; Curtin & Lorenzi, 2010; Gehrke, Gandhirajan, & Kreuzer, 2009; Herr, Hausmann, & Basler, 2012; Monroe, McGee-Lawrence, Oursler, & Wesetndorf, 2012; Niehrs, 2006; Niehrs & Shen, 2010; Pinzone et al., 2009). In accordance with the positive effect of Wnt3a on the FOXM1 protein stability, DKK1 down-regulated not only the endogenous FOXM1 protein expression (Fig. 3.10), but also decreased the protein level of ectopically expressed FOXM1, indicating that DKK1 diminishes the stability of the FOXM1 protein (Zhang et al., 2011).

Hence, the FOXM1 protein is stabilized by Wnt3a but destabilized by the Wnt antagonist DKK1 (Fig. 3.10; Zhang et al., 2011).

6.8.5 Stabilization of FOXM1 by OGT

The enzyme OGT (O-GlcNAc (O-linked b-N-acetylglucosamine) transferase), which uses UDP-GlcNAc (uridine diphosphate-N-acetylglucosamine) as a donor substrate, catalyzes the covalent addition of a single monosaccharide (GlcNAc) onto serine or threonine residues of cytosolic and nuclear proteins (Butkinaree, Park, & Hart, 2010; Comer & Hart, 1999, 2000; Hart, Housley, & Slawson, 2007; Hart, Slawson, Ramirez-Correa, & Lagerlof, 2011; Issad & Kuo, 2008; Kamemura & Hart, 2003; Kudlow, 2006; Love, Krause, & Hanover, 2010; Özcan et al., 2010; Slawson & Hart, 2011; Wells, Vosseller, & Hart, 2001, 2003; Zachara & Hart, 2006).

The FOXM1 protein level was decreased by shRNA against OGT and by the OGT inhibitors 9D or ST060266, which suggests that OGT increases the FOXM1 protein expression (Fig. 3.10; Caldwell et al., 2010; Lynch et al., 2012). However, FOXM1 did not show any O-GlcNAc modification so that FOXM1 is presumably not O-GlcNAcated by OGT (Caldwell et al., 2010; Lynch et al., 2012). Therefore, OGT seems to upregulate the FOXM1 protein expression only indirectly (Caldwell et al., 2010; Lynch et al., 2012). OGT may elevate the FOXM1 protein expression (Fig. 3.10) by preventing the proteasome-dependent degradation of FOXM1 because the proteasome inhibitor lactacystin abrogated the reduction of the FOXM1 protein level by shRNA against OGT (Lynch et al., 2012) and because shRNA against OGT failed to affect the expression of a proteasomal degradation-resistant N-terminal FOXM1 deletion mutant, which lacked the D-boxes and the KEN-box (Caldwell et al., 2010; Lynch et al., 2012).

6.8.6 Stabilization of FOXM1 by PHGDH

PHGDH (phosphoglycerate dehydrogenase) is a key enzyme for serine biosynthesis and has also been implicated in tumorigenesis due to its oncogenic effects and its amplification or overexpression in some human cancers (DeBerardinis, 2011; Liu, Guo, Li, et al., 2013; Locasale et al., 2011; Luo, 2011; Mullarky, Mattaini, Vander Heiden, Cantley, & Locasale, 2011; Possemato et al., 2011; Seton-Rogers, 2011).

PHGDH interacts possibly indirectly with FOXM1, namely with the N-terminus (aa 1–234) of FoxM1B (corresponding to aa 1–233 of

FOXM1c) (Liu, Guo, Li, et al., 2013). PHGDH increases the protein stability of FOXM1 (Fig. 3.10) by protecting FOXM1 against ubiquitin-dependent degradation by the 26S proteasome (see below) (Liu, Guo, Li, et al., 2013). Since the two D-boxes (aa 6–18) and the KEN-box (aa 206–208) in the N-terminus of FOXM1, which mediate the recognition of FOXM1 by the E3 ubiquitin ligase APC/C^{Cdh1} (Laoukili, Alvarez-Fernandez, et al., 2008; Park, Costa, Lau, Tyner, & Raychaudhuri, 2008), are located within FOXM1's interaction domain for PHGDH (aa 1–233) (Liu, Guo, Li, et al., 2013), it is tempting to speculate that PHGDH might mask these two recognition motifs for Cdh1 thereby preventing the ubiquitination of FOXM1 by APC/C^{Cdh1} and thus the subsequent proteasomal destruction of FOXM1.

ShRNA-mediated knockdown of PHGDH decreased the FOXM1 protein level, but not the *foxm1* mRNA level in U87 and U251 glioma cells, which indicates that PHGDH increases the FOXM1 protein expression (Fig. 3.10) through stabilization of the FOXM1 protein (Liu, Guo, Li, et al., 2013). In fact, shRNA against PHGDH accelerated the decay of FOXM1 protein in the presence of cycloheximide in U251 cells (Liu, Guo, Li, et al., 2013). Moreover, the decline of the FOXM1 protein expression, which was caused by shRNA against PHGDH, was attenuated by the proteasome inhibitor MG132 in U87 cells (Liu, Guo, Li, et al., 2013). These two findings demonstrate that PHGDH protects FOXM1 from degradation by the 26S proteasome (Liu, Guo, Li, et al., 2013). Actually, shRNA-mediated depletion of PHGDH induced the ubiquitylation of FOXM1 in U87 cells, showing that PHDGH prevents the ubiquitination of FOXM1 (Liu, Guo, Li, et al., 2013). Thus, PHGDH stabilizes the FOXM1 protein (Fig. 3.10) because PHGDH protects FOXM1 against proteasomal destruction by preventing its ubiquitylation (Liu, Guo, Li, et al., 2013).

6.9. Control of the *foxm1* promoter by transcription factors

The *foxm1* promoter was shown to be *in vivo* occupied by the transcription factors C/EBPα (Chen et al., 2010; Wang, Salisbury, et al., 2008; Wang, Shi, et al., 2007), CREB (Xia, Huang, et al., 2012), E2F-1 (Chen et al., 2013; Millour et al., 2011; Wang, Shi, et al., 2007), E2F-4 (Wang, Salisbury, et al., 2008; Wang, Shi, et al., 2007), ERα (Horimoto et al., 2011; Millour et al., 2010), ERβ1 (Horimoto et al., 2011), ETV5 (Llaurado et al., 2012), FXR/RXRα (retinoid X receptor α) (Chen et al., 2010; Meng et al., 2011), HIF-1α (hypoxia-inducible factor α) (Xia et al., 2009; Xia, Mo, et al., 2012), HSF1 (Dai et al., 2013), KLF4

(Li, Jia, et al., 2012), c-Myc (Blanco-Bose et al., 2008; Delpuech et al., 2007; Fernandez et al., 2003; Huynh et al., 2010), Mad1 (Huynh et al., 2010), Mxi1 (Delpuech et al., 2007), B-Myb (Down et al., 2012; Lorvellec et al., 2010), c-Myb (Lefebvre et al., 2010), c-Rel (Gieling et al., 2010), p53 (Kurinna et al., 2013), and Sp1 (specificity protein 1) (Chen et al., 2013; Petrovich et al., 2010) as well as by the coactivator YAP (Table 3.2; Mizuno et al., 2012), which binds to DNA in complex with the transcription factor TEAD (TEA domain family member) (Badouel, Garg, & McNeill, 2009; Badouel & McNeill, 2011; Halder & Johnson, 2011; Mauviel, Nallet-Staub, & Varelas, 2012; Saucedo & Edgar, 2007; Sudol & Harvey, 2010; Zhao, Li, & Guan, 2010; Zhao, Li, Lei, & Guan, 2010) and activates the *foxm1* promoter in the YAP–TEAD complex (Fig. 3.10; Table 3.2; Mizuno et al., 2012). Moreover, the *foxm1* promoter is *in vivo* occupied by FOXM1 itself (Table 3.2; Down et al., 2012) because the *foxm1* gene is subject to positive autoregulation (Fig. 3.10; Halasi & Gartel, 2009; Wang, Teh, Ji, et al., 2010; Xie et al., 2010).

Thus, *foxm1* is a direct target gene of at least these 21 transcription factors (Fig. 3.10; Table 3.2). Among them are transcriptional activators of the *foxm1* promoter (E2F-1, ERα, FOXM1, HIF-1α, YAP–TEAD, STAT3, Sp1) (de Olano et al., 2012; Mencalha et al., 2012; Millour et al., 2010, 2011; Mizuno et al., 2012; Petrovich et al., 2010; Wang, Teh, Ji, et al., 2010; Xia et al., 2009; Xia, Mo, et al., 2012) as well as transcriptional repressors of the *foxm1* promoter (C/EBPα, ERβ1, KLF4) (Fig. 3.10; Horimoto et al., 2011; Li, Jia, et al., 2012; Wang, Shi, et al., 2007). Furthermore, the *foxm1* mRNA expression was increased by ETV5 (Llaurado et al., 2012), HSF1 (Dai et al., 2013), c-Myc (Blanco-Bose et al., 2008; Delpuech et al., 2007; Huynh et al., 2010; Keller et al., 2007), B-Myb (Down et al., 2012; Lorvellec et al., 2010; Sadavisam et al., 2012), c-Myb (Lefebvre et al., 2010), and c-Rel (Gieling et al., 2010) but decreased by E2F-4 (Bellelli et al., 2012), Mad1 (Huynh et al., 2010), and Mxi1 (Fig. 3.10; Delpuech et al., 2007).

p53 decreased the *foxm1* mRNA expression in ten studies (Barsotti & Prives, 2009; Bellelli et al., 2012; Bhonde et al., 2006; Chen et al., 2013; Lok et al., 2011; Millour et al., 2011; Pandit et al., 2009; Qu et al., 2013; Rovillain et al., 2011; Spurgers et al., 2006), but it surprisingly increased the *foxm1* expression in one study (Table 3.2; Kurinna et al., 2013).

Hence, *foxm1* is a direct target gene of the E2F family (Barsotti & Prives, 2009; Bellelli et al., 2012; Chen et al., 2013; de Olano et al., 2012; Hallstrom et al., 2008; Millour et al., 2011; Rovillain et al., 2011; Wang, Salisbury, et al., 2008; Wang, Shi, et al., 2007) and the Myc/Max/Mad

Table 3.2 Transcription factors, which regulate the foxm1 expression

Transcription factor	Binding to foxm1 promoter		Regulation of endogenous foxm1 expression		Regulation of foxm1 promoter		References
	Method[a]	Comment[b]	Expression[c]	Manipulation of transcription factor[d]	Manipulation of transcription factor-binding site[e]	Manipulation of transcription factor[d]	
AP-2γ			T	↑ siRNA			Woodfield, Horan, Chen, and Weigel (2007)
CAR			T	↑ KO			Huang et al. (2006)
β-catenin[f]			T, P	↓ KO			Bräuning et al. (2011)
C/EBPα	C		P	↓ P	wt, P	↓ OE, P	Wang, Shi, et al. (2007), Wang, Salisbury, et al. (2008), Chen et al. (2010), Jin et al. (2010)
CREB[g]	C				wt, del, P	↑ B	Xia, Huang, et al. (2012)
E2F-1	C		T, P	↑ OE, dn, siRNA	wt, P	↑ OE, P	Wang, Shi, et al. (2007), Hallstrom, Mori, and Nevins (2008), Barsotti and Prives (2009), Millour et al. (2011), de Olano et al. (2012), Chen et al. (2013)
E2F-4	E, C	S, P	T	↓ OE			Wang, Shi, et al. (2007), Wang, Salisbury, et al. (2008), Bellelli et al. (2012)

Continued

Table 3.2 Transcription factors, which regulate the *foxm1* expression—cont'd

Transcription factor	Binding to *foxm1* promoter		Regulation of endogenous *foxm1* expression		Regulation of *foxm1* promoter			References
	Method	Comment	Expression	Manipulation of transcription factor	Manipulation of transcription factor-binding site	Manipulation of transcription factor		
E2F			T	↓ dn	P	↓		Wang, Shi, et al. (2007), Millour et al. (2011), Rovillain et al. (2011)
ERα	E, C, D	S, C, P	T, P	↑ siRNA, L, A	wt, del, P	↑ OE, L, A		Millour et al. (2010), Horimoto et al. (2011)
ERβ1[h]	C		T, P	↓ OE	wt, P	↓ OE		Horimoto et al. (2011)
ETV5	C		T, P	↑ siRNA				Llaurado et al. (2012)
FOXD3			T	↑ KO				Plank, Frist, LeGrone, Magnuson, and Labosky (2011)
FOXM1	C		T, P	↑ OE	wt	↑ OE		Halasi and Gartel (2009), Wang, Teh, Ji, et al. (2010), Xie et al. (2010), Down et al. (2012)
FOXO3a			T, P	↓ OE, ca, siRNA	wt	↓ ca		Laoukili et al. (2005), Delpuech et al. (2007), Zou et al. (2008), McGovern et al. (2009), Karadedou et al. (2012)
FOXO1 +FOXO3[i]			T	↓ DKO				Sengupta et al. (2013)

FXR/RXRα[j]	E, C	IV, S, C	T	↑ ca, KO, L	P, H	↑ L	Huang et al. (2006), Chen et al. (2010), Meng et al. (2010, 2011), Zhang, Li, et al. (2012), Zhang, Wang, et al. (2012)
GATA3			T	↓ siRNA			Carr et al. (2012)
GLI-1			T, P	↑ OE, siRNA			Teh et al. (2002), Calvisi et al. (2009)
HIF-1α	C		T, P	↑ siRNA	wt, del, P	↑ siRNA	Xia et al. (2009), Xia, Mo, et al. (2012), Raghavan et al. (2012)
HIF-2α			T	↑ siRNA			Raghavan et al. (2012)
HNF-6			T	↑ OE			Tan et al. (2006)
HSF1	C		T, P	↑ ca, siRNA, KO	wt, del, P	↑ HS	Dai et al. (2013)
KLF4	C		P	↓ OE, KO	wt	↓ OE, siRNA	Li, Jia, et al. (2012)
Mad1	C		T	↓ OE			Huynh et al. (2010)
Mxi1	C		T	↓ siRNA			Delpuech et al. (2007)
B-Myb	C		T	↑ siRNA, KO			Lorvellec et al. (2010), Down et al. (2012), Sadavisam et al. (2012)

Continued

Table 3.2 Transcription factors, which regulate the *foxm1* expression—cont'd

Transcription factor	Binding to *foxm1* promoter		Regulation of endogenous *foxm1* expression			Regulation of *foxm1* promoter		References
	Method	Comment	Expression		Manipulation of transcription factor	Manipulation of transcription factor-binding site	Manipulation of transcription factor	
c-Myb	C		↑	T, P	siRNA			Lefebvre et al. (2010)
c-Myc	C, D		↑	T, P	siRNA, tg, KO			Fernandez et al. (2003), Delpuech et al. (2007), Keller et al. (2007), Blanco-Bose et al. (2008), Huynh et al. (2010)
Notch1			↑	T, P	OE, siRNA			Wang, Zhang, Snyder, et al. (2010)
p53[k]	C		↓	T, P	OE, P, siRNA, KO, CC, SM, GSE	wt, P	↓ OE, P	Bhonde et al. (2006), Spurgers et al. (2006), Barsotti and Prives (2009), Pandit, Halasi, and Gartel (2009), Lok et al. (2011), Millour et al. (2011), Rovillain et al. (2011), Chen et al. (2013), Kurinna et al. (2013), Qu et al. (2013)
p53			↑	T, P	siRNA, KO			Tan et al. (2010), Kurinna et al. (2013)
c-Rel	C		↑	T, P	KO			Gieling et al. (2010)

Sox2		T	↑ tg	I?	↑ OE	Tompkins et al. (2011)
Sox17		T	↑ OE			Lange, Keiser, Wells, Zorn, and Whitsett (2009)
Spl1	C	P	↑ siRNA	wt	↑ OE	Petrovich, Costa, Lau, Raychaudhuri, and Tyner (2010), Chen et al. (2013)
SPDEF		T	↓ KO			Ren et al. (2013)
STAT3	E, C S, C	T	↑ IS3	wt	↑ IS3	Mencalha et al. (2012)
TEAD				wt	↑ OE, del	Mizuno et al. (2012)
YAP	C	T	↑ siRNA	wt	↑ OE, ca, P	Mizuno et al. (2012)

AP-2γ, activator protein-2γ; CAR, constitutive androstane receptor; C/EBPα, CCAAT/enhancer-binding protein α; CREB, cAMP (cyclic adenosine monophosphate) response element-binding protein; ER, estrogen receptor; ETV5, ETS variant 5, ERM, Ets-related molecule; FOX, Forkhead box; FXR, farnesoid X receptor; GATA3, GATA-binding protein 3; GLI-1, glioma transcription factor-1; HIF, hypoxia-inducible factor; HNF-6, hepatocyte nuclear factor-6; HSF1, heat shock factor 1; KLF4, Krüppel-like factor 4; Mad1, Max dimerizer 1, MXD1, Max dimerization protein 1; Mxi1, Max interactor 1; c-Myb, MYB; B-Myb, MYBL2, MYB-like 2; c-Myc, MYC; p53, TP53, tumor protein p53; c-Rel, REL; RXRα, retinoid X receptor α; Sox, SRY (sex-determining region Y) box, SRY-related HMG (high mobility group)-box; Sp1, specificity protein 1; SPDEF, SAM pointed domain-containing ETS transcription factor, PDEF, prostate-derived ETS transcription factor; STAT3, signal transducer and activator of transcription 3; TEAD, TEAD1, TEA domain family member 1; YAP, Yes-associated protein.

[a]**Method:** E=EMSA=electrophoretic mobility shift assay; C=ChIP=chromatin immunoprecipitation assay; D=DNAP=DNA precipitation assay=DAPA=DNA affinity pull-down assay.

[b]**Comment:** IV=in vitro=purified or in vitro transcribed/translated transcription factor; S=supershift experiments; C=competition experiments; P=binding site was point-mutated.

[c]**Expression:** T=transcript=mRNA=endogenous mRNA level affected; P=protein=endogenous protein level affected.

[d]**Manipulation of transcription factor:** OE=overexpression of wild-type; dn=dominant-negative form; ca=constitutively active form; del=analyzed with deletion mutant of the transcription factor; P=analyzed with point mutant of the transcription factor; siRNA=siRNA-mediated or shRNA-mediated knockdown; tg=transgenic cells/mice; KO=knockout cells/mice; CC=p53-deficient colon carcinoma cell lines; SM=small-molecule activator of p53 (nutlin-3); L=activation by ligand; A=ERα antagonist (4-OHT (4-hydroxytamoxifen), ICI (ICI182780, fulvestrant, Faslodex)); GSE=genetic suppressor element (100–300 nt (nucleotide) long rat p53 cDNA segment); B=HBX=HBV (hepatitis B virus) X protein; DKO=cardiomyocyte-specific conditional foxo1/foxo3 double knockout; HS=heat shock; IS3=small-molecule inhibitor of STAT3 (LLL-3).

^a**Manipulation of transcription factor-binding site:** wt = "wild-type" *foxm1* promoter; del = analyzed with deletion mutant of *foxm1* promoter; P = transcription factor-binding site was point-mutated; H = transcription factor-binding site in context of heterologous core promoter; I? = 0.78 kb (kilobase) *foxm1* reporter construct containing a regulatory sequence from the first intron.

^fThe negative effect of β-catenin on the *foxm1* mRNA and protein expression was observed in female mice in the presence of TCPOBOP (1,4-bis[2-(3,5-dichloropyridyloxy)]benzene), but β-catenin had no negative effect on the *foxm1* mRNA level in the absence of TCPOBOP (Bräuning et al., 2011).

^gThe activation of the *foxm1* promoter by HBX (hepatitis B virus X protein) was considerably reduced if a putative CREB-binding site in the *foxm1* promoter was point-mutated, but an effect of CREB on the *foxm1* promoter or the *foxm1* expression was not shown (Xia, Huang, et al., 2012).

^hERβ1, which displaces ERα from the *foxm1* promoter, represses the *foxm1* promoter and downregulates the *foxm1* mRNA and protein expression only in ERα-positive but not in ERα-negative breast carcinoma cell lines (Horimoto et al., 2011).

ⁱThe *foxm1* mRNA level was increased in the neonatal hearts of mice with a cardiomyocyte-specific conditional *foxo1*/*foxo3* double knockout compared to the neonatal hearts of control mice, indicating that FOXO1 + FOXO3 decrease the *foxm1* mRNA expression (Sengupta et al., 2013).

^jThe FXR/RXRα binding site is located in intron 3 of the *foxm1* gene.

^kThe point-mutated versions of the *foxm1* promoter had point mutations in the two putative E2F-binding sites (but not in any putative p53-binding site).

^lOverexpression of MnSOD (manganese superoxide dismutase) increased the FOXM1 protein level in H460 cells lung cancer cells and shRNA against Sp1 abolished this MnSOD-induced increase in the FOXM1 protein level, which indicates an upregulation of the FOXM1 protein expression by Sp1. However, the effect of the Sp1 knockdown on the FOXM1 protein expression in the absence of exogenous MnSOD was not analyzed.

network (Blanco-Bose et al., 2008; Delpuech et al., 2007; Fernandez et al., 2003; Huynh et al., 2010; Keller et al., 2007) and in both cases the *foxm1* gene is antagonistically regulated by individual members of the same transcription factor family (Fig. 3.10; Table 3.2) (see below).

Among the transcription factors, which regulate the *foxm1* expression, are protooncoproteins (β-catenin, ETV5, GLI-1, c-Myc, c-Myb, Notch1, STAT3, c-Rel, YAP) as well as tumor suppressors (GATA3, KLF4, p53), and potential tumor suppressors (FOXO3a, Mad1) (Fig. 3.10; Table 3.2). Again, this given fact offers the opportunity for uncontrolled *foxm1* expression in such human tumors, where these transcription factors are deregulated either because of mutations in their genes or because of mutations in the many cancer-related signal transduction pathways that control their expression or activity (Wierstra & Alves, 2007c).

In addition, several other transcription factors, which regulate the *foxm1* expression, play important proliferative roles (E2F-1, ERα, HSF1, B-Myb) or antiproliferative roles (C/EBPα, Mxi1) (Fig. 3.10; Table 3.2) and are known to be deregulated in cancer so that they could contribute to uncontrolled expression of *foxm1* in tumors, too (Wierstra & Alves, 2007c).

Since the transcription factors, which control the *foxm1* expression (Fig. 3.10; Table 3.2), encompass not only key players in proliferation and tumorigenesis (see above), but also central players in other cellular processes, which in turn are regulated by various signal transduction chains, it seems obvious that the expression of *foxm1* might be subject to control by very different kinds of signaling inputs.

6.10. Regulation of the FOXM1 expression by miRNAs

So far, five miRNAs have been described to reduce the FOXM1 expression (Fig. 3.10): miRNA-370 (Zhang, Zeng, et al., 2012) and miR-31 (Lin et al., 2013) downregulated the *foxm1* mRNA and FOXM1 protein levels. miR-34 decreased the *foxm1* mRNA level (Chang et al., 2007; Christoffersen et al., 2010; He, He, Lim, et al., 2007; He, He, Lowe, & Hannon, 2007). miR-134 (Li, Wang, et al., 2012) and miR-200b (Bao et al., 2011) diminished the FOXM1 protein level.

MiR-370 (Zhang, Zeng, et al., 2012), miR-31 (Lin et al., 2013), and miR-134 (Li, Wang, et al., 2012) target the 3′-UTR of the *foxm1* transcript, which was demonstrated through the fusion of the coding sequence for luciferase to the wild-type *foxm1* 3′-UTR and to a point-mutated version of the *foxm1* 3′-UTR that eliminates the predicted miRNA-370 (Zhang, Zeng, et al., 2012), miR-31 (Lin et al., 2013), or miR-134 (Li, Wang, et al., 2012) target sequences.

6.11. Selected pathways that regulate the expression of FOXM1

6.11.1 Regulation of the FOXM1 expression by the RB tumor-suppressor pathway

6.11.1.1 *foxm1* is a direct target gene of the E2F family and pocket proteins
foxm1 is a direct E2F target gene (Fig. 3.10; Table 3.2; Chen et al., 2013; Millour et al., 2011; Wang, Salisbury, et al., 2008; Wang, Shi, et al., 2007). ChIP assays demonstrated that the *foxm1* promoter is *in vivo* occupied by the E2F family members E2F-1 (Chen et al., 2013; Millour et al., 2011; Wang, Shi, et al., 2007) and E2F-4 (Wang, Salisbury, et al., 2008; Wang, Shi, et al., 2007) as well as by the pocket proteins RB (Millour et al., 2011; Wang, Salisbury, et al., 2008; Wang, Shi, et al., 2007) and p130 (RBL2 (retinoblastoma-like 2)) (Down et al., 2012). Accordingly, EMSAs showed that E2F-4 and the E2F-4/p130 complex bind to an E2F-binding site in the *foxm1* promoter *in vitro* (Wang, Shi, et al., 2007).

E2F-1 transactivates the *foxm1* promoter (de Olano et al., 2012; Millour et al., 2011), depending on an E2F-binding site in the *foxm1* promoter (Millour et al., 2011). Accordingly, E2F-1 overexpression increased the *foxm1* mRNA expression (Hallstrom et al., 2008) whereas knockdown of E2F-1 with siRNA/shRNA decreased the *foxm1* mRNA and protein expression (Barsotti & Prives, 2009; Chen et al., 2013; de Olano et al., 2012). In contrast, overexpression of E2F-4 diminished the *foxm1* mRNA level (Bellelli et al., 2012). Thus, the *foxm1* expression is upregulated by E2F-1 (Barsotti & Prives, 2009; Chen et al., 2013; de Olano et al., 2012; Hallstrom et al., 2008; Millour et al., 2011) but downregulated by E2F-4 (Fig. 3.10; Table 3.2; Bellelli et al., 2012). This opposite regulation of *foxm1* by E2F-1 and E2F-4 (Barsotti & Prives, 2009; Bellelli et al., 2012; Chen et al., 2013; de Olano et al., 2012; Hallstrom et al., 2008) corresponds with their general roles as activating or repressing E2F family members, respectively (Attwooll, Lazzerini Denchi, & Helin, 2004; Blais & Dynlacht, 2004; Bracken, Ciro, Cocito, & Helin, 2004; Calzone, Gelay, Zinovyev, Radvanyi, & Barillot, 2008; Cam & Dynlacht, 2003; Chen, Chien, et al., 2009; Chen, Dominguez-Brauer, et al., 2009; Chen, Korfhagen, et al., 2009; Chen, Tsai, & Leone, 2009; DeGregori, 2002; DeGregori & Johnson, 2006; Dimova & Dyson, 2005; Iaquinta & Lees, 2007;

McClellan & Slack, 2007; Morgan, 2008b; Mundle & Saberwal, 2003; Rowland & Bernards, 2006; Stevens & LaThangue, 2003; Trimarchi & Lees, 2002; Tsantoulis & Gorgoulis, 2005).

Loss of RB raised the *foxm1* mRNA level in asynchronously proliferating MEFs, indicating that RB represses the *foxm1* mRNA expression (Fig. 3.10; Markey et al., 2007). Importantly, the RB loss-induced increase in the *foxm1* mRNA expression was observed not only following chronic RB loss (i.e., in $Rb^{-/-}$ MEFs from *Rb* knockout mice harboring germline loss of the *Rb* gene), but also following acute RB loss (i.e., in adult fibroblasts from $Rb^{loxP/loxP}$ mice with floxed *Rb* alleles, where the *Rb* gene was acutely knocked out through infection with adenovirus encoding Cre recombinase), the latter system of which excludes long-term adaptive changes in gene expression that occur during development of the $Rb^{-/-}$ embryos (Markey et al., 2007). In accordance with this negative effect of RB on the *foxm1* mRNA expression (Markey et al., 2007), RB knockdown elevated the FOXM1 protein level in MDA-MB-231 breast cancer cells (Dean et al., 2012). Hence, the key tumor suppressor RB represses the *foxm1* expression (Fig. 3.10; Dean et al., 2012; Markey et al., 2007).

Accordingly, siRNA to the three pocket proteins (RB, p107, p130) partially abrogated the downregulation of the *foxm1* mRNA levels by the DNA-damaging agent daunorubicin in MCF-7 cells, suggesting a role of (one of) the pocket proteins for repression of the *foxm1* expression in response to daunorubicin-induced DNA damage (Barsotti & Prives, 2009).

Together, these findings point to a repression of the *foxm1* promoter by E2F/RB or E2F/p130 complexes (Fig. 3.10). Accordingly, overexpression of a dn E2F protein, which interferes with the repression of E2F target genes by pocket proteins, enhanced the *foxm1* mRNA expression in conditionally immortalized HMFs (Rovillain et al., 2011) and the E2F site(s) in the *foxm1* promoter exerted a negative effect on the *foxm1* transcription in MCF-7 breast cancer cells (Millour et al., 2011), C33A cervix cancer cells, and HEK293 human embryonic kidney cells (Table 3.2; Wang, Shi, et al., 2007).

6.11.1.2 A contradictory finding for RB

Surprisingly, a single study showed an upregulation of the *foxm1* mRNA expression by RB because shRNA-mediated silencing of RB in MCF-7 breast cancer cells reduced not only the endogenous *foxm1* mRNA expression, but also the mRNA level of ectopically overexpressed FOXM1 (Fig. 3.10; Carr et al., 2012). In clear contrast, three other studies demonstrated a downregulation of the FOXM1 mRNA and protein expression by RB because chronic loss of RB in $RB^{-/-}$ MEFs increased the *foxm1* mRNA expression (Markey et al., 2007), because acute loss of RB in conditional *Rb* knockout MEFs enhanced the *foxm1* mRNA expression (Markey et al., 2007), because RB knockdown elevated the FOXM1 protein level in MDA-MB-231 breast cancer cells (Dean et al., 2012) and because siRNA to the three pocket proteins (RB, p107, p130) partially abrogated the downregulation of the *foxm1* mRNA level by daunorubicin in MCF-7 cells (Barsotti & Prives, 2009). Consequently, the significance of the singular positive effect of RB on the *foxm1* mRNA level in MCF-7 cells (Carr et al., 2012) remains unclear.

6.11.1.3 Astonishing results for an exceptional RB mutant protein

Since RB downregulates the *foxm1* mRNA and protein expression (Fig. 3.10; Dean et al., 2012; Markey et al., 2007) the *foxm1* mRNA level is expected to be decreased by a ca RB mutant and increased by a dn RB mutant.

In transgenic mice, which expressed in the liver a mutant RB protein (mt-Rb) with 12 of the 16 Cdk consensus sites mutated to alanine residues, the *foxm1* mRNA level was increased in tumorous transgenic livers (with macroscopic tumors) (Fig. 3.10) but not in nontumorous transgenic livers (without macroscopic tumors) (Wang, Hikosaka et al., 2012). Due to the alanine substitutions mt-Rb should represent a ca RB mutant. However, surprisingly, mt-RB transgenic mice developed liver tumors (Wang, Hikosaka et al., 2012), indicating that mt-RB behaves as a dn RB mutant in this context. Thus, the unexpected upregulation of the *foxm1* mRNA level by mt-Rb (Fig. 3.10; Wang, Hikosaka, et al., 2012) might be attributed to its exceptional dominant-negative behavior in this context. Furthermore, the failure of mt-Rb to affect the *foxm1* mRNA expression in nontumorous transgenic livers suggests that the rise in the *foxm1* mRNA level in tumorous mt-Rb transgenic livers (Wang, Hikosaka, et al., 2012) is not a primary effect of mt-Rb but results from a secondary deregulation event in these liver tumors.

6.11.2 Regulation of the FOXM1 expression by the p53 tumor-suppressor pathway

6.11.2.1 *foxm1* is a direct target gene of p53

foxm1 is a direct p53 target gene because the *foxm1* promoter was *in vivo* occupied by p53 in adult mouse livers, namely, in untreated quiescent livers and during the proliferative phase of liver regeneration after PHx (Table 3.2;

Kurinna et al., 2013). The key tumor suppressor p53 represses the *foxm1* promoter (Chen et al., 2013; Millour et al., 2011) and thus reduces the *foxm1* mRNA and protein expression (Fig. 3.10; Table 3.2; Barsotti & Prives, 2009; Bellelli et al., 2012; Bhonde et al., 2006; Chen et al., 2013; Lok et al., 2011; Millour et al., 2011; Pandit et al., 2009; Qu et al., 2013; Rovillain et al., 2011; Spurgers et al., 2006).

6.11.2.2 The *foxm1* mRNA level is reduced by p21 and the miR34 family, which are encoded by p53 target genes

The products of three direct p53 target genes decrease the *foxm1* mRNA expression, namely, on the one hand, the CKI (Cdk inhibitor) p21 (Barsotti & Prives, 2009; Bellelli et al., 2012; Millour et al., 2011; Rovillain et al., 2011; Stoyanova et al., 2012; Weymann et al., 2009) and, on the other, the miR-34 family (Fig. 3.10; Chang et al., 2007; Christoffersen et al., 2010; He, He, Lim, et al., 2007), which is encoded by the *miR-34a* and *miR-34b/miR-34c* loci (Babashah & Soleimani, 2011; Bueno, Perez de Castro, & Malumbres, 2008; Chivukula & Mendell, 2008; Corney & Nikitin, 2008; He, He, & Hannon, 2007; He, He, Lowe, et al., 2007; Hermeking, 2010, 2012; Lee & Dutta, 2009; Lotterman, Kent, & Mendell, 2008; Medina & Slack, 2008; Sato, Tshuchiya, Meltzer, & Shimizu, 2011; Ventura & Jacks, 2009). Consequently, p53 may indirectly downregulate the *foxm1* mRNA level through either p21 or the miR-34 family (Fig. 3.10), which are the products of the direct p53 target genes *p21* (El-Deiry et al., 1993) and *miR-34a* or *miR-34b/miR34c* (Bommer et al., 2007; Cannell et al., 2010; Chang et al., 2007; Corney, Flesken-Nikitin, Godwin, Wang, & Nikitin, 2007; He, He, Lim, et al., 2007; Raver-Shapira et al., 2007; Tarasov et al., 2007).

Additionally, p53 can repress the *foxm1* mRNA expression independent of p21 (Barsotti & Prives, 2009) in accordance with the *in vivo* occupancy of the *foxm1* promoter by p53 (Table 3.2; Kurinna et al., 2013).

6.11.2.3 p53 and p21 are involved in the downregulation of FOXM1 by daunorubicin and nutlin-3

Daunorubicin, a DNA-damaging anthracyclin, reduces the *foxm1* mRNA and protein levels dependent on p53 and p21 (Fig. 3.10; Barsotti & Prives, 2009). Also nutlin-3, a small-molecule MDM2 (mouse double minute 2) inhibitor and thus p53 activator, decreases the *foxm1* mRNA and protein expression dependent on p53 and p21 (Fig. 3.10; Barsotti & Prives, 2009).

However, both p53 (Barsotti & Prives, 2009; Bellelli et al., 2012; Bhonde et al., 2006; Chen et al., 2013; Lok et al., 2011; Millour et al., 2011; Pandit et al., 2009; Qu et al., 2013; Rovillain et al., 2011; Spurgers et al., 2006) and p21 (Barsotti & Prives, 2009; Bellelli et al., 2012; Millour et al., 2011; Rovillain et al., 2011; Stoyanova et al., 2012; Weymann et al., 2009) can diminish the *foxm1* mRNA and protein expression in the absence of nutlin-3 and in the absence of daunorubicin-induced DNA damage. Accordingly, p53 can repress the *foxm1* promoter independent of nutlin-3 and daunorubicin-induced DNA damage (Table 3.2; Chen et al., 2013; Millour et al., 2011).

6.11.2.4 p53, but not p21, is involved in the downregulation of FOXM1 by oxaliplatin

The DNA-damaging, alkylating agent oxaliplatin decreases the *foxm1* mRNA and protein expression dependent on p53 but independent of p21 (Qu et al., 2013).

Yet, p53 can also repress the *foxm1* promoter (Table 3.2; Chen et al., 2013; Millour et al., 2011) and downregulate the *foxm1* mRNA and protein levels (Barsotti & Prives, 2009; Bellelli et al., 2012; Bhonde et al., 2006; Chen et al., 2013; Lok et al., 2011; Millour et al., 2011; Pandit et al., 2009; Qu et al., 2013; Rovillain et al., 2011; Spurgers et al., 2006) in the absence of oxaliplatin-induced DNA damage.

6.11.2.5 p53 blocks the binding of E2F-1 and Sp1 to the *foxm1* promoter

The *foxm1* mRNA expression was reduced by p53 but not by the p53 mutants p53(L194R) and p53(R249S) with point mutations in the DBD (Chen et al., 2013). Accordingly, the *foxm1* promoter was repressed by p53 but not by the p53 mutants p53(L194R) and p53(R249S) (Chen et al., 2013).

p53 blocked the binding of E2F-1 and Sp1 to the *foxm1* promoter because the *in vivo* occupancy of the *foxm1* promoter with E2F-1 and Sp1 was decreased by p53 overexpression but increased by shRNA-mediated knockdown of p53 (Chen et al., 2013). Since the *foxm1* promoter is transactivated by both E2F-1 (de Olano et al., 2012; Millour et al., 2011) and Sp1 (Table 3.2; Petrovich et al., 2010) p53 may indirectly repress the *foxm1* promoter by blocking the binding of E2F-1 and Sp1 to the *foxm1* promoter and thus preventing the transactivation of the *foxm1* promoter by E2F-1 and Sp1 (Fig. 3.10; Chen et al., 2013). This assumption is supported by the finding that the *in vivo* occupancy of the *foxm1* promoter with E2F-1 and Sp1 was unaffected by the p53 mutants p53(L194R) and p53(R249S), which neither repressed the *foxm1* promoter nor reduced the *foxm1* mRNA expression (Chen et al., 2013).

The antioxidant enzyme MnSOD (manganese superoxide dismutase) seems to activate the *foxm1* expression via the pathway MnSOD ⊣ p53 ⊣ E2F-1 + Sp1 → *foxm1* (Fig. 3.10; Chen et al., 2013):

MnSOD, which activated the *foxm1* promoter and upregulated the FOXM1 protein level (Fig. 3.10), increased the *in vivo* occupancy of the *foxm1* promoter with E2F-1 and Sp1 (Chen et al., 2013). Conversely, the *in vivo* occupancy of the *foxm1* promoter with E2F-1 and Sp1 was decreased by shRNA against MnSOD, which repressed the *foxm1* promoter and downregulated the FOXM1 protein level (Chen et al., 2013). The MnSOD-induced increase in the FOXM1 protein expression was eliminated by shRNA-mediated depletion of either E2F-1 or Sp1, demonstrating that MnSOD upregulates the FOXM1 protein expression through E2F-1 and Sp1 (Fig. 3.10; Chen et al., 2013). Hence, MnSOD seems to activate the *foxm1* promoter by enhancing the binding of E2F-1 and Sp1 to the *foxm1* promoter and thus promoting the transactivation of the *foxm1* promoter by E2F-1 and Sp1 (Fig. 3.10; Chen et al., 2013).

Moreover, MnSOD downregulated the p53 protein expression (Fig. 3.10) but did not affect the E2F-1 or Sp1 protein expression (Chen et al., 2013). Conversely, p53 diminished the MnSOD protein level (Fig. 3.10) but not the E2F-1 or Sp1 protein levels (Chen et al., 2013). The negative effects of shRNA against MnSOD on the FOXM1 protein expression and on the *in vivo* occupancy of the *foxm1* promoter with E2F-1 and Sp1 were abolished by shRNA-mediated silencing of p53, indicating that MnSOD activates the *foxm1* promoter via the pathway MnSOD ⊣ p53 ⊣ E2F-1 + Sp1 → *foxm1* promoter (Fig. 3.10; Chen et al., 2013).

6.11.2.6 Contradictory findings for p53

Surprisingly, two studies reported a positive effect of p53 on the FOXM1 mRNA (Kurinna et al., 2013) and protein (Tan et al., 2010) expression (Fig. 3.10): first, siRNA-mediated depletion of p53 downregulated the FOXM1 protein level in U2OS osteosarcoma cells (Table 3.2; Tan et al., 2010). Second, the *foxm1* mRNA expression was reduced in the livers of adult $p53^{-/-}$ mice compared to those of wild-type mice, namely, in untreated quiescent livers and in proliferating livers during PHx-induced liver regeneration (Table 3.2; Kurinna et al., 2013).

In contrast, 10 other studies reported a negative effect of p53 on the FOXM1 mRNA and protein expression (Fig. 3.10) because they demonstrated that p53 represses the *foxm1* promoter (Chen et al., 2013; Millour et al., 2011) and decreases the FOXM1 mRNA and protein levels (Table 3.2) in various cell lines (BJ human foreskin fibroblasts, $p53^{-/-}$ MEFs, HMFs, U2OS cells, MCF-7 breast cancer cells, PC3 and LNCaP prostate cancer cells, H1299 NSCLC (non-small cell lung cancer) cells, H460 large lung cancer cells, A549 lung adenocarcinoma cells, HepG2 and SMMC-7721 HCC cells, HCT116 and other CRC (colorectal cancer) cells, HTH74 and 8505C ATC (anaplastic thyroid carcinoma) cells), among them even U2OS cells (Barsotti & Prives, 2009; Bellelli et al., 2012; Bhonde et al., 2006; Chen et al., 2013; Lok et al., 2011; Millour et al., 2011; Pandit et al., 2009; Qu et al., 2013; Rovillain et al., 2011; Spurgers et al., 2006).

Therefore, the significance of the exceptional positive effects of p53 on the FOXM1 expression (Kurinna et al., 2013; Tan et al., 2010) remains unclear although the downregulation of the *foxm1* mRNA and protein levels in $p53^{-/-}$ mouse livers (Kurinna et al., 2013) or p53 knockdown cells (Tan et al., 2010), respectively, indicates an upregulation of the *foxm1* mRNA and protein expression by p53 (Table 3.2).

6.11.3 The two major human tumor-suppressor pathways control the expression of FOXM1

The two major tumor-suppressor pathways in human cells are the RB pathway and the p53 pathway (Hanahan & Weinberg, 2000, 2011; Sherr, 2004; Sherr & McCormick, 2002; Stein & Pardee, 2004; Vogelstein & Kinzler, 2004; Weinberg, 2006). The RB tumor-suppressor pathway p16 ⊣ cyclin D1/Cdk4 ⊣ RB leads from the tumor suppressor p16 via cyclin D1/Cdk4, a complex of two protooncoproteins, to the tumor suppressor RB (Campisi, 2003; Fanciulli, 2006; Hahn & Weinberg, 2002a, 2002b; Hall & Peters, 1996; Polager & Ginsberg, 2009; Sherr, 2000, 2001, 2002, 2004, 2006; Sherr & McCormick, 2002; Stein & Pardee, 2004; Vogelstein & Kinzler, 2004; Weinberg, 2006; Yu & Hahn, 2004). Similarly, the p53 tumor-suppressor pathway ARF ⊣ MDM2 ⊣ p53 leads from the tumor-suppressor ARF (alternative reading frame, CDKN2A (Cdk inhibitor 2A), p14, p19) via the protooncoprotein MDM2 to the tumor suppressor p53 (Campisi, 2003; Hahn & Weinberg, 2002a, 2002b; Polager & Ginsberg, 2009; Sherr, 1998, 2000, 2001, 2002, 2004, 2006; Sherr & McCormick, 2002; Sherr & Weber, 2000; Stein & Pardee, 2004; Vogelstein & Kinzler, 2004; Weinberg, 2006; Yu & Hahn, 2004).

The RB tumor-suppressor pathway is inactivated in virtually every human cancer cell owing to either mutations in the *RB1* gene itself or alterations of its upstream regulators (Fanciulli, 2006; Hall & Peters, 1996; Stein & Pardee, 2004; Weinberg, 2006).

Likewise, the p53 tumor-suppressor pathway is inactivated in virtually every human cancer cell (Sherr & McCormick, 2002; Stein & Pardee, 2004; Weinberg, 2006). Actually, *p53* is the most frequently mutated gene in human tumors (Levine, Momand, & Finlay, 1991; Stein & Pardee, 2004; Weinberg, 2006). Mutations in the *p53* gene occur in about half of all human cancers (Greenblatt, Bennett, Hollstein, & Harris, 1994; Hainaut & Hollstein, 2000; Hainaut et al., 1997; Hollstein et al., 1994; Hollstein, Sidransky, Vogelstein, & Harris, 1991; Olivier, Hollstein, & Hainaut, 2010) and the other half is thought to contain alterations in other components of the p53 pathway (Stein & Pardee, 2004; Weinberg, 2006).

foxm1 is a direct target gene of the RB tumor-suppressor pathway because *foxm1* is activated by E2F-1 (Barsotti & Prives, 2009; Chen et al., 2013; de Olano et al., 2012; Hallstrom et al., 2008; Millour et al., 2011) but repressed by RB (Fig. 3.10; Dean et al., 2012; Markey et al., 2007), which both bind to the *foxm1* promoter (Table 3.2; Chen et al., 2013; Millour et al., 2011; Wang, Salisbury, et al., 2008; Wang, Shi, et al., 2007). Additionally, *foxm1* is a direct target gene of the p53 tumor-suppressor pathway because *foxm1* is repressed by p53 (Fig. 3.10; Barsotti & Prives, 2009; Bellelli et al., 2012; Bhonde et al., 2006; Chen et al., 2013; Lok et al., 2011; Millour et al., 2011; Pandit et al., 2009; Qu et al., 2013; Rovillain et al., 2011; Spurgers et al., 2006), which binds to (Kurinna et al., 2013) and represses (Chen et al., 2013; Millour et al., 2011) the *foxm1* promoter (Table 3.2). Hence, the expression of *foxm1* is controlled by both the RB pathway and the p53 pathway, so that *foxm1* represents a target gene of the two major human tumor-suppressor pathways.

Since the RB and p53 tumor-suppressor pathways are inactivated in virtually every human cancer cell (Hanahan & Weinberg, 2000, 2011; Sherr, 2004; Sherr & McCormick, 2002; Stein & Pardee, 2004; Vogelstein & Kinzler, 2004; Weinberg, 2006), overexpression of FOXM1 is predicted to occur in a large percentage of human tumors. In fact, FOXM1 is overexpressed in a vast number of human cancers (Banz et al., 2009; Bektas et al., 2008; Bellelli et al., 2012; Calvisi et al., 2009; Chan et al., 2008; Chen, Yang, et al., 2012; Chu et al., 2012; Craig et al., 2013; Dibb et al., 2012; Elgaaen et al., 2012; Francis et al., 2009; Garber et al., 2001; Gemenetzidis et al., 2009; Gialmanidis et al., 2009; Guerra et al., 2013; He et al., 2012; Hodgson et al., 2009; Hui et al., 2012;

Iacobuzio-Donahue et al., 2003; Janus et al., 2011; Jiang et al., 2011; Kalin et al., 2006; Kalinina et al., 2003; Kim et al., 2006; Kretschmer et al., 2011; Laurendeau et al., 2010; Li, Peng, et al., 2013; Li et al., 2009; Lin, Guo, et al., 2010; Lin, Madan, et al., 2010; Liu, Dai, et al., 2006; Llaurado et al., 2012; Lok et al., 2011; Nakamura, Hirano, et al., 2010; Nakamura, Thakore, et al., 2004; Nakamura, Yamashita, et al., 2010; Newick et al., 2012; Obama et al., 2005; Okabe et al., 2001; Park et al., 2012; Pellegrino et al., 2010; Perou et al., 2000; Pignot et al., 2012; Pilarsky et al., 2004; Qu et al., 2013; Rickman et al., 2001; Rodriguez et al., 2005; Romagnoli et al., 2009; Rosty et al., 2005; Salvatore et al., 2007; Santin et al., 2005; Schüller et al., 2006; Sorlie et al., 2001; Sun et al., 2011; Takahashi et al., 2006; Teh et al., 2002; The Cancer Genome Atlas Research Network, 2011; Uddin et al., 2011, 2012; van den Boom et al., 2003; Waseem et al., 2010; Wonsey & Follettie, 2005; Wu, Liu, et al., 2010; Xia, Huang, et al., 2012; Xia, Wang, et al., 2012; Xue et al., 2012; Yokomine et al., 2009; Yoshida et al., 2007; Zeng et al., 2009; Zhang, Zeng, et al., 2012).

6.11.4 Regulation of the FOXM1 expression by the Myc/Max/Mad network
6.11.4.1 The Myc/Max/Mad network
The Myc/Max/Mad network of bHLHLZ (basic region/helix–loop–helix/leucine zipper) transcription factors consists of the Myc family (c-Myc, N-Myc (MYCN, Myc oncogene neuroblastoma-derived), L-Myc (MYCL1, Myc oncogene lung carcinoma-derived)), the Mad (MXD) family (Mad1, Mxi1, Mad3 (MXD3 (Max dimerization protein 3)), Mad4 (MXD4 (Max dimerization protein 4)), Mnt (MNT (Max's next tango), Rox), and Max, the latter of which is the common heterodimerization partner for Myc proteins, Mad proteins, and Mnt but also forms Max/Max homodimers (Baudino & Cleveland, 2001; Eisenman, 2001b; Foley & Eisenman, 1999; Grandori, Cowley, James, & Eisenman, 2000; Hooker & Hurlin, 2005; Hurlin & Dezfouli, 2004; Hurlin & Huang, 2006; Hurlin et al., 2004; Lüscher, 2001, 2012; McArthur et al., 1998; Rottmann & Lüscher, 2006; Wahlstrom & Henriksson, 2007; Zhou & Hurlin, 2001).

The Mad family and Mnt represent antagonists for the Myc family because the Myc proteins stimulate proliferation, promote cell cycle entry and progression, inhibit differentiation, induce apoptosis, and cause transformation and tumorigenesis whereas the Mad proteins and Mnt inhibit proliferation, prevent cell cycle entry and progression, promote differentiation, suppress apoptosis, and interfere with transformation and tumorigenesis

(Adhikary & Eilers, 2005; Amati, Frank, Donjerkovic, & Taubert, 2001; Baudino & Cleveland, 2001; Dang, 2012; Eisenman, 2001b; Foley & Eisenman, 1999; Grandori et al., 2000; Hooker & Hurlin, 2005; Hurlin & Dezfouli, 2004; Hurlin & Huang, 2006; Hurlin et al., 2004; Larsson & Henriksson, 2010; Leon, Ferrandiz, Acosta, & Delgado, 2009; Lüscher, 2001, 2012; McArthur et al., 1998; Meyer & Penn, 2008; Oster, Ho, Soucle, & Penn, 2002; Pelengaris, Khan, & Evan, 2002; Pirity, Blanck, & Schreiber-Agus, 2006; Rottmann & Lüscher, 2006; Wahlstrom & Henriksson, 2007; Zhou & Hurlin, 2001).

The E-box is the shared DNA-binding site for Myc/Max, Mad/Max, Mnt/Max, and Max/Max dimers, the latter of which are transcriptionally inert (Eisenman, 2001b; Grandori et al., 2000; Lüscher & Larsson, 1999; Nair & Burley, 2006). Via their E-boxes, the common target genes of the Myc/Mad/Max network are activated by Myc/Max heterodimers but repressed by Mad/Max and Mnt/Max heterodimers, which includes either Myc-mediated HAT (histone acetyltransferase) recruitment and histone acetylation or Mad- and Mnt-mediated HDAC (histone deacetylase) recruitment and histone deacetylation (Amati et al., 2001; Baudino & Cleveland, 2001; Cole & McMahon, 1999; Cole & Nikiforov, 2006; Cowling & Cole, 2006; Dang et al., 2006; Eisenman, 2001a, 2001b; Foley & Eisenman, 1999; Grandori et al., 2000; Hooker & Hurlin, 2005; Hurlin & Dezfouli, 2004; Hurlin & Huang, 2006; Hurlin et al., 2004; Lüscher, 2001, 2012; Lüscher & Vervoorts, 2012; McArthur et al., 1998; Rottmann & Lüscher, 2006; Schreiber-Agus & DePinho, 1998; van Riggelen, Yetil, & Felsher, 2010; Zhou & Hurlin, 2001).

In accordance with their antagonism, the Myc and Mad families exhibit opposite expression patterns so that the Myc family is expressed in proliferating cells but not in quiescent or differentiated cells, whereas, conversely, the Mad family (i.e., Mad1, Mxi1, Mad4) is expressed in quiescent and differentiated cells but not in proliferating cells (Amati & Land, 1994; Bouchard, Staller, & Eilers, 1998; Grandori et al., 2000; Henriksson & Lüscher, 1996; Hooker & Hurlin, 2005; Hurlin & Dezfouli, 2004; Kelly & Siebenlist, 1986; Marcu, Bossone, & Patel, 1992; Meichle, Philipp, & Eilers, 1992; Obaya, Mateyak, & Sedivy, 1999; Pelengaris et al., 2002; Rottmann & Lüscher, 2006). The expression patterns of Mad3 and Mnt deviate from this scheme (Hooker & Hurlin, 2005; Hurlin & Dezfouli, 2004; Larsson & Henriksson, 2010; Rottmann & Lüscher, 2006; Wahlstrom & Henriksson, 2007).

The Myc/Max/Mad network can be viewed as a functional module, which integrates numerous environmental signals and converts them into specific gene-regulatory programs that relate to cell cycle progression and cell growth (Eisenman, 2001a, 2001b; Grandori et al., 2000; Levens, 2002, 2003; Oster et al., 2002).

6.11.4.2 *foxm1* is a direct target gene of the Myc/Max/Mad network

foxm1 is a direct target gene of the Myc/Max/Mad network because the *foxm1* promoter was *in vivo* occupied by c-Myc (http://www.myccancergene.org/site/mycTargetDB.asp) (Blanco-Bose et al., 2008; Delpuech et al., 2007; Fernandez et al., 2003; Huynh et al., 2010), Mad1 (Huynh et al., 2010) and Mxi1 (Table 3.2; Delpuech et al., 2007). Yet, N-Myc did not occupy the *foxm1* promoter in mouse liver (Blanco-Bose et al., 2008).

In accordance with the antagonism between the Myc and Mad families (see above), the *foxm1* mRNA expression is upregulated by c-Myc (Blanco-Bose et al., 2008; Delpuech et al., 2007; Huynh et al., 2010; Keller et al., 2007) but downregulated by Mad1 (Huynh et al., 2010) and Mxi1 (Fig. 3.10; Table 3.2; Delpuech et al., 2007), two c-Myc antagonists.

However, it should be noted that a regulation of the *foxm1* promoter by c-Myc, Mad1, or Mxi1 has not been demonstrated so far and that the control of the *foxm1* mRNA expression by c-Myc and Mxi1 was shown rather indirectly (see below) (Table 3.2).

Nonetheless, direct evidence for the repression of the *foxm1* mRNA expression by Mad1 has been obtained because overexpression of Mad1 diminished the *foxm1* mRNA level in HO-1 cells (Fig. 3.10; Table 3.2; Huynh et al., 2010).

A repression of the *foxm1* mRNA expression by Mxi1 is suggested by the observation that siRNA-mediated knockdown of Mxi1 increased the *foxm1* mRNA level in DL23 colon carcinoma cells (Fig. 3.10; Table 3.2), which overexpressed ca FOXO3a (ca FOXO3a decreased the *foxm1* mRNA level in the absence of siRNA against Mxi1) (Table 3.2; Delpuech et al., 2007).

The activation of the *foxm1* mRNA expression by c-Myc is indicated by the following findings (Fig. 3.10; Table 3.2): first, silencing of c-Myc by siRNA reduced the *foxm1* mRNA expression in DLD-1 colon carcinoma cells (Delpuech et al., 2007) and HO-1 metastatic melanoma cells (Huynh et al., 2010). Second, the *foxm1* mRNA level was increased in bone marrow cells from transgenic $E\mu$-*Myc* mice compared to bone marrow cells from wild-type animals (Keller et al., 2007). Third, exogenous c-Myc prevented the repression of the *foxm1* promoter by IFN-β+mezerein during the IFN-β/mezerein-induced terminal differentiation of HO-1 cells (Huynh et al., 2010). Fourth, in adult mouse liver, the knockout of *c-myc* abolished the induction of *foxm1* mRNA expression by the hepatomitogen TCPOBOP (Blanco-Bose et al., 2008). The halogenated hydrocarbon TCPOBOP is a xenobiotic, nongenotoxic carcinogen that induces acute hepatocyte proliferation and growth (hepatomegaly) (Costa, Kalinichenko, Tan, et al., 2005; Oliver & Roberts, 2002; Qatanani & Moore, 2005; Swales & Negishi, 2004).

6.11.4.3 IFN-β+mezerein utilize the Myc/Max/Mad network to repress the foxm1 transcription during terminal differentiation

During the IFN-β/mezerein-induced terminal differentiation of HO-1 cells, the *foxm1* mRNA and protein expression is downregulated dramatically (Huynh et al., 2009, 2010). Accordingly, IFN-β+MEZ treatment strongly represses both the *foxm1* promoter and the *foxm1* transcription (Huynh et al., 2010).

This repression of the *foxm1* promoter by IFN-β+mezerein depends on an E-box in the *foxm1* promoter, which is bound by c-Myc, and the overexpression of c-Myc prevents the repression of the *foxm1* promoter by IFN-β+mezerein (Huynh et al., 2010). Since IFN-β+mezerein treatment results in the removal of the transactivator c-Myc from the *foxm1* promoter as well as in the arrival of the transrepressor Mad1 at the *foxm1* promoter, IFN-β+mezerein bring about the repression of the *foxm1* promoter by causing the exchange of c-Myc for Mad1 at the E-box in the *foxm1* promoter (Fig. 3.10; Huynh et al., 2010). Accordingly, IFN-β+mezerein downregulate the c-Myc mRNA and protein levels but upregulate the Mad1 mRNA and protein levels in HO-1 cells (Fig. 3.10; Huynh et al., 2010).

In summary, IFN-β+mezerein exploit the antagonistic control of the *foxm1* promoter by members of the Myc/Max/Mad network for repression of *foxm1* transcription because they finish the occupancy of the *foxm1* promoter by c-Myc and induce its occupancy by Mad1 (Huynh et al., 2010).

6.11.4.4 The putative tumor-suppressor FOXO3a employs the Myc/Max/Mad network for repression of foxm1 transcription

The antiproliferative transcription factor FOXO3a represses the *foxm1* promoter (McGovern et al., 2009) and thus decreases the *foxm1* mRNA and protein expression (Fig. 3.10; Table 3.2; Delpuech et al., 2007; He et al., 2013; Karadedou et al., 2012; Laoukili et al., 2005; McGovern et al.,

2009; Zou et al., 2008). However, a binding of FOXO3a to the *foxm1* promoter has not been shown so far (Table 3.2).

Instead, FOXO3a exploits the antagonistic control of *foxm1* transcription by members of the Myc/Max/Mad network for repression of the *foxm1* promoter because, in DL23 colon carcinoma cells, a ca form of FOXO3a induced the occupancy of the *foxm1* promoter with the repressor Mxi1 but finished the occupancy of the *foxm1* promoter with the activator c-Myc (Fig. 3.10; Delpuech et al., 2007). In accordance with this ca FOXO3a-mediated switch in *foxm1* promoter occupancy from c-Myc to Mxi1, ca FOXO3a decreased the c-Myc mRNA and protein expression but increased the Mxi1 mRNA and protein expression in DL23 cells (Fig. 3.10; Delpuech et al., 2007). The repression of the *foxm1* mRNA expression by ca FOXO3a in DL23 cells requires Mxi1 because siRNA-mediated depletion of Mxi1 alleviated the reduction of the *foxm1* mRNA level by ca FOXO3a (Delpuech et al., 2007). Hence, FOXO3a represses the transcription of *foxm1* indirectly through Mxi1 and c-Myc, which both bind to an E-box in the *foxm1* promoter (Fig. 3.10; Delpuech et al., 2007).

6.11.5 Regulation of the FOXM1 expression by the RAS/MAPK pathway

The expression of *foxm1* is activated by the RAS-RAF-MEK1/2-ERK1/2 pathway (Fig. 3.10):

First, oncogenic H-Ras upregulated the *foxm1* mRNA and protein expression (Behren et al., 2010; Park, Carr, et al., 2009; Pellegrino et al., 2010). Accordingly, the FOXM1 protein level was also increased by wild-type Ha-Ras (Pellegrino et al., 2010) and by an activated form of N-Ras (Ho et al., 2012).

Second, ERK2 raised the FOXM1 protein expression (Calvisi et al., 2009). Accordingly, the *foxm1* mRNA level was reduced by the ERK1/2 inhibitor A6355 (Berger, Vega, Vidal, & Geloen, 2012).

Third, the MEK1/2 inhibitor U0126 decreased the *foxm1* mRNA and protein levels, indicating that MEK1/2 elevates the *foxm1* mRNA and protein expression (Bonet et al., 2012; Chan et al., 2012; Faust et al., 2012; Lok et al., 2011; Madureira et al., 2006; Wang, Snyder, et al., 2012; Xia, Huang, et al., 2012).

Thus, the classical RAS-RAF-MEK1/2-ERK1/2 cascade enhances the *foxm1* expression (Fig. 3.10).

RAS disposes of multiple downstream effectors (Ahearn, Haigis, Bar-Sagi, & Philips, 2012; Campbell, Khosravi-Far, Rossman, Clark, &

Der, 1998; Chakrabarty & Heumann, 2008; Cully & Downward, 2008; Der & Van Dyke, 2007; Downward, 2003; Friday & Adjei, 2005; Fuentes & Valencia, 2009; Giehl, 2005; Karnoub & Weinberg, 2008; Kern, Niault, & Baccarini, 2011; Malumbres & Barbacid, 2003; Marshall, 1996; McCormick, 1999, 2011; Moon, 2006; Overmeyer & Maltese, 2011; Perez-Sala & Rebello, 1999; Pylayeva-Gupta, Grabocka, & Bar-Sagi, 2011; Rajalimgam, Schreck, Rapp, & Albert, 2007; Repasky, Chenette, & Der, 2004; Reuther & Der, 2000; Schubbert, Shannon, & Bollag, 2007; Shields, Pruitt, McFall, Shaub, & Der, 2000; Vakiani & Solit, 2011; Vojtek & Der, 1998; Yamamoto, Taya, & Kaibuchi, 1999). In fact, p38 (Behren et al., 2010), JNK, and ROS (reactive oxygen species) (Park, Carr, et al., 2009) each have been implicated in the RAS-induced FOXM1 expression (Fig. 3.10):

The p38 inhibitor SB203580 decreased the oncogenic H-Ras-induced *foxm1* mRNA expression and *foxm1* promoter activity in NIH3T3 fibroblasts indicating that p38 is required for activation of the *foxm1* promoter and induction of *foxm1* mRNA expression by H-Ras (Behren et al., 2010).

The oncogenic H-Ras-induced FOXM1 protein expression was reduced by the JNK inhibitor SB600125 indicating that it depends on JNK activity (Park, Carr, et al., 2009).

The oncogenic H-Ras-induced FOXM1 mRNA and/or protein expression was also reduced by the antioxidants Tempol, MnTM-2-PyP (Mn(III) *meso*-tetrakis(*N*-methyl-2-pyridyl)porphyrin Pentachloride, Manganese (III)-5,10,15,20-tetrakis(*N*-methylpyridinium-2-yl)porphyrin Pentachloride), catalase, and NAC (*N*-acetyl-cysteine) indicating that it depends on ROS (Park, Carr, et al., 2009).

Interestingly, the conditional deletion of *foxm1* from the distal respiratory epithelium in $epFoxm1^{-/-}$ mouse embryos attenuated those defects in branching lung morphogenesis and lung sacculation, which were caused by the transgenic expression of activated K-Ras^{G12D} in these distal lung epithelial cells, indicating that FOXM1 is an important downstream target of K-Ras during embryonic lung morphogenesis (Wang, Snyder, et al., 2012). However, an effect of K-Ras on the expression or the transcriptional activity of FOXM1 has not been reported so far.

6.11.6 The foxm1 gene is subject to positive autoregulation

The *foxm1* gene is subject to positive autoregulation because FOXM1 itself *in vivo* occupies (Down et al., 2012) and transactivates (Wang, Teh, Ji, et al.,

2010) the *foxm1* promoter (Fig. 3.10; Table 3.2). Consequently, overexpression of exogenous FOXM1 enhanced the endogenous *foxm1* mRNA and protein expression (Halasi and Gartel, 2009; Xie et al., 2010). Thus, FOXM1 activates its own expression (Halasi and Gartel, 2009; Wang, Teh, Ji, et al., 2010; Xie et al., 2010) in a positive autoregulatory feedback loop (Fig. 3.10).

This positive FOXM1 autoregulation (Down et al., 2012; Halasi and Gartel, 2009; Wang, Teh, Ji, et al., 2010; Xie et al., 2010) entails that a small initial rise in the *foxm1* expression will ultimately result in a tremendous increase in the FOXM1 level. Therefore, once induced by a triggering signal, abundant FOXM1 expression will ensue. Accordingly, the expression of *foxm1* is tightly controlled by a vast number of regulators (Fig. 3.10), presumably in order to limit the momentous outcome of the positive *foxm1* feedback loop to appropriate cellular settings, where a high FOXM1 level is desired and not harmful.

6.11.7 Repression of the foxm1 gene by the C/EBPα–RB–E2F4–Brm–HDAC1 complex in old-aged mouse livers

After PHx-induced liver injury, *foxm1* is reexpressed in the adult liver during tissue repair, that is, during the regenerative hepatocyte proliferation (Chen et al., 2010; Gieling et al., 2010; Krupczak-Hollis et al., 2003; Kurinna et al., 2013; Lehmann et al., 2012; Matsuo et al., 2003; Meng et al., 2011; Mukhopadhyay et al., 2011; Tan et al., 2006; Wang et al., 2003; Wang, Kiyokawa, et al., 2002; Wang, Krupczak-Hollis, et al., 2002; Wang, Quail, et al., 2001; Weymann et al., 2009; Ye et al., 1999; Zhang, Wang, et al., 2012). Yet, the PHx-induced *foxm1* mRNA and protein expression is reduced in old-aged mice (Chen et al., 2010; Wang, Krupczak-Hollis, et al., 2002; Wang, Quail, et al., 2001) because, in the livers of old mice, a repressive C/EBPα–RB–E2F4–Brm (Brahma, SMARCA4)–HDAC1 complex forms on the *foxm1* promoter (Fig. 3.10; Wang, Salisbury, et al., 2008; Wang, Shi, et al., 2007), which is not disassembled in response to PHx and, therefore, prevents the transcription of *foxm1* not only before, but also after PHx (Wang, Shi, et al., 2007).

In contrast, in the livers of young mice, the *foxm1* promoter is repressed by a RB–E2F4 complex (and a RB–E2F1 complex) prior to PHx (Wang, Salisbury, et al., 2008; Wang, Shi, et al., 2007) and PHx causes the disappearance of RB from the *foxm1* promoter so that the *foxm1* transcription is activated by E2F-1 (and E2F-4) following PHx (Fig. 3.10; Wang, Shi, et al., 2007).

Notably, GH restores the PHx-induced *foxm1* mRNA and protein expression in old-aged mouse livers (Krupczak-Hollis et al., 2003) because GH eliminates the repressive C/EBPα–RB–E2F4–Brm–HDAC1 complex that occupies the *foxm1* promoter in the absence of GH (Fig. 3.10; Wang, Salisbury, et al., 2008).

Multiple regulatory mechanisms (Jin, Wang, Salisbury, Timchenko, & Timchenko, 2009; Jin, Wang, Timchenko, & Timchenko, 2009; Jones, Timchenko, & Timchenko, 2012; Schmucker & Sanchez, 2011; Timchenko, 2009; Willis-Martinez, Richards, Timchenko, & Medrano, 2010) mediate this age-specific switch at the *foxm1* promoter from its PHx-reversible repression by a RB–E2F4 complex (and a RB–E2F1 complex) in the livers of young mice to its irreversible repression by the C/EBPα–RB–E2F4–Brm–HDAC1 complex in the livers of old mice (Wang, Salisbury, et al., 2008; Wang, Shi, et al., 2007). Actually, this age-specific switch in gene repression from a RB–E2F complex in young livers to the C/EBP–RB–E2F4–Brm–HDAC1 complex in old livers is not limited to the *foxm1* promoter, but it represents a general mechanism that takes place at the promoters of several proliferation-stimulating E2F target genes, the expression of which is induced by PHx (Jin, Wang, Salisbury, et al., 2009; Jin, Wang, Timchenko, et al., 2009; Jones et al., 2012; Schmucker & Sanchez, 2011; Timchenko, 2009), for example, at the *c-myc* and *cdk1* promoters (Iakova, Awad, & Timchenko, 2003; Wang, Salisbury, et al., 2008; Wang, Shi, et al., 2007).

6.11.8 Is FOXM1 an immediate-early gene in FSH-treated Sertoli cells?

In contrast to the classification of *foxm1* as a delayed-early gene during the serum-induced cell cycle reentry from quiescence (Costa et al., 2003; Mackey et al., 2003; Ye et al., 1999, 1997), *foxm1* is suggested to act as an immediate-early gene in FSH (follicle-stimulating hormone)-stimulated Sertoli cells (Chaudhary et al., 2000). The gonadotropin FSH controls the function of testis Sertoli cells, which form the seminiferous tubule and respond to FSH treatment with either proliferation (embryonic and prepubertal Sertoli cells) or differentiation (postpubertal Sertoli cells) (Eddy, 2002; Loss, Jacobus, & Wassermann, 2007; Plant & Marshall, 2001; Walker & Cheng, 2005). FSH treatment of cultured Sertoli cells, which were isolated from the testis of mid-pubertal (20 day old) rats, induced *foxm1* mRNA expression (Fig. 3.10) and the *foxm1* mRNA level peaked already 30 min after FSH addition, indicating that *foxm1* represents an immediate-early gene (Chaudhary et al., 2000). Also cAMP, a downstream effector of

FSH (Gloaguen, Crepieux, Heitzler, Poupon, & Reiter, 2011; Gorcynska-Fjälling, 2004; Hansson, Skalhegg, & Tasken, 2000; Ulloa-Aguirre, Zarinan, Pasapera, Casa-Gonzalez, & Dias, 2007), induced *foxm1* mRNA expression in these cultured Sertoli cells (Fig. 3.10) and the *foxm1* mRNA level peaked already 60 min after cAMP addition, again pointing to a role of *foxm1* as an immediate-early gene (Chaudhary et al., 2000). This assumption is supported by the observation that treatment of Sertoli cells with cycloheximide for 2 h superinduced the *foxm1* expression (Chaudhary et al., 2000). Thus, in FSH-treated Sertoli cells, *foxm1* seems to act as an immediate-early gene (Chaudhary et al., 2000).

6.12. The expression of FOXM1 in human disease
6.12.1 Hutchinson–Gilford progeria
Hutchinson–Gilford progeria is a rare genetic disorder characterized by accelerated aging, including loss or graying of hair, skeletal abnormalities, cardiovascular disease, and diminished subcutaneous fat (Burtner & Kennedy, 2010; Dominguez-Gerpe & Araujo-Vilar, 2008; Hennekam, 2006; Kieran, Gordon, & Kleinman, 2007; Kudlow, Kennedy, & Monnat, 2007).

foxm1 mRNA expression is considerably reduced in Hutchinson–Gilford progeria (Fig. 3.10; Ly et al., 2000). The analysis of transcripts in actively dividing, early-passage fibroblasts, which were obtained from either healthy young humans or patients with Hutchinson–Gilford progeria, revealed a strong reduction of the *foxm1* RNA expression in Hutchinson–Gilford progeria because the *foxm1* mRNA level was 8.7-fold lower in fibroblasts from progeria patients than in fibroblasts from healthy young humans (Ly et al., 2000).

This finding is consistent with the observed decrease in the *foxm1* mRNA expression during normal human aging (see above) (Ly et al., 2000) because progeria patients show at a very early age physical features and cellular characteristics typically associated with natural old age, for example, an increase in the number of polyploid cells with 4N and 8N DNA content (Ly et al., 2000), genomic instability, telomere attrition, premature senescence, and defective stem cell homeostasis (Burtner & Kennedy, 2010; Dominguez-Gerpe & Araujo-Vilar, 2008; Hennekam, 2006; Kieran et al., 2007; Kudlow et al., 2007).

6.12.2 Trisomy 21
Trisomy 21 or the Down syndrome is defined by three copies of chromome 21 per body cell instead of the normal two copies (Kahlem, 2006;

Roubertoux & Kerdelhue, 2006; Sommer & Henrique-Silva, 2008). This congenital aneuploidy afflicts about 1 of every 700 newborn children (Megarbane et al., 2009).

The Ts65Dn mouse model of the Down syndrome (trisomy 21) showed a downregulation of the *foxm1* mRNA level in trisomic neuropheres, that is, in trisomic neural precursor cells (Trazzi et al., 2011). Ts65Dn mice are trisomic for a segment of mouse chromosome 16, highly homologous to the long arm of human chromosome 21 (Akeson et al., 2001; Davisson, Schmidt, & Akeson, 1990).

6.12.3 Pathological cardiac hypertrophy

Pathological cardiac hypertrophy (heart growth) is defined as an abnormal increase in heart muscle mass (ventricular remodeling) as a result of an increased size of the individual cardiac myocytes (cardiomyocyte enlargement) (Catalucci, Latronico, Ellingsen, & Condorelli, 2008; Dorn, 2007; Heineke & Molkentin, 2006; Luedde, Katus, & Frey, 2006). Pathological cardiac hypertrophy represents a compensatory response to enhanced hemodynamic stress and workload, but it is rather a maladaptive process that ultimately leads to heart failure (cardiac dysfunction) and sudden death (Catalucci et al., 2008; Dorn, 2007; Heineke & Molkentin, 2006; Luedde et al., 2006).

foxm1 is barely expressed in normal hearts, but the *foxm1* expression is strongly increased in a model of PAH (pathological hypertrophy) of the heart, that is, upon PO (pressure overload) by TAC (traverse aortic constriction) (Bolte et al., 2012; Song, Hong, Kim, & Kim, 2012). In the heart, application of TAC, that is, PO, results in cardiac hypertrophy and cardiac dysfunction (heart failure) (Catalucci et al., 2008; Dorn, 2007; Heineke & Molkentin, 2006; Luedde et al., 2006).

Although the *foxm1* mRNA expression in the heart increased in the course of PO-induced cardiac hypertrophy (Bolte et al., 2012; Song et al., 2012) FOXM1 is not critical for this model of PAH because no differences in the timeline or extent of TAC-induced cardiac hypertrophy were found between conditional *foxm1* knockout mice (α-*MHC-Cre*/*Foxm1*$^{fl/fl}$) with a selective deletion of *foxm1* from cardiomyocytes in the late postnatal period and control mice (Bolte et al., 2012).

6.12.4 Asthma and COPD (chronic obstructive pulmonary disease)

FOXM1 protein is not expressed in normal human lungs, but abundant FOXM1 protein expression was detected in the lungs of patients with severe asthma and COPD, namely, in bronchiolar epithelium and inflammatory cells (Ren et al., 2013).

6.12.5 Idiopathic pulmonary fibrosis

IPF (idiopathic pulmonary fibrosis) is an interstitial lung disease, which is characterized by inflammation, excessive deposition of ECM proteins (e.g., collagen), and permanent fibrotic remodeling of lung tissue, culminating in replacement of normal lung parenchyma with fibrotic tissue (American Thoracic Society, 2000; Bringardner, Bara, Eubank, & Marsh, 2008; Corvol, Flamein, Epaud, Clement, & Guillot, 2009; Coward, Saini, & Jenkins, 2010; Crystal et al., 2002; Gharaee-Khermani, Hu, Phan, & Gyetko, 2009; Günther et al., 2012; Hardie et al., 2010; King, Pardo, & Selman, 2011; Scotton & Chambers, 2007; Selman et al., 2001; Wynn, 2011). IPF patients suffer from progressive and irreversible destruction of the lung architecture, pulmonary volume restriction, disruption of alveolar gas exchange, and death from respiratory failure (American Thoracic Society, 2000; Wynn, 2011). Pulmonary fibrosis is often regarded as a dysregulated wound repair process of damaged lung tissue, in which fibrotic scar formation ultimately leads to organ malfunction (Kisseleva & Brenner, 2008a, 2008b; Selman et al., 2001; Thannickal, Toews, White, Lynch, & Martinez, 2004; Wynn, 2007). Pulmonary fibrosis can be triggered by recurrent and persistent injury to the respiratory epithelium and alveolar type II epithelial cells participate in the pathogenesis of pulmonary fibrosis (Chapman, 2011; Corvol et al., 2009; Coward et al., 2010; Günther et al., 2012; Karge & Borok, 2012; King et al., 2011; Sisson et al., 2010).

FOXM1 protein expression was increased in the fibrotic lesions of patients with IPF compared to the lung tissue from control organ donors (Balli et al., 2013). In particular, the FOXM protein levels were elevated in alveolar type II epithelial cells of the IPF fibrotic lesions (Balli et al., 2013). Accordingly, in a mouse model of radiation-induced pulmonary fibrosis, in which pulmonary fibrosis is caused by thoracic irradiation, *foxm1* mRNA expression in the lung was progressively upregulated following IR (Balli et al., 2013). Again, a raised FOXM1 protein level was detected in alveolar type II epithelial cells within fibrotic lesions of the irradiated mouse lungs (Balli et al., 2013).

In accordance with its increased expression in this mouse model, FOXM1 plays a critical role in radiation-induced pulmonary fibrosis (see below) (Balli et al., 2013).

6.13. Additional regulators of the FOXM1 expression

Figure 3.10 summarizes the control of the FOXM1 expression. This chapter provides brief additional information for some regulators of the FOXM1 expression.

6.13.1 Regulators of the foxm1 mRNA expression
6.13.1.1 Hypoxia
Hypoxia activates the *foxm1* promoter and upregulates *foxm1* mRNA and protein levels (Fig. 3.10; Raghavan et al., 2012; Xia et al., 2009). In HepG2 HCC cells, hypoxia activated the *foxm1* promoter and increased FOXM1 protein expression through HIF-1α (Fig. 3.10), which binds to the *foxm1* promoter (Table 3.2; Xia et al., 2009). In contrast, in HPASMC (human pulmonary artery SMCs), hypoxia raised the *foxm1* mRNA expression through HIF-2α (Fig. 3.10) but not through HIF-1α (Raghavan et al., 2012).

6.13.1.2 ERβ1
ERβ1, which displaces ERα from the *foxm1* promoter, represses the *foxm1* promoter and decreases the *foxm1* mRNA and protein expression only in ERα-positive but not in ERα-negative breast carcinoma cell lines (Fig. 3.10; Table 3.2; Horimoto et al., 2011).

6.13.1.3 STAT3
The binding of STAT3 to the *foxm1* promoter was shown *in vivo* in ChIP assays and *in vitro* in EMSAs (Table 3.2; Mencalha et al., 2012). The small-molecule STAT3 inhibitor LLL-3, which interferes with STAT3 dimerization, STAT3 DNA binding, and STAT3 transcriptional activity (Fossey et al., 2009; Fuh et al., 2009), repressed the *foxm1* promoter and reduced the *foxm1* mRNA expression, indicating that the *foxm1* promoter is activated by STAT3 (Table 3.2; Mencalha et al., 2012). However, STAT3 has not been demonstrated to activate the *foxm1* promoter or the *foxm1* expression so far.

6.13.1.4 FOXO1+FOXO3
The *foxm1* mRNA level was increased in the neonatal hearts (at postnatal day 1 (pd1)) of mice with a cardiomyocyte-specific conditional *foxo1/foxo3* double knockout compared to the neonatal hearts of control mice, indicating that FOXO1+FOXO3 decrease the *foxm1* mRNA expression (Fig. 3.10; Table 3.2; Sengupta et al., 2013).

6.13.1.5 LIN9

foxm1 is a direct target gene of LIN9, a subunit of the MuvB complex, because the *foxm1* promoter was occupied *in vivo* by LIN9 in immortalized MEFs (Down et al., 2012), because shRNA against LIN9 decreased the *foxm1* mRNA level in murine F9 EC cells (Knight et al., 2009) and because siRNA against LIN9 decreased the *foxm1* protein level in U2OS osteosarcoma cells (Chen, Müller, et al., 2013), indicating that LIN9 increases the *foxm1* mRNA and protein expression (Fig. 3.10). *foxm1* is also a direct target gene of the transcription factor B-Myb (Lorvellec et al., 2010; Down et al., 2012), which enhances the *foxm1* mRNA expression, too (Fig. 3.10; Table 3.2; Down et al., 2012; Lorvellec et al., 2010; Sadavisam et al., 2012). In immortalized MEFs, both LIN9 and B-Myb *in vivo* occupied the *foxm1* promoter in late S-phase (Down et al., 2012). This *in vivo* association of LIN9 with the *foxm1* promoter in late S-phase depends on B-Myb because it was abolished when *B-Myb* was knocked out in conditional *B-Myb* knockout MEFs (Down et al., 2012). Since LIN9 interacts with B-Myb (Calvisi et al., 2011; Down et al., 2012; Knight et al., 2009; Mannefeld, Klassen, & Gaubatz, 2009; Osterloh et al., 2007; Pilkinton, Sandoval, & Colamonici, 2007; Pilkinton, Sandoval, Song, Ness, & Colamonici, 2007; Sadavisam et al., 2012; Schmit et al., 2007) one might speculate that B-Myb and the MuvB complex cooperate in the activation of *foxm1* transcription in late S-phase and that B-Myb is required for recruitment of the MuvB complex to the *foxm1* promoter.

6.13.1.6 p130

In late S-phase, the *foxm1* promoter was occupied *in vivo* by the pocket protein p130 in *B-Myb* knockout MEFs but not in control MEFs with B-Myb expression (Down et al., 2012).

6.13.1.7 c-Met

During pregnancy (gestational day 15), *foxm1* mRNA levels were increased in mouse pancreatic islets (Demirci et al., 2012; Huang, 2013), but this increase was abolished in conditional knockout mice lacking *c-Met* in the pancreas, which indicates that c-Met upregulates the *foxm1* mRNA expression (Fig. 3.10; Demirci et al., 2012). However, the lower *foxm1* mRNA level in pancreatic islets of nonpregnant mice was unaffected by this *c-Met* knockout, probably because also the c-Met expression in pancreatic islets is strongly induced during pregnancy but weak in nonpregnant mice (Demirci et al., 2012).

6.13.1.8 BMP signaling

Inhibition of BMP (bone morphogenetic protein) signaling by either Noggin or Chordin or a dn BMPR (BMP receptor) induced *foxm1* mRNA expression, which points to a repression of the *foxm1* mRNA expression by BMP signaling (Fig. 3.10; Ueno et al., 2008). The BMP antagonists Noggin and Chordin are ligand traps, which inhibit BMPs by blocking their surfaces that are required for their interaction with the type I and type II BMP receptors (Balemans & Van Hul, 2002; Gordon & Blobe, 2008; Massagué, 2008; Schmierer & Hill, 2007; Shi & Massagué, 2003; Wu & Hill, 2009; Yanagita, 2005).

6.13.1.9 PKCα

foxm1 mRNA and protein expression is downregulated during the terminal differentiation of HO-1 human metastatic melanoma cells induced by IFN-β+mezerein (Huynh et al., 2009, 2010). Accordingly, IFN-β+mezerein repress the *foxm1* promoter (Huynh et al., 2010). The IFN-β/mezerein-induced repression of the *foxm1* promoter and the IFN-β/mezerein-induced downregulation of the *foxm1* mRNA and protein levels depend on PKCα (protein kinase Cα) because they were abrogated by siRNA against PKCα and/or by the PKC inhibitors Gö6983 and Ro-31-8220 (Huynh et al., 2010). Likewise, the *foxm1* mRNA expression is downregulated during the differentiation of N/Tert-1 keratinocytes (hTERT- and p16 loss-immortalized normal human skin keratinocytes) induced by suspension culture in methylcellulose (anchorage deprivation) (Bose et al., 2012). Also the methylcellulose suspension culture-induced downregulation of the *foxm1* mRNA level depends on PKC because it was prevented by the PKC inhibitor GF109203 (Bose et al., 2012). Hence, PKC seems to be involved in the repression of the *foxm1* expression during the terminal differentiation of several cell types and in response to several differentiation stimuli (Fig. 3.10).

6.13.1.10 PI3K inhibitors

The PI3K (phosphatidylinositol-3 kinase) inhibitor LY204002 repressed the *foxm1* promoter in HTH74 and 8505C primary ATC cells, which suggests an activation of the *foxm1* promoter by PI3K (Bellelli et al., 2012). Accordingly, the FOXM1 protein level was downregulated by the PI3K inhibitors LY204002 and Wortmannin in human PC-3 and C4-2B prostate cancer cells, suggesting a positive effect of PI3K on the FOXM1 protein expression (Wang, Li, Ahmad, et al., 2010).

6.13.1.11 EPS8

The PI3K inhibitor LY294002 abolished the EPS8 (EGFR pathway substrate 8)-induced increase in the *foxm1* mRNA expression, indicating that PI3K is implicated in the upregulation of the *foxm1* mRNA level by EPS8 (Fig. 3.10; Wang, Teh, Ji, et al., 2010).

6.13.1.12 Calcium

foxm1 mRNA levels were upregulated in DMH (1,2-dimethylhydrazine)-induced CRC, and calcium downregulated this heightened *foxm1* mRNA expression in DMH-induced CRC (Fig. 3.10; Wang, Lin, et al., 2011). Yet, the effect of calcium alone on the *foxm1* expression was not analyzed.

6.13.1.13 Allergen stimulation

Allergen stimulation with either HDM (house dust mite) extract or OVA (ovalbumin) induced abundant FOXM1 protein expression in mouse lungs (Fig. 3.10), which otherwise do not express FOXM1 protein (Ren et al., 2013). In these two mouse models of asthma, the FOXM1 protein was expressed in bronchiolar epithelial cells and inflammatory cells (Ren et al., 2013). The HDM challenge also induced *foxm1* mRNA expression in mouse lungs (Fig. 3.10; Ren et al., 2013).

6.13.1.14 SPDEF

The *foxm1* mRNA level after HDM challenge was higher in airway epithelial cells of $Spdef^{-/-}$ mouse lungs than in the airway epithelium of wild-type lungs, indicating that SPDEF represses the *foxm1* mRNA expression (Fig. 3.10; Table 3.2; Ren et al., 2013). However, whether SPDEF affects the expression of *foxm1* without HDM treatment was not analyzed.

6.13.1.15 Dextrose

Dextrose, which inhibits liver regeneration after PHx, suppresses the induction of *foxm1* mRNA expression by PHx (Fig. 3.10; Weymann et al., 2009). This negative effect of dextrose on the PHx-induced *foxm1* mRNA expression depends on the presence of p21 (Fig. 3.10; Weymann et al., 2009).

6.13.1.16 Cry1 and Cry2

Thirty-six and forty hours after PHx, that is, at the peak of the PHx-induced *foxm1* mRNA expression, the *foxm1* mRNA level was lower in *Cry* (*Cryptochrome*)-deficient mice ($Cry1^{-/-}$ $Cry2^{-/-}$) than in wild-type mice, indicating that the circadian clock proteins Cry1 and/or Cry2 have a positive effect on the induction of *foxm1* expression by PHx (Fig. 3.10; Matsuo et al., 2003).

6.13.1.17 c-Rel
PHx strongly induced *foxm1* mRNA expression in wild-type mouse livers, but the PHx-induced *foxm1* mRNA expression was blunted and delayed in the livers of $c\text{-}rel^{-/-}$ mice, indicating that c-Rel is required for the induction of *foxm1* expression by PHx (Fig. 3.10; Gieling et al., 2010). In the livers of wild-type animals, PHx induced the binding of c-Rel to the *foxm1* promoter (Table 3.2), which was not occupied *in vivo* by c-Rel in sham-operated mice (Gieling et al., 2010).

6.13.1.18 β-Catenin
The halogenated hydrocarbon TCPOBOP is xenobiotic, nongenotoxic carcinogen, which induces acute hepatocyte proliferation and growth (hepatomegaly) (Costa, Kalinichenko, Tan, et al., 2005; Oliver & Roberts, 2002; Qatanani & Moore, 2005; Swales & Negishi, 2004).

The *foxm1* mRNA and protein levels were higher in TCPOBOP-treated livers of female mice with a hepatocyte-specific knockout of *β-catenin* than in TCPOBOP-treated livers of female control mice, which points to a negative effect of β-catenin on the *foxm1* expression (Fig. 3.10; Table 3.2; Bräuning et al., 2011). However, without TCPOBOP treatment, β-catenin did not affect the *foxm1* expression because the *foxm1* mRNA level in untreated female livers was similar in *β-catenin* knockout and control mice (Bräuning et al., 2011).

6.13.1.19 SRp20
The splicing factor SRp20 (SFRS3 (splicing factor, arginine/serine-rich 3)) increased not only the FOXM1 protein level (Fig. 3.10), but also the *foxm1* mRNA level (Jia et al., 2010). This suggests that SRp20 may indirectly stimulate *foxm1* transcription through a positive effect on the splicing of a transcription factor, which activates the *foxm1* promoter (Jia et al., 2010). This transcription factor could be FOXM1 itself because FOXM1 binds to (Down et al., 2012) and transactivates (Wang, Teh, Ji, et al., 2010) the *foxm1* promoter so that the SRp20-mediated increase in the FOXM1 protein level (Fig. 3.10) could explain the upregulation of the *foxm1* mRNA expression by SRp20 (Jia et al., 2010).

6.13.1.20 Fibronectin

The *foxm1* mRNA expression is downregulated during the differentiation of N/Tert-1 skin keratinocytes induced by methylcellulose suspension culture (Bose et al., 2012). This suspension-induced downregulation of the *foxm1* mRNA level was alleviated by fibronectin and by a RGD peptide (Bose et al., 2012), which mimics fibronectin (Lash, Linask, & Yamada, 1987), presumably because fibronectin, more precisely engaging β1-integrin with fibronectin, suppresses the methylcellulose-induced terminal differentiation of keratinocytes (Adams & Watt, 1989, 1990).

6.13.1.21 AMPK

The *foxm1* mRNA level in rat neonatal cardiomyocytes was increased by the AMPK (AMP (adenosine monophosphate)-activated protein kinase) inhibitor compound C but decreased by the AMP-mimetic AMPK activator AICAR (5-aminoimidazole-4-carboxamide ribonucleoside), which suggests a downregulation of the *foxm1* mRNA expression by AMPK (Sengupta et al., 2013). Accordingly, compound C also elevated the FOXM1 protein level and enhanced the *in vivo* occupancy of the *IGF-1* promoter by FOXM1 in rat neonatal cardiomyocytes (Sengupta et al., 2013).

6.13.2 Regulators of the FOXM1 protein expression
6.13.2.1 IGF-I

The PI3K inhibitor LY294002 mitigated the IGF-I-induced increase in the FOXM1 protein expression, indicating that PI3K is implicated in the upregulation of the FOXM1 protein level by IGF-I (Fig. 3.10; Aburto et al., 2012).

6.13.2.2 GRB7

The MEK1/2 inhibitors U0126 and PD98059 abolished the GRB7-induced increases in the FOXM1 protein level, which indicates an involvement of MEK1/2 in the upregulation of the FOXM1 protein expression by GRB7 (Fig. 3.10; Chan et al., 2012).

6.13.2.3 Sonic hedgehog (Shh)

The transcription factor GLI-1 augments *foxm1* mRNA and protein expression (Fig. 3.10; Table 3.2; Calvisi et al., 2009; Teh et al., 2002). Also the extracellular ligand Shh elevates the FOXM1 protein level (Fig. 3.10; Schüller et al., 2007). Since Shh is an activating upstream signal for GLI-1 in the

Hedgehog pathway (Hui & Angers, 2011; Ingham, 2008; Ingham, Nakano, & Seger, 2011; Jiang & Hui, 2008; Ribes & Briscoe, 2009; Riobo & Manning, 2007; Rohatgi & Scott, 2007; Ruiz i Altaba, Mas, & Stecca, 2007; Ryan & Chiang, 2012; Varjaluso & Taipale, 2007, 2008; Wang, McMahon, & Allen, 2007), Shh may increase the FOXM1 expression via activation of GLI-1, which suggests that *foxm1* may be a target gene of the Hedgehog pathway (Calvisi et al., 2009; Schüller et al., 2007; Teh et al., 2002).

6.13.2.4 ConA/PMA

In addition to its prominent function in apoptosis, FADD (Fas (CD95, DR2 (death receptor 2))-associated death domain) plays a role in T-cell proliferation (Budd, 2002; Strasser & Newton, 1999; Tourneur & Chiocchia, 2010). Human FADD is phosphorylated by CKIα (casein kinase Iα) at S-194 (Alappat et al., 2005), which is equivalent to S-191 in murine FADD. The Ser-to-Asp point mutation at S-191 (S191D) mimics constitutive phosphorylation of murine FADD. FADD-D transgenic mice, which bear the S191D point mutation in FADD, exhibit proliferative T-cell defects but no apoptotic T-cell defects (Hua, Sohn, Kang, Cado, & Winoto, 2003; Osborn et al., 2007).

Murine naïve T-cells from wild-type animals express a low level of FOXM1 protein and activation of these naïve T-cells with ConA (concanavalin A)/PMA leads to a rise in the FOXM1 protein level (Fig. 3.10; Osborn et al., 2007). In contrast, both the basal and the ConA/PMA-induced FOXM1 protein expression were abolished in naïve or activated T-cells from FADD-D transgenic mice, respectively, which indicates that the phosphorylation of FADD at S-191 is detrimental to FOXM1 expression in peripheral T-cells at any point before and after their activation by ConA/PMA (Osborn et al., 2007). However, FOXM1 protein was highly expressed in thymocytes from both wild-type and FADD-D transgenic mice, suggesting that the phosphomimetic S191D point mutation in FADD specifically impairs mature T-cells in their ability to express FOXM1 (Osborn et al., 2007).

6.13.3 Regulators of the FOXM1 expression, which have contradictory effects on the expression of FOXM1

6.13.3.1 H_2O_2

H_2O_2, which causes oxidative stress, increased the *foxm1* mRNA expression in adult HMECs (human microvascular endothelial cells) (Ye et al., 1997) and in primary cultures of endothelial cells isolated from mouse lungs (Fig. 3.10; Zhao et al., 2006). Accordingly, H_2O_2 elevated the FOXM1 protein level in immortalized MEFs (Park, Carr, et al., 2009) and U2OS osteosarcoma cells (Fig. 3.10; de Olano et al., 2012).

In contrast, the *foxm1* mRNA and protein levels were decreased by H_2O_2 in human MCF-7 breast cancer cells (Fig. 3.10; Chua et al., 2009). Accordingly, the *foxm1* mRNA expression was reduced by H_2O_2 in human dermal fibroblasts (Fig. 3.10; Kim et al., 2011).

6.13.3.2 TNF-α

TNF-α (tumor necrosis factor-α) and its downstream effector IKKβ (IκB (inhibitor of NF-κB (nuclear factor-κB)) kinase β) repressed the *foxm1* mRNA expression in immortalized MEFs during a transient cell proliferation arrest associated with a G1/S cell cycle block (Fig. 3.10; Penzo et al., 2009).

In contrast, TNF-α induced *foxm1* mRNA expression in primary cultures of endothelial cells isolated from mouse lungs (Fig. 3.10; Zhao et al., 2006). Likewise, TNF-α activated the *foxm1* promoter via its downstream effectors ROS and HIF-1α in HepG2 and Huh-7 HCC cells, where TNF-α enhanced the *foxm1* mRNA and protein expression (Fig. 3.10; Xia, Mo, et al., 2012).

6.13.3.3 TGF-β1

TGF-β1 (transforming growth factor-β1) increased the FOXM1 protein expression in human A549 NSCLC cells during their TGF-β1-induced EMT (Fig. 3.10; Li, Wang, et al., 2012).

In contrast, TGF-β1 decreased the *foxm1* mRNA expression in PrSC (prostate stromal cells) during their TGF-β1-induced transdifferentiation into myofibroblasts/SMCs (Fig. 3.10; Untergasser et al., 2005).

6.13.3.4 IGF-I

The *foxm1* mRNA level was upregulated in the cochlea of embryonic $Igf1^{-/-}$ mice, indicating a downregulation the *foxm1* mRNA expression by IGF-I (Fig. 3.10; Sanchez-Calderon et al., 2010).

In contrast, IGF-I increased the FOXM1 protein level in quiescent chicken HH18 otic vesicles, namely, in a PI3K- and AKT-dependent manner (Fig. 3.10; Aburto et al., 2012).

6.13.3.5 AKT

Activated AKT1 decreased the FOXM1 protein expression in human IMR90 primary fibroblasts (Fig. 3.10; Park, Carr, et al., 2009).

In contrast, the FOXM1 protein level was increased by AKT overexpression but decreased by siRNA directed against AKT in human PC-3 prostate cancer cells (Wang, Li, Ahmad, et al., 2010) and human HLF HCC cells (Ho et al., 2012). Accordingly, AKT overexpression raised the FOXM1 protein level in mouse liver tumor cells (Ho et al., 2012). Likewise, the *foxm1* mRNA and protein expression was diminished in both *AKT1* knockout MEFs and *AKT1/AKT2* double knockout MEFs, showing that AKT1 enhances the *foxm1* mRNA and protein expression in MEFs (Fig. 3.10; Wang, Li, Ahmad, et al., 2010). Similarly, the AKT inhibitor Merck 124005 reduced the FOXM1 protein level in EPS8-overexpressing HN4 HNSCC (head and neck squamous cell carcinoma) cells, which express more FOXM1 protein than control H4 cells in the absence of this AKT inhibitor, suggesting that the EPS8-induced upregulation of the FOXM1 protein expression requires AKT activity (Wang, Teh, Ji, et al., 2010). Accordingly, the AKT inhibitor AKTi VIII alleviated the IGF-I-induced rise in the FOXM1 protein level in chicken otic vesicles, indicating a requirement of AKT activity for the upregulation of the FOXM1 protein expression by IGF-I (Aburto et al., 2012).

Finally, in contrast to the positive effect of AKT on the endogenous FOXM1 protein expression (Aburto et al., 2012; Ho et al., 2012; Wang, Li, Ahmad, et al., 2010; Wang, Teh, Ji, et al., 2010), the protein level of ectopically expressed FoxM1B was unaffected by a pharmacological AKT kinase inhibitor in U2OS cells (Major et al., 2004).

Since a possibly indirect interaction between FOXM1 and phosphorylated AKT was observed after PHx in mouse liver (Mukhopadhyay et al., 2011) the serine/threonine kinase AKT might posttranslationally control the FOXM1 protein level through phosphorylation of FOXM1. However, the failure of the AKT inhibitor to affect the expression of exogenous FoxM1B (Major et al., 2004) would argue against such a posttranslational regulation of FOXM1 by AKT. Mukhopadhyay et al. (2011) did not specify the phosphorylated residue of AKT and they neither analyzed whether FOXM1 interacts with unphosphorylated AKT nor whether AKT phosphorylates FOXM1 nor whether AKT regulates the FOXM1 expression.

6.13.3.6 JNK

Both siRNA against JNK1 and the JNK inhibitor S600125 decreased the FOXM1 protein level in human U2OS osteosarcoma cells, which points to an upregulation of the FOXM1 protein expression by JNK1 (Fig. 3.10; Wang, Chen, et al., 2008). Accordingly, in human IMR90 primary fibroblasts, the oncogenic H-Ras-induced FOXM1 protein expression was reduced by the JNK inhibitor SB600125, indicating a requirement of JNK activity for the upregulation of the FOXM1 protein expression by H-Ras (Fig. 3.10; Park, Carr, et al., 2009).

In contrast, the epirubicin-induced increase in the FOXM1 protein expression was enhanced in $JNK1/2^{-/-}$ MEFs compared to wild-type MEFs, which indicates a downregulation of the FOXM1 protein level by JNK1/2 (Fig. 3.10; de Olano et al., 2012). Accordingly, siRNA against JNK1/2 and the JNK inhibitor SP600125 augmented the epirubicin-induced FOXM1 protein expression in U2OS cells, again pointing to a negative effect of JNK1/2 on the FOXM1 protein level (Fig. 3.10; de Olano et al., 2012).

6.13.3.7 p38

Dn p38α and the p38 inhibitor SB203580 increased the *foxm1* mRNA level in "spontaneous" dormant D-HEp3 human epidermoid carcinoma cells, indicating that p38 decreases the *foxm1* mRNA expression (Fig. 3.10; Adam et al., 2009).

In contrast, p38 is required for the induction of FOXM1 protein expression by epirubicin because siRNA against p38 and SB203580 prevented the epirubicin-induced upregulation of the FOXM1 protein level in human U2OS osteosarcoma and/or MCF-7 breast carcinoma cells (Fig. 3.10; de Olano et al., 2012). Accordingly, SB203580 reduced both the oncogenic H-Ras-induced and the ca MKK3-induced *foxm1* promoter activity and *foxm1* mRNA expression in NIH3T3 fibroblasts which indicates that p38 activity is required for the activation of the *foxm1* promoter and the induction of *foxm1* mRNA expression by H-Ras and MKK3 (Fig. 3.10; Behren et al., 2010). Likewise, SB203580 inhibited the transactivation of the *foxm1* promoter by E2F-1 in U2OS cells, indicating a requirement of p38 activity for the E2F-1-mediated transactivation of the *foxm1* promoter (Fig. 3.10; de Olano et al., 2012).

7. FOXM1 MOUSE MODELS

7.1. FOXM1 knockout mice

7.1.1 General knockout of foxm1

The knockout of *foxm1* in mice is embryonically lethal (Kim et al., 2005; Korver et al., 1998; Krupczak-Hollis et al., 2004; Ramakrishna et al.,

2007) with defective development of heart (Bolte et al., 2011; Korver et al., 1998; Krupczak-Hollis et al., 2004; Ramakrishna et al., 2007), lung (Kalin et al., 2008; Kim et al., 2005; Ustiyan et al., 2009, 2012) and liver (Krupczak-Hollis et al., 2004). Thus, FOXM1 is essential for embryonic development (Fig. 3.2; Kim et al., 2005; Korver et al., 1998; Krupczak-Hollis et al., 2004; Ramakrishna et al., 2007). However, the presence of heart, lung, and liver in $foxm1^{-/-}$ mouse embryos (Kim et al., 2005; Korver et al., 1998; Krupczak-Hollis et al., 2004; Ramakrishna et al., 2007) indicates that FOXM1 is dispensable for early embryogenesis and for initiation of morphogenesis in these organs, but that it is required for later stages of heart, lung, and liver organogenesis (Laoukili et al., 2007; Wierstra & Alves, 2007c).

The hearts of $foxm1^{-/-}$ mouse embryos displayed a reduced number of cardiomyocytes, a decreased heart size, a thinned myocardium, irregularly orientated cardiomyocytes, dilatation of ventricles, and ventricular hypoplasia (Korver et al., 1998; Ramakrishna et al., 2007). Accordingly, embryonic $foxm1^{-/-}$ mice are most likely dying of circulatory failure (Korver et al., 1998). Enlarged polyploid nuclei with a 30- to 50-fold increased DNA content were found in embryonic $foxm1^{-/-}$ cardiomyocytes (Korver et al., 1998; Ramakrishna et al., 2007), whereas the natural cardiomyocyte polyploidization in wild-type mice takes place postnatally (Lacroix & Maddox, 2012; Normand & King, 2010; Winkelman, Pfitzer, & Schneider, 1987). This polyploidy of cardiomyocytes in $foxm1^{-/-}$ mouse embryos (Korver et al., 1998; Ramakrishna et al., 2007) points to an importance of FOXM1 for prevention of DNA replication reinitiation or for mitotic entry and progression (Laoukili et al., 2007; Wierstra & Alves, 2007c).

The lungs of $foxm1^{-/-}$ mouse embryos exhibited a reduced number of mesenchymal lung cells, a decrease in the number of large blood vessels, defective microvasculature formation, and VSMC (vascular SMC) hypertrophy in pulmonary arteries (Kim et al., 2005).

The livers of $foxm1^{-/-}$ mouse embryos showed a reduced number of hepatoblasts, a disrupted organization of hepatic cords and sinusoids, a decreased number of large hepatic veins, and a failure in the formation of intrahepatic bile ducts (Krupczak-Hollis et al., 2004). Like cardiomyocytes (Korver et al., 1998; Ramakrishna et al., 2007), embryonic $foxm1^{-/-}$ hepatoblasts had enlarged polyploid nuclei with a sixfold increased DNA content (Krupczak-Hollis et al., 2004) whereas the natural hepatocyte polyploidization occurs after birth in wild-type mice (Celton-Morizur & Desdouets, 2010; Centric, Celton-Morizur, & Desdouets, 2012a, 2012b; Gupta, 2000).

7.1.2 Pancreas-specific deletion of foxm1 in Foxm1$^{\Delta panc}$ mice

Although *Foxm1*$^{\Delta panc}$ mice with a pancreas-specific deletion of the *foxm1* gene were born normal, their β-cell mass was reduced in comparison to control mice after birth because the β-cell mass of control mice increased continuously in the postnatal period, whereas the β-cell mass of *Foxm1*$^{\Delta panc}$ mice expanded during only the first four postnatal weeks and afterward remained the same or decreased slightly (Zhang, Ackermann, et al., 2006). Thus, FOXM1 is required for postnatal expansion and maintenance of β-cell mass (Fig. 3.2; Zhang, Ackermann, et al., 2006). Accordingly, in the *Foxm1*$^{\Delta panc}$ mice, the average islet size was decreased, the pancreatic insulin content was reduced, and the glucose tolerance was impaired so that part of them developed overt diabetes mellitus by 9 weeks of age (Zhang, Ackermann, et al., 2006). The postnatal reduction of β-cell mass in *Foxm1*$^{\Delta panc}$ mice compared to control mice was caused by a decrease in β-cell proliferation but not by increased apoptosis of β-cells (Zhang, Ackermann, et al., 2006). In addition, the pancreata of *Foxm1*$^{\Delta panc}$ mice showed cellular necrosis and premature senescence (Zhang, Ackermann, et al., 2006).

During pregnancy, the maternal pancreatic β-cell mass expands by ∼50% mainly due to a hormone-induced increase in β-cell proliferation (Butler, Meier, Butler, & Bhushan, 2007; Dhawan, Georgia, & Bhushan, 2007; Parsons, Brelje, & Sorenson, 1992; Rieck & Kaestner, 2010; Sorensen & Brelje, 1997). Also the maternal *foxm1* mRNA expression in pancreatic islets is elevated more than 50% during pregnancy (Zhang et al., 2010). Virgin mice with a pancreas-wide *foxm1* deletion (FoxM1$^{\Delta panc}$) displayed reduced β-cell proliferation and reduced β-cell mass compared to wild-type animals (Zhang et al., 2010). In contrast to the increase in β-cell proliferation and the expansion of β-cell mass during pregnancy in wild-type animals, neither an increase in β-cell proliferation nor an expansion of β-cell mass was observed in pregnant FoxM1$^{\Delta panc}$ mice indicating that FOXM1 is absolutely essential for maternal β-cell mass expansion during pregnancy (Zhang et al., 2010). However, there was no increase in apoptotic β-cells in pregnant FoxM1$^{\Delta panc}$ mice (Zhang et al., 2010). Pregnant FoxM1$^{\Delta panc}$ females exhibited a reduced total pancreatic insulin content, impaired glucose tolerance, and overt GDM (gestational diabetes mellitus) showing that FOXM1 is absolutely essential for the maintenance of glucose homeostasis during pregnancy (Zhang et al., 2010). In summary, FOXM1 plays a critical role in maternal β-cell compensation during pregnancy, at least partially because it is important for increased maternal β-cell proliferation during pregnancy (Zhang et al., 2010).

7.1.3 Conditional deletion of foxm1 in the developing pulmonary epithelium in epFoxm1$^{-/-}$ mice

Conditional deletion of *foxm1* in the developing pulmonary epithelium (*epFoxm1*$^{-/-}$) from embryonic day 7.5 (E7.5) onward inhibited the morphological and biochemical maturation of the lung, causing respiratory failure at birth and perinatal lethality within the first 24 h after birth (Kalin et al., 2008). Thus, FOXM1 was not required for fetal survival *in utero*, but it was critical for survival immediately after birth and for adaptation to air breathing (Kalin et al., 2008). The *epFoxm1*$^{-/-}$ newborn mice displayed pulmonary congestion, lung atelectasis, bronchial occlusion, and hyaline membranes lining terminal airways, findings consistent with severe RDS (respiratory distress syndrome) (Kalin et al., 2008). Yet, in *epFoxm1*$^{-/-}$ newborn mice, the number of lung lobes, body weight, and lung to body weight ratio were unchanged (Kalin et al., 2008). The lung epithelial-specific *foxm1* knockout did also not alter lung growth, branching lung morphogenesis, or epithelial cell proliferation (Kalin et al., 2008). Instead, the cell-selective deletion of *foxm1* from the developing respiratory epithelium delayed lung maturation and impaired lung sacculation leading to a diminished size of peripheral saccules (Kalin et al., 2008). Consistent with this delayed lung maturation, the pulmonary mesenchyme failed to thin (mesenchymal thickening) and the peripheral lung tubules remained closed in *epFoxm1*$^{-/-}$ lungs (Kalin et al., 2008). *Foxm1* deficiency did not influence the differentiation of ciliated and Clara cells but delayed the differentiation of type I and type II epithelial cells (Kalin et al., 2008). Accordingly, the numbers of squamous type I epithelial cells were decreased and the lamellar bodies in type II cells were significantly smaller in *epFoxm1*$^{-/-}$ embryos (Kalin et al., 2008).

In contrast, postnatal *foxm1* deletion from postnatal day 3 (P03) onward did not influence overall lung morphology (Kalin et al., 2008).

In summary, FOXM1 plays a role in lung morphogenesis (Kalin et al., 2008). It is required for perinatal lung function and for proper development of lung epithelial cells (Kalin et al., 2008). It is critical for lung maturation before birth and for surfactant homeostasis (Kalin et al., 2008).

When the deletion of *foxm1* from the distal respiratory epithelium in *epFoxm1*$^{-/-}$ mouse embryos was combined with the expression of activated K-Ras^{G12D} in these distal lung epithelial cells, it was found that the deletion of *foxm1* attenuates those defects in branching lung morphogenesis and lung sacculation, which were caused by activated K-Ras^{G12D} (Wang, Snyder, et al., 2012).

In the developing lung epithelium of such $epKras^{G12D}/epFoxm1^{-/-}$ mouse embryos, $foxm1$ was conditionally deleted from E7.5–E15.5 and, simultaneously, activated K-Ras^{G12D} was expressed (Wang, Snyder, et al., 2012). For comparison, the ectopic overexpression of activated K-Ras^{G12D} in the distal respiratory epithelium from E7.5–E15.5 without conditional deletion of $foxm1$ was performed in $epKras^{G12D}$ transgenic mice (Wang, Snyder, et al., 2012).

The embryonic lungs of $epKras^{G12D}$ mice displayed enlarged distal epithelial tubes (cysts) at E15.5, but the deletion of $foxm1$ in $epKras^{G12D}/epFoxm1^{-/-}$ embryos significantly reduced the size of Kras^{G12D}-expressing distal epithelial tubules (Wang, Snyder, et al., 2012). The FOXM1 protein expression was maintained in the dilated lung tubules of $epKras^{G12D}$ embryos and the epithelial cells lining these dilated lung cysts in $epKras^{G12D}$ mice were distal lung cells (Wang, Snyder, et al., 2012). In the lungs of $epKras^{G12D}/epFoxm1^{-/-}$ mice, the pulmonary capillary endothelium was present (Wang, Snyder, et al., 2012). The transgenic overexpression of activated Kras^{G12D} ($epKras^{G12D}$) caused large intrapulmonary cysts at E17.5, but the additional deletion of $foxm1$ in Kras^{G12D}-expressing embryos ($epKras^{G12D}/epFoxm1^{-/-}$) significantly decreased the number and size of intrapulmonary cysts and increased the number of small peripheral saccules (alveolar sacs) (Wang, Snyder, et al., 2012). Moreover, the deletion of $foxm1$ from Kras^{G12D}-expressing embryonic lungs ($epKras^{G12D}/epFoxm1^{-/-}$) improved their morphogenetic abnormalities in lung epithelium and mesenchyme because it normalized the ratio between epithelial and mesenchymal areas, which was increased in $epKras^{G12D}$ mice but redecreased in $epKras^{G12D}/epFoxm1^{-/-}$ animals (Wang, Snyder, et al., 2012).

Hence, the expression of activated Kras^{G12D} disrupts branching lung morphogenesis and causes severe sacculation defects during embryonic lung development (Shaw et al., 2007; Wang, Snyder, et al., 2012), whereas the simultaneous deletion of $foxm1$ from distal respiratory epithelial cells attenuates these branching and sacculation defects caused by activated Kras^{G12D} (Wang, Snyder, et al., 2012). The finding that $foxm1$ deficiency prevents the Kras^{G12D}-induced structural abnormalities in the developing respiratory epithelium indicates that FOXM1 is an important downstream target of K-Ras during embryonic lung morphogenesis (Wang, Snyder, et al., 2012).

So far, an effect of K-Ras on the expression or activity of FOXM1 has not been reported, but the Ras family members H-Ras (Behren et al., 2010; Park, Carr, et al., 2009; Pellegrino et al., 2010) and N-Ras (Ho et al., 2012) are known to upregulate the $foxm1$ mRNA and/or protein expression

(Fig. 3.10). In addition, Ras stimulates the transcriptional activity of FOXM1 because dn Ras suppressed the transcriptional activity of FoxM1B (Major et al., 2004).

The conditional lung epithelial-specific *foxm1* knockout attenuated the defects in branching lung morphogenesis and lung sacculation caused by activated Kras^{G12D} without influencing cell proliferation or proximal–distal patterning in the lung epithelium (Wang, Snyder, et al., 2012). This observation suggests that K-Ras and FOXM1 may influence branching lung morphogenesis by regulating cell migration, cell-to-cell contacts, and epithelial polarity (Wang, Snyder, et al., 2012), all of which are critical for epithelial branching (Morrisey & Hogan, 2010).

Canonical Wnt/β-catenin signaling is critical for proper lung morphogenesis (Goss et al., 2009; Mucenski et al., 2003).

The study of Wang, Snyder, et al. (2012b) revealed that the deletion of *foxm1* from the distal respiratory epithelium may attenuate the defects in branching lung morphogenesis and lung sacculation caused by activated K-Ras^{G12D} because both K-Ras and FOXM1 inhibit the canonical Wnt/β-catenin signaling pathway in the developing lung epithelium. Ectopic overexpression of either activated K-Ras^{G12D} (*epKras^{G12D}*) or a ca mutant of FoxM1B (*epFoxm1*), which lacks the autoinhibitory FOXM1 N-terminus (FoxM1B-ΔN), in distal respiratory epithelial cells of transgenic mouse embryos decreased the activity of the β-catenin/TCF-4 reporter construct TOPGAL, which is driven by TCF-4-binding sites, whereas conditional knockout of FOXM1 from the distal respiratory epithelium in mouse embryos (*epFoxm1$^{-/-}$*) increased the TOPGAL activity showing that the canonical Wnt/β-catenin signaling pathway is inhibited by both K-Ras and FOXM1 (Wang, Snyder, et al., 2012). In this context, the upregulation of the protein expression of the β-catenin inhibitor AXIN2 (axis inhibitor 2, conductin) by both K-Ras and FOXM1 seems to play a crucial role in their inhibitory effects on canonical Wnt/β-catenin signaling because the AXIN2 protein level was increased in distal (cystic) lung epithelium after overexpression of Kras^{G12D} (*epKras^{G12D}*) and because *foxm1* deletion (*epFoxm1$^{-/-}$*) selectively reduced the AXIN2 protein level in distal lung epithelial tubules without influencing proximal epithelial tubules (Wang, Snyder, et al., 2012). Therefore, Wang, Snyder, et al. (2012b) suggested that FOXM1 mediates a crosstalk between K-Ras/MAPK and canonical Wnt/β-catenin signaling pathways during development of the respiratory epithelium and they postulated the existence of a pathway

K-Ras → FOXM1 → Axin2 ⊣ β-catenin → lung morphogenesis in the embryonic lung epithelium.

7.1.4 Conditional deletion of foxm1 from airway Clara cells in CCSP-Foxm1$^{-/-}$ mice

Conditional deletion of *foxm1* from airway Clara cells in *CCSP-Foxm1*$^{-/-}$ mice from E16.5 onward inhibited Clara cell proliferation and disrupted the normal patterning of epithelial cell differentiation in the bronchioles of the developing mouse lung, demonstrating that FOXM1 is required for proliferation and differentiation of Clara cells as well as for proper epithelial differentiation during development of conducting airways (Ustiyan et al., 2012).

The selective deletion of *foxm1* from Clara cells (*CCSP-Foxm1*$^{-/-}$) from E16.5 to P5 significantly decreased the total numbers of proliferating bronchiolar epithelial cells as well as the percentages of proliferating Clara cells at P5 (Ustiyan et al., 2012). Thus, the conditional Clara cell-specific *foxm1* knockout in *CCSP-Foxm1*$^{-/-}$ mouse embryos reduced the proliferation of airway Clara cells during the early postnatal period of lung development indicating that FOXM1 is required for Clara cell proliferation during development of conducting airways (Ustiyan et al., 2012).

In accordance with the reduced proliferation of epithelial cells in *CCSP-Foxm1*$^{-/-}$ airways at P5, the number of epithelial cells lining the *CCSP-Foxm1*$^{-/-}$ bronchioles was dramatically decreased at P30 when *foxm1* was conditionally deleted from Clara cells from E16.5 to P30 (Ustiyan et al., 2012). Moreover, the distribution of epithelial cells in *CCSP-Foxm1*$^{-/-}$ bronchioles with conditional *foxm1* knockout from E16.5 to P30 was considerably altered at P30 such that they displayed a decrease in the number of Clara cells and ciliated cells but an increase in the number of squamous cells, goblet cells, and alveolar type II cells (Ustiyan et al., 2012). Since squamous, goblet, and alveolar type II cells were absent from the pulmonary bronchioles of control mice the Clara cell-specific deletion of *foxm1* led to the appearance of ectopic squamous, goblet, and alveolar type II cells in the conducting airway epithelium (Ustiyan et al., 2012). The ectopic goblet cells and alveolar type II cells were directly derived from *foxm1*-deficient Clara cells, whereas the ectopic squamous cells did not originate from Clara cells and their origin remained unclear (Ustiyan et al., 2012). This transdifferentiation of *foxm1*-deficient Clara cells into alveolar type II cells suggests that FOXM1 expression in Clara cell progenitors may be required to prevent their differentiation into alveolar type II cells in the bronchiolar epithelium (Ustiyan et al., 2012). Hence, FOXM1 is required not only for proliferation of Clara cells, but also for normal differentiation of the airway epithelium during the late embryonic/postnatal period of lung development (Ustiyan et al., 2012).

In accordance with the loss of Clara and ciliated cells and with the squamous and globlet cell metaplasia in the bronchiolar epithelium of *CCSP-Foxm1*$^{-/-}$ mice at P30, the loss of *foxm1* in Clara cells resulted in dramatic airway remodeling in *CCSP-Foxm1*$^{-/-}$ lungs at P30, including increased thickness of stromal tissue and peribronchiolar fibrosis (Ustiyan et al., 2012): extensive regions of the *CCSP-Foxm1*$^{-/-}$ bronchioles lacked the normal cuboidal epithelium consisting of Clara and ciliated cells, but instead they were lined by a squamous epithelium, which was not found in control lungs and which often contained enlarged, elongated, extremely thin epithelial cells with enlarged nuclei (Ustiyan et al., 2012). Moreover, both Clara cells and ciliated cells were elongated and lacked the normal columnar/cuboidal morphology (Ustiyan et al., 2012). The abnormally large Clara cells contained numerous characteristic mitochondria but no secretory granules (Ustiyan et al., 2012). The number of cilia on the surface of the dramatically altered ciliated cells was reduced (Ustiyan et al., 2012). Yet, apoptotic cells were not detected in *CCSP-Foxm1*$^{-/-}$ bronchiolar epithelium (Ustiyan et al., 2012). This dramatically altered morphology of the airway epithelium in *CCSP-Foxm1*$^{-/-}$ mice manifested in significantly higher airway resistance and hysteresivity at P30 (Ustiyan et al., 2012).

Furthermore, *CCSP-Foxm1*$^{-/-}$ bronchioles exhibited disrupted epithelial junctions (Ustiyan et al., 2012). At P30, E-cadherin and β-catenin were present in epithelial junctions of control bronchioles whereas epithelial junctions were disrupted in *foxm1*-deficient bronchiolar epithelium, causing abnormal localization of E-cadherin and β-catenin in basal membranes (Ustiyan et al., 2012).

In addition to its above-mentioned importance for Clara cell proliferation and differentiation during late embryonic and postnatal lung development, FOXM1 plays a critical role in proliferation and differentiation of Clara cells in the adult lung, where it is critical for long-term maintenance of bronchiolar epithelium and airway structure (Ustiyan et al., 2012): extended deletion of *foxm1* from Clara cells in adult mice from 6 to 8 weeks until 5 months after birth caused extensive peribronchial fibrosis, enlargement of peripheral respiratory airspaces as well as disrupted airway structure and it resulted in airway hyperreactivity (Ustiyan et al., 2012). This long-term conditional Clara cell-specific *foxm1* knockout in the adult lung decreased the number of Clara cells dramatically and caused abnormal accumulation of ectopic squamous cells and ectopic alveolar type II cells in the bronchiolar epithelium, which indicates a requirement of FOXM1 for maintenance of airway epithelial differentiation and supports a role of FOXM1 in self-renewal of airway epithelium (Ustiyan et al., 2012).

Also after naphthalene-induced lung injury, *foxm1*-deficient bronchioles displayed a reduced proliferation of Clara cells and an accumulation of ectopic alveolar type II cells in the airway epithelium at day 5 and day 14 after injury (Ustiyan et al., 2012). To study the role of FOXM1 in naphthalene lung injury, *foxm1* was selectively deleted from Clara cells in adult (6–8 week old) $CCSP$-$Foxm1^{-/-}$ mice for 7 days before naphthalene administration (Ustiyan et al., 2012). This short *foxm1* knockout was insufficient to disrupt airway epithelium prior to naphthalene-induced lung injury (Ustiyan et al., 2012).

The conditional deletion of *foxm1* from airway Clara cells did not influence the tracheal epithelium although the number of Clara cells was reduced in trachea of $CCSP$-$Foxm1^{-/-}$ lungs (Ustiyan et al., 2012). However, neither alveolar type II cells nor squamous cells were found in $CCSP$-$Foxm1^{-/-}$ trachea and also the numbers and distribution of basal cells were unchanged in $CCSP$-$Foxm1^{-/-}$ mice (Ustiyan et al., 2012). Therefore, FOXM1 expression in Clara cells is required for development of bronchiolar (columnar) epithelium but dispensable for development of tracheal (pseudostratified) epithelium (Ustiyan et al., 2012).

Foxm1 deletion from Clara cells in $CCSP$-$Foxm1^{-/-}$ mice did not influence the morphology of the distal lung region, alveolar structure, the number of alveolar type I and type II cells in distal lung regions, or the number of proliferative cells in distal lung saccules (Ustiyan et al., 2012). Also the numbers of BASCs (CCSP+/proSP-C+ cells) lining bronchoalveolar duct junctions were not altered in $CCSP$-$Foxm1^{-/-}$ lungs (Ustiyan et al., 2012). Moreover, neither pulmonary inflammation nor significant changes in total or differential counts of inflammatory cells in BALF (bronchoalveolar lavage fluid) were observed in $CCSP$-$Foxm1^{-/-}$ mice (Ustiyan et al., 2012).

In summary, deletion of *foxm1* from bronchiolar Clara cells decreased Clara cell proliferation and dramatically altered the cellular composition of the bronchiolar epithelium, demonstrating that FOXM1 is required for proliferation and differentiation of Clara cells as well as for proper epithelial differentiation not only during late embryonic and postnatal development of conducting airways, but also in the adult mouse lung, where FOXM1 is required for maintenance of epithelial cells lining conducting airways (Ustiyan et al., 2012). FOXM1 regulates progenitor properties of Clara cells in the conducting airways and FOXM1 is critical for maintenance of normal bronchiolar epithelial cell differentiation as well as for prevention of trans-differentiation of Clara cells into alveolar type II cells in the bronchiolar epithelium (Ustiyan et al., 2012). Notably, loss of *foxm1* in Clara cells resulted in their transdifferentiation into ectopic alveolar type II cells in the bronchioles of $CCSP\text{-}Foxm1^{-/-}$ mice, which indicates that FOXM1 restricts differentiation of alveolar type II cells from conducting airway epithelial cells so that it plays a critical role in the differentiation and maintenance of proximal versus peripheral respiratory epithelial cell fate in the perinatal and adult lung (Ustiyan et al., 2012). Thus, in Clara cells, FOXM1 plays a critical role in the restriction of bronchiolar cells to conducting versus alveolar regions of the lung (Ustiyan et al., 2012).

7.1.5 Conditional deletion of foxm1 in developing smooth muscle cells (SMCs) in smFoxm1$^{-/-}$ mice

The conditional deletion of *foxm1* in developing SMCs caused a perinatal lethality in the majority (87%) of these $smFoxm1^{-/-}$ mice within the first 24 h after birth indicating that FOXM1 function in SMC is not required for fetal survival *in utero* but is critical for survival immediately after birth (Ustiyan et al., 2009). Mice with this smooth muscle-specific conditional *foxm1* knockout ($smFoxm1^{-/-}$) exhibited severe pulmonary hemorrhage as well as structural defects in arterial wall and esophagus (Ustiyan et al., 2009). *Foxm1* deficiency did not influence the differentiation of SMC during embryonic development, but it decreased the proliferation of SMC in muscle layers of embryonic blood vessels and esophagus (Ustiyan et al., 2009). The structural abnormalities of the esophagus in $smFoxm1^{-/-}$ embryos included significantly diminished thickness of esophageal muscle and loss of mesenchymal cells (Ustiyan et al., 2009). $smFoxm1^{-/-}$ newborn mice displayed extensive perinatal pulmonary hemorrhage with red blood cells present in bronchioles and peripheral pulmonary saccules (Ustiyan et al., 2009). Since lung hemorrhage was not detected in $smFoxm1^{-/-}$ mutants prior to birth the hemorrhaging phenotype was concomitant with the increased pulmonary arterial blood flow that occurs following birth (Ustiyan et al., 2009). Thus, the structural abnormalities of the arterial wall led to vascular leakage from perforated pulmonary arteries in $smFoxm1^{-/-}$ newborn mice (Ustiyan et al., 2009). In summary, FOXM1 is critical for proliferation of SMC and it is required for proper embryonic development of blood vessels and esophagus (Ustiyan et al., 2009).

7.1.6 Conditional deletion of foxm1 from cardiomyocytes early during heart development

Conditional deletion of *foxm1* from cardiomyocytes early during heart development disrupts heart morphogenesis in mice and results in abnormal cardiac morphology and decreased cardiomyocyte proliferation, culminating in embryonic lethality in late gestation (Bolte et al., 2011). Deletion of *foxm1* from cardiomyocytes caused chamber dilation, thinning of the ventricular walls and the interventricular septum, disorganization of the myocardium (cardiomyocyte disarray), and decreased cardiomyocyte proliferation (Bolte et al., 2011). The myocardial thinning and the ventricular hypoplasia were due to the diminished cardiomyocyte proliferation (Bolte et al., 2011). Despite the *foxm1* deletion in developing cardiomyocytes the hearts possessed all four chambers and the overall heart size was unaltered (Bolte et al., 2011). Mouse hearts with the cardiomyocyte-specific deletion of *foxm1* exhibited decreased capillary density but no change in the number or morphology of coronary vessels (Bolte et al., 2011). Moreover, mice with the cardiomyocyte-specific *foxm1* deletion displayed significant cardiac fibrosis in the postnatal heart (Bolte et al., 2011). Although an increased deposition of extracellular matrix in the atrioventricular valves was observed the valve size was unaltered (Bolte et al., 2011). There was also no difference in the size of the leaflets in either the mitral or tricuspid valves (Bolte et al., 2011). In summary, FOXM1 expression in cardiomyocytes is critical for proper heart development and required for cardiomyocyte proliferation and myocardial growth (Bolte et al., 2011). Thus, FOXM1 plays a cell-autonomous role in cardiomyocytes during cardiac development (Bolte et al., 2011).

7.1.7 Conditional deletion of foxm1 from cardiomyocytes in the late postnatal period in α-MHC-Cre/Foxm1$^{fl/fl}$ mice

Conditional deletion of *foxm1* from cardiomyocytes in the late postnatal period (α-*MHC-Cre/Foxm1$^{fl/fl}$*) predisposes to cardiac hypertrophy and fibrosis late in life because old α-*MHC-Cre/Foxm1$^{fl/fl}$* mice developed cardiac hypertrophy and fibrosis, whereas young adult α-*MHC-Cre/Foxm1$^{fl/fl}$*

mice exhibited normal cardiac morphology and function (Bolte et al., 2012). Thus, FOXM1 is dispensable for postnatal heart development, but it is essential for proper long-term maintenance of cardiac structure and function in the older heart (Bolte et al., 2012).

α-*MHC-Cre/Foxm1*$^{fl/fl}$ mice with a postnatal cardiomyocyte-specific *foxm1* deletion were viable and healthy (Bolte et al., 2012). They developed normally into adulthood displaying no heart abnormalities in either cardiac structure (left or right ventricular anatomy, heart weight-to-body weight ratio, cardiomyocyte size, capillary density, coronary vessel formation) or cardiac function (cardiac inotrophy and lusitrophy, sinus rhythm, occurrence of arrhythmia) until old age (Bolte et al., 2012). However, old α-*MHC-Cre/Foxm1*$^{fl/fl}$ mice (16 month) suffered from late onset cardiac hypertrophy because they exhibited a significant increase in the heart weight-to-body weight ratio compared to control mice and because their cardiomyocyte size (area) in both ventricles was increased by 60% (Bolte et al., 2012). Additionally, the hearts of old α-*MHC-Cre/Foxm1*$^{fl/fl}$ mice displayed increased myocardial fibrosis (as shown by Masson's Trichome staining) (Bolte et al., 2012).

In summary, the postnatal deletion of *foxm1* from cardiomyocytes causes late onset cardiac hypertrophy and fibrosis in old α-*MHC-Cre/Foxm1*$^{fl/fl}$ mice, which demonstrates a critical role of FOXM1 in maintenance of myocardial structure during aging (Bolte et al., 2012). In contrast, FOXM1 is not important for heart development during the late postnatal period into adulthood because of the normal cardiac structure and function in young adult α-*MHC-Cre/Foxm1*$^{fl/fl}$ mice (Bolte et al., 2012).

7.1.8 Deletion of foxm1 *in epithelial cells as they differentiate during pregnancy in* FoxM1 *FL/FL, WAP-Cre mice*

Deletion of *foxm1* in epithelial cells as they differentiate during pregnancy in *FoxM1* FL/FL, WAP-Cre mice led to a delay in the development of lobuloalveolar structures in the mammary gland during the second pregnancy, indicating that FOXM1 promotes lobuloalveolar differentiation (Carr et al., 2012). During early stages of pregnancy, the mammary epithelium expands and begins to form lobuloalveolar units (Henninghausen & Robinson, 2005), which were found in wild-type mice at pregnancy day 6 (early pregnancy), whereas the *FoxM1* FL/FL, WAP-Cre mice did not develop lobuloalveolar structures (Carr et al., 2012). The pregnancy day 6 mammary glands in *FoxM1* FL/FL, WAP-Cre mice appeared similar to those of 8-week-old virgin controls and failed to form appreciable alveoli (Carr et al., 2012). However, alveoli were clearly visible, though in reduced number, in *FoxM1* FL/FL, WAP-Cre mice at pregnancy day 18 (late pregnancy), which indicates a delay in lobuloalveolar differentiation in the mammary glands of these conditional *foxm1* knockout mice (Carr et al., 2012).

In lactating *FoxM1* FL/FL, WAP-Cre mammary glands, which appeared flattened, the number of milk globules in the alveoli was reduced (Carr et al., 2012). Nonetheless, *FoxM1* FL/FL, WAP-Cre mice produced sufficient milk to support their litters because their pups survived and did not differ in weight from pups born to control animals (Carr et al., 2012). During pregnancy, *FoxM1* FL/FL, WAP-Cre mice exhibited a delay in accumulation of the key milk proteins α-casein and β-casein, but, on lactation day 1, they showed equivalent expression of milk proteins as wild-type animals (Carr et al., 2012).

In summary, the loss of *foxm1* in *FoxM1* FL/FL, WAP-Cre mice leads to a delay in lobuloalveolar differentiation in the mammary gland, indicating that FOXM1 stimulates lobuloalveolar differentiation during pregnancy (Carr et al., 2012). Accordingly, the *foxm1* RNA expression in the mammary gland increases during pregnancy, a period of ductal growth and expansion, but it re-decreases again during lactation so that the lowest *foxm1* mRNA level is observed upon involution, which is characterized by apoptosis and remodeling (Carr et al., 2012).

7.1.9 Conditional deletion of foxm1 in the mammary tissue of adult virgin FoxM1 *FL/FL, WAP-rtTA-Cre mice*

Conditional deletion of *foxm1* in the mammary tissue of adult (8 week old) virgin *FoxM1* FL/FL, WAP-rtTA-Cre mice resulted in an expansion of differentiated luminal cells, indicating that FOXM1 inhibits mammary luminal differentiation (Carr et al., 2012). This conditional *foxm1* knockout reduced the *foxm1* mRNA level by 80% in luminal progenitors and by 90% in differentiated luminal cells, whereas mammary stem cells displayed no significant reduction of the *foxm1* mRNA expression (Carr et al., 2012). The mammary glands of *FoxM1* FL/FL, WAP-rtTA-Cre mice exhibited narrow ductal branching, but their number of branches was unchanged (Carr et al., 2012). *FoxM1* FL/FL, WAP-rtTA-Cre mammary glands were not composed of a single layer of epithelial cells and lumens were filled with cells that expanded beyond the myoepithelial layer (Carr et al., 2012). These cells were mature luminal epithelium, suggesting an expansion of the differentiated pool (Carr et al., 2012). Yet, the rates of proliferation and apoptosis were almost unaltered in *FoxM1* FL/FL, WAP-rtTA-Cre mammary glands (Carr et al., 2012). After the conditional *foxm1* deletion, there was an approx. 20% increase in differentiated luminal cells with a concomitant loss of the mammary stem cell and luminal progenitor populations in *FoxM1* FL/FL, WAP-rtTA-Cre mice, so that this acute loss of *foxm1* in the mammary gland resulted in a shift toward the differentiated state (Carr et al., 2012). Since the *foxm1* knockdown affected the mammary stem cell population although *foxm1* was not deleted in this population, the negative effect on the stem cell pool may be secondary to changes in the differentiated luminal cell population (Carr et al., 2012) because hormonally mediated paracrine signaling by differentiated luminal cells is known to regulate the mammary stem cell pool (Asselin-Labat et al., 2010; Joshi et al., 2010). In summary, acute loss of *foxm1* in the adult mammary gland of *FoxM1* FL/FL, WAP-rtTA-Cre mice leads to an increase in differentiated luminal cells and a loss of luminal progenitor pools, that is, to a shift toward the differentiated state, which indicates that FOXM1 inhibits mammary luminal differentiation in nonpregnant mice (Carr et al., 2012). Hence, FOXM1 is a critical regulator of mammary luminal cell fate because it plays an important role in prevention of terminal differentiation of luminal epithelial progenitors and thus in the maintenance of the mammary luminal progenitor pool (Carr et al., 2012). Accordingly, the *foxm1* mRNA expression is highest in luminal progenitors and decreases upon differentiation because the *foxm1* mRNA level was 10-fold and nearly 50-fold higher in mammary stem cells or luminal progenitors than in differentiated luminal cells, respectively, so that less differentiated mammary cells display a higher *foxm1* mRNA expression (Carr et al., 2012).

FOXM1 inhibits mammary luminal differentiation at least in part through the RB-dependent repression of the direct FOXM1 target gene *GATA3* because the *foxm1* knockout phenotype of *FoxM1* FL/FL, WAP-rtTA-Cre mice was reversed by simultaneous knockdown of GATA3 (Carr et al., 2012): mammary glands were regenerated from *FoxM1* FL/FL, WAP-rtTA-Cre mammospheres, which had been infected with retrovirus-expressing GATA3-targeting shRNA, so that these repopulated glands lacked both FOXM1 and GATA3 (Carr et al., 2012). The shRNA-mediated knockdown of GATA3 was sufficient to reverse the loss of luminal progenitors and the expansion of differentiated luminal cells observed after loss of *foxm1* in *FoxM1* FL/FL, WAP-rtTA-Cre mice (Carr et al., 2012).

The transcription factor GATA3 is a key regulator of mammary luminal differentiation, which is required for proper mammary gland development and maintenance of mature luminal cells (Asselin-Labat et al., 2007; Kouros-Mehr, Slorach, Sternlicht, & Werb, 2006). The tumor-suppressor GATA3 maintains the differentiated state of the mammary gland through activation of genes, which are necessary for the transition from the progenitor to the differentiated state (Asselin-Labat et al., 2007; Kouros-Mehr et al., 2006).

The above-mentioned results for the inhibition of mammary luminal differentiation by FOXM1, which were obtained with conditional *foxm1* knockout mice (*FoxM1* FL/FL, WAP-rtTA-Cre), are confirmed by the reverse findings for FOXM1-overexpressing repopulated mammary glands, which were regenerated from mammospheres that had been infected with retrovirus-expressing FOXM1 (Carr et al., 2012). This overexpression of FOXM1 in regenerated mammary glands leads to expansion of the luminal progenitor pool, inhibition of mammary luminal differentiation and aberrant ductal morphology (Carr et al., 2012).

The FOXM1-overexpressing mammary glands showed narrowing and distinct hyperplastic regions of excessive cell infiltration, where epithelial cells filled the lumen or spread beyond the basal layer (Carr et al., 2012). The altered architecture of FOXM1-overexpressing mammary glands revealed a startling phenotype (Carr et al., 2012). Control mammary glands contained a layer of SMA (smooth muscle actin)-positive cells surrounding luminal cells whereas the FOXM1-overexpressing mammary glands contained the expected pattern as well as SMA-positive cells surrounded by luminal cells, which were not misplaced myoepithelial cells (Carr et al., 2012). However, the FOXM1 overexpression in repopulated mammary glands did not affect the rates of proliferation and apoptosis (Carr et al., 2012). Control mammary glands displayed a uniform luminal restricted KRT18 (cytokeratin 18) staining pattern, whereas the FOXM1-overexpressing mammary glands exhibited a punctate pattern distinct from differentiated luminal cells (Carr et al., 2012). The expanded cells indicated the expansion of an undifferentiated cell of luminal origin (Carr et al., 2012). This expansion of an undifferentiated cell type in FOXM1-overexpressing mammary glands manifested in a shift away from the differentiated state because, in the FOXM1-overexpressing mammary glands, the luminal progenitor pool expanded by approx. 20%, with a similar reduction in the percentage of differentiated luminal cells, suggesting that the FOXM1 overexpression results in a failure of cells to properly exit the luminal progenitor pool and differentiate fully (Carr et al., 2012). Accordingly, also the mammary stem cell population was increased in the FOXM1-overexpressing mammary glands (Carr et al., 2012). Thus, FOXM1 overexpression in the mammary gland inhibits mammary luminal differentiation and results in an expansion of mammary progenitors (Carr et al., 2012). In summary, FOXM1 regulates mammary luminal cell fate through inhibition of mammary luminal differentiation (Carr et al., 2012).

Again, FOXM1 inhibits mammary luminal differentiation at least partially through the RB-dependent repression of the direct FOXM1 target gene GATA3 because the simultaneous overexpression of GATA3 reversed the effects of FOXM1 overexpression on luminal differentiation when mammary glands were regenerated from FOXM1-overexpressing mammospheres, which had been infected with retrovirus-expressing GATA3, so that the repopulated glands co-overexpressed GATA3 along with FOXM1 (Carr et al., 2012): these reconstituted mammary glands overexpressing both FOXM1 and GATA3 had visible lumens and no extensive cellular hyperplasia (Carr et al., 2012). The expression of differentiation markers was corrected too (Carr et al., 2012). Moreover, co-overexpression of GATA3 was sufficient to reverse the expansion of luminal progenitors and the loss of differentiated luminal cells observed in FOXM1-overexpressing mammary glands (Carr et al., 2012). Hence, FOXM1 downregulates the GATA3 expression in the progenitor population and this repression of *GATA3* transcription seems to be the dominant mechanism, by which FOXM1 inhibits the differentiation of luminal epithelial progenitors (Carr et al., 2012).

Notably, the repression of the *GATA3* mRNA expression by FOXM1 depends on RB because FOXM1 itself upregulates the *GATA3* mRNA level in RB-deficient cells (Carr et al., 2012). In order to repress the *GATA3* transcription, the transcriptional activator FOXM1 recruits the corepressors RB and DNMT3b to the *GATA3* promoter so that FOXM1 represses the transcription of *GATA3* through DNMT3b-mediated, RB-dependent DNA methylation of the *GATA3* promoter (Carr et al., 2012). Accordingly, FOXM1 interacts directly with RB (Carr et al., 2012; Major et al., 2004; Wierstra, 2013a; Wierstra & Alves, 2006a) and possibly indirectly with DNMT3b (Carr et al., 2012).

FOXM1 inhibits mammary luminal differentiation through the RB-dependent repression of *GATA3* transcription so that RB is required for the inhibition of mammary luminal differentiation by FOXM1 because the simultaneous knockdown of RB reversed the effects of FOXM1 overexpression on luminal differentiation (Carr et al., 2012): mammary glands were regenerated from FOXM1-overexpressing mammospheres, which had been infected with retrovirus-expressing RB-targeting shRNA, so that these repopulated glands overexpressed FOXM1 and lacked RB (Carr et al., 2012). The shRNA-mediated knockdown of RB was sufficient to reverse the loss of differentiated luminal cells as well as the expansion of luminal progenitors and mammary stem cells observed in FOXM1-overexpressing mammary glands (Carr et al., 2012). Hence, FOXM1 overexpression inhibits mammary luminal differentiation, but knockdown of RB alleviates this FOXM1-mediated inhibition of differentiation, because FOXM1 functions in a complex with RB to inhibit *GATA3* expression and thereby mammary luminal differentiation (Carr et al., 2012). Therefore, both RB and GATA3 play pivotal roles in the inhibition of mammary luminal differentiation by FOXM1 (Carr et al., 2012).

7.2. FOXM1 transgenic mouse models

7.2.1 Transgenic epFoxm1 *mice with conditional expression of an N-terminal FoxM1B mutant in respiratory epithelial cells*

Transgenic *epFoxm1* mice with conditional expression of a FoxM1B mutant, which lacks the 231 N-terminal aa (FoxM1B-ΔN), in respiratory epithelial cells showed that the precise control of FOXM1 expression in the lung epithelium is critical for normal lung morphogenesis, in particular for lung sacculation and proper development of the airway epithelium (Wang, Zhang, Snyder, et al., 2010).

Expression of this N-terminally truncated FoxM1B mutant (FoxM1B-ΔN) during embryogenesis induced epithelial hyperplasia, which was characterized by the presence of a pseudostratified epithelium in contrast to a single columnar epithelium in wild-type lungs, indicating that FoxM1B-ΔN expression is sufficient to accelerate cellular proliferation in the undifferentiated respiratory epithelium (Wang, Zhang, Snyder, et al., 2010). Moreover, FoxM1B-ΔN expression impaired lung sacculation and abnormal, dilated epithelial cysts were observed throughout peripheral *epFoxm1* lungs (Wang, Zhang, Snyder, et al., 2010). The severe sacculation defects were characterized by thickening of mesenchyme and increased size of peripheral saccules (Wang, Zhang, Snyder, et al., 2010). These results suggest that the expression of FoxM1B-ΔN might prevent the differentiation of epithelial progenitor cells toward the type II lineage (Wang, Zhang, Snyder, et al., 2010).

Expression of FoxM1B-ΔN during postnatal lung development or in adult *epFoxm1* mice caused focal airway hyperplasia, which was characterized by a pseudostratified epithelium at sites normally lined by a single columnar epithelium (Wang, Zhang, Snyder, et al., 2010). Furthermore, FoxM1B-ΔN expression increased the proliferation of Clara cells and led to Clara cell hyperplasia (Wang, Zhang, Snyder, et al., 2010).

7.2.2 Transgenic CCSP-Foxm1 *mice with conditional expression of the N-terminal FoxM1B mutant in Clara cells*

Transgenic *CCSP-Foxm1* mice, which conditionally express the N-terminally truncated FoxM1B mutant (FoxM1B-ΔN) in Clara cells, showed that the expression of FoxM1B-ΔN in adult *CCSP-Foxm1* lungs caused airway hyperplasia and was sufficient to induce hyperplasia of Clara cells (Wang, Zhang, Snyder, et al., 2010).

7.3. Xenopus

In early *Xenopus* embryos, FOXM1 is required for neural development, namely, for proliferation of primary neuronal precursors and for primary neuronal differentiation but not specification (Ueno et al., 2008).

8. BIOLOGICAL FUNCTIONS OF FOXM1

8.1. Cell proliferation and cell cycle progression

8.1.1 Stimulation of cell proliferation

FOXM1 stimulates cell proliferation and promotes cell cycle progression at both the G1/S-transition and the G2/M-transition (Fig. 3.2; Costa, 2005; Costa et al., 2003; Costa, Kalinichenko, Major, et al., 2005; Costa, Kalinichenko, Tan, et al., 2005; Kalin et al., 2011; Koo et al., 2011; Laoukili et al., 2007; Mackey et al., 2003; Myatt & Lam, 2007; Raychaudhuri & Park, 2011; Wierstra & Alves, 2007c). The stimulation of proliferation is one main function of FOXM1 and forms the basis of many of its biological roles (see below).

In accordance with its proliferation-stimulating effect, FOXM1 displays a strictly proliferation-specific expression pattern (see above), and its expression is upregulated by proliferation signals but downregulated by antiproliferation signals (Fig. 3.10; Koo et al., 2011; Laoukili et al., 2007; Raychaudhuri & Park, 2011; Wang, Ahmad, Li, et al., 2010; Wierstra & Alves, 2007c). Also the transcriptional activity of FOXM1 is increased and decreased by proliferation or antiproliferation signals, respectively (Costa, 2005; Costa, Kalinichenko, Major, et al., 2005; Koo et al., 2011; Laoukili et al., 2007; Murakami et al., 2010; Myatt & Lam, 2007; Raychaudhuri & Park, 2011; Wang, Ahmad, Li, et al., 2010; Wierstra & Alves, 2007c), which correlates with the role of FOXM1 in stimulation of cell proliferation too.

Thus, FOXM1 promotes proliferation and is antagonistically regulated by growth versus antigrowth stimuli so that it exhibits the characteristics of a typical proliferation-associated transcription factor (Kalin et al., 2011; Koo et al., 2011; Laoukili et al., 2007; Raychaudhuri & Park, 2011; Wierstra & Alves, 2007c).

8.1.2 Promotion of cell cycle progression

FOXM1 stimulates cell proliferation (Ahmad et al., 2011, 2010; Bao et al., 2011; Bergamaschi et al., 2011; Bhat et al., 2011; Carr et al., 2010; Chen et al., 2011; Chu et al., 2012; Dai et al., 2013; Faust et al., 2012; Ho et al., 2012; Li et al., 2011; Lok et al., 2011; Mencalha et al., 2012; Millour et al., 2010, 2011; Monteiro et al., 2012; Ning et al., 2012; Park et al., 2012; Park, Wang, et al., 2008; Pellegrino et al., 2010; Raghavan et al., 2012; Uddin et al., 2011; Ueno et al., 2008; Wang, Ahmad, Banerjee, et al., 2010; Wang, Lin, et al., 2011; Wierstra & Alves, 2006d; Xia et al., 2009; Xia, Mo, et al., 2012; Xie et al., 2010; Xue et al., 2012; Zeng et al., 2009; Zhang, Wu, et al., 2012) by promoting cell cycle progression at both the G1/S- and G2/M-transitions (Fig. 3.2). First, FOXM1 stimulates the entry into S-phase (Ackermann Misfeldt et al., 2008; Anders et al., 2011; Brezillon et al., 2007; Calvisi et al., 2009; Chan et al., 2008; Chen, Müller, et al., 2013; Chen et al., 2010; Davis et al., 2010; Gusarova et al., 2007; Huang & Zhao, 2012; Kalin et al., 2006; Kalinichenko et al., 2003, 2004; Kim et al., 2005, 2006; Krupczak-Hollis et al., 2003, 2004; Lefebvre et al., 2010; Liu, Gampert, et al., 2006; Liu et al., 2011; Park, Carr, et al., 2009; Park, Costa, Lau, Tyner, & Raychaudhuri, 2008; Ramakrishna et al., 2007; Wang, Banerjee, et al. 2007; Wang, Chen, et al., 2008; Wang, Hung, et al., 2001; Wang, Kiyokawa, et al., 2002; Wang, Krupczak-Hollis, et al., 2002; Wang, Meliton, et al., 2008; Wang, Quail, et al., 2001; Wang, Teh, Ji, et al., 2010; Wang et al., 2005; Wonsey & Follettie, 2005; Wu, Liu, et al., 2010; Xue et al., 2010; Ye et al., 1999; Yoshida et al., 2007; Zhang, Ackermann, et al., 2006, 2009; Zhao et al., 2006). Second, FOXM1 promotes the entry into M-phase (Barsotti & Prives, 2009; Bolte et al., 2011; Brezillon et al., 2007;

Calvisi et al., 2009; Chan et al., 2008; Fu et al., 2008; Kalinichenko et al., 2003; Kim et al., 2006; Krupczak-Hollis et al., 2003; Laoukili, Alvarez, et al., 2008; Laoukili et al., 2005; Leung et al., 2001; Nakamura, Hirano, et al., 2010; Ramakrishna et al., 2007; Schüller et al., 2007; Sengupta et al., 2013; Ustiyan et al., 2009, 2012; Wang, Chen, et al., 2008; Wang et al., 2005; Wang, Kiyokawa, et al., 2002; Wang, Krupczak-Hollis, et al., 2002; Wang, Quail, et al., 2001; Wonsey & Follettie, 2005; Ye et al., 1999; Zhang et al., 2011; Zhao et al., 2006). Accordingly, FOXM1 enhances the kinase activity of both Cdk2 and Cdk1 (Wang et al., 2005; Wang, Kiyokawa, et al., 2002; Wang, Krupczak-Hollis, et al., 2002; Zhao et al., 2006).

In accordance with its positive effect on cell cycle progression, FOXM1 controls the expression of key regulators of the entry into S-phase and mitosis (Fig. 3.2). In particular, FOXM1 not only activates the expression of those factors, which induce S- and M-phase entry (e.g., cyclins and Cdks (cyclin-dependent kinases)), but it also represses the expression of those factors, which prevent S- and M-phase entry (i.e., CKIs) (Fig. 3.11).

Since FOXM1 augments the expression of Cdk2, cyclin A1 and cyclin A2 (Fig. 3.2; Table 3.1), which together form cyclin A/Cdk2 that is essential for S-phase progression (Desdouets, Sobczak-Thepot, Murphy, & Brechot, 1995; Gopinathan, Ratnacaram, & Kaldis, 2011; Hochegger, Takeda, & Hunt, 2008; Malumbres, 2011; Malumbres & Barbacid, 2005, 2009; Morgan, 2007, 2008a; Nigg, 1995; Pines, 1995; Satyanarayana & Kaldis, 2009; Yam, Fung, & Poon, 2002), FOXM1 might also promote the progression through S-phase (Fig. 3.11). This supposition is in line with the given fact that several FOXM1 target genes encode proteins for DNA replication (Fig. 3.2).

Figure 3.11 shows how FOXM1 can activate cyclin D/Cdk4,6, cyclin E/Cdk2, cyclin A/Cdk2, cyclin A/Cdk1, and cyclin B/Cdk 1 through activation and repression of its target genes (Costa et al., 2003; Costa, Kalinichenko, Tan, et al., 2005; Kalin et al., 2011; Minamino & Komuro, 2006; Wierstra & Alves, 2007c). Hence, FOXM1 should be capable of activating all five cyclin/Cdk complexes, which successively drive the progression through the cell cycle (Fig. 3.11). Therefore, FOXM1 promotes cell cycle progression and proliferation (Fig. 3.2).

The entry into S-phase from quiescence requires three major steps, namely, the induction of *cyclin D1* and *c-myc* expression, which both are

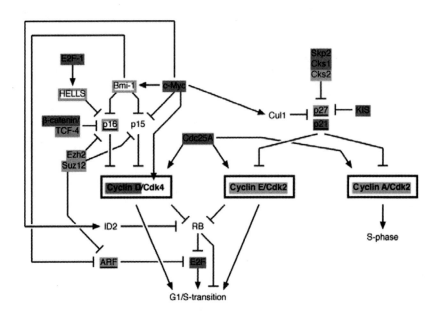

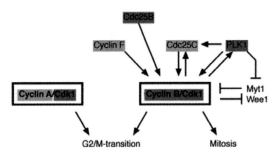

Figure 3.11 The control of the cell cycle by the protein products of FOXM1 target genes. Direct (dark gray), possibly indirect (light gray), and probably indirect (light gray border) FOXM1 target genes are indicated. FOXM1 upregulates the expression of most of its target genes. Those four genes (*p16*, *p21*, *p27*, *ARF*), whose expression is downregulated by FOXM1, are marked (underlined). Cdk4, cyclin-dependent kinase 4; Cul1, Cullin1; ID2, inhibitor of DNA binding 2; Myt1, membrane-associated and tyrosine/threonine-specific 1; p15, CDKN2B, Cdk inhibitor 2B, INK4B, inhibitor of Cdk4 B; RB, retinoblastoma protein, RB1, retinoblastoma 1, p105.

not expressed in quiescent cells, as well as the downregulation of the p27 level, which is high in quiescent cells (Chambard, Lefloch, Pouysségur, & Lenormand, 2007; Chang et al., 2003; Coleman, Marshall, & Olson, 2004; Crespo & Leon, 2000; Ewen, 2000; Frame & Balmain, 2000; Jones & Kazlauskas, 2001a, 2001b; Kerkhoff & Rapp, 1998; Liang &

Slingerland, 2003; Malumbres & Pellicier, 1998; Marshall, 1999a, 1999b; Massagué, 2004; Pruitt & Der, 2001; Sears & Nevins, 2002; Shapiro, 2001; Takuwa & Takuwa, 2001; Wilkinson & Millar, 2000). FOXM1 accomplishes each of these three steps so that it stimulates proliferation and S-phase entry (Fig. 19). On the one hand, FOXM1 induces the expression of its direct target genes *cyclin D1* and *c-myc* by transactivating the *cyclin D1* and *c-myc* promoters (Fig. 3.2; Table 3.1). On the other hand, FOXM1 downregulates p27 levels not only by reducing the *p27* mRNA expression (Fig. 3.2; Table 3.1), but also by triggering the degradation of p27 protein via the ubiquitin–proteasome pathway (Fig. 3.11). For the latter purpose, FOXM1 increases the expression of three subunits of the p27-targeting E3 ubiquitin ligase SCF^{Skp2} (Skp1-Cullin1-F-box protein) (Fig. 3.11), namely, that of Skp2, Cks1 (Cdk subunit 1, CDC2-associated protein CKS1), and Cks2 (Fig. 3.2; Table 3.1). Moreover, FOXM1 upregulates the expression of KIS (Fig. 3.2; Table 3.1), a kinase that can also trigger the proteasomal destruction of p27 (Fig. 3.11) because the KIS-mediated p27 phosphorylation results in nuclear export of p27 into the cytoplasm, where p27 is ubiquitylated by the ubiquitin ligase KPC (Kip1 ubiquitylation-promoting complex) (Alkarain & Slingerland, 2004; Besson, Assoian, & Roberts, 2004; Besson, Dowdy, & Roberts, 2008; Borriello, Cucciolla, Oliva, Zappia, & Della Ragione, 2007; Chu, Hengst, & Slingerland, 2008; Denicourt & Dowdy, 2004; le Sage, Nagel, & Agami, 2007; Le, Pruefer, & Bast, 2005; Lu & Hunter, 2010; Nakayama & Nakayama, 2006; Reed, 2002; Starostina & Kipreos, 2011; Vervoorts & Lüscher, 2008; Viglietto, Motti, & Fusco, 2002).

8.1.3 Proper execution of mitosis
FOXM1 plays a role not only for M-phase entry, but also for proper execution of mitosis (Fig. 3.2; Chan et al., 2008; Fu et al., 2008; Gusarova et al., 2007; Laoukili et al., 2005; Priller et al., 2011; Ramakrishna et al., 2007; Schüller et al., 2007; Wang et al., 2005; Wonsey & Follettie, 2005; Yoshida et al., 2007). Therefore, FOXM1 deficiency leads to pleiotropic mitotic defects, including (Chan et al., 2008; Fu et al., 2008; Gusarova et al., 2007; Laoukili et al., 2005; Priller et al., 2011; Ramakrishna et al., 2007; Schüller et al., 2007; Wang et al., 2005; Wonsey & Follettie, 2005; Yoshida et al., 2007):
– aneuploidy and polyploidy
– delay in progression through mitosis
– failure to progress beyond the prophase stage of mitosis
– aberrant furrow formation

- defects in cytokinesis
- mitotic spindle defects
- misalignment of chromosomes at the metaphase plate
- chromosome missegregation
- defective spindle assembly checkpoint
- propensity to be binucleated
- centrosome amplification
- cell death by mitotic catastrophe

This requirement of FOXM1 for proper execution of mitosis implies the necessity of FOXM1 for preservation of mitotic fidelity and thus for maintenance of chromosome integrity and genomic stability (see below). In particular, FOXM1-deficient cells generally suffer from polyploidy and aneuploidy (Chen, Müller, et al., 2013; Fu et al., 2008; Kalinichenko et al., 2004; Kim et al., 2005; Korver et al., 1998; Krupczak-Hollis et al., 2004; Laoukili et al., 2005; Nakamura, Hirano, et al., 2010; Ramakrishna et al., 2007; Ustiyan et al., 2009; Wan et al., 2012; Wang et al., 2005; Wonsey & Follettie, 2005; Zhao et al., 2006).

The function of FOXM1 in orderly progression through M-phase is reflected by the FOXM1 target genes, which include many key players in mitosis (Fig. 3.2), above all Cdk1, cyclin B1, cyclin B2, and cyclin B3 (Table 3.1), which together form cyclin B/Cdk1, the master regulator of M-phase (Ferrari, 2006; Gopinathan et al., 2011; Hochegger et al., 2008; Malumbres, 2011; Malumbres & Barbacid, 2005, 2009; Morgan, 2007, 2008a; Nigg, 1995, 2001; Pines, 1995; Satyanarayana & Kaldis, 2009). Furthermore, the mitotic target genes of FOXM1 encode other mitotic kinases, both activator proteins of the APC/C, subunits of the chromosomal passenger complex, and SAC (spindle assembly checkpoint) components (Fig. 3.2).

8.1.4 Cell cycle-dependent regulation of the expression and the transcriptional activity of FOXM1

FOXM1 stimulates proliferation (Ahmad et al., 2011, 2010; Bao et al., 2011; Bergamaschi et al., 2011; Bhat et al., 2011; Carr et al., 2010; Chen et al., 2011; Chu et al., 2012; Dai et al., 2013; Faust et al., 2012; Ho et al., 2012; Li et al., 2011; Lok et al., 2011; Mencalha et al., 2012; Millour et al., 2010, 2011; Monteiro et al., 2012; Ning et al., 2012; Park et al., 2012; Park, Wang, et al., 2008; Pellegrino et al., 2010; Raghavan et al., 2012; Uddin et al., 2011; Ueno et al., 2008; Wang, Ahmad, Banerjee, et al., 2010; Wang, Park, et al., 2011; Wierstra & Alves, 2006d; Xia et al., 2009; Xia, Mo, et al., 2012; Xie et al., 2010; Xue et al., 2012; Zeng et al., 2009; Zhang, Wu, et al., 2012) and cell cycle progression by promoting the entry into both S-phase and M-phase (Fig. 3.2; Ackermann Misfeldt et al., 2008; Anders et al., 2011; Barsotti & Prives, 2009; Bolte et al., 2011; Brezillon et al., 2007; Calvisi et al., 2009; Chan et al., 2008; Chen, Müller, et al., 2013; Chen et al., 2010; Davis et al., 2010; Fu et al., 2008; Gusarova et al., 2007; Huang & Zhao, 2012; Kalin et al., 2006; Kalinichenko et al., 2003, 2004; Kim et al., 2005, 2006; Krupczak-Hollis et al., 2003, 2004; Laoukili, Alvarez, et al., 2008; Laoukili et al., 2005; Lefebvre et al., 2010; Leung et al., 2001; Liu, Gampert, et al., 2006; Liu et al., 2011; Nakamura, Hirano, et al., 2010; Park, Carr, et al., 2009; Park, Costa, Lau, Tyner, & Raychaudhuri, 2008; Ramakrishna et al., 2007; Schüller et al., 2007; Sengupta et al., 2013; Ustiyan et al., 2009, 2012; Wang, Banerjee, et al. 2007; Wang, Chen, et al., 2008; Wang, Hung, et al., 2001; Wang, Kiyokawa, et al., 2002; Wang, Krupczak-Hollis, et al., 2002; Wang, Meliton, et al., 2008; Wang, Quail, et al., 2001; Wang, Teh, Ji, et al., 2010; Wang et al., 2005; Wonsey & Follettie, 2005; Wu, Liu, et al., 2010; Xue et al., 2010; Ye et al., 1999; Yoshida et al., 2007; Zhang, Ackermann, et al., 2006; Zhang et al., 2011, 2010; Zhao et al., 2006). Moreover, FOXM1 is crucial for proper execution of mitosis (Fig. 3.2; Chan et al., 2008; Fu et al., 2008; Gusarova et al., 2007; Laoukili et al., 2005; Priller et al., 2011; Ramakrishna et al., 2007; Schüller et al., 2007; Wang et al., 2005; Wonsey & Follettie, 2005; Yoshida et al., 2007).

Accordingly, the expression (Alvarez-Fernandez et al., 2011; Bar-Joseph et al., 2008; Chaudhary et al., 2000; Chen, Dominguez-Brauer, et al., 2009; Dibb et al., 2012; Down et al., 2012; Fu et al., 2008; Gemenetzidis et al., 2010; Grant et al., 2012; Korver, Roose, & Clevers, 1997; Korver, Roose, Heinen, et al., 1997; Laoukili, Alvarez, et al., 2008; Laoukili, Alvarez-Fernandez, et al., 2008; Laoukili et al., 2005; Leung et al., 2001; Li et al., 2008; Ma, Tong, et al., 2005; Park, Costa, Lau, Tyner, & Raychaudhuri, 2008; Petrovic et al., 2008; Sadavisam et al., 2012; Sullivan et al., 2012; Wan et al., 2012; Wang, Chen, et al., 2008; Wang, Hung, et al., 2001; Whitfield et al., 2002; Ye et al., 1999, 1997) as well as the transcriptional activity (Chen, Dominguez-Brauer, et al., 2009; Laoukili, Alvarez, et al., 2008; Laoukili et al., 2005; Littler et al., 2010; Major et al., 2004; Park, Wang, et al., 2008) of FOXM1 increase strongly during G1-phase, persists throughout S-phase, and reach their maxima in G2/M-phase.

In particular, (almost) all cyclin/Cdk complexes, which drive the cell cycle, enhance the transcriptional activity of FOXM1 because FOXM1 is activated by cyclin D1/Cdk4, cyclin D3/Cdk6, cyclin E/Cdk2, cyclin A/Cdk2, cyclin A/Cdk1 (Alvarez-Fernandez et al., 2011; Alvarez-Fernandez, Halim, et al., 2010; Alvarez-Fernandez, Medema, et al., 2010; Anders et al., 2011; Chen, Dominguez-Brauer, et al., 2009; Laoukili, Alvarez, et al., 2008; Laoukili, Alvarez-Fernandez, et al., 2008; Laoukili et al., 2005; Lüscher-Firzlaff, Lilischkis, & Lüscher, 2006; Major et al., 2004; Park, Wang, et al., 2008; Sullivan et al., 2012; Wierstra, 2013a; Wierstra & Alves, 2006b, 2006c, 2008), and possibly cyclin B/Cdk1. Conversely, the transcriptional activity of FOXM1 is repressed by some CKIs (p16, p21, p27) (Kalinichenko et al., 2004; Laoukili et al., 2005; Lüscher-Firzlaff et al., 2006; Wierstra, 2013a; Wierstra & Alves, 2006b, 2006c, 2008) and by the tumor suppressor RB (Wierstra, 2013a; Wierstra & Alves, 2006b, 2008), which controls the restriction point in late G1-phase (Blagosklonny & Pardee, 2002; Lundberg & Weinberg, 1999; Planas-Silva & Weinberg, 1997; Weinberg, 1995; Zetterberg, Larsson, & Wiman, 1995), so that FOXM1 is inactivated by the key inhibitors of cell cycle entry and progression.

Additionally, cyclin D1/Cdk4 and cyclin D3/Cdk6 increase the FOXM1 protein expression by stabilizing FOXM1 (Anders et al., 2011) whereas RB (Dean et al., 2012; Markey et al., 2007) and p21 (Barsotti & Prives, 2009; Bellelli et al., 2012; Millour et al., 2011; Rovillain et al., 2011; Stoyanova et al., 2012; Weymann et al., 2009) decreases the *foxm1* mRNA expression (Fig. 3.10).

The parallel culmination of both the expression (Bar-Joseph et al., 2008; Dibb et al., 2012; Down et al., 2012; Gemenetzidis et al., 2010; Grant et al., 2012; Laoukili, Alvarez, et al., 2008; Laoukili, Alvarez-Fernandez, et al., 2008; Leung et al., 2001; Li et al., 2008; Ma, Tong, et al., 2005; Park, Costa, Lau, Tyner, & Raychaudhuri, 2008; Sadavisam et al., 2012; Sullivan et al., 2012; Wan et al., 2012; Whitfield et al., 2002) and the transcriptional activity (Chen, Dominguez-Brauer, et al., 2009; Laoukili, Alvarez, et al., 2008; Laoukili et al., 2005; Littler et al., 2010; Major et al., 2004; Park, Wang, et al., 2008) of FOXM1 in G2/M-phase points to a central function of FOXM1 in G2/M-phase. Indeed, the entirety of all studies suggests that FOXM1 may be more important for G2/M-transition and mitosis than for G1/S-transition although FOXM1 plays an undisputed role in S-phase entry, too (see below).

8.1.5 Experimental findings for the function of FOXM1 in M-phase entry and mitotic progression

The results obtained with $foxm1^{-/-}$ MEFs (Laoukili et al., 2005; Wang et al., 2005) demonstrate the central importance of FOXM1 for G2/M-transition and progression through mitosis:

Laoukili et al. (2005) reported that $foxm1^{-/-}$ MEFs exhibited a defect in mitotic entry because of a significant delay in G2-phase whereas they entered S-phase as efficiently as wild-type MEFs. Likewise, Wang et al. (2005) found that $foxm1^{-/-}$ MEFs accumulated in G2/M-phase but not in G1-phase, and that they displayed a block in mitotic progression due to a failure to progress beyond the prophase stage of mitosis.

Thus, FOXM1 deficiency has more severe consequences for the entry into M-phase and for progression through mitosis than for the entry into S-phase.

This conclusion is supported by several studies with siRNA/shRNA-mediated depletion of FOXM1 (Bellelli et al., 2012; Chen, Yang, et al., 2012; Fu et al., 2008; Green et al., 2011; Mencalha et al., 2012; Millour et al., 2011; Nakamura, Hirano, et al., 2010; Ustiyan et al., 2009; Wang, Chen, et al., 2008; Wang et al., 2005; Xiang, Liu, Quan, Cao, & Lv, 2012), which mainly increased the portion of cells in G2/M phase but (almost) not that of cells in G1-phase. Accordingly, knockout of *foxm1* in MEFs (Wang et al., 2005) and conditional knockout of *foxm1* in CGNP (cerebellar granule neuron precursors) (Schüller et al., 2007) had the same effect. Accordingly, siRNA to FOXM1 caused a p53-independent G2-arrest in MCF-7 mammary carcinoma cells (Barsotti & Prives, 2009) and shRNA to FOXM1 arrested TMZ (temozolomide)-treated recurrent GBM1 (glioblastoma multiforme) cells in G2-phase, which are otherwise resistant to TMZ-induced G2-arrest (Zhang, Wu, et al., 2012). Accordingly, the heat shock-induced cell cycle arrest in G2/M-phase was exacerbated by siRNA against FOXM1 and alleviated by FOXM1 overexpression in human U-87MG or Hs683 glioma cells, respectively (Dai et al., 2013).

In fact, progression through mitosis, in particular prometaphase progression, is delayed by FOXM1 deficiency (Chen, Dominguez-Brauer, et al., 2009; Fu et al., 2008; Wang et al., 2005; Wonsey & Follettie, 2005):

Both $foxm1^{-/-}$ MEFs and U2OS cells depleted of FOXM1 by siRNA showed a block in mitotic progression because they failed to proceed beyond the prophase stage of mitosis (Wang et al., 2005). Similarly, another study reported that siRNA-mediated knockdown of FOXM1 in U2OS cells resulted in a prolonged progression from prophase to prometaphase (Fu et al., 2008). Accordingly, the prometaphase arrest caused by inhibition of PLK1 was partially rescued by overexpression of a phosphomimetic FOXM1 mutant, in which two PLK1-phosphorylated serines were replaced by glutamic acid (Fu et al., 2008). Likewise, after release from a nocodazole-induced arrest at prometaphase U2OS cells expressing wild-type FOXM1 progressed into metaphase, whereas U2OS cells expressing a FOXM1 point mutant with impaired transcriptional activity remained in prometaphase (Chen, Dominguez-Brauer, et al., 2009).

Furthermore, mitotic BT-20 breast cancer cells (Wonsey & Follettie, 2005) and mitotic DAOY medulloblastoma cells (Priller et al., 2011) depleted of FOXM1 by shRNA died by mitotic catastrophe.

In summary, FOXM1 plays an important role in M-phase entry and mitotic progression (Fig. 3.2).

Accordingly, the expression of numerous direct FOXM1 target genes peaks in late G2- and M-phase (Chen, Müller, et al., 2013). Moreover, siRNA-mediated knockdown of FOXM1 curtailed the strong induction of *cyclin B1*, *Cdh1*, *KPNA2* (karyopherin α-2), and *CENP-A* (centrosomal protein-A) mRNA expression in G2/M-phase after release of G1/S-arrested U2OS from a double-thymidine block, which underscores the importance of FOXM1 for the cell cycle-dependent activation of these four direct FOXM1 target genes (Fig. 3.2; Table 3.1) during G2- and M-phase (Chen, Müller, et al., 2013).

Actually, FOXM1, which interacts with B-Myb (Chen, Müller, et al., 2013) and three subunits (LIN9, LIN37, LIN53) of the MuvB complex (Chen, Müller, et al., 2013; Sadavisam et al., 2012), seems to cooperate with the MuvB complex and B-Myb in the G2/M-phase-specific activation of their shared target genes (Chen, Müller, et al., 2013; Down et al., 2012; Sadavisam et al., 2012). Thus, the pivotal function of FOXM1 during the G2/M-transition and mitosis also manifests in its protein–protein interactions and its gene activation mechanisms.

8.1.6 Experimental findings for the function of FOXM1 in S-phase entry
The function of FOXM1 in stimulation of cell cycle progression is not limited to M-phase entry and progression though mitosis but extends to S-phase entry as well (Fig. 3.2):

Several studies reported an increase in the portion of cells in S-phase upon FOXM1 overexpression (Chan et al., 2008; Liu, Gampert, et al., 2006; Uddin et al., 2011; Wang, Banerjee, et al., 2007) and a decrease in S-phase cells upon depletion of FOXM1 by siRNA (Bellelli et al., 2012; Calvisi et al., 2009; Chan et al., 2008; Kalin et al., 2006; Liu, Gampert, et al., 2006; Qu et al., 2013; Wan et al., 2012; Wang, Banerjee, et al., 2007; Wu, Liu, et al., 2010; Xue et al., 2012). Also the percentage of *foxm1*-deficient mature T-cells in S-phase was reduced compared to their wild-type counterparts (Xue et al., 2010). Accordingly, *foxm1* knockout MEFs displayed a delay in the serum-induced entry into S-phase after serum starvation (Anders et al., 2011). All these results demonstrate a pivotal role of FOXM1 for the entry into S-phase.

This importance of FOXM1 for the G1/S-transition is underscored by the finding that during the serum-induced entry of starved MEFs into S-phase the induction of *cyclin E2*, *MSH6* (*mutS homolog 6*), and *c-Myb* mRNA expression by serum was blunted in *foxm1* knockout MEFs, which indicates a requirement of FOXM1 for the serum-induced transcription of these three FOXM1 target genes (Fig. 3.2; Table 3.1) during the G0/G1/S-transition (Anders et al., 2011).

Moreover, siRNA-mediated depletion of FOXM1 decreased the *cyclin D1* and *c-myc* mRNA expression more strongly in the G0/G1 population than in the G2/M population of sorted HAVSMCs (human aortic vascular SMCs), indicating that FOXM1 regulates the expression of the direct FOXM1 target genes *cyclin D1* and *c-myc* (Fig. 3.2; Table 3.1) mainly prior to DNA replication (Ustiyan et al., 2009).

In summary, FOXM1 plays an important role in S-phase entry (Fig. 3.2).

8.1.7 FOXM1 is a typical proliferation-associated transcription factor
The following overall picture arises:

FOXM1 represents a typical proliferation-associated transcription factor, which stimulates proliferation and cell cycle progression by promoting the entry into both S-phase and M-phase (Fig. 3.2; Costa, 2005; Costa et al., 2003; Costa, Kalinichenko, Major, et al., 2005; Costa, Kalinichenko, Tan, et al., 2005; Kalin et al., 2011; Koo et al., 2011; Laoukili et al., 2007; Mackey et al., 2003; Myatt & Lam, 2007; Raychaudhuri & Park, 2011; Wierstra & Alves, 2007c). In addition, FOXM1 is required for proper execution of mitosis (Fig. 3.2). Accordingly, FOXM1 controls the expression of genes, whose protein products regulate the G1/S- and G2/M-transitions as well as the orderly progression through mitosis (Figs. 3.2 and 3.11; Costa, 2005; Costa et al., 2003; Costa, Kalinichenko, Tan, et al., 2005; Davis et al., 2010; Kalin et al., 2011; Koo et al., 2011; Laoukili et al., 2007; Minamino & Komuro, 2006; Murakami et al., 2010; Raychaudhuri & Park, 2011; Wierstra & Alves, 2007c).

This proliferative function of FOXM1 explains many of its physiological effects in normal cells (see below) and many of its pathological effects in cancer cells (see Part II of this two-part review, that is see Wierstra, 2013b).

The proliferation-stimulating function of FOXM1 underlies at least in part its roles in (see below):
— embryonic development
— interference with contact inhibition
— prevention of cellular senescence
— adult tissue repair after liver, lung, and pancreas injury
— liver repopulation by transplanted hepatocytes in a model of chronic liver injury
— postnatal expansion and maintenance of pancreatic β-cell mass
— pregnancy-induced expansion of maternal β-cell mass
— maintenance of the proliferative capacity of cells

8.2. Embryonic development
8.2.1 Embryonic development of Mus musculus
8.2.1.1 foxm1 knockout mice
FOXM1 is required for the embryonic development of heart, liver, and lung (Fig. 3.2; Bolte et al., 2011; Kalin et al., 2008; Kim et al., 2005; Korver et al., 1998; Krupczak-Hollis et al., 2004; Ramakrishna et al., 2007; Ustiyan et al., 2009, 2012).

$Foxm1^{-/-}$ mouse embryos died *in utero*, which demonstrates the essential function of FOXM1 during embryonic development (Kim et al., 2005; Korver et al., 1998; Krupczak-Hollis et al., 2004; Ramakrishna et al., 2007). Also a hepatoblast-specific *foxm1* knockout (Krupczak-Hollis et al., 2004) and a cardiomyocyte-specific *foxm1* knockout (Bolte et al., 2011) resulted in embryonic lethality of mice, whereas newborn mice with a lung epithelial-specific *foxm1* knockout (Kalin et al., 2008) or a smooth muscle-specific *foxm1* knockout (Ustiyan et al., 2009) died perinatally within the first 24 h after birth. In contrast, no mortality was reported for mice with a Clara cell-specific *foxm1* knockout from E16.5 onward (Ustiyan et al., 2012).

All these *foxm1* knockout mice suffered from severe defects in heart (Bolte et al., 2011; Korver et al., 1998; Krupczak-Hollis et al., 2004; Ramakrishna et al., 2007), lung (Kalin et al., 2008; Kim et al., 2005; Ustiyan et al., 2009, 2012), and liver (Krupczak-Hollis et al., 2004) development. In particular, *foxm1* knockout embryos displayed a decrease in the number of hepatoblasts, cardiomyocytes, and mesenchymal lung cells because the proliferation of these cell types was reduced, whereas apoptosis was not increased (Kim et al., 2005; Krupczak-Hollis et al., 2004; Ramakrishna et al., 2007).

The Clara cell-specific *foxm1* knockout in $CCSP\text{-}Foxm1^{-/-}$ mouse embryos significantly decreased the proliferation of airway Clara cells and bronchiolar epithelial cells in the developing mouse lung during the late embryonic/early postnatal period of development of conducting airways, but apoptotic cells were not detected in $CCSP\text{-}Foxm1^{-/-}$ bronchiolar epithelium (Ustiyan et al., 2012).

$smFoxm1^{-/-}$ embryos with a smooth muscle-specific *foxm1* knockout exhibited reduced proliferation of SMCs in blood vessels and esophagus with no significant increase in apoptosis prior to birth, whereas $smFoxm1^{-/-}$ newborn mice displayed increased apoptosis in lung, pulmonary blood vessels, and esophagus (Ustiyan et al., 2009).

The lung epithelial-specific knockout of *foxm1* from embryonic day 7.5 (E7.5) onward in $epFoxm1^{-/-}$ mice did not alter epithelial cell proliferation although the number of squamous type I epithelial cells was reduced, possibly due to delayed differentiation of type I epithelial cells (Kalin et al., 2008).

The conditional deletion of *foxm1* from the distal respiratory epithelium in $epFoxm1^{-/-}$ mouse embryos attenuated those defects in branching lung morphogenesis and lung sacculation, which were caused by the transgenic expression of activated K-Ras^{G12D} in these distal lung epithelial cells (Wang, Snyder, et al., 2012).

The conditional deletion of *foxm1* from airway Clara cells ($CCSP\text{-}Foxm1^{-/-}$) demonstrated that, during late embryonic and postnatal development of conducting airways, FOXM1 is required not only for Clara cell proliferation and proliferation of bronchiolar epithelial cells, but also for Clara cell differentiation and proper epithelial cell differentiation in the bronchiolar epithelium, in particular for prevention of transdifferentiation of Clara cells into alveolar type II cells (Ustiyan et al., 2012). These roles of FOXM1 extend to the adult lung so that FOXM1 is critical for long-term maintenance of bronchiolar epithelium and conducting airway structure (Ustiyan et al., 2012).

Additionally, this Clara cell-specific *foxm1* knockout revealed a requirement of FOXM1 for the preservation of epithelial junctions in bronchioles of the developing mouse lung (Ustiyan et al., 2012).

8.2.1.2 *foxm1* transgenic mice

Transgenic mice, which conditionally express an N-terminally truncated FoxM1B mutant (FoxM1B-ΔN) in respiratory epithelial cells, showed that the expression of this FoxM1B mutant during embryogenesis impaired lung sacculation and induced epithelial hyperplasia, indicating that FoxM1B-ΔN expression is sufficient to accelerate cellular proliferation in the undifferentiated lung epithelium (Wang, Zhang, Snyder, et al., 2010). Accordingly, expression of FoxM1B-ΔN during postnatal lung development or in adult mice caused focal airway hyperplasia, increased the proliferation of Clara cells, and was sufficient to induce Clara cell hyperplasia (Wang, Zhang, Snyder, et al., 2010).

8.2.2 Embryonic development of Xenopus laevis

In early *Xenopus* embryos, FOXM1 is required for neural development (Fig. 3.2), namely, for proliferation of primary neuronal precursors and for primary neuronal differentiation (Ueno et al., 2008).

8.3. Interference with contact inhibition

FOXM1 protein expression is rapidly downregulated upon contact inhibition, that is, when cells reach confluence (see above) (Coller et al., 2006; Faust et al., 2012; Küppers et al., 2010; Mirza et al., 2010). Stable overexpression of FoxM1B in NIH3T3 fibroblasts, which abolished this normal FOXM1 downregulation at increasing cell density, caused a twofold increase in saturation density indicating loss of contact inhibition (Faust et al., 2012). Hence, FOXM1 interferes with contact inhibition (Fig. 3.2) so that the downregulation of FOXM1 is required for contact inhibition (Faust et al., 2012). As a result, ectopic FoxM1B expression caused the loss of contact inhibition, which manifested in an increased saturation density (Faust et al., 2012). Thus, overexpression of FOXM1 is sufficient to overcome contact inhibition (Fig. 3.2; Faust et al., 2012).

Confluent cells are arrested in G0/G1-phase (Coller et al., 2006; O'Farrell, 2011; Valcourt et al., 2012; Yanagida, 2009). Since FOXM1 deficiency was reported to result in a G2/M-arrest (Bellelli et al., 2012; Chen, Yang, et al., 2012; Dai et al., 2013; Fu et al., 2008; Green et al., 2011; Laoukili et al., 2005; Mencalha et al., 2012; Millour et al., 2011; Nakamura, Hirano, et al., 2010; Schüller et al., 2007; Ustiyan et al., 2009; Wang, Chen, et al., 2008; Wang et al., 2005; Xiang et al., 2012; Zhang, Wu, et al., 2012) Faust et al. (2012) suggested that the decrease in FOXM1 upon contact inhibition is not responsible for induction of G0/G1 arrest but is required to prevent bypass of contact inhibition by FOXM1.

Contact inhibition of NIH3T3 fibroblasts includes not only the downregulation of the FOXM1 protein expression, but also the downregulation of the cyclin A and PLK1 protein levels (Faust et al., 2012). Ectopic expression of FoxM1B prevented this decrease in cyclin A and PLK1 at increasing NIH3T3 cell densities (Faust et al., 2012). Since both *cyclin A* and *PLK1* are direct FOXM11 target genes (Fig. 3.2; Table 3.1) the rapid downregulation of FOXM1 upon contact inhibition seems to be necessary to block the transcription of *cyclin A* and *PLK1* and to decrease the cyclin A and PLK1 expression thereby preventing S-phase entry and further cell cycle progression (Faust et al., 2012).

In summary, the downregulation of FOXM1 is crucial for contact inhibition because FOXM1 interferes with contact inhibition (Fig. 3.2; Faust et al., 2012). The ability of FOXM1 overexpression to overcome contact inhibition in NIH3T3 fibroblasts (Faust et al., 2012) underscores the potency of FOXM1 for stimulation of cell proliferation and cell cycle progression.

8.4. Cellular senescence
8.4.1 Prevention of cellular senescence
Cellular senescence is an irreversible growth arrest (Adams, 2009; Ben-Porath & Weinberg, 2004, 2005; Blagosklonny, 2006; Campisi, 2005a; Campisi, 2011; Campisi & d'Adda di Fagagna, 2007; Campisi, Kim, Lim, & Rubio, 2001; Coppe, Desprez, Krtolica, & Campisi, 2010; d'Adda di Fagagna, 2008; Dimri, 2005; Evan & d'Adda di Fagagna, 2009; Freund, Orjalo, Desprez, & Campisi, 2010; Fridman & Tainsky, 2008; Itahana, Campisi, & Dimri, 2004; Kiyokawa, 2006; Kuilman, Michloglou, Mooi, & Peeper, 2010; Kuilman & Peeper, 2009; Lanigan, Geraghty, & Bracken, 2011; McDuff & Turner, 2011; Ogrunc & d'Adda di Fagana, 2011; Pazolli & Stewart, 2008; Serrano & Blasco, 2001; Sharpless, 2004; Sherr & DePinho, 2000; Stewart & Weinberg, 2002, 2006). This permanent exit from

the cell cycle represents an important tumor suppression mechanism (Braig & Schmitt, 2006; Caino, Meshki, & Kazanietz, 2009; Campisi, 2000, 2001, 2003, 2005b; Cichowski & Hahn, 2008; Collado, Blasco, & Serrano, 2007; Collado & Serrano, 2005, 2006, 2010; Courtois-Cox, Jones, & Cichowski, 2008; Di Micco, Fumagalli, & d'Adda di Fagagna, 2007; Evan et al., 2005; Ewald, Desotelle, Wilding, & Jarrard, 2010; Hanahan & Weinberg, 2011; Hemann & Narita, 2007; Lleonart, Artero-Castro, & Kondoh, 2009; Lowe, Cepero, & Evan, 2004; Mathon & Lloyd, 2001; Mooi & Peeper, 2006; Nardella, Clohessy, Alimonti, & Pandolfi, 2011; Ohtani, Mann, & Hara, 2009; Prieur & Peeper, 2008; Rangarajan & Weinberg, 2003; Roninson, 2003; Sage, 2005; Schmitt, 2003, 2007; Sharpless & DePinho, 2004; Shay & Roninson, 2004).

In accordance with its proliferation-stimulating function, FOXM1 prevents both oncogene-induced and oxidative stress-induced premature senescence (Fig. 3.2; Anders et al., 2011; Li et al., 2008; Park, Carr, et al., 2009; Qu et al., 2013; Wang et al., 2005; Zeng et al., 2009; Zhang, Ackermann, et al., 2006). In fact, a ca FOXM1c mutant bypassed senescence in conditionally immortalized HMFs, where senescence was induced by reactivation of the p16–RB and p53–p21 pathways (Rovillain et al., 2011), and another ca FOXM1c mutant attenuated the senescence of U2OS cells, which was induced by the CDK4/6 inhibitor PD0332991 (Anders et al., 2011).

In accordance with its ability to suppress premature senescence (Anders et al., 2011; Li et al., 2008; Park, Carr, et al., 2009; Qu et al., 2013; Rovillain et al., 2011; Wang et al., 2005; Zeng et al., 2009; Zhang, Ackermann, et al., 2006), the expression of FOXM1 ceases when cells undergo cellular senescence so that *foxm1* is not expressed in senescent cells (Hardy et al., 2005; Rovillain et al., 2011).

Early-passage $foxm1^{-/-}$ MEFs (Wang et al., 2005), immortalized $foxm1^{-/-}$ MEFs (Anders et al., 2011), and pancreata from mice with a pancreas-specific *foxm1* knockout (Zhang, Ackermann, et al., 2006) displayed premature senescence, which indicates that FOXM1 prevents premature senescence. Accordingly, silencing of FOXM1 by siRNA in U2OS cells resulted in cellular senescence, too (Anders et al., 2011). Also in HepG2 and SMMC-7721 HCC cells, siRNA-mediated knockdown of FOXM1 led to cellular senescence whereas the number of senescent cells was decreased by FOXM1 overexpression (Qu et al., 2013).

H_2O_2 treatment, that is oxidative stress, caused senescence of NIH3T3 mouse fibroblasts (Li et al., 2008) and human IMR90 primary fibroblasts

(Park, Carr, et al., 2009). Overexpression of FOXM1c in NIH3T3 fibroblasts suppressed this oxidative stress-induced premature senescence (Li et al., 2008). Conversely, siRNA-mediated knockdown of FOXM1 in IMR90 fibroblasts reinforced the H_2O_2-induced senescence (Park, Carr, et al., 2009), verifying that FOXM1 hampers this oxidative stress-induced premature senescence.

Treatment of immortalized MEFs with the ROS-inducing drug Imexon revealed that Imexon caused senescence in $foxm1^{-/-}$ MEFs whereas it barely did so in wild-type MEFs (Anders et al., 2011). This finding shows again that FOXM1c counteracts oxidative stress-induced premature senescence (Anders et al., 2011). Accordingly, the Imexon-induced senescence of $foxm1^{-/-}$ MEFs was lowered by reexpression of exogenous FOXM1c, confirming the suppression of oxidative stress-induced premature senescence by FOXM1 (Anders et al., 2011).

Activated (myristylated) mAKT1 triggered senescence of human IMR90 primary fibroblasts and human U2OS osteosarcoma cells (Park, Carr, et al., 2009). Overexpression of FOXM1 in IMR90 fibroblasts suppressed this oncogene-induced premature senescence (Park, Carr, et al., 2009). Conversely, silencing of FOXM1 with siRNA in U2OS cells amplified the mAKT1-induced senescence, which confirms that FOXM1 impedes this oncogene-induced premature senescence (Park, Carr, et al., 2009).

Likewise, the oncogenic H-RasV12-induced senescence of immortalized MEFs was exacerbated by siRNA against FOXM1, which indicates that FOXM1 attenuates this oncogene-induced premature senescence (Park, Carr, et al., 2009). Hence, FOXM1 prevents oxidative stress-induced premature senescence (Anders et al., 2011; Li et al., 2008; Park, Carr, et al., 2009) as well as oncogene-induced premature senescence (Fig. 3.2; Park, Carr, et al., 2009).

Acute inhibition of CDK4/6 with the CDK4/6 inhibitor PD0332991 induced senescence of U2OS cells (Anders et al., 2011). The PD0332991-induced senescence was attenuated by overexpression of a ca FOXM1c mutant but not by overexpression of wild-type FOXM1c (Anders et al., 2011). Thus, FOXM1c has the potential to protect U2OS cells from PD0332991-induced senescence (Anders et al., 2011), but this ability is restricted by negative-regulatory mechanisms, which keep wild-type FOXM1c inactive.

Accordingly, the CDK4/6 inhibitor PD0332991 downregulates the endogenous FOXM1 protein level by triggering the proteasomal degradation of FOXM1 because the FOXM1 protein is stabilized through its phosphorylation by cyclin D1/Cdk4 and cyclin D3/Cdk6 (Fig. 3.10; Anders et al., 2011). Moreover, PD0332991 represses the transcriptional activity of exogenous FOXM1c independent of an effect on the FOXM1 protein stability (Anders et al., 2011), because both cyclin D1/Cdk4 (Anders et al., 2011; Wierstra, 2013a; Wierstra & Alves, 2006b, 2006c, 2008) and cyclin D3/Cdk6 (Anders et al., 2011) enhance the transcriptional activity of FOXM1c.

When primary HMFs were immortalized with hTERT and SV40 large T antigen the resulting conditionally immortalized HMFs could be induced to senesce by inactivation of SV40 large T (Hardy et al., 2005; Rovillain et al., 2011). Since the viral oncoprotein SV40 large T inhibits the cellular tumor suppressors RB and p53 (Ahuja et al., 2005; Ali & DeCaprio, 2001; Caracciolo et al., 2006; Cheng et al., 2009; DeCaprio, 2009; Felsani et al., 2006; Helt & Galloway, 2003; Lee & Cho, 2002; Levine, 2009; Liu & Marmorstein, 2006; Moens et al., 2007; O'Shea & Fried, 2005; Pipas, 2009; Pipas & Levine, 2001; Saenz-Robles & Pipas, 2009; Saenz-Robles et al., 2001; Sullivan & Pipas, 2002; White & Khalili, 2004, 2006) the senescence of these conditionally immortalized HMFs upon inactivation of SV40 large T is induced by reactivation of the p16–RB and p53–p21 pathways (Hardy et al., 2005; Rovillain et al., 2011).

Overexpression of a ca FOXM1c mutant readily bypassed the induced senescence of these HMFs, whereas overexpression of wild-type FOXM1c had no effect (Rovillain et al., 2011). Hence, FOXM1c has the potential to protect conditionally immortalized HMFs from senescence induced by reactivation of the p16–RB and p53–p21 pathways (Rovillain et al., 2011), but this ability is restricted by negative-regulatory mechanisms which keep wild-type FOXM1c inactive.

Accordingly, the *foxm1* mRNA expression in the conditionally immortalized HMFs decreased during their senescence upon reactivation of the p16–RB and p53–p21 pathways (Hardy et al., 2005; Rovillain et al., 2011).

The ca FOXM1c mutant lacked the autoinhibitory FOXM1c N-terminus (Rovillain et al., 2011), which directly inhibits the TAD of FOXM1c (Fig. 3.3; Wierstra, 2011a, 2011b, 2013a; Wierstra & Alves, 2006a, 2006b, 2006c, 2006d, 2007b, 2008) and mediates the degradation of FOXM1 via the ubiquitin–proteasome pathway (Laoukili, Alvarez-Fernandez, et al., 2008; Park, Costa, Lau, Tyner, & Raychaudhuri, 2008).

The DNA-alkylating agent oxaliplatin, a third-generation platinum-derived chemotherapeutic drug, causes DNA damage and intracellular ROS generation (Chaney, Campbell, Bassett, & Wu, 2005; Kelland, 2007; Kweekel, Gelderblom, & Guchelaar, 2005; Lim, Choi, Kang, & Han, 2010; Qu et al., 2013; Raymond, Faivre, Chaney, Waynarowski, & Cvitkovic, 2002). Oxaliplatin induced senescence of HepG2 and SMMC-7721 HCC cells (Qu et al., 2013). The oxaliplatin-induced senescence was attenuated by FOXM1 overexpression but enhanced by siRNA-mediated knockdown of FOXM1, which demonstrates that FOXM1 impairs oxaliplatin-induced senescence (Qu et al., 2013).

Accordingly, oxaliplatin treatment decreased the *foxm1* mRNA and protein expression in HepG2 and SMMC-7721 cells (Qu et al., 2013).

The senescence caused by oxaliplatin might be classified as oxidative stress-induced senescence or DNA damage-induced senescence or chemotherapy-induced senescence (Qu et al., 2013). siRNA-mediated depletion of FOXM1 resulted in cellular senescence of gastric cancer cells (AGS, BGC-823, HGC-27, KATO-III), indicating that FOXM1 counteracts senescence (Zeng et al., 2009).

FOXM1 is known to repress the *p27* promoter (Zeng et al., 2009) and to downregulate the *p27* mRNA and protein expression (Fig. 3.2; Table 3.1). Accordingly, silencing of FOXM1 by siRNA in gastric cancer cells led to an upregulation of the *p27* mRNA and protein levels (Zeng et al., 2009). The FOXM1 knockdown-induced senescence of gastric cancer cells depends on this p27 upregulation because siRNA against p27 reduced the percentage of senescent FOXM1 knockdown cells (Zeng et al., 2009). Thus, FOXM1 protects gastric cancer cells from cellular senescence through repression of the *p27* expression (Zeng et al., 2009).

siRNA-mediated depletion of FOXM1 in IMR90 fibroblasts caused cellular senescence, which indicates that FOXM1 halts senescence (Park, Carr, et al., 2009).

FOXM1 inhibits the production or accumulation of ROS because the intracellular ROS level was decreased by FOXM1 overexpression but increased by siRNA to FOXM1 (Park, Carr, et al., 2009). In addition, FOXM1 leads to the inactivation of p38 because the level of phosphorylated, activated p38 was augmented by FOXM1-targeting siRNA (Park, Carr, et al., 2009). The serine/threonine kinase p38 is a known downstream effector of ROS (Bonner & Arbiser, 2012; Hayakawa, Hayakawa, Takeda, & Ichijo, 2012; Matsuzawa & Ichijo, 2008; Nagai, Noguchi, Takeda, & Ichijo, 2007; Pan, Hong, & Ren, 2009; Reuter, Gupta,

Chaturvedi, & Aggarwal, 2010; Shao et al., 2011; Torres & Forman, 2003; Weinberg & Chandel, 2009).

The FOXM1 knockdown-induced senescence of IMR90 fibroblasts depends on both ROS and p38 activity because the percentage of senescent FOXM1 knockdown cells was reduced by the p38 inhibitor SB203580 and the antioxidant detoxifying enzyme catalase, a ROS scavenger (Park, Carr, et al., 2009). Therefore, FOXM1 counteracts cellular senescence of IMR90 fibroblasts by protecting them from oxidative stress through downregulation of the intracellular ROS level and through inactivation of p38 (Park, Carr, et al., 2009).

The antioxidant detoxifying enzymes catalase, MnSOD, and PRDX3 (peroxiredoxin 3) are encoded by two direct (*catalase, MnSOD*) and one possibly indirect (*PRDX3*) FOXM1 target gene (Fig. 3.2; Table 3.1) so that FOXM1 can inhibit the accumulation of ROS by activating the expression of these three ROS scavengers (Park, Carr, et al., 2009).

In summary, FOXM1 suppresses premature cellular senescence (Fig. 3.2; Anders et al., 2011; Li et al., 2008; Park, Carr, et al., 2009; Qu et al., 2013; Rovillain et al., 2011; Wang et al., 2005; Zeng et al., 2009; Zhang, Ackermann, et al., 2006). For this purpose, FOXM1 represses the expression of the CKI p27 (Zeng et al., 2009), downregulates the intracellular ROS level, and mediates the inactivation of p38 (Park, Carr, et al., 2009). The ability of FOXM1 to prevent oncogene-induced and oxidative stress-induced premature senescence (Anders et al., 2011; Li et al., 2008; Park, Carr, et al., 2009; Qu et al., 2013; Rovillain et al., 2011; Wang et al., 2005; Zeng et al., 2009; Zhang, Ackermann, et al., 2006) underscores the potency of FOXM1 for stimulation of cell proliferation and cell cycle progression.

8.4.2 An exceptional positive effect of FOXM1 on cellular senescence
All the eight above-mentioned studies demonstrated a negative effect of FOXM1 on cellular senescence and they reported that FOXM1 prevents both oxidative stress-induced premature senescence and oncogene-induced premature senescence (Anders et al., 2011; Li et al., 2008; Park, Carr, et al., 2009; Qu et al., 2013; Rovillain et al., 2011; Wang et al., 2005; Zeng et al., 2009; Zhang, Ackermann, et al., 2006). In contrast, a single study showed a positive effect of FOXM1 on cellular senescence because Tan et al. (2010) observed that primary *FoxM1B* transgenic MEFs exhibited an earlier onset of premature senescence during MEF culture (i.e., appearance of the senescent phenotype at earlier passages) compared to primary wild-type MEFs.

Cell type-specific differences cannot explain the opposite effects of FOXM1 on cellular senescence because they were also observed in the same cell type, namely, primary MEFs: on the one hand, early-passage $foxm1^{-/-}$ MEFs exhibited premature senescence (Wang et al., 2005). On the other, *FoxM1B* transgenic MEFs with a twofold elevated FOXM1 protein expression displayed an earlier onset of cellular senescence (i.e., at earlier passages) (Tan et al., 2010). One possible interpretation of these two contradictory findings might be that primary MEFs are exquisitely sensitive to their FOXM1 protein levels so that any deviation from the optimum results in senescence, regardless whether FOXM1 is overexpressed or underexpressed.

The primary *FoxM1B* transgenic MEFs showed not only a surprising positive effect of FOXM1 on cellular senescence (Tan et al., 2010), but also two other astonishing properties, which are in contradiction to many other studies:

First, the *p21* mRNA level was increased in *FoxM1B* transgenic MEFs compared to wild-type MEFs (Tan et al., 2010), whereas FOXM1 was found to decrease the *p21* mRNA expression in seven other studies (Ahmad et al., 2010; Bolte et al., 2011; Sengupta et al., 2013; Tan et al., 2007; Wang, Banerjee, et al., 2007; Wang, Hung, et al., 2001; Xue et al., 2012) with various cell types (U2OS osteosarcoma cells, BxPC-3, HPAC and PANC-1 pancreatic cancer cells, MDA-MB-231 and SUM149 breast cancer cells, Caki-1 and 786-O RCC (renal cell carcinoma), regenerating mouse liver after CCl_4 injury and neonatal mouse heart) (Fig. 3.2; Table 3.1).

Second, the p21 protein level was upregulated in *FoxM1B* transgenic MEFs compared to wild-type MEFs (Tan et al., 2010), whereas FOXM1 was found to downregulate the p21 protein expression in 14 other studies (Bolte et al., 2011; Chan et al., 2008; Kalin et al., 2006; Kalinichenko et al., 2003; Li et al., 2008; Nakamura, Hirano, et al., 2010; Qu et al., 2013; Ramakrishna et al., 2007; Tan et al., 2007; Wang, Banerjee, et al., 2007; Wang et al., 2005; Wang, Kiyokawa, et al., 2002; Xia et al., 2009; Xue et al., 2012) with various cell types (NIH3T3 fibroblasts, HL-1 cardiomyocyte cells, U2OS osteosarcoma cells, PC-3 prostate cancer cells, SiHa cervical cancer cells, HepG2 and SMMC-7721 HCC cells, BxPC-3, HPAC and PANC-1 pancreatic cancer cells, Caki-1 and 786-O RCC cells, KG-1, Kasumi-1, U937 and YRK2 leukemia cells, mouse heart, regenerating mouse liver after PHx, regenerating mouse lung after BHT injury) (Fig. 3.2; Table 3.1).

In conclusion, the acceleration of premature senescence in primary *FoxM1B* transgenic MEFs, which points to a positive effect of FOXM1 on cellular senescence (Tan et al., 2010), represents a remarkable exception that demands further analysis to clarify its significance.

8.5. Adult tissue repair after injury
8.5.1 Injury-induced regeneration of adult liver, lung, and pancreas
Adult liver, lung, and pancreas are capable of regenerative cell proliferation and therefore of tissue repair following injury (Poss, 2010; Stoick-Cooper et al., 2007; Zaret & Grompe, 2008). The expression of *foxm1* is strongly induced in the course of this adult tissue repair (see above) (Ackermann Misfeldt et al., 2008; Chen et al., 2010; Gieling et al., 2010; Huang & Zhao, 2012; Kalinichenko et al., 2003, 2001; Krupczak-Hollis et al., 2003; Kurinna et al., 2013; Lehmann et al., 2012; Liu et al., 2011; Matsuo et al., 2003; Meng et al., 2011, 2010; Mukhopadhyay et al., 2011; Ren et al., 2010; Tan et al., 2006; Wang et al., 2003; Wang, Hung, et al., 2001; Wang, Kiyokawa, et al., 2002; Wang, Krupczak-Hollis, et al., 2002; Wang, Quail, et al., 2001; Weymann et al., 2009; Ye et al., 1999, 1997; Zhang, Li, et al., 2012; Zhang, Wang, et al., 2012; Zhao et al., 2006) because FOXM1 plays a pivotal role for liver, lung and pancreas regeneration after injury (Fig. 3.2), in particular for proliferation and cell cycle reentry of the otherwise quiescent cells in adult liver, lung and pancreas (Ackermann Misfeldt et al., 2008; Huang & Zhao, 2012; Kalinichenko et al., 2003; Krupczak-Hollis et al., 2003; Liu et al., 2011; Wang, Hung, et al., 2001; Wang, Kiyokawa, et al., 2002; Wang, Krupczak-Hollis, et al., 2002; Wang, Quail, et al., 2001; Ye et al., 1999; Zhao et al., 2006).

The importance of FOXM1 for adult tissue repair following injury (Fig. 3.2) has been demonstrated for:
- liver regeneration after PHx (Krupczak-Hollis et al., 2003; Wang, Kiyokawa, et al., 2002; Wang, Krupczak-Hollis, et al., 2002; Wang, Quail, et al., 2001; Ye et al., 1999),
- liver regeneration after CCl_4 liver injury (Wang, Hung, et al., 2001),
- pancreas regeneration after PPx (Ackermann Misfeldt et al., 2008),
- lung regeneration after BHT lung injury (Kalinichenko et al., 2003),
- lung regeneration after LPS-induced acute vascular lung injury (Zhao et al., 2006),
- lung regeneration after CLP-induced lung vascular injury (Huang & Zhao, 2012),

— lung regeneration after *Pseudomonas aeruginosa*-induced lung alveolar injury (Liu et al., 2011).

In general, FOXM1 overexpression accelerates and FOXM1 deficiency attenuates the regenerative cell proliferation during injury-induced tissue repair (Ackermann Misfeldt et al., 2008; Huang & Zhao, 2012; Kalinichenko et al., 2003; Krupczak-Hollis et al., 2003; Liu et al., 2011; Wang, Hung, et al., 2001; Wang, Kiyokawa, et al., 2002; Wang, Krupczak-Hollis, et al., 2002; Wang, Quail, et al., 2001; Ye et al., 1999; Zhao et al., 2006).

During liver regeneration after PHx, cell proliferation was reduced in mice with a hepatocyte-specific *foxm1* knockout (Wang, Kiyokawa, et al., 2002). Conversely, FOXM1 transgenic mice with ectopic expression of FoxM1B in hepatocytes displayed an acceleration in regenerative proliferation following PHx (Ye et al., 1999) and following CCl_4 liver injury (Wang, Hung, et al., 2001).

Cell proliferation was decreased during lung regeneration after LPS-induced acute vascular lung injury in mice with an endothelial cell-specific *foxm1* knockout (Zhao et al., 2006) and after *Pseudomonas aeruginosa*-induced lung alveolar injury in mice with an alveolar type II cell-specific *foxm1* knockout (Liu et al., 2011). Conversely, FOXM1 transgenic mice with ectopic FoxM1B expression in all tissues exhibited an acceleration in regenerative proliferation following BHT lung injury (Kalinichenko et al., 2003) and following CLP-induced lung vascular injury (Huang & Zhao, 2012).

During pancreas regeneration after PPx, cell proliferation was diminished in mice with a pancreas-specific *foxm1* knockout (Ackermann Misfeldt et al., 2008).

Hence, FOXM1 is important for regenerative cell proliferation during adult tissue repair after liver, lung, and pancreas injury (Fig. 3.2; Costa et al., 2003; Gartel, 2010; Kalin et al., 2011; Koo et al., 2011; Laoukili et al., 2007; Minamino & Komuro, 2006; Wierstra & Alves, 2007c).

8.5.2 PHx-induced liver regeneration in old-aged mice

The liver regeneration after PHx is impaired in old-aged mice, which show not only a reduced proliferation of hepatocytes following PHx (Schmucker, 2005; Schmucker & Sanchez, 2011; Timchenko, 2009), but also a diminished induction of *foxm1* expression in response to PHx (see above) (Chen et al., 2010; Wang, Krupczak-Hollis, et al., 2002; Wang, Quail, et al., 2001).

However, the reduced PHx-induced hepatocyte proliferation in the injured livers of old mice can be restored to the level in young animals if the decreased PHx-induced *foxm1* expression is re-increased, which can be achieved either by expression of a *foxm1b* transgene in hepatocytes (Wang, Quail, et al., 2001) or by acute adenovirus-mediated delivery of FoxM1B (via tail vein injection) (Wang, Krupczak-Hollis, et al., 2002) or by human GH administration (via intraperitoneal injection) (Krupczak-Hollis et al., 2003) or by acute adenovirus-mediated delivery of a ca FXRα2-VP16 fusion protein (via tail vein injection) (Chen et al., 2010). Thus, the impaired liver regeneration after PHx in old-aged mice is restored to the level in young animals if the age-dependent downregulation of the PHx-induced *foxm1* expression is compensated by exogenous FoxM1B (Wang, Krupczak-Hollis, et al., 2002; Wang, Quail, et al., 2001) or by upregulation of the endogenous *foxm1* expression through treatment with GH (Krupczak-Hollis et al., 2003) or FXRα2-VP-16 (Chen et al., 2010).

This ability of exogenous FoxM1B to cancel the age-dependent decline in the PHx-induced hepatocyte proliferation (Wang, Krupczak-Hollis, et al., 2002; Wang, Quail, et al., 2001) confirms the central importance of FOXM1 for regenerative hepatocyte proliferation during liver repair following PHx. Moreover, it points to a requirement of FOXM1 for maintenance of the proliferative capacity of cells (Costa et al., 2003; Costa, Kalinichenko, Tan, et al., 2005; Mackey et al., 2003).

8.5.3 Special functions of FOXM1 during adult liver and lung repair following injury

In the course of adult tissue regeneration after injury, FOXM1 not only has a general proliferation-stimulating function in all organs analyzed (i.e., liver, lung, pancreas), but it also fulfills some special nonproliferative functions in individual tissues:

During liver repair following CCl_4-mediated liver injury, FOXM1 is required for recruitment of monocytes to the injured liver but dispensable for neutrophil infiltration of the injured liver (Ren et al., 2010). Mice with a knockout of *foxm1* in myeloid cells (including macrophages, monocytes, and neutrophils) displayed a delay in liver repair after CCl_4 injury coinciding with reduced numbers of monocytes and later on macrophages in the CCl_4-injured liver, whereas the number of neutrophils in the CCl_4-injured liver remained unchanged (Ren et al., 2010). Surprisingly, this myeloid-specific deletion of *foxm1* affected neither the proliferation of hepatic myeloid-derived inflammatory cells (macrophages, monocytes, neutrophils) nor the percentages of monocytes and neutrophils in the blood and the bone marrow of untreated mice (Ren et al., 2010). Adoptive transfer of control monocytes restored liver repair after CCl$_4$ injury in mice with the myeloid-specific *foxm1* knockout demonstrating that the observed delay in liver repair is caused by the impaired recruitment of monocytes, which is due to *foxm1* deficiency in myeloid cells (Ren et al., 2010).

During lung regeneration after acute vascular lung injury, FOXM1 plays a dual role in endothelial repair: first, FOXM1 promotes endothelial regeneration by inducing endothelial cell proliferation (Zhao et al., 2006). Second, FOXM1 mediates the reannealing of endothelial adherens junctions in order to restore the restrictive endothelial barrier (Fig. 3.2; Mirza et al., 2010). The VE-cadherin (vascular endothelial-cadherin, CDH5 (cadherin 5)) interaction partner β-catenin is part of the intercellular adherens junctions (Bazzoni & Dejana, 2004; Dejana, 2004; Dejana, Orsenigo, & Lampugnani, 2008; Dejana, Orsenigo, Molendini, Baluk, & McDonald, 2009; Mehta & Malik, 2006; Vestweber, Broermann, & Schulte, 2010; Wallez & Huber, 2008) so that the activation of the direct FOXM1 target gene *β-catenin* (Fig. 3.2; Table 3.1) by FOXM1 is important for the FOXM1-mediated recovery of endothelial barrier integrity and function, that is, for endothelial barrier repair (Mirza et al., 2010).

During lung repair after *Pseudomonas aeruginosa*-induced lung alveolar injury, FOXM1 is important not only for proliferation of adult alveolar type II epithelial cells but also for transdifferentiation of type II cells into type I cells so that FOXM1 plays a dual role in reannealing of the alveolar epithelial barrier and recovery of alveolar epithelial barrier function (Liu et al., 2011).

During lung repair following CLP-induced lung vascular injury, which is a model of polymicrobial sepsis, FOXM1 plays a role in restoration of endothelial barrier integrity and resolution of lung inflammation (Huang & Zhao, 2012):
Lung injury induced by CLP challenge manifests in increased lung vascular permeability and lung inflammation, which both re-decrease to baseline during endothelial repair (Huang & Zhao, 2012). Lung vascular permeability was measured as EBA (Evans Blue-conjugated albumin) extravasation (Huang & Zhao, 2012). Lung inflammation was assessed by

leukocyte sequestration and MPO (myeloperoxidase) activity, an indicator of neutrophil infiltration (Huang & Zhao, 2012).

FoxM1B overexpression accelerates both the restoration of endothelial barrier integrity and the resolution of lung inflammation because vascular permeability and MPO activity returned earlier to basal levels in *foxm1b* transgenic mice than in wild-type mice (Huang & Zhao, 2012). Accordingly, FoxM1B overexpression in *foxm1b* transgenic mice accelerated the resolution of CLP-induced lung edema, reduced the leukocyte sequestration post-CLP, and improved the survival rate of mice after CLP challenge. Moreover, FoxM1B overexpression caused endothelial cell proliferation in *foxm1b* transgenic mice thereby promoting rapid endothelial repair (Huang & Zhao, 2012).

Vice versa, *foxm1* deficiency in *FoxM1 CKO* mice with an endothelial cell-restricted disruption of *foxm1* impaired endothelial repair following CLP challenge so that vascular permeability and MPO activity remained elevated in *FoxM1 CKO* lungs even 72 h post-CLP when both had recovered to nearly baseline in wild-type mice (Huang & Zhao, 2012).

Hence, FOXM1 is necessary and sufficient for restoration of endothelial barrier integrity and resolution of lung inflammation during endothelial repair upon lung regeneration after CLP-induced lung vascular injury (Huang & Zhao, 2012). Accordingly, CLP challenge induced strong endogenous *foxm1* mRNA expression in wild-type mouse lungs (Huang & Zhao, 2012).

8.6. Liver repopulation by transplanted hepatocytes in a model of chronic liver injury

In primary cell cultures, FoxM1B-overexpressing hepatocytes, which were isolated from FoxM1B-transgenic mice, progressed faster through the cell cycle than control cells as evidenced by an acceleration of their entry into both S-phase and M-phase (Brezillon et al., 2007). Thus, overexpression of FoxM1B accelerates the hepatocyte cell cycle in a cell-autonomous fashion (Brezillon et al., 2007).

Because of this proliferative advantage, FoxM1B-overexpressing hepatocytes repopulated the liver more efficiently than control cells after hepatocyte transplantation into chronically injured mouse liver (Brezillon et al., 2007).

Transplantation of donor hepatocytes into recipient liver and following repopulation of the liver by the engrafted donor hepatocytes is considered an appealing alternative to orthotopic liver transplantation (Allen & Bhatia,

2002; Grompe, 2006; Gupta, 2002; Gupta, Bhargava, & Novikoff, 1999; Gupta, Malhi, Gagandeep, & Novikoff, 1999; Gupta, Rajvanshi, Bhargava, & Kerr, 1996; Malhi & Gupta, 2001). Since transplanted hepatocytes do not proliferate in the recipient liver a continuous liver regeneration stimulus is needed (Alison, Islam, & Lim, 2009; Gilgenkrantz, 2010; Grompe, 1999, 2001; Grompe, Laconi, & Shafritz, 1999; Guha et al., 2001; Gupta & Chowdhury, 2002; Gupta & Rogler, 1999; Kawashita et al., 2005; Laconi & Laconi, 2002; Mizuguchi, Mitaka, Katsuramaki, & Hirata, 2005; Shafritz & Oertel, 2011; Wu & Gupta, 2009). Additionally, a selective proliferative advantage of the transplanted hepatocytes over those in the recipient liver is required for efficient liver repopulation by the donor hepatocytes (Alison et al., 2009; Gilgenkrantz, 2010; Grompe, 1999, 2001; Grompe et al., 1999; Guha et al., 2001; Gupta & Chowdhury, 2002; Gupta & Rogler, 1999; Kawashita et al., 2005; Laconi & Laconi, 2002; Mizuguchi et al., 2005; Shafritz & Oertel, 2011; Wu & Gupta, 2009). Therefore, transplanted hepatocytes can efficiently repopulate the liver provided that they dispose of a proliferative advantage over resident ones (Alison et al., 2009; Gilgenkrantz, 2010; Grompe, 1999, 2001; Grompe et al., 1999; Guha et al., 2001; Gupta & Chowdhury, 2002; Gupta & Rogler, 1999; Kawashita et al., 2005; Laconi & Laconi, 2002; Mizuguchi et al., 2005; Shafritz & Oertel, 2011; Wu & Gupta, 2009). Strikingly, FoxM1B overexpression confers such a proliferative advantage on transplanted hepatocytes, in particular on old donor hepatocytes (Brezillon et al., 2007).

uPA/SCID (urokinase-type plasminogen activator/severe combined immunodeficiency) mice represent a model of chronic liver injury (Knetemann & Mercer, 2005; Meuleman & Leroux-Roels, 2008; Meuleman et al., 2008; Strom, Davila, & Grompe, 2010). As a result of the continous selective pressure induced by the cytotoxicity of uPA transgene expression, normal transplanted hepatocytes have a strong survival advantage and repopulate the liver of the uPA/SCID mice (Meuleman & Leroux-Roels, 2008; Meuleman et al., 2008; Rhim, Sandgren, Degen, Palmiter, & Brinster, 1994; Strom et al., 2010).

Both FoxM1B-overexpressing and control hepatocytes from young donor mice (2 month old) repopulated the liver after transplantation into uPA/SCID mice (Brezillon et al., 2007). Liver repopulation was twofold more efficient with FoxM1B-overexpressing hepatocytes than with control cells, demonstrating that the overexpression of FoxM1B confers a selective proliferative advantage on the transplanted hepatocytes (Brezillon et al., 2007). This advantage can be attributed to the FoxM1B-mediated acceleration of cell cycle progression, that is, to the faster S- and M-phase entry observed in FoxM1B-overexpressing hepatocytes (see above) (Brezillon et al., 2007).

The difference in the liver repopulation efficiency between FoxM1B-overexpressing and control hepatocytes becomes significantly more pronounced when donor hepatocytes from old mice (14 month old) are transplanted into the livers of young uPA/SCID mice (18 day old) (Brezillon et al., 2007). FoxM1B-overexpressing hepatocytes from old donors repopulated the liver sevenfold more efficiently than control hepatocytes from old donors (Brezillon et al., 2007). Thus, FoxM1B overexpression confers a dramatic proliferative advantage on old donor hepatocytes (Brezillon et al., 2007).

As a result, the liver repopulation efficiency is similar for FoxM1B-overexpressing hepatocytes from both old and young donors, whereas the liver repopulation efficiency of control hepatocytes from old donors is reduced by approx. 70% compared to those from young donors (Brezillon et al., 2007). Hence, the proliferative capacity of transplanted donor hepatocytes decreases during aging, and FoxM1B overexpression is sufficient to prevent this age-dependent decline in their proliferation capacity (Brezillon et al., 2007). Therefore, in contrast to nonmodified old donor hepatocytes, FoxM1B-overexpressing hepatocytes from old donors retain an undiminished potency for liver repopulation after transplantation into the chronically injured livers of uPA/SCID mice (Brezillon et al., 2007).

These findings point to an importance of FOXM1 for maintenance of the proliferative capacity of cells (Costa et al., 2003; Costa, Kalinichenko, Tan, et al., 2005; Mackey et al., 2003).

In summary, FoxM1B overexpression in transplanted hepatocytes allows efficient liver repopulation in a model of chronic liver injury, because FoxM1B confers a proliferative advantage on the donor hepatocytes, especially on those from old donors (Brezillon et al., 2007). Therefore, the overexpression of FoxM1B in old donor hepatocytes preserves their proliferative capacity, which is otherwise strongly decreased due to aging (Brezillon et al., 2007).

8.7. Insulin-producing β-cells in pancreatic islets
8.7.1 Proliferation of pancreatic β-cells after birth

Overexpression of FoxM1B in isolated pancreatic islets stimulates the proliferation of β-cells, which was evidenced by increased DNA replication rates of β-cells in FoxM1B-overexpressing islets from human and murine pancreata (Davis et al., 2010). In accordance with this stimulation of β-cell proliferation by FoxM1B (Davis et al., 2010), mice with a conditional *foxm1* knockout in the developing embryonic pancreas (FoxM1$^{\Delta panc}$) revealed a postnatal requirement of FOXM1 for β-cell proliferation and expansion of β-cell mass in two different biological settings, namely, during the period after birth (Zhang, Ackermann, et al., 2006) and during pregnancy (see below) (Zhang et al., 2010).

FOXM1 is highly expressed in the pancreas of the developing embryo (Kalin et al., 2011; Yao et al., 1997; Zhang, Ackermann, et al., 2006), whereas the overall FOXM1 mRNA and protein levels are relatively low or undetectable in the adult pancreas (Kalinichenko et al., 2003; Yao et al., 1997; Ye et al., 1997). However, FOXM1 remains highly expressed in subpopulations of endocrine cells in adult pancreatic islets, in accordance with its necessity for postnatal β-cell proliferation (Zhang, Ackermann, et al., 2006). Accordingly, abundant FOXM1 expression was found in all endocrine cell lines analyzed, including rat 38 cells and rat INS-1insuloma cells, which both produce insulin (Yao et al., 1997).

8.7.2 Postnatal expansion and maintenance of pancreatic β-cell mass

FOXM1 is required for postnatal expansion and maintenance of pancreatic β-cell mass (Fig. 3.2; Zhang, Ackermann, et al., 2006). Male FoxM1$^{\Delta panc}$ mice with a pancreas-specific deletion of *foxm1* were born normal, but after birth they had less β-cell mass than control animals (Zhang, Ackermann, et al., 2006). This lack of β-cell mass was due to diminished β-cell proliferation whereas no increase in β-cell apoptosis was detected (Zhang, Ackermann, et al., 2006). Since the Foxm1$^{\Delta panc}$ mice are born normal FOXM1 is specifically required for the proliferation of β-cells after birth (Zhang, Ackermann, et al., 2006). The Foxm1$^{\Delta panc}$ mice revealed a dual requirement of FOXM1 for postnatal β-cell proliferation (Zhang, Ackermann, et al., 2006). First, FOXM1 is required for β-cell proliferation in the course of normal β-cell turnover and thus for postnatal maintenance of the existing β-cell mass (Fig. 3.2; Zhang, Ackermann, et al., 2006). Second, FOXM1 is required for the normal continuous postnatal expansion of β-cell mass (Fig. 3.2), which occurs after birth in control animals (Zhang, Ackermann, et al., 2006).

Because of their lack of insulin-producing β-cells in comparison to control animals, 13% of the male FoxM1$^{\Delta panc}$ mice developed overt diabetes mellitus by nine weeks of age (Zhang, Ackermann, et al., 2006).

8.7.3 Pregnancy-induced expansion of maternal β-cell mass

FOXM1 is required for the expansion of maternal β-cell mass during pregnancy because it is essential for the pregnancy-induced increase in maternal β-cell proliferation (Zhang et al., 2010). In contrast to wild-type animals, female FoxM1$^{\Delta panc}$ mice with a pancreas-wide *foxm1* deletion failed to increase β-cell proliferation and thus to expand β-cell mass during pregnancy (Zhang et al., 2010). Yet, there was no increase in apoptotic β-cells in pregnant FoxM1$^{\Delta panc}$ females (Zhang et al., 2010).

Due to their scarcity of insulin-producing β-cells compared to control animals, pregnant female FoxM1$^{\Delta panc}$ mice generally developed overt GDM at GD (gestational day) 15.5 (Zhang et al., 2010).

Hence, the requirement of FOXM1 for postnatal β-cell proliferation manifests in the development of diabetes in the case of pancreatic *foxm1* deficiency (Zhang, Ackermann, et al., 2006; Zhang et al., 2010). Thus, diabetes mellitus is one example for the implication of FOXM1 in disease.

8.8. Maintenance of the proliferative capacity of cells

The reexpression of FOXM1 in PHx-injured livers of old-aged mice (Wang, Krupczak-Hollis, et al., 2002; Wang, Quail, et al., 2001) and in transplanted hepatocytes from old donors (Brezillon et al., 2007) is able to overcome their age-related proliferative defects, which suggests an importance of FOXM1 for maintenance of the proliferative capacity of cells (Costa et al., 2003; Costa, Kalinichenko, Tan, et al., 2005; Mackey et al., 2003). This view is supported by the capability of FOXM1 to suppress cellular senescence (Anders et al., 2011; Li et al., 2008; Park, Carr, et al., 2009; Rovillain et al., 2011; Wang et al., 2005; Zeng et al., 2009; Zhang, Ackermann, et al., 2006).

8.9. Homologous recombination (HR) repair of DNA damage
8.9.1 FOXM1 promotes HR repair

HR (homologous recombination) and NHEJ (nonhomologous end joining) are two DNA repair mechanisms for DSBs (DNA-double-strand breaks) (Amunugama & Fishel, 2012; Bordeianu, Zugun-Eloae, & Rusu, 2011; Burma, Chen, & Chen, 2006; Cerbinskaite, Mukhopadhyay, Plummer, Curtin, & Edmondson, 2012; Ciaccia & Elledge, 2010; Dever, White, Hartman, & Valerie, 2012; Evers, Helleday, & Jonkers, 2010; Ferguson et al., 2000; Hakem, 2008; Hasty, 2008; Helleday, 2010; Heyer, Ehmsen, & Liu, 2010; Hoeijmakers, 2001; Jeggo, Geuting, & Löbrich, 2011; Kass & Jasin, 2010; Khanna & Jackson, 2001; Krejci, Altmannova,

Spirek, & Zhao, 2012; Lieber, 1999, 2010; Ma, Lu, et al., 2005; Moynahan & Jasin, 2010; Peterson & Cote, 2004; Polo & Jackson, 2011; Raassool, 2003; Raassool & Tomkinson, 2010; San Filippo, Sung, & Klein, 2008; Scott & Pandita, 2006; Symington & Gautier, 2011; van Gent, Hoeijmakers, & Kanaar, 2001; Wyman & Kanaar, 2006).

FOXM1 is required for HR (Fig. 3.2; Park et al., 2012) but not for NHEJ (Monteiro et al., 2012) because siRNA-mediated knockdown of FOXM1 significantly impaired the HR repair of defined genomic DSBs (Monteiro et al., 2012; Park et al., 2012), whereas the NHEJ activity was only marginally affected (Monteiro et al., 2012). The role of FOXM1 for HR is verified by the finding that FOXM1 overexpression enhanced the DSB repair via HR (Monteiro et al., 2012).

Since FOXM1 promotes HR repair (Fig. 3.2; Monteiro et al., 2012; Park et al., 2012), FOXM1-deficient cells exhibit an increase in the number of DSBs without external infliction of DNA damage (Chetty et al., 2009; Kwok et al., 2010; Monteiro et al., 2012; Tan et al., 2007). This observation was reported for $foxm1^{-/-}$ MEFs (Tan et al., 2007) as well as for various cell types depleted of FOXM1 by siRNA, namely, for U2OS osteosarcoma cells (Tan et al., 2007), A549 lung cancer cells (Chetty et al., 2009), epirubin-resistant MCF-7-EPI^R breast cancer cells (Monteiro et al., 2012), and cisplatin-resistant MCF-7-CIS^R cells (Kwok et al., 2010).

The importance of FOXM1 for HR repair of DSBs (Fig. 3.2; Monteiro et al., 2012; Park et al., 2012) is corroborated by the ability of FOXM1 overexpression to prevent the accumulation of DSBs in response to treatment with DNA-damaging chemotherapeutic drugs (epirubicin, cisplatin) (Kwok et al., 2010; Monteiro et al., 2012):

Epirubicin-treated human MCF-7 mammary carcinoma cells accumulated a high number of DSBs, but overexpression of FOXM1 abolished the formation of epirubicin-induced DSBs, which indicates that FOXM1 overexpression is sufficient to render MCF-7 cells resistant to induction of DSBs by epirubicin (Monteiro et al., 2012). Likewise, a high number of DSBs is formed in cisplatin-treated MCF-7 cells, but the accumulation of cisplatin-induced DSBs was abolished by overexpression of a ca FOXM1 mutant, indicating that ca FOXM1 overexpression can confer resistance to induction of DSBs by cisplatin (Kwok et al., 2010). Probably, the (ca) FOXM1-overexpressing MCF-7 cells are more effective in repairing the sustained DSBs rather than less susceptible to induction of

DSBs so that epirubicin and cisplatin probably still induce DSBs, but these DSBs do not progressively accumulate on prolonged treatment because they are very efficiently repaired via HR due to the improvement of HR activity by the overexpressed (ca) FOXM1 (Monteiro et al., 2012).

Hence, FOXM1 plays a role in HR repair of DSBs (Fig. 3.2; Monteiro et al., 2012; Park et al., 2012).

Accordingly, FOXM1 activates the expression of the HR proteins BRCA2 (breast cancer-associated gene 2, FANCD1 (Fanconi anemia complementation group D1)), XRCC2 (X-ray cross-completing group 2), RAD51, EXO1 (exonuclease 1), and BRIP1 (BRCA1-interacting protein 1, Bach1 (BRCA1-associated C-terminal helicase)) as well as the expression of the DNA replication proteins RFC4 (replication factor C), POLE2 (DNA polymerase ε 2), and PCNA (proliferating cell nuclear antigen) (Fig. 3.2; Table 3.1), which are involved in HR too (Amunugama & Fishel, 2012; Bordeianu et al., 2011; Cerbinskaite et al., 2012; Ciaccia & Elledge, 2010; Dever et al., 2012; Evers et al., 2010; Hakem, 2008; Helleday, 2010; Heyer et al., 2010; Hoeijmakers, 2001; Jeggo et al., 2011; Kass & Jasin, 2010; Khanna & Jackson, 2001; Krejci et al., 2012; Moynahan & Jasin, 2010; Peterson & Cote, 2004; Polo & Jackson, 2011; Raassool & Tomkinson, 2010; San Filippo et al., 2008; Scott & Pandita, 2006; Symington & Gautier, 2011; van Gent et al., 2001; Wyman & Kanaar, 2006). These eight proteins are encoded by five direct FOXM1 target genes (*brca2*, *rad51*, *exo1*, *brip1*, *rfc4*) and three possibly indirect FOXM1 target genes (*xrcc2*, *pole2*, *pcna*) (Fig. 3.2; Table 3.1). Also cyclin D1, the protein product of a direct FOXM1 target gene (Fig. 3.2; Table 3.1), is required for an efficient HR DNA repair process (Bartek & Lukas, 2011; Jirawatnotai et al., 2011).

8.9.2 FOXM1 expression during the DNA damage response

The checkpoint kinase Chk2, a key player in the DNA damage response to DSBs (Abraham, 2001; Ahn, Urist, et al., 2004; Antoni et al., 2007; Bartek et al., 2001; Bartek & Lukas, 2003, 2007; Bartek et al., 2004; Branzei & Foiani, 2008; Chen & Poon, 2008; Deckbar et al., 2011; Finn et al., 2012; Harper & Elledge, 2007; Harrison & Haber, 2006; Ishikawa et al., 2006; Jackson, 2009; Jackson & Bartek, 2009; Kastan, 2008; Kastan & Bartek, 2004; Li & Zou, 2005; Liang et al., 2009; Lukas et al., 2004; McGowan, 2002; Nakanishi et al., 2006; Nyberg et al., 2002; Perona et al., 2008; Reinhardt & Yaffe, 2009; Rouse & Jackson, 2002; Samuel et al., 2002;

Sancar et al., 2004; Smith et al., 2010; Stracker et al., 2009; Su, 2006; Walworth, 2000; Wang, Ji, et al., 2009; Yang & Zou, 2006; Zhou & Elledge, 2000), increases the FOXM1 protein expression by stabilizing the FOXM1 protein through its phosphorylation (see above) (Tan et al., 2007). This Chk2-mediated FOXM1 stabilization has been observed after DNA damage induced by IR, UV, or etoposide (Fig. 3.10; Tan et al., 2007).

The *foxm1* mRNA and protein levels are also upregulated by BRCA1 (Bae et al., 2005) or RAD51 (Fig. 3.10; Zhang, Wu, et al., 2012), respectively, which both have crucial functions in the HR repair of DSBs (Baumann & West, 1998; Caestecker & Van de Walle, 2013; Daboussi, Dumay, Delacote, & Lopez, 2002; D'Andrea, 2003; Forget & Kowalczykowski, 2010; Henning & Stürzbecher, 2003; Huen, Sy, & Chen, 2010; Kawabata, Kawabata, & Nishibori, 2005; Kennedy & D'Andrea, 2005; Li & Greenberg, 2012; Masson & West, 2001; McKinnon & Caldecott, 2007; Morris, 2010; Mullan, Quinn, & Harkin, 2006; Murray, Mullan, & Harkin, 2007; Ohta, Sato, & Wu, 2011; Richardson, 2005; Roy, Chun, & Powell, 2011; Shinohara & Ogawa, 1999; Silver & Livingston, 2012; Sung, Krejci, Van Komen, & Sehorn, 2003; Thacker, 2005; Tutt & Ashworth, 2002; Venkitaraman, 2004; Vispe & Defais, 1997; Wang, 2007; Wu, Lu, et al., 2010; Yang & Xia, 2010; Yarden & Papa, 2006; Yoshida & Miki, 2004).

Although the DNA damage-induced stabilization of FOXM1 by Chk2 (Tan et al., 2007) and the upregulation of FOXM1 by BRCA1 and RAD51 (Bae et al., 2005; Zhang, Wu, et al., 2012) fit the role of FOXM1 for HR repair (Monteiro et al., 2012; Park et al., 2012), the situation is far more complex because the FOXM1 expression was also found to be downregulated in response to DNA damage (see above) induced by doxorubicin (Alvarez-Fernandez, Halim, et al., 2010; Halasi & Gartel, 2012; Pandit et al., 2009), daunorubicin (Barsotti & Prives, 2009), epirubicin (Millour et al., 2011; Monteiro et al., 2012), cisplatin (Kwok et al., 2010), oxaliplatin (Qu et al., 2013) and 5-FU (Monteiro et al., 2012), which all can cause DSBs (Adamsen, Kravik, & DeAngelis, 2011; Chiu, Lee, Hsu, & Chen, 2009; El-Awady, Saleh, & Dahm-Daphi, 2010; Gewirtz, 1999; Kizek et al., 2012; Minotti, Menna, Salvatorelli, Cairo, & Gianni, 2004; Müller et al., 1998; Nowosielska & Marinus, 2005; Perego et al., 2001; Rabbani, Finn, & Ausio, 2005; Saleh, El-Awady, & Anis, 2013; Siddik, 2003; Taatjes, Fenick, Gaudiana, & Koch, 1998; Takahashi, Koi, Balaguer, Boland, & Goel, 2011). Since the DNA damage-induced changes in the FOXM1 expression depend on both the cell type and the p53 status of the cell (see above) (Barsotti & Prives, 2009; Halasi & Gartel, 2012; Millour et al., 2011; Pandit et al., 2009) FOXM1 could play a role in DSB repair via HR (Monteiro et al., 2012; Park et al., 2012) only in those scenarios, where DNA damage induces an increase in the FOXM1 level.

Regarding the general role of FOXM1 during the DNA damage response, it has been suggested that the p53-mediated downregulation of FOXM1 (Fig. 3.10; Table 3.2) might be important for the long-term maintenance of a stable G2-arrest (Barsotti & Prives, 2009) and that, conversely, FOXM1 activity might be important for the entry into mitosis after checkpoint silencing, that is, for checkpoint recovery (Alvarez-Fernandez, Halim, et al., 2010; Alvarez-Fernandez, Medema, et al., 2010).

8.10. Maintenance of genomic stability

The role of FOXM1 in HR repair of DSBs (Monteiro et al., 2012; Park et al., 2012) suggests that FOXM may be important for the maintenance of chromosomal integrity and genomic stability (Laoukili et al., 2007; Wierstra & Alves, 2007c; Wilson, Brosens, Schwenen, & Lam, 2011; Zhao & Lam, 2012).

In fact, FOXM1 is required for prevention of polyploidy (and aneuploidy) (Fig. 3.2) because FOXM1-deficient cells generally suffer from polyploidy (and aneuploidy) (Chen, Yang, et al., 2012; Fu et al., 2008; Kalinichenko et al., 2004; Kim et al., 2005; Korver et al., 1998; Krupczak-Hollis et al., 2004; Laoukili et al., 2005; Nakamura, Hirano, et al., 2010; Ramakrishna et al., 2007; Ustiyan et al., 2009; Wan et al., 2012; Wang et al., 2005; Wonsey & Follettie, 2005; Zhao et al., 2006). This is explained by the necessity of FOXM1 for proper execution of mitosis (Fig. 3.2; Chan et al., 2008; Fu et al., 2008; Gusarova et al., 2007; Laoukili et al., 2005; Priller et al., 2011; Ramakrishna et al., 2007; Schüller et al., 2007; Wang et al., 2005; Wonsey & Follettie, 2005; Yoshida et al., 2007) and thus for preservation of mitotic fidelity (see above).

Moreover, FOXM1 protects cells from DSBs because it promotes DSB repair via HR (Fig. 3.2; Monteiro et al., 2012; Park et al., 2012). Consequently, FOXM1-deficient cells display an increased number of DSBs (Chetty et al., 2009; Kwok et al., 2010; Monteiro et al., 2012; Tan et al., 2007). Actually, epirubicin and cisplatin induce DSBs in MCF-7 cells but not in MCF-7 cells overexpressing FOXM1 (Monteiro et al., 2012) or a ca FOXM1 mutant (Kwok et al., 2010), respectively, showing that FOXM1 confers resistance to epirubicin-induced and cisplatin-induced DSBs.

Among the FOXM1 target genes are several DNA repair genes (Fig. 3.2), namely not only the five above-mentioned HR genes (*brca2, xrcc2, exo1, rad51, brip1*), but also two MMR (mismatch repair) genes (*msh6, exo1*) and one BER (base-excision repair) gene (*xrcc1*) (Fig. 3.2; Table 3.1). *Xrcc1* is a direct FOXM1 target gene and *msh6* is a possibly indirect FOXM1 target gene (Fig. 3.2; Table 3.1).

Another possibly indirect FOXM1 target gene encodes the SRF (substrate recognition factor) DTL (denticleless, CDT2 (chromatin licensing and DNA replication factor 1), DCAF2 (Ddb1 (damage-specific DNA-binding protein 1)- and Cul4 (Cullin 4)-associated factor 2)) of the E3 ubiquitin ligase CRL4 (Cullin 4-RING E3 ubiquitin ligase) (Fig. 3.2; Table 3.1), which plays a vital role in DNA repair pathways, normal cell cycle progression, prevention of DNA re-replication, and preservation of genomic stability and genome integrity (Abbas & Dutta, 2011; Havnes & Walter, 2011; Scrima, Fischer, Lingaraju, Cavadini, & Thomä, 2011).

Furthermore, FOXM1 upregulates the mRNA or protein levels of the checkpoint kinases Chk1 and Chk2, respectively (Fig. 3.2; Table 3.1), two key factors in the DNA damage response (Abraham, 2001; Ahn, Urist, et al., 2004; Antoni et al., 2007; Bartek et al., 2001; Bartek & Lukas, 2003, 2007; Bartek et al., 2004; Branzei & Foiani, 2008; Chen & Poon, 2008; Chen & Sanchez, 2004; Deckbar et al., 2011; Finn et al., 2012; Goto, Izawa, Li, & Inagaki, 2012; Harper & Elledge, 2007; Harrison & Haber, 2006; Ishikawa et al., 2006; Jackson, 2009; Jackson & Bartek, 2009; Kastan, 2008; Kastan & Bartek, 2004; Li & Zou, 2005; Liang et al., 2009; Lukas et al., 2004; McGowan, 2002; Merry, Fu, Wang, Yeh, & Zhang, 2010; Meuth, 2010; Nakanishi et al., 2006; Nyberg et al., 2002; Perona et al., 2008; Reinhardt & Yaffe, 2009; Rouse & Jackson, 2002; Samuel et al., 2002; Sancar et al., 2004; Smith et al., 2010; Stracker et al., 2009; Su, 2006; Tapia-Alveal, Calonge, & O'Connell, 2009; Walworth, 2000; Wang, Ji, et al., 2009; Yang & Zou, 2006; Zhou & Elledge, 2000). *Chk1* is a direct FOXM1 target gene (Fig. 3.2; Table 3.1).

Together, these findings point to the importance of FOXM1 for the maintenance of chromosomal integrity and genomic stability (Laoukili et al., 2007; Wierstra & Alves, 2007c; Wilson et al., 2011; Zhao & Lam, 2012).

Since genomic stability seems to be preserved by FOXM1 but perturbed by FOXM1 deficiency, FOXM1 might somehow function as a caretaker and thus might also dispose of a tumor suppressing side in addition to its predominant tumorigenic activities (see Part II of this two-part review, that is see Wierstra, 2013b).

8.11. Maintenance of stem cell pluripotency

FOXM1 plays a role in the maintenance of stem cell pluripotency (Fig. 3.2; Xie et al., 2010).

Accordingly, FOXM1 activates its target genes *oct-4*, *sox2*, and *nanog* (Fig. 3.2; Table 3.1), the protein products of which are essential for pluripotency and self-renewal of ES (embryonic stem) cells (Amabile & Meissner, 2009; Brumbaugh, Rose, Phanstiel, Thomson, & Coon, 2011; de Vries et al., 2008; Dejosez & Zwaka, 2012; Geoghegan & Byrnes, 2008; Gonzales & Ng, 2011; Hanna, Saha, & Jaenisch, 2010; Hochedlinger & Plath, 2009; Jaenisch & Young, 2008; Li, 2010; Loh, Ng, & Ng, 2008; Loh et al., 2011; MacArthur, Ma'ayan, & Lemischka, 2009; Ng & Surani, 2011; Niwa, 2007a, 2007b; Ohtsuka & Dalton, 2008; Orkin & Hochedlinger, 2011; Pauklin, Pedersn, & Vallier, 2011; Pei, 2009; Ralston & Rossant, 2010; Rossant, 2008; Scheper & Copray, 2009; Silva & Smith, 2008; Yamanaka, 2007, 2008; Young, 2011; Yu & Thomson, 2008; Zhao & Daley, 2008). Oct-4, Sox2, and Nanog together form the core regulatory circuitry of pluripotency, in which each of these three transcription factors induces its own expression and that of the two other genes (Brumbaugh et al., 2011; Gonzales & Ng, 2011; Hanna et al., 2010; Jaenisch & Young, 2008; Loh et al., 2011; MacArthur et al., 2009; Orkin & Hochedlinger, 2011; Scheper & Copray, 2009; Young, 2011). Notably, overexpression of FOXM1 can not only prevent the downregulation of the *oct-4*, *sox2*, and *nanog* levels during RA-induced differentiation of mouse P19 EC cells, but FOXM1 overexpression is also capable of restarting the expression of *oct-4*, *sox2*, and *nanog* in RA-differentiated P19 EC cells and in differentiated human newborn fibroblasts, both of which do not normally express the pluripotency factors Oct-4, Sox2, or Nanog (Xie et al., 2010).

The involvement of FOXM1 in the maintenance of stem cell pluripotency (Fig. 3.2) was analyzed in three experimental systems (Xie et al., 2010):

Mouse P19 EC cells represent pluripotent stem cells and they are capable of forming teratomas with tissues of all three germlayers (i.e., endoderm, mesoderm, ectoderm) upon subcutaneous inoculation into nude mice (McBurney, 1993; van der Heyden & Defize, 2003; Xie et al., 2010). When teratoma formation was performed with P19 cells, which had been infected with adenovirus-expressing siRNA against FOXM1 (AdFoxm1siRNA), the AdFoxm1siRNA-infected P19 cells produced significantly smaller teratomas and these teratomas expressed only the mesodermal marker Brachyury but neither the ectodermal marker MAP2 (microtubule-associated protein 2) nor the endodermal marker GATA4 (Xie et al., 2010). This finding suggests that the P19 cells with knockdown of FOXM1 failed to form endodermal or ectodermal structures but instead underwent spontaneous differentiation to mesodermal derivatives (Xie et al., 2010). In fact, hematoxylin and eosin staining revealed mesoderm-derived muscle and adipose connective tissue in the teratomas formed by AdFoxm1siRNA-infected P19 cells (Xie et al., 2010). Also the expression of the cardiac muscle-specific marker ACTC1 (actin α cardiac muscle, SMA) only in teratomas produced with AdFoxm1siRNA-infected P19 cells, but not in their control counterparts, points to a spontaneous differentiation of FOXM1-deficient P19 cells to cardiomyocytes (Xie et al., 2010). Thus, the reduced teratoma size and the failure to differentiate into all three germlayers indicate a loss of pluripotency in FOXM1 knockdown P19 cells, which implies a role of FOXM1 in maintaining the pluripotency of stem cells (Fig. 3.2; Xie et al., 2010).

Pluripotent P19 EC cells differentiate into neural cells in response to RA treatment when aggregated to form EBs (embryoid bodies) (Bain, Ray, Yao, & Gottlieb, 1994; Jones-Villeneuve, McBurney, Rogers, & Kalnins, 1982; McBurney, Jones-Villeneuve, & Edwards, 1982; Soprano, Teets, & Soprano, 2007; Ulrich & Majumder, 2006; Xie et al., 2010). During the RA-induced neural differentiation of P19 cells, *foxm1* mRNA and protein expression is downregulated at early time points and also the mRNA and protein expression of the key pluripotency genes *oct-4*, *nanog*, and *sox2* decreases so that the differentiated P19 cells in EB express neither *foxm1* nor *oct-4* nor *nanog* nor *sox2* (Xie et al., 2010).

Both *oct-4* and *sox2* are direct FOXM1 target genes and *nanog* is a possibly indirect FOXM1 target gene (Fig. 3.2; Table 3.1). FOXM1 transactivates the *oct-4* and *sox2* promoters (Table 3.1). Moreover, FOXM1 increases the *nanog* mRNA and protein expression (Fig. 3.2; Table 3.1).

FOXM1 overexpression **during** the RA-induced P19 cell differentiation prevented the downregulation of the *oct-4*, *sox2*, and *nanog* mRNA levels (Xie et al., 2010). Strikingly, FOXM1 overexpression in EB **after** the neural differentiation of P19 cells induced the reexpression of *oct-4*, *sox2*, and *nanog* mRNA and/or protein in the differentiated P19 cells in EB (Xie et al.,

2010). Hence, the reexpression of FOXM1 in RA-differentiated P19 cells in EB is sufficient to restart the already closed expression of the key pluripotency genes *oct-4*, *sox2*, and *nanog* (Xie et al., 2010).

The ability of FOXM1 to reactivate the silenced *oct-4*, *sox2*, and *nanog* expression was also demonstrated in well differentiated somatic cells because FOXM1 overexpression could reinduce the mRNA expression of *oct-4*, *sox2*, and *nanog* in human newborn fibroblasts, which otherwise did not express these three pluripotency genes (Xie et al., 2010).

This potency of FOXM1 in driving the expression of the transcription factors Oct-4, Sox2, and Nanog, which are the key mediators of ES cell pluripotency and self-renewal (Amabile & Meissner, 2009; Brumbaugh et al., 2011; de Vries et al., 2008; Dejosez & Zwaka, 2012; Geoghegan & Byrnes, 2008; Gonzales & Ng, 2011; Hanna et al., 2010; Hochedlinger & Plath, 2009; Jaenisch & Young, 2008; Li, 2010; Loh et al., 2008, 2011; MacArthur et al., 2009; Ng & Surani, 2011; Niwa, 2007a, 2007b; Ohtsuka & Dalton, 2008; Orkin & Hochedlinger, 2011; Pauklin et al., 2011; Pei, 2009; Ralston & Rossant, 2010; Rossant, 2008; Scheper & Copray, 2009; Silva & Smith, 2008; Yamanaka, 2007, 2008; Young, 2011; Yu & Thomson, 2008; Zhao & Daley, 2008), is a clue to FOXM1's critical involvement in the maintenance of stem cell pluripotency (Fig. 3.2; Xie et al., 2010).

SSEA-1 is a cell surface marker of pluripotent stem cells (Hanna et al., 2010; Pera & Tam, 2010; Yanagisawa, 2011). Another marker of pluripotent stem cells is alkaline phosphatase activity (Hanna et al., 2010; Pera & Tam, 2010).

The role of FOXM1 in maintaining the pluripotency of stem cells was confirmed by its effects on the alkaline phosphatase activity and on the cell surface level of SSEA-1 (Xie et al., 2010):

Silencing of FOXM1 with siRNA in undifferentiated P19 EC cells resulted in decreased activity of alkaline phosphatase and in a decline of the SSEA-1 surface level (Xie et al., 2010). Conversely, reexpression of FOXM1 in RA-differentiated P19 cells in EB caused both a higher alkaline phosphatase activity and a significant increase in the SSEA-1 surface level (Xie et al., 2010). Accordingly, FOXM1 overexpression in human newborn fibroblasts led to higher alkaline phosphatase activity (Xie et al., 2010).

Thus, the effects of FOXM1 on the alkaline phosphatase activity and the cell surface level of SSEA-1 indicate that FOXM1 knockdown causes undifferentiated P19 EC cells to lose their pluripotency whereas FOXM1-overexpressing differentiated cells acquire some characteristics similar to pluripotent stem cells, as observed in RA-differentiated P19 cells in EB and in well-differentiated somatic cells (Xie et al., 2010). These results suggest that FOXM1 is important for the maintenance of stem cell pluripotency (Fig. 3.2; Xie et al., 2010).

In summary, FOXM1 plays a role in the maintainance of stem cell pluripotency (Fig. 3.2) (Xie et al., 2010).

8.12. Stem cell self-renewal

FOXM1 may be important for stem cell self-renewal because it had a positive effect on the self-renewal capacity of stem cells and cancer cells (Fig. 3.2; Bao et al., 2011; Wang, Park, et al., 2011; Zhang et al., 2011).

FOXM1 overexpression increased the formation of pancreatospheres in pancreatic cancer cells (Bao et al., 2011) and augmented the formation of neurospheres in glioma cells (Zhang et al., 2011). Conversely, *foxm1* deficiency decreased the formation of neurospheres in neural cortical stem/progenitor cells (Wang, Park, et al., 2011), and FOXM1 knockdown reduced the formation of neurospheres in GIC (GBM-initiating cell) cells, which could be rescued by FOXM1 reexpression (Zhang et al., 2011).

Thus, a positive effect of FOXM1 on the self-renewal capacity was reported for several types of stem cells or cancer cells (Fig. 3.2), namely for: AsPC-1 human pancreatic cancer cells:

Stable overexpression of FOXM1 in AsPC-1 cells enhanced the formation of pancreatospheres, indicating a positive effect of FOXM1 on their self-renewal capacity (Bao et al., 2011).

Accordingly, stable FOXM1 overexpression increased the protein levels of the CSC (cancer stem cell) surface markers CD44 and EpCAM in AsPC-1 cells (Bao et al., 2011).
SW1783 human glioma cells:

Stable overexpression of FOXM1 in SW1783 cells augmented the neurosphere formation efficiency, which points to a positive effect of FOXM1 on their self-renewal capacity (Zhang et al., 2011).

Accordingly, stable FOXM1 overexpression elevated the protein level of the neuroprogenitor markers Nestin and SSEA-1 in SW1783 cells (Zhang et al., 2011).
Neural cortical stem/progenitor cells:

The conditional deletion of *foxm1* in neural cortical stem/progenitor cells reduced the frequency of newly formed primary and secondary neurospheres, showing that *foxm1* loss impairs their self-renewal capacity (Wang, Park, et al., 2011).

Accordingly, this conditional *foxm1* knockout decreased the protein expression of Nestin, a NSC marker (Wang, Park, et al., 2011).

MD11 GIC cells:

ShRNA-mediated knockdown of FOXM1 in MD11 cells substantially decreased the size and number of neurospheres formed in primary and secondary sphere formation assays, whereas overexpression of shRNA-resistant FOXM1 restored the secondary neurosphere formation ability of these MD11 cells with FOXM1 knockdown (Zhang et al., 2011). Thus, FOXM1 is important for the GIC self-renewal capacity (Zhang et al., 2011).

Accordingly, FOXM1 knockdown by shRNA decreased the protein expression of the neuroprogenitor markers CD133, Nestin, Musashi-1, and Sox2 but increased the protein levels of the differentiation markers GFAP (astrocytic marker) and Tuj1 (neuronal class III β-tubulin) (neuronal marker), whereas overexpression of shRNA-resistant FOXM1 in these FOXM1 knockdown cells restored the protein expression of CD133, Nestin, Musashi-1, and Sox2 but suppressed the GFAP protein expression (Zhang et al., 2011).

The effect of FOXM1 on the self-renewal potential of GIC cells was also determined by NCFCA (neural colony-forming cell assay) (Zhang et al., 2011), a more stringent test for the presence of self-renewing cells (Louis et al., 2008). shRNA-mediated knockdown of FOXM1 reduced the efficiency of neural colony formation in GIC cells, whereas overexpression of shRNA-resistant FOXM1 rescued the NCFCA ability of these GIC cells with FOXM1 knockdown (Zhang et al., 2011). Hence, FOXM1 is important for the GIC self-renewal capacity (Zhang et al., 2011).

In summary, FOXM1 has a positive effect on the self-renewal capacity of stem cells and cancer cells, which suggests that FOXM1 may be important for stem cell self-renewal (Fig. 3.2; Bao et al., 2011; Wang, Park, et al., 2011; Zhang et al., 2011).

8.13. Balance between progenitor cell renewal and the commitment to differentiation

Zhang et al. (2011) concluded that FOXM1 perturbs the balance between progenitor cell renewal and the commitment to differentiation by favoring GIC self-renewal and disfavoring GIC differentiation.

Accordingly, Gemenetzidis et al. (2010) suggested that FOXM1 might favor the clonal expansion of epithelial stem/progenitor cells at the expense of their commitment to terminal differentiation.

However, FOXM1 was unable to revert the differentiated phenotype of committed keratinocytes and the onset of epithelial differentiation rendered primary oral keratinocytes refractory to FOXM1-induced proliferation (Gemenetzidis et al., 2010). This suggests that FOXM1 may not inhibit the differentiation process but that the ability of FOXM1 to drive cell cycle progression may interfere with the final exit from the cell cycle, which is a prerequisite for terminal differentiation (Brown, Hughes, & Michell, 2003; Buttitta & Edgar, 2007; Lipinski & Jacks, 1999; Miller, Yeh, Vidal, & Koff, 2007; Myster & Duronio, 2000; Scott, Tzen, Witte, Blatti, & Wang, 1993; Yee, Shih, & Tevosian, 1998; Zhu & Skoultchi, 2001).

Since FOXM1 promotes both stem cell self-renewal (Bao et al., 2011; Wang, Park, et al., 2011; Zhang et al., 2011) and the maintenance of stem cell pluripotency (Xie et al., 2010) it is expected to influence the balance between progenitor cell renewal and their commitment to differentiation in favor of clonal expansion of stem/progenitor cells but in disfavor of terminal differentiation (Fig. 3.2), as described (Carr et al., 2012; Gemenetzidis et al., 2010; Zhang et al., 2011). The same result was also obtained during mammary luminal differentiation, where FOXM1 increased the numbers of luminal progenitors and mammary stem cells but decreased the number of differentiated luminal cells (see below) (Carr et al., 2012).

8.14. Cellular differentiation
8.14.1 Inhibition of cellular differentiation
The roles of FOXM1 in maintenance of stem cell pluripotency (Xie et al., 2010), in stem cell self-renewal (Bao et al., 2011; Wang, Park, et al., 2011; Zhang et al., 2011), and in balancing progenitor cell renewal with commitment to differentiation (Fig. 3.2; Carr et al., 2012; Gemenetzidis et al., 2010; Zhang et al., 2011) suggest that FOXM1 may counteract cellular differentiation. In particular, FOXM1 activates the expression of the key pluripotency factors Oct-4, Sox2, and Nanog (Fig. 3.2; Table 3.1), which generally inhibit the expression of lineage-specific genes that are expressed in terminally differentiated cells (Brumbaugh et al., 2011; Colman & Dreesen, 2009; de Vries et al., 2008; Hochedlinger & Jaenisch, 2006; Hochedlinger & Plath, 2009; Jaenisch & Young, 2008; Keith & Simon, 2007; Loh et al., 2008, 2011; MacArthur et al., 2009; Niwa, 2007a; Orkin & Hochedlinger, 2011; Ralston & Rossant, 2010; Scheper & Copray, 2009; Silva & Smith, 2008; Yamanaka, 2008; Young, 2011; Zhao & Daley, 2008).

Three observations point to an inhibitory effect of FOXM1 on cellular differentiation:

First, teratomas formed with pluripotent P19 EC cells, which had been infected with siRNA against FOXM1 (AdFoxm1siRNA), revealed that knockdown of FOXM1 results in the spontaneous differentiation of P19 cells to mesodermal derivatives, such as muscle, adipose connective tissue, and cardiomyocytes (see above) (Xie et al., 2010). This finding suggests that FOXM1 suppresses the differentiation of mesoderm-derived tissues (Xie et al., 2010).

Second, siRNA-mediated depletion of FOXM1 caused a significant increase of the neuronal differentiation phenotype in BE(2)-C neuroblastoma cells so that FOXM1 deficiency leads to their spontaneous differentiation (Wang, Park, et al., 2011). Accordingly, silencing of FOXM1 by siRNA raised the protein levels of the neuronal differentiation markers NF-M and tubulin β III in BE(2)-C cells (Wang, Park, et al., 2011). These findings indicate that FOXM1 impedes neuronal differentiation (Wang, Park, et al., 2011).

Third, during the IFN-β/mezerein-induced terminal differentiation of human HO-1 metastatic melanoma cells, overexpression of N-terminally truncated FOXM1 interfered with both the G1 cell cycle arrest and the downregulation of some genes, for example, *CDC45L* (CDC45-like, CDC45), *CKS1B* (Cks1), *GINS2* (GINS (Go, Ichi, Nii, San) complex subunit 2), *GMNN* (geminin), *KIAA0101* (PAF (PCNA-associated factor)), *PEG10* (paternally expressed gene 10), *H2AFZ* (H2A histone family, member Z), and *TMPO* (thymopoietin) (Huynh et al., 2010). However, the sustained expression of these mostly cell cycle-related genes could be a secondary effect resulting from the ongoing cell cycle in the presence of overexpressed, N-terminally truncated FOXM1. Nonetheless, the ability of FOXM1 to prevent the IFN-β/mezerein-induced cell cycle arrest upon terminal differentiation of HO-1 cells (Huynh et al., 2010) represents an obstacle to differentiation because permanent withdrawal from the cell cycle is a prerequisite for terminal differentiation (Brown et al., 2003; Buttitta & Edgar, 2007; Lipinski & Jacks, 1999; Miller et al., 2007; Myster & Duronio, 2000; Scott et al., 1993; Yee et al., 1998; Zhu & Skoultchi, 2001). Thus, the potency of FOXM1 in stimulation of proliferation and cell cycle progression might manifest in an inhibitory effect of FOXM1 on terminal differentiation because FOXM1 hampers the required final cell cycle exit.

8.14.2 Inhibition of mammary luminal differentiation

FOXM1 inhibits mammary luminal differentiation in virgin mice (Carr et al., 2012): conditional deletion of *foxm1* in adult mammary tissue increased the number of differentiated luminal cells but decreased the numbers of luminal progenitors and mammary stem cells, indicating that FOXM1 inhibits mammary luminal differentiation (Carr et al., 2012). Conversely, overexpression of FOXM1 in mammary glands expanded the luminal progenitor pool and the mammary stem cell population but reduced the percentage of differentiated luminal cells, again pointing to an inhibition of mammary luminal differentiation by FOXM1 (Carr et al., 2012). Accordingly, *foxm1* mRNA expression declines during mammary luminal differentiation so that the *foxm1* mRNA level was approximately 50-fold and 10-fold higher in luminal progenitors or mammary stem cells than in differentiated luminal cells, respectively (Carr et al., 2012).

The mRNA expression of the direct FOXM1 target gene *GATA3* is repressed by FOXM1 through a RB-dependent and DNMT-dependent mechanism, which involves the FOXM1-induced methylation of the *GATA3* promoter as well as the FOXM1-mediated recruitment of both RB and DNMT3b to the *GATA3* promoter (Carr et al., 2012).

The transcription factor GATA3 is a master regulator of luminal differentiation, which is required for proper mammary gland development, for differentiation of the luminal epithelial lineage, and for maintenance of the differentiated state of the adult mammary gland (Asselin-Labat et al., 2007; Kouros-Mehr et al., 2006).

Accordingly, the FOXM1-mediated repression of the *GATA3* expression plays a crucial role in the inhibition of mammary luminal differentiation by FOXM1 (Carr et al., 2012): shRNA-mediated knockdown of GATA3 was sufficient to reverse the loss of luminal progenitors and expansion of differentiated luminal cells observed after loss of *foxm1* (Carr et al., 2012). Vice versa, coexpression of GATA3 reversed the shift away from the differentiated luminal state and toward the luminal progenitor state found after overexpression of FOXM1 (Carr et al., 2012). Hence, repression of *GATA3* transcription seems to be the dominant mechanism by which FOXM1 inhibits the differentiation of luminal epithelial progenitors (Carr et al., 2012).

In addition, RB has a central role in the inhibition of mammary luminal differentiation by FOXM1 because shRNA-mediated silencing of RB alleviated the expansion of the luminal progenitor pool and the loss of differentiated luminal cells observed after FOXM1 overexpression (Carr et al., 2012). Thus, FOXM1 relies on RB for the inhibition of mammary luminal differentiation (Carr et al., 2012).

In summary, FOXM1 regulates mammary luminal cell fate through inhibition of mammary luminal differentiation (Carr et al., 2012). FOXM1 prevents the terminal differentiation of mammary luminal progenitor cells and is required to maintain the mammary luminal progenitor pool (Carr et al., 2012).

Accordingly, it is highly expressed in luminal epithelial progenitors (Carr et al., 2012).

8.14.3 Stimulation of lobuloalveolar differentiation

In contrast to the inhibition of mammary luminal differentiation by FOXM1 in virgin mice, FOXM1 promotes alveolar differentiation during pregnancy (Carr et al., 2012): deletion of *foxm1* in epithelial cells as they differentiate during pregnancy results in a delay in lobuloalveolar differentiation during the second pregnancy (Carr et al., 2012). This delay in lobuloalveolar differentiation in *foxm1*-deficient mammary glands suggests that FOXM1 stimulates lobuloalveolar differentiation during pregnancy (Carr et al., 2012). Thus, FOXM1 participates in alveolar cell differentiation so that loss of *foxm1* leads to defects in mammary development during pregnancy, namely, to a delay in the development of lobuloalveolar structures (Carr et al., 2012).

Accordingly, the *foxm1* RNA and protein expression in the mammary gland increases during pregnancy but redecreases during lactation (Carr et al., 2012).

8.14.4 Versatile regulation of cellular differentiation

As outlined above, FOXM1 exerts opposite effects on cellular differentiation in the mammary gland because it inhibits mammary luminal differentiation in virgin mice but promotes lobuloalveolar differentiation during pregnancy (Carr et al., 2012).

Consequently, FOXM1 cannot be considered a general inhibitor of differentiation as one would expect at first glance. Rather, FOXM1 seems to affect differentiation in a context-dependent manner so that it might be better described as a versatile regulator of cellular differentiation (Fig. 3.2).

This view is supported by some findings in *foxm1* knockout mice (Kalin et al., 2008; Ueno et al., 2008; Ustiyan et al., 2012), in a mouse model of asthma (Ren et al., 2013), and during injury-induced adult lung regeneration (Liu et al., 2011), where FOXM1 has been reported either to play a role in the transdifferentiation between two cell types (Liu et al., 2011; Ustiyan et al., 2012) or to facilitate the differentiation of a cell type (Kalin et al., 2008; Ren et al., 2013; Ueno et al., 2008).

8.15. Apoptosis and cellular survival
8.15.1 General conclusions
So far, a clear-cut function for FOXM in either apoptosis or cell survival has not been established (Wierstra & Alves, 2007c). On the one hand, a few studies described a positive effect of FOXM1 on apoptosis (Gusarova et al., 2007; Kalinichenko et al., 2004; Teh et al., 2010). On the other hand, several studies reported a negative effect of FOXM1 on apoptosis (Ahmed et al., 2012; Bhat et al., 2009a, 2009b; Calvisi et al., 2009; Carr et al., 2010; Halasi & Gartel, 2012; Huynh et al., 2010; Kwok et al., 2010; Lefebvre et al., 2010; McGovern et al., 2009; Mencalha et al., 2012; Millour et al., 2011; Ning et al., 2012; Pandit & Gartel, 2011a, 2011b; Park, Carr, et al., 2009; Priller et al., 2011; Radhakrishnan et al., 2006; Ren et al., 2010; Uddin et al., 2011, 2012; Ustiyan et al., 2009; Wan et al., 2012; Wang, Ahmad, Banerjee, et al., 2010; Wonsey & Follettie, 2005; Xia, Mo, et al., 2012; Xu, Zhang, et al., 2012; Xue et al., 2010; Zhao et al., 2006). In contrast, many other studies found that FOXM1 overexpression did not promote cellular survival (Bhat et al., 2009b; Carr et al., 2012; Huynh et al., 2010; Kalinina et al., 2003; McGovern et al., 2009; Ning et al., 2012; Xu, Zhang, et al., 2012; Yoshida et al., 2007) and that FOXM deficiency did not cause cell death (Balli et al., 2012; Carr et al., 2012; Chen, Yang, et al., 2012; Dibb et al., 2012; Gusarova et al., 2007; Halasi & Gartel, 2012; Kalin et al., 2008; Kalinichenko et al., 2004; Kim et al., 2005; Krupczak-Hollis et al., 2004; Millour et al., 2011; Nakamura, Hirano, et al., 2010; Pandit & Gartel, 2011a, 2011b; Priller et al., 2011; Ramakrishna et al., 2007; Schüller et al., 2007; Tan et al., 2007; Ustiyan et al., 2009, 2012; Wang, Meliton, et al., 2009; Wang, Park, et al., 2011; Wang et al., 2005; Wonsey & Follettie, 2005; Xie et al., 2010; Xu, Zhang, et al., 2012; Xue et al., 2010; Yoshida et al., 2007; Zhang, Ackermann, et al., 2006, 2009; Zhao et al., 2006). Thus, most studies showed that FOXM1 affects neither apoptosis nor cell survival.

Often the positive effects of FOXM1 on cellular survival were detected in tumor cells treated with anticancer drugs (Ahmed et al., 2012; Bhat et al., 2009a, 2009b; Carr et al., 2010; Halasi & Gartel, 2012; Kwok et al., 2010;

McGovern et al., 2009; Millour et al., 2011; Ning et al., 2012; Pandit & Gartel, 2011a, 2011b; Uddin et al., 2011; Xu, Zhang, et al., 2012), suggesting that FOXM1 overexpression in cancer cells contributes to their chemotherapeutic drug resistance (see Part II of this two-part review, that is see Wierstra, 2013b) (Koo et al., 2011; Raychaudhuri & Park, 2011; Teh, 2012; Wilson et al., 2011; Zhao & Lam, 2012).

Since the intracellular signaling circuitry of cancer cells differs from that of normal cells (Weinstein, 2000) the totality of all studies leads to the conclusion that FOXM1 has no strong net effect on cell viability and that the few effects observed may rather be due to the intrinsic balance of pro- and antiapoptotic signals in the individual system analyzed, in particular in the cancer cell lines often used. This view is underscored by the spectrum of known FOXM1 target genes (Fig. 3.2; Table 3.1) that includes proapoptotic as well as antiapoptotic factors (Fig. 3.12), which may reciprocally negate each other so that the apoptotic threshold is never exceeded and thus apoptosis does not occur, just as found in most studies with FOXM1 (Balli et al., 2012; Bhat et al., 2009b; Carr et al., 2012; Chen, Yang, et al., 2012; Dibb et al., 2012; Gusarova et al., 2007; Halasi & Gartel, 2012; Huynh et al., 2010; Kalin et al., 2008; Kalinichenko et al., 2004; Kalinina et al., 2003; Kim et al., 2005; Krupczak-Hollis et al., 2004; McGovern et al., 2009; Millour et al., 2011; Nakamura, Hirano, et al., 2010; Ning et al., 2012; Pandit & Gartel, 2011a, 2011b; Priller et al., 2011; Ramakrishna et al., 2007; Schüller et al., 2007; Tan et al., 2007; Ustiyan et al., 2009, 2012; Wang, Meliton, et al., 2009; Wang, Park, et al., 2011; Wang, Zhang, Snyder, et al., 2010; Wang et al., 2005; Wonsey & Follettie, 2005; Xie et al., 2010; Xu, Zhang, et al., 2012; Xue et al., 2010; Yoshida et al., 2007; Zhang, Ackermann, et al., 2006, 2009; Zhao et al., 2006).

8.15.2 Stimulation of apoptosis

The following results indicate a positive effect of FOXM1 on apoptosis, suggesting that FOXM1 may promote cell death (Gusarova et al., 2007; Kalinichenko et al., 2004; Teh et al., 2010):

TUNEL (TdT (terminal deoxynucleotidyl transferase)-mediated dUTP (desoxy-uridine-triphosphate) nick end labeling) staining on liver sections from either 6-week-old male conditional *foxm1* knockout mice with a hepatocyte-specific *foxm1* deletion or control mice revealed that the percentage of apoptotic *foxm1*-deficient hepatocytes was lower than the

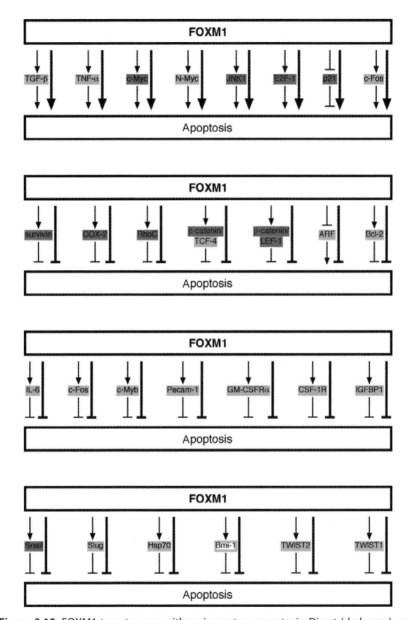

Figure 3.12 FOXM1 target genes with an impact on apoptosis. Direct (dark gray), possibly indirect (light gray), and probably indirect (light gray border) FOXM1 target genes are indicated. Thin arrows and thin perpendicular ends depict the known FOXM1-mediated regulation of FOXM1 target genes and the known proapoptotic or antiapoptotic functions of the encoded proteins. Thick arrows and thick perpendicular ends indicate the hypothetic ensuing effects of FOXM1 on apoptosis. c-Fos is shown twice because it can promote either apoptosis or cellular survival depending on the cellular context. RhoC suppresses apoptosis in a cell type-specific manner.

percentage of apoptotic control hepatocytes, indicating a positive effect of FOXM1 apoptosis (Kalinichenko et al., 2004).

UVB irradiation induced apoptosis in primary NHEK (normal human epidermal keratinocytes) as evidenced by PARP (poly(ADP (adenosine diphosphate)-ribose)polymerase) cleavage and a larger sub-G1 population (Teh et al., 2010). Overexpression of FoxM1B exacerbated the UVB-induced apoptosis of primary NHEK because it increased both the sub-G1 population and the level of cleaved PARP, which points to a positive effect of FoxM1B on apoptosis (Teh et al., 2010).

However, without UVB exposure, apoptosis rate of primary NHEK was (almost) unaltered by FoxM1B overexpression, suggesting that FoxM1B alone has no (significant) negative effect on cellular survival (Teh et al., 2010).

A cell-penetrating ARF peptide (aa 26–44 of ARF plus nine N-terminal D-arginine residues), which encompasses the FOXM1 interaction domain of ARF (Kalinichenko et al., 2004), relocalizes FOXM1 to the nucleolus (Gusarova et al., 2007; Kalinichenko et al., 2004) and represses the transcriptional activity of exogenous FoxM1B (Kalinichenko et al., 2004).

This ARF peptide induced apoptosis (measured as TUNEL staining) in human hepatoma HepG2 cells, whereas the extent of ARF peptide-induced apoptosis was considerably reduced by siRNA-mediated depletion of FOXM1 (Gusarova et al., 2007). Since the ARF-dependent apoptosis of HepG2 cells requires FOXM1, FOXM1 seems to promote apoptosis (Gusarova et al., 2007).

Yet, the FOXM1 knockdown did not change the percentage of apoptotic HepG2 cells without ARF peptide treatment, suggesting that FOXM1 alone has no negative effect on cellular survival (Gusarova et al., 2007).

DEN/PB (diethylnitrosamine/phenobarbital) exposure creates HCC and hepatic adenomas (Lee, 2000; Pitot, Dragan, Teeguarden, Hsia, & Campbell, 1996; Sargent et al., 1996; Tamano, Merlino, & Ward, 1994). The ARF peptide caused apoptosis in many tumor cells of the DEN/PB-induced HCC and hepatic adenomas, whereas no apoptotic cells were found in the *foxm1*-deficient DEN/PB-induced liver tumors of *foxm1*$^{-/-}$ mice, in which the *foxm1* gene had been conditionally deleted in preexisting, DEN/PB-induced HCC and hepatic adenomas (Gusarova et al., 2007). Hence, also the ARF-dependent apoptosis of liver tumor cells *in vivo* requires FOXM1, which again indicates that FOXM1 promotes apoptosis (Gusarova et al., 2007).

8.15.3 Stimulation of cell survival

The following results, which were obtained without anticancer drug treatment, indicate a negative effect of FOXM1 on apoptosis, suggesting that FOXM1 may promote cellular survival (Ahmed et al., 2012; Calvisi et al., 2009; Halasi & Gartel, 2012; Huynh et al., 2010; Lefebvre et al., 2010; Mencalha et al., 2012; Ning et al., 2012; Park, Carr, et al., 2009; Priller et al., 2011; Ren et al., 2010; Uddin et al., 2011, 2012; Ustiyan et al., 2009; Wan et al., 2012; Wang, Ahmad, Banerjee, et al., 2010; Wonsey & Follettie, 2005; Xia, Mo, et al., 2012; Xue et al., 2010; Zhao et al., 2006):

FOXM1 depletion by shRNA resulted in reduced cell survival of BT-20 cells (Wonsey & Follettie, 2005).

Primary mouse lung endothelial *foxm1* knockout cells died after several days in culture (Zhao et al., 2006).

In human HCC cell lines, apoptosis was reduced by FOXM1 overexpression (HuH6 and HLE cells) but increased by siRNA-mediated depletion of FOXM1 (SNU-182 cells) (Calvisi et al., 2009).

FOXM1-overexpressing mouse NIH3T3 fibroblasts were more resistant to H_2O_2-induced cell death and, conversely, human IMR90 primary fibroblasts depleted of FOXM1 by siRNA were more sensitive to H_2O_2-induced cell death (Park, Carr, et al., 2009).

Newborn $smFoxm1^{-/-}$ mice with a smooth muscle-specific conditional *foxm1* knockout displayed an increase in apoptosis in lung, pulmonary blood vessels, and esophagus (Ustiyan et al., 2009).

Cultured *foxm1*-deficient DP (double positive) thymocytes were slightly more sensitive to apoptosis (Xue et al., 2010). However, no changes in differential apoptosis were observed when apoptotic stimuli (α-CD3/CD28 antibodies, α-Fas antibody, phorbol ester PMA, dexamethasone) were applied to these *foxm1*-deficient DP thymocytes, indicating that *foxm1*-deficient DP thymocytes display normal levels of apoptosis (Xue et al., 2010).

In human MIA PaCa-2 pancreatic cancer cells, apoptosis was decreased by FOXM1 overexpression but increased by siRNA-mediated silencing of FOXM1 expression (Wang, Ahmad, Banerjee, et al., 2010).

Silencing of FOXM1 expression by shRNA increased the percentage of apoptotic ST468 cells, a Burkitt's lymphoma cell line (Lefebvre et al., 2010).

IFN-β+mezerein induced massive apoptosis in human HO-1 metastatic melanoma cells and a ca (N-terminally truncated) FOXM1 mutant, but not full-length FOXM1, diminished the percentage of apoptotic HO-1 cells induced by IFN-β+mezerein treatment (Huynh et al., 2010).

Mice with a knockout of *foxm1* in myeloid cells showed increased hepatic apoptosis in pericentral liver regions after CCl_4-mediated liver injury (Ren et al., 2010).

siRNA-mediated depletion of FOXM1 increased the cleavage of both caspase-3 and PARP in HCT-15 cells (Uddin et al., 2011).

shRNA-mediated knockdown of FOXM1 increased the cleavage of caspase-3 and decreased the cell viability in four medulloblastoma cell lines (D425med, DAYO, R300, UW228) (Priller et al., 2011).

siRNA against FOXM1 increased the cleavage of caspase-3 and PARP in BCPAP and TPC-1 PTC (papillary thyroid carcinoma) cells (Ahmed et al., 2012). Accordingly, siRNA against FOXM1 increased the percentage of cell death in TPC-1 cells (Ahmed et al., 2012).

Knockdown of FOXM1 by siRNA increased the percentage of apoptotic SUDHL4 and HBL-1 human DLBCL (diffuse large B-cell lymphoma) cells (Uddin et al., 2012). Accordingly, siRNA against FOXM1 increased the cleavage of caspase-3 and caspase-9 but decreased the level of inactivated, phosphorylated Bad (Bcl-2 antagonist of cell death, BCL2L8 (Bcl-2-like 8)), a proapoptotic Bcl-2 (B-cell lymphoma-2) family member, in SUDHL4 and OCI-LY19 cells (Uddin et al., 2012).

shRNA-mediated knockdown of FOXM1 increased the IR-induced apoptosis in MDA-MB-231 breast cancer cells and MIA Paca-2 pancreatic cancer cells, indicating that FOXM1 deficiency sensitizes these cells to DNA damage-induced apoptosis (Halasi & Gartel, 2012). Accordingly, shRNA against FOXM1 increased the IR-induced caspase-3 cleavage in MIA Paca-2 and MDA-MB-231 cells (Halasi & Gartel, 2012). However, without IR exposure of MDA-MB-23 and MIA Paca-2 cells, that is, in the absence of IR-induced DNA damage, shRNA against FOXM1 had no effect on cellular survival and caspase-3 cleavage (Halasi & Gartel, 2012).

Silencing of FOXM1 by siRNA increased the percentage of apoptotic human SKOV3 and CoC1 ovarian cancer cells (Ning et al., 2012).

The percentage of apoptotic HepG2 HCC cells was decreased by FOXM1 overexpression but increased by shRNA-mediated knockdown of FOXM1 (Xia, Mo, et al., 2012).

Fas-induced apoptosis of HepG2 cells was exacerbated by shRNA against FOXM1 but mitigated by FOXM1 overexpression (Xia, Mo, et al., 2012).

Knockdown of FOXM1 by shRNA in embryonal RD RMS (rhabdomyosarcoma) cells resulted in a significant accumulation of sub-G1 cells, indicating that FOXM1 deficiency leads to apoptosis (Wan et al., 2012).

siRNA-mediated silencing of FOXM1 increased the percentage of apoptotic human K562 CML (chronic myeloid leukemia) cells (Mencalha et al., 2012).

8.16. Autophagy

The proteasome inhibitors MG132, bortezomib, and thiostrepton induced autophagy in MIA Paca-2 pancreatic cancer cells, MDA-MB-231 breast cancer cells, and HCT-116 colon cancer cells (Pandit & Gartel, 2011a). The percentage of these cells undergoing autophagy following treatment with these proteasome inhibitors was (almost) unchanged by shRNA-mediated knockdown of FOXM1, indicating that FOXM1 does not affect proteasome inhibitor-induced autophagy (Pandit & Gartel, 2011a). Thus, FOXM1 knockdown does not sensitize human cancer cells to autophagy induced by proteasome inhibitors (Pandit & Gartel, 2011a).

8.17. Global DNA hypomethylation and focal DNA hypermethylation

Analysis of genome-wide promoter methylation in primary NOKs (normal oral human keratinocytes) overexpressing FoxM1B revealed that FoxM1B causes global hypomethylation as well as focal hypermethylation (affecting individual genes) (Fig. 3.2; Teh et al., 2012).

The FoxM1B-induced methylation pattern in NOKs resembled that of the HNSCC cell line SCC15, suggesting that FOXM1 causes a DNA methylation pattern similar to those found in cancer cells (Teh et al., 2012). This possibility is supported by the fact that both global hypomethylation and focal hypermethylation are typical DNA methylation patterns found in cancer (Baylin & Jones, 2011; Dawson & Kouzarides, 2012; Pujadas & Feinberg, 2012; Tsai & Baylin, 2011). It is tempting to speculate that FOXM1 might upregulate the expression of oncogenes through their promoter hypomethylation and downregulate the expression of tumor-suppressor genes through their promoter hypermethylation (Teh et al., 2012).

In fact, FOXM1 causes hypermethylation at the promoters of two tumor-suppressor genes, namely, *p16* (Teh et al., 2012) and *GATA3* (Carr et al., 2012). Accordingly, overexpression of FOXM1 led to the expected downregulation of the expression of *p16* (Teh et al., 2012) and *GATA3* (Carr et al., 2012).

First, FoxM1B overexpression induced hypermethylation of the *p16* promoter in NOKs and reduced the *p16* mRNA and protein levels in primary human oral keratinocytes (Teh et al., 2012).

Second, overexpression of FOXM1 induced methylation of the *GATA3* promoter in MCF-7 mammary tumor cells, and the *GATA3* mRNA and/or protein expression was decreased by FOXM1 overexpression in MCF-7 cells and mouse mammary glands but increased in *foxm1*-deficient mammary glands of conditional *foxm1* knockout mice (Carr et al., 2012).

Thus, the tumor-suppressor genes *p16* (Teh et al., 2012) and *GATA3* (Carr et al., 2012) are two examples for gene repression through FOXM1-induced promoter methylation.

There are two principal possibilities of how FOXM1 could cause the global hypomethylation and focal hypermethylation (Fig. 3.2), which were found in the genome-wide promoter methylation profiling in FoxM1B-overexpressing primary NOKs (Teh et al., 2012):

First, FOXM1 could bind to an individual promoter and recruit either a DNMT or an appropriate enzyme/protein, which triggers DNA demethylation (possibility 1).

Second, FOXM1 could activate those genes, which encode DNMTs and enzymes/proteins for DNA demethylation (possibility 2). Moreover, FOXM1 could activate the genes for transcription factors and other factors, which recruit DNA-methylating or DNA-demethylating enzymes to target genes (possibility 2).

Recently, genome-wide ChIP-seq assays with an α-FOXM1 antibody in human U2OS osteosarcoma cells accumulated in late G2- and M-phase showed that FOXM1 *in vivo* occupies only 270 binding regions (Chen, Müller, et al., 2013). The majority of these FOXM1-binding regions is located in close proximity to promoters, with 74% either in the 5′-UTR or within 1 kb upstream of the transcription start site (Chen, Müller, et al., 2013). In comparison to other transcription factors, the number of FOXM1-binding sites in the human genome is extremely low, at least in late G2- and M-phase (Chen, Müller, et al., 2013).

Therefore, at least in late G2- and M-phase, FOXM1 could not bring about global promoter hypomethylation (Teh et al., 2012) by binding to the affected genes and recruiting enzymes/proteins for DNA demethylation (possibility 1) so that FOXM1 is most likely to activate those genes, which encode DNMTs and DNA-demethylating enzymes as well as (transcription) factors for their recruitment (possibility 2). Nevertheless, FOXM1 could effectuate focal promoter hypermethylation (Teh et al., 2012) at some genes by binding to their promoters and recruiting DNMTs (possibility 1), namely, maximally at 200 promoters in late G2- and M-phase (i.e., 74% of the 270 FOXM1-binding regions (Chen, Müller, et al., 2013)).

The *p16* promoter represents an example for possibility 2 because it was not occupied *in vivo* by FOXM1 in oral keratinocytes, not even in FoxM1B-overexpressing oral keratinocytes (Gemenetzidis et al., 2009).

In contrast, the *GATA3* promoter represents an example for the first possibility 1 because FOXM1 *in vivo* occupied the *GATA3* promoter in mammary glands and MDA-MB-453 breast cancer cells and because FOXM1 recruited DNMT3b to the *GATA3* promoter in MDA-MB-453 cells (Carr et al., 2012). Hence, FOXM1 represses the direct FOXM1 target gene *GATA3* by inducing its DNMT3b-mediated promoter hypermethylation (Carr et al., 2012).

However, RB is essentially required for both the repression of the *GATA3* mRNA expression by FOXM1 and the FOXM1-induced methylation of the *GATA3* promoter so that actually RB represses the *GATA3* transcription and triggers the *GATA3* promoter methylation, presumably through its interaction with DNTM3b, whereas FOXM1 is only necessary for the recruitment of RB and DNMT3b to the *GATA3* promoter (Carr et al., 2012). In fact, in RB-deficient cells, FOXM1 alone enhances the *GATA3* mRNA expression and reduces the methylation at the *GATA3* promoter, clearly demonstrating that FOXM1 itself is not capable of mediating gene repression or promoter methylation of *GATA3* (Carr et al., 2012).

Thus, FOXM1 represses the *GATA3* transcription through RB-dependent, DNMT3b-mediated methylation of the *GATA3* promoter (Carr et al., 2012).

In summary, FOXM1 causes global DNA hypomethylation and focal DNA hypermethylation in promoter regions (Fig. 3.2; Teh et al., 2012). Yet, the very limited number of FOXM1-binding sites in the human genome in late G2- and M-phase (Chen, Müller, et al., 2013) excludes the direct association of FOXM1 with most of the affected loci so that FOXM1 must act indirectly through activation of genes, which encode DNMTs and DNA-demethylating enzymes as well as (transcription) factors for their recruitment.

8.18. Heat shock response

The heat shock response, which is induced by heat shock and other types of stress, represents a specific gene expression program that enables cells to survive and recover from otherwise lethal conditions, e.g. severe heat stress (thermotolerance) (Buchberger, Bukau, & Sommer, 2010; Craig, 1985; Craig & Gross, 1991; Daugaard, Rohde, & Jäättela, 2007; Diller, 2006; Feder & Hofman, 1999; Gabai & Sherman, 2002; Garrido et al., 2006; Garrido, Gurguxani, Ravagnan, & Kroemer, 2001; Garrido et al., 2003; Gidalevitz, Prahlad, & Morimoto, 2011; Gupta, Sharma, Mishra, Mishra, & Chowdhuri, 2010; Hartl, Bracher, & Hayer-Hartl, 2011; Hightower, 1991; Jäättelä, 1999; Jäättelä & Wissing, 1992; Jolly & Morimoto, 2000; Kampinga et al., 2009; Kao & Nevins, 1983; Kiang & Tsokos, 1998; Kim, Hwang, & Lee, 2007; Kregel, 2002; Kültz, 2005; Lanks, 1986; Lindquist, 1986; Lindquist & Craig, 1988; Liu, Gampert, et al., 2006; Morimoto, 1993, 1998, 2008, 2011; Morimoto, Sarge, & Abravaya, 1992; Mosser & Morimoto, 2004; Parsell & Lindquist, 1993; Phillips & Morimoto, 1991; Powers & Workman, 2007; Richter, Haslbeck, & Buchner, 2010; Schmitt, Gehrmann, Brunet, Multhoff, & Garrido, 2007; Shamovsky & Nudler, 2008; Sherman & Goldberg, 2001; Somero, 1995; Tavaria, Gabriele, Kola, & Anderson, 1996; Tyedmers, Mogk, & Bukau, 2010; Voellmy, 1994, 2004; Welch, 1992; Wu, Hunt, & Morimoto, 1985; Wu, Kingston, & Morimoto, 1986). Heat shock activates the latent transcription factor HSF1 that in turn induces the transcription of heat shock proteins and other stress proteins, which protect the cell against damage caused by heat stress (Akerfelt, Morimoto, & Sistonen, 2010; Anckar & Sistonen, 2007, 2011; Björk & Sistonen, 2010; Calderwood et al., 2010; Fujimoto & Nakai, 2010; Powers & Workman, 2007; Sakurai & Enoki, 2010; Shamovsky & Nudler, 2008; Vigh, Horvath, Maresca, & Harwood, 2007; Voellmy & Boellmann, 2007).

Foxm1 is a direct HSF1 target gene (Table 3.2) (Dai et al., 2013). Heat shock induces *foxm1* transcription and thus FOXM1 mRNA and protein expression through HSF1 (Fig. 3.10; Dai et al., 2013). Hence, *foxm1* is a heat shock response gene (Dai et al., 2013).

In fact, FOXM1 protects cells from heat shock stress (Fig. 3.2) by promoting cell cycle progression and cell survival under heat stress conditions (Dai et al., 2013).

Mild heat shock (42 °C) caused a cell cycle arrest in G2/M-phase of human U-87MG and Hs683 glioma cells (Dai et al., 2013). This heat shock-induced G2/M-arrest was exacerbated by siRNA against FOXM1 and alleviated by FOXM1 overexpression in U-87MG or Hs683 cells, respectively, demonstrating that FOXM1 promotes G2/M-phase progression under mild heat stress conditions (Dai et al., 2013).

Lethal heat shock stress (44 °C) led to death of U-87MG cells and siRNA-mediated knockdown of FOXM1 dramatically reinforced this heat shock-induced cell death, suggesting that FOXM1 promotes glioma cell survival under lethal heat stress conditions (Dai et al., 2013).

In summary, FOXM1 protects cells from heat shock stress (Fig. 3.2) because it enables G2/M-phase cell cycle progression and prevents cell death under mild or lethal heat shock conditions, respectively (Dai et al., 2013). Thus, FOXM1 is implicated in the heat shock response (Fig. 3.2) and contributes to thermotolerance downstream of HSF1 (Dai et al., 2013).

8.19. Allergen-induced lung inflammation

The chronic respiratory disease asthma is characterized by persistent pulmonary inflammation, goblet cell metaplasia, airway hyperresponsiveness, and lung remodeling (Bai & Knight, 2005; Carter, Heinly, Yates, & Lieberman, 1997; Durrani, Viswanathan, & Busse, 2011; Ober & Yao, 2011; Sumi & Hamid, 2007; Tagaya & Tamaoki, 2007; Warner & Knight, 2008; Yamauchi, 2006).

The HDM sensitization/challenge protocol is a mouse model of asthma, in which allergen stimulation, that is, exposure to HDM extract, results in lung inflammation, increased airway resistance, T-cell activation, and goblet cell metaplasia (Cates, Fattouh, Johnson, Llop-Guevara, & Jordana, 2007; Fuchs & Braun, 2008; Gregory & Lloyd, 2011; Le Cras et al., 2011; Lewkovich et al., 2008; Plopper & Hyde, 2008; Ren et al., 2013; Stevenson & Birrell, 2011).

HDM challenge induced *foxm1* mRNA and protein expression in mouse lungs (Fig. 3.10), which do otherwise not express FOXM1 (Ren et al., 2013). In HDM-treated lungs, FOXM1 protein was expressed in bronchiolar epithelial cells and inflammatory cells (Ren et al., 2013). Airway epithelial cells expressing FOXM1 protein in response to HDM included Clara cells and subsets of goblet cells (Ren et al., 2013). *foxm1* mRNA expression in inflammatory cells of HDM-challenged lungs was high in macrophages, monocytes, and neutrophils but low in lymphocytes and mDCs (myeloid dendritic cells) as well as undetectable in eosinophils (Ren et al., 2013).

Accordingly, the lungs of patients with severe asthma displayed abundant FOXM1 protein expression in bronchiolar epithelium and inflammatory cells whereas FOXM1 protein is not expressed in normal human lungs (Ren et al., 2013).

In HDM-treated mouse lungs, FOXM1 promotes allergen-induced lung inflammation (Fig. 3.2) and goblet cell metaplasia (Ren et al., 2013).

Since HDM exposure induced FOXM1 expression in bronchiolar epithelial and inflammatory cells the *foxm1* gene was conditionally deleted in *foxm1* knockout mice from either myeloid cells (*LysM-Cre Foxm$^{fl/fl}$*) or airway Clara cells (*CCSP-Foxm1$^{-/-}$*) (Ren et al., 2013). HDM-treated lungs of *LysM-Cre Foxm$^{fl/fl}$* mice revealed efficient deletion of *foxm1* in myeloid cell lineage (macrophages, monocytes, neutrophils, mDCs) and alveolar type II cells but unchanged *foxm1* mRNA levels in lymphocytes (Ren et al., 2013).

HDM challenge caused pulmonary inflammation and accordingly increased airway resistance after methacoline treatment (Ren et al., 2013). Deletion of *foxm1* from either myeloid cells or Clara cells in *LysM-Cre Foxm$^{fl/fl}$* and *CCSP-Foxm1$^{-/-}$* mice, respectively, reduced both lung inflammation and airway resistance after HDM exposure, demonstrating that FOXM1 promotes allergen-induced lung inflammation (Fig. 3.2; Ren et al., 2013).

In response to HDM, both the total number of inflammatory cells and the expression of proinflammatory cytokines were lower in lungs of *LysM-Cre Foxm$^{fl/fl}$* mice compared to control mice (Ren et al., 2013). In more detail, the numbers of macrophages and eosinophils were decreased in HDM-treated *LysM-Cre Foxm$^{fl/fl}$* lungs, whereas those of neutrophils, mDCs, and pDCs (plasmacytoid dendritic cells) remained unchanged (Ren et al., 2013).

Similarly, the total number of inflammatory cells in BALF and the expression of proinflammatory cytokines in lung tissue were diminished in HDM-challenged *CCSP-Foxm1$^{-/-}$* mice in comparison to control mice (Ren et al., 2013). In more detail, HDM-treated *CCSP-Foxm1$^{-/-}$* lungs exhibited decreased numbers of macrophages, eosinophils, and lymphocytes in BALF (Ren et al., 2013).

Hence, FOXM1 expression in both myeloid inflammatory cells and Clara cells of the airway epithelium is critical for allergen-induced lung inflammation in response to HDM exposure (Fig. 3.2; Ren et al., 2013).

Specialized DC populations are critical for antigen presentation and T-cell activation during the pathogenesis of asthma (Bharadwaj, Bewtra, & Agrawal, 2007; Hammad et al., 2010; Holgate, 2012; Lambrecht, 2001a, 2001b; Minnicozzi, Sawyer, & Fenton, 2011; von Garnier & Nicod, 2009; Willart & Hammad, 2010; Willart & Lambrecht, 2009).

FOXM1, which is expressed in mDCs after HDM challenge, plays a role in antigen presentation by mDCs and for the resulting T-cell activation in response to HDM because the *foxm1*-deficient mDCs in HDM-treated *LysM-Cre Foxm1$^{fl/fl}$* lungs were impaired in the activation of T-cells due to decreased cell surface expression of the receptor MHC II and the costimulatory molecule CD86 on these *foxm1*-deficient mDCs (Ren et al., 2013). This diminished T-cell activation by antigen-presenting mDCs contributes to the decreased allergen-induced lung inflammation in *LysM-Cre Foxm1$^{fl/fl}$* mice following HDM exposure (Ren et al., 2013).

HDM challenge caused goblet cell metaplasia (Ren et al., 2013). The number of goblet cells in the airway epithelium of HDM-treated *CCSP-Foxm1$^{-/-}$* mice was reduced in comparison to control mice, showing that FOXM1 promotes allergen-mediated goblet cell metaplasia.

Since the proliferation of airway epithelial cells after HDM exposure was unchanged in *CCSP-Foxm1$^{-/-}$* lungs compared to control lungs FOXM1 may have a cell cycle-independent function in formation of goblet cell metaplasia. In fact, FOXM1 activates the transcription of its direct target gene *spdef* (Fig. 3.2; Table 3.1) so that the conditional deletion of *foxm1* from Clara cells led to decreased SPDEF mRNA and protein expression in HDM-treated *CCSP-Foxm1$^{-/-}$* lungs. Since the transcription factor SPDEF is both necessary and sufficient for differentiation of goblet cells (Chen, Korfhagen, et al., 2009; Curran & Cohn, 2010; Park et al., 2007) FOXM1 may trigger the differentiation of Clara cells into goblet cells by activating SPDEF expression in Clara cells (Ren et al., 2013).

Thus, FOXM1 expression in airway Clara cells is critical for allergen-mediated goblet cell metaplasia following HDM exposure (Ren et al., 2013).

In contrast, FOXM1 expression in myeloid cells is dispensable for differentiation of goblet cells because the deletion of *foxm1* from myeloid cells in *LysM-Cre Foxm1$^{fl/fl}$* mice did not alter goblet cell metaplasia in HDM-treated lungs (Ren et al., 2013).

In summary, FOXM1 promotes allergen-induced lung inflammation (Fig. 3.2) and goblet cell metaplasia in response to HDM challenge in a mouse model of asthma (Ren et al., 2013). The expression of FOXM1 in both myeloid inflammatory cells and Clara cells of the airway epithelium is important for the pathogenesis of this asthma-like phenotype after allergen exposure (Ren et al., 2013).

8.20. Radiation-induced pulmonary fibrosis

Radiation-induced pulmonary fibrosis is a mouse model, in which pulmonary fibrosis is caused by thoracic irradiation (Moore & Hogaboam, 2008). Ionizing irradiation (IR) to the lung damages the epithelium so that epithelial cells release inflammatory mediators that attract inflammatory cells, which in turn secrete profibrotic cytokines and chemokines thereby amplifying the inflammatory response (Chapman, 2011; Hardie et al., 2010; King et al., 2011; Thannickal et al., 2004; Wynn, 2004, 2007, 2008, 2011; Wynn & Barron, 2010). These profibrotic mediators lead to the recruitment, proliferation, and activation of (myo)fibroblasts, which produce ECM proteins (e.g., collagen) resulting in the excess deposition of ECM components (Chapman, 2011; Hardie et al., 2010; King et al., 2011; Thannickal et al., 2004; Wynn, 2004, 2007, 2008, 2011; Wynn & Barron, 2010).

FOXM1 plays a critical role in radiation-induced pulmonary fibrosis by increasing inflammation, inducing proliferation of myofibroblasts, and promoting EMT (Balli et al., 2013).

Radiation-induced pulmonary fibrosis was exacerbated in transgenic *epiFoxm1-ΔN* mice, which conditionally overexpress a ca N-terminally truncated FoxM1B mutant (FoxM1B-ΔN) in the respiratory epithelium (Balli et al., 2013). The *epiFoxm1-ΔN* mice displayed an increased deposition of collagen in lungs after irradiation compared to control animals (Balli et al., 2013). Conversely, the radiation-induced pulmonary fibrosis was mitigated in *epiFoxm1 KO* knockout mice with a conditional deletion of *foxm1* from the pulmonary epithelium (Balli et al., 2013). The irradiated lungs of *epiFoxm1 KO* mice exhibited a decreased collagen deposition in comparison to the lungs of control animals (Balli et al., 2013). Thus, the expression of FOXM1 in the respiratory epithelium plays a role in radiation-induced pulmonary fibrosis so that ablation of *foxm1* protects from lung fibrosis, whereas lung fibrosis is enhanced by ectopic expression of a ca FoxM1B mutant (Balli et al., 2013).

Accordingly, in the mouse model of radiation-induced pulmonary fibrosis, the *foxm1* mRNA expression in lung was progressively upregulated following IR and the FOXM1 protein expression was elevated in alveolar type II epithelial cells within fibrotic lesions of the irradiated mouse lungs (Balli et al., 2013). Accordingly, the FOXM1 protein level was increased in the fibrotic lesions of human IPF patients, namely, in alveolar type II epithelial cells of the IPF fibrotic lesions (Balli et al., 2013).

The enhanced fibrosis in *epiFoxm1-ΔN* lungs was associated with increased and sustained pulmonary inflammation because the total number of inflammatory cells in BAL (brochoalveolar lavage) as well as the numbers of interstitial F4/80-positive macrophages and perivascular CD3-positive lymphocytes in lung sections were increased in irradiated *epiFoxm1-ΔN* mice compared to control mice (Balli et al., 2013). In addition, the mRNA expression of proinflammatory cytokines (*CCL2* (*chemokine, CC motif, ligand 2*), *CXCL5, IL-1β*) was higher in *epiFoxm1-ΔN* lungs than in control lungs (Balli et al., 2013). Conversely, the mRNA levels of proinflammatory genes (*IL-1β, CXCL5, TGF-β*) were lower in irradiated *epiFoxm1* KO lungs than in control lungs (Balli et al., 2013).

The enhanced fibrosis in *epiFoxm1-ΔN* animals was also associated with an increased proliferation rate as evidenced by a higher number of Ki-67 positive cells in *epiFoxm1-ΔN* lungs than in control lungs following irradiation (Balli et al., 2013). Interestingly, in the irradiated *epiFoxm1-ΔN* lungs, which express ca FoxM1B-ΔN specifically in alveolar epithelial type II cells, only the number of proliferating myofibroblasts (α-SMA-positive) was increased but not that of proliferating alveolar epithelial type II cells (pro-SP-C (pro-surfactant protein C)-positive) (Balli et al., 2013). Hence, the overexpression of FoxM1B-ΔN in alveolar epithelial type II cells augmented the proliferation of myofibroblasts, which might be the result of an increased release of profibrotic mediators (IL-1β, CCL2, CXCL5) from alveolar epithelial type II cells because the mRNA expression of *IL-1β*, *CCL2*, and *CXCL5* was elevated in irradiated *epiFoxm1-ΔN* lungs compared to control lungs (Balli et al., 2013).

Importantly, without IR, the *epiFoxm1-ΔN* mice developed neither pulmonary fibrosis nor lung inflammation, and neither collagen content in lung nor the proliferation of lung cells were altered in comparison to control mice (Balli et al., 2013). Consequently, the conditional overexpression of a ca FoxM1B mutant in the respiratory epithelium of *epiFoxm1-ΔN* mice reinforced the induction of pulmonary fibrosis and lung inflammation (pneumonitis) by IR, but FoxM1B-ΔN alone was unable to cause these defects (Balli et al., 2013).

After irradiation, *epiFoxm1-ΔN* mice showed a higher *COL1A1* (collagen, type I, α-1), *COL3A1* (collagen, type III, α-1), and α-*SMA* mRNA expression in lung than control mice, indicating increased ECM production and myofibroblast activation (Balli et al., 2013). Conversely, the *COL1A1*, *COL3A1*, and α-*SMA* mRNA levels in lung were reduced in irradiated *epiFoxm1 KO* mice compared to control animals, which indicates decreased ECM production and myofibroblast activation (Balli et al., 2013).

Following IR, *epiFoxm1-ΔN* lungs exhibited increased mRNA levels of mesenchymal markers (*vimentin*, *fibronectin*) and EMT transcription factors (*Snail1* (*SNAI1*, *Snail*), *Slug* (*SNAI2*, *Snail2*), *TWIST2* (*Dermo1* (*Dermis-expressed protein 1*)), *ZEB-1* (*zinc finger E-box-binding homeobox-1*, δ*EF1* (δ-*crystallin enhancer-binding factor 1*)), *ZEB-2* (*SIP1* (*Smad-interacting protein 1*)) but a decreased mRNA level of the epithelial marker *E-cadherin* compared to control lungs, suggesting that FoxM1B-ΔN-expressing epithelial cells underwent EMT after irradiation (Balli et al., 2013). Vice versa, a decreased mRNA expression of the EMT transcription factors (*Snail1*, *TWIST1*, *TWIST2*, *ZEB-1*, *ZEB-2*) was found in irradiated *epiFoxm1 KO* lungs in comparison to control lungs (Balli et al., 2013).

Together, these findings suggest that ca FoxM1B enhances radiation-induced pulmonary fibrosis by increasing inflammation, inducing proliferation of myofibroblasts, and promoting EMT (Balli et al., 2013). Accordingly, *foxm1* deficiency protects against radiation-induced pulmonary fibrosis (Balli et al., 2013). In summary, FOXM1 plays a critical role in radiation-induced pulmonary fibrosis (Balli et al., 2013).

Bleomycin-induced pulmonary fibrosis is another mouse model, in which lung fibrosis is caused by intratracheal administration of the glycopeptide antibiotic bleomycin (Chen & Stubbe, 2005; Moeller, Ask, Warburton, Gauldie, & Kolb, 2008; Mouratis and Aidini, 2011).

However, after bleomycin-induced lung injury, the collagen content in lung was similar in *epiFoxm1-ΔN* mice and control mice (Balli et al., 2013).

Nevertheless, when primary type II lung epithelial cells were isolated followed by flow cytometry-based cell sorting for GFP (green fluorescent protein), the FoxM1B-ΔN-expressing cell population (GFP-positive) of bleomycin-treated *epiFoxm1-ΔN* lungs contained increased *vimentin* and *Snail* mRNA levels but a decreased *E-cadherin* mRNA level, in comparison to control cells (GFP-negative) of control lungs (Balli et al., 2013). This suggests that the FoxM1B-ΔN-expressing population of alveolar type II epithelial cells underwent EMT in response to bleomycin treatment (Balli et al., 2013).

REFERENCES

Abbas, T., & Dutta, A. (2011). CRL4^{Cdt2}: Master coordinator of cell cycle progression and genome stability. *Cell Cycle*, *10*, 241–249.

Abraham, R. T. (2001). Cell cycle checkpoint signaling through the ATM and ATR kinases. *Genes & Development*, *15*, 2177–2196.

Aburto, M. R., Magarinos, M., Leon, Y., Varela-Nieto, I., & Sanchez-Calderon, H. (2012). AKT signaling mediates IGF-I survival actions on otic neural progenitors. *PLoS One*, *7*, e30790.

Ackermann Misfeldt, A., Costa, R. H., & Gannon, M. (2008). β-Cell proliferation, but not neogenesis, following 60% partial pancreatectomy is impaired in the absence of FoxM1. *Diabetes*, *57*, 3069–3077.

Adam, A. P., George, A., Schewe, D., Bragado, P., Iglesias, B. V., Ranganathan, A. C., et al. (2009). Computational identification of a p38^{SAPK}-regulated transcription factor network required for tumor cell quiescence. *Cancer Research*, *69*, 5664–5672.

Adams, P. D. (2009). Healing and hurting: Molecular mechanisms, functions, and pathologies of cellular senescence. *Molecular Cell*, *36*, 2–14.

Adams, J. C., & Watt, F. M. (1989). Fibronectin inhibits the terminal differentiation of human keratinocytes. *Nature*, *340*, 307–309.

Adams, J. C., & Watt, F. M. (1990). Changes in keratinocyte adhesion during terminal differentiation: Reduction in fibronectin binding precedes a5b1 integrin loss from the cell surface. *Cell*, *63*, 424–435.

Adamsen, B. L., Kravik, K. L., & DeAngelis, P. M. (2011). DNA damage signaling in response to 5-fluorouracil in three colorectal cancer cell lines with different mismatch repair and TP53 status. *International Journal of Oncology*, *39*, 673–682.

Adhikary, S., & Eilers, M. (2005). Transcriptional regulation and transformation by Myc proteins. *Nature Reviews. Molecular Cell Biology*, *6*, 635–645.

Ahearn, I. M., Haigis, K., Bar-Sagi, D., & Philips, M. R. (2012). Regulating the regulator: Post-translational modifications of RAS. *Nature Reviews. Molecular Cell Biology*, *13*, 39–51.

Ahmad, A., Ali, S., Wang, Z., Ali, A. S., Sethi, S., Sakr, W. A., et al. (2011). 3,3′-Diindolylmethane enhances Taxotere-induced growth inhibition of breast cancer cells through down-regulation of FoxM1. *International Journal of Cancer*, *129*, 1781–1791.

Ahmad, A., Wang, Z., Kong, D., Ali, S., Li, Y., Banerjee, S., et al. (2010). FoxM1 down-regulation leads to inhibition of proliferation, migration and invasion of breast cancer cells through the modulation of extra-cellular matrix degrading factors. *Breast Cancer Research and Treatment*, *122*, 337–346.

Ahmed, M., Uddin, S., Hussain, A. R., Alyan, A., Jehan, Z., Al-Dayel, F., et al. (2012). FoxM1 and its association with matrix metalloproteinases (MMP) signaling pathway in papillary thyroid carcinoma. *Journal of Clinical Endocrinology and Metabolism*, *97*, E1–E13.

Ahn, J. I., Lee, K. H., Shin, D. M., Shim, J. W., Kim, C. M., Kim, H., et al. (2004). Temporal expression changes during differentiation of neural stem cells derived from mouse embryonic stem cells. *Journal of Cellular Biochemistry*, *93*, 563–578.

Ahn, J., Urist, M., & Prives, C. (2004). The Chk2 protein kinase. *DNA Repair*, *3*, 1039–1047.

Ahuja, D., Saenz-Robles, M. T., & Pipas, J. M. (2005). SV40 large T antigen targets multiple cellular pathways to elicit cellular transformation. *Oncogene*, *24*, 7729–7745.

Akerfelt, M., Morimoto, R. I., & Sistonen, L. (2010). Heat shock factors: Integrators of cell stress, development and life span. *Nature Reviews. Molecular Cell Biology*, *11*, 545–556.

Akeson, E. C., Lambert, J. P., Narayanswami, S., Gardiner, K., Bechtel, L. J., & Davisson, M. T. (2001). Ts65Dn—Localization of the translocation breakpoint and trisomic gene content in a mouse model for Down syndrome. *Cytogenetics and Cell Genetics*, *93*, 270–276.

Alappat, E. C., Feig, C., Boyerinas, B., Volkland, J., Samuels, M., Murmann, A. E., et al. (2005). Phosphorylation of FADD at serine 194 by CKIα regulates its nonapoptotic activities. *Molecular Cell, 19*, 321–332.

Ali, S. H., & DeCaprio, J. A. (2001). Cellular transformation by SV40 large T antigen: Interaction with host proteins. *Seminars in Cancer Biology, 11*, 15–23.

Alison, M. R., Islam, S., & Lim, S. M. (2009). Cell therapy for liver disease. *Current Opinion in Molecular Therapeutics, 11*, 364–374.

Alkarain, A., & Slingerland, J. (2004). Deregulation of p27 by oncogenic signaling and its prognostic significance in breast cancer. *Breast Cancer Research, 6*, 13–21.

Allen, J. W., & Bhatia, S. N. (2002). Engineering liver therapies for the future. *Tissue Engineering, 8*, 725–737.

Alvarez-Fernandez, M., Halim, V. A., Aprelia, M., Mohammed, S., & Medema, R. H. (2011). Protein phosphatase 2A (B55α) prevents premature activation of transcription factor FOXM1 by antagonizing cyclin A/cyclin dependent kinase mediated phosphorylation. *Journal of Biological Chemistry, 286*, 33029–33036.

Alvarez-Fernandez, M., Halim, V. A., Krenning, L., Aprelia, M., Mohammed, S., Heck, A. J., et al. (2010). Recovery from a DNA-damage-induced G2 arrest requires Cdk-dependent activation of FoxM1. *EMBO Reports, 11*, 452–458.

Alvarez-Fernandez, M., Medema, R. H., & Lindqvist, A. (2010). Transcriptional regulation underlying recovery from a DNA-damage-induced arrest. *Transcription, 1*, 1–4.

Amabile, G., & Meissner, A. (2009). Induced pluripotent stem cells: Current progress and potential for regenerative medicine. *Trends in Molecular Medicine, 15*, 59–68.

Amati, B., Frank, S. R., Donjerkovic, D., & Taubert, S. (2001). Function of the c-Myc oncoprotein in chromatin remodeling and transcription. *Biochimica et Biophysica Acta, 1471*, M135–M145.

Amati, B., & Land, H. (1994). Myc-Max-Mad: A transcription factor network controlling cell cycle progression, differentiation and death. *Current Opinion in Genetics & Development, 4*, 102–108.

American Thoracic Society (2000). Idiopathic pulmonary fibrosis: Diagnosis and treatment. International consensus statement. American Thoracic Society (ARS), and the European Respiratory Society (ERS). *American Journal of Respiratory and Critical Care Medicine, 161*, 646–664.

Amunugama, R., & Fishel, R. (2012). Homologous recombination in eukaryotes. *Progress in Molecular Biology and Translational Science, 110*, 155–206.

Anckar, J., & Sistonen, L. (2007). Heat shock factor 1 as a coordinator of stress and developmental pathways. *Advances in Experimental Medicine and Biology, 594*, 78–88.

Anckar, J., & Sistonen, L. (2011). Regulation of HSF1 function in the heat shock response: Implications in aging and disease. *Annual Review of Biochemistry, 80*, 1089–1115.

Anders, L., Ke, N., Hydbring, P., Choi, Y. J., Widlund, H. R., Chick, J. M., et al. (2011). A systematic screen for CDK4/6 substrates links FOXM1 phosphorylation to senescence suppression in cancer cells. *Cancer Cell, 20*, 620–634.

Antoni, L., Sodha, N., Collins, I., & Garrett, M. D. (2007). CHK2 kinase: Cancer susceptibility and cancer therapy—Two sides of the same coin? *Nature Reviews. Cancer, 7*, 925–936.

Asselin-Labat, M. L., Sutherland, K. D., Barker, H., Thomas, R., Shackleton, M., Forrest, N. C., et al. (2007). Gata-3 is an essential regulator of mammary-gland morphogenesis and luminal-cell differentiation. *Nature Cell Biology, 9*, 201–209.

Asselin-Labat, M. L., Vaillant, F., Sheridan, J. M., Pal, B., Wu, D., Simpson, E. R., et al. (2010). Control of mammary stem cell function by steroid hormone signalling. *Nature, 465*, 798–802.

Attwooll, C., Lazzerini Denchi, E., & Helin, K. (2004). The E2F family: Specific functions and overlapping interests. *EMBO Journal, 23*, 4709–4716.

Babashah, S., & Soleimani, M. (2011). The oncogenic and tumour suppressive roles of microRNAs in cancer and apoptosis. *European Journal of Cancer, 47*, 1127–1137.

Badouel, C., Garg, A., & McNeill, H. (2009). Herding Hippos: Regulating growth in flies and man. *Current Opinion in Cell Biology, 21*, 837–843.

Badouel, C., & McNeill, H. (2011). SnapShot: The Hippo signaling pathway. *Cell, 145*, 484–484.e1.

Bae, I., Rih, J. K., Kim, H. J., Kang, H. J., Haddad, B., Kirilyuk, A., et al. (2005). BRCA1 regulates gene expression for orderly mitotic progression. *Cell Cycle, 4*, 1641–1666.

Bai, T. R., & Knight, D. A. (2005). Structural changes in the airways in asthma: Observations and consequences. *Clinical Science, 108*, 463–477.

Bain, G., Ray, W. J., Yao, M., & Gottlieb, D. I. (1994). From embryonal carcinoma cells to neurons: The P19 pathway. *Bioessays, 16*, 343–348.

Balemans, W., & Van Hul, W. (2002). Extracellular regulation of BMP signaling in vertebrates: A cocktail of modulators. *Developmental Biology, 250*, 231–250.

Balli, D., Ren, X., Chou, F. S., Cross, E., Zhang, Y., Kalinichenko, V. V., et al. (2012). Foxm1 transcription factor is required for macrophage migration during lung inflammation and tumor formation. *Oncogene, 31*, 3875–3888.

Balli, D., Ustiyan, V., Zhang, Y., Wang, I. C., Masino, A. J., Ren, X., et al. (2013). Foxm1 transcription factor is required for lung fibrosis and epithelial-to mesenchymal transition. *EMBO Journal, 32*, 231–244.

Balli, D., Zhang, Y., Snyder, J., Kalinichenko, V. V., & Kalin, T. V. (2011). Endothelial-specific deletion of transcription factor FoxM1 increases urethane-induced lung carcinogenesis. *Cancer Research, 71*, 40–50.

Balmain, A., Gray, J., & Ponder, B. (2003). The genetics and genomics of cancer. *Nature Genetics, 33*, 238–244.

Bandukwala, H. S., Wu, Y., Feuerer, M., Chen, Y., Barboza, B., Ghosh, S., et al. (2011). Structure of a domain-swapped FOXP3 dimer on DNA and its function in regulatory T cells. *Immunity, 34*, 479–491.

Banz, C., Ungethuem, U., Kuban, R. J., Diedrich, K., Lengyel, E., & Hornung, D. (2009). The molecular signature of endometriosis-associated endometrioid ovarian cancer differs significantly from endometriosis-independent endometrioid ovarian cancer. *Fertility and Sterility, 94*, 1212–1217.

Bao, B., Wang, Z., Ali, S., Kong, D., Banerjee, S., Ahamd, A., et al. (2011). Over-expression of FoxM1 leads to epithelial-mesenchymal transition and cancer stem cell phenotype in pancreatic cancer cells. *Journal of Cellular Biochemistry, 112*, 2296–2306.

Bar-Joseph, Z., Siegfried, Z., Brandeis, M., Brors, B., Lu, Y., Eils, R., et al. (2008). Genome-wide transcriptional analysis of the human cell cycle identifies genes differentially regulated in normal and cancer cells. *Proceeding of the National Academy of Sciences of the United States of America, 105*, 955–960.

Barsotti, A. M., & Prives, C. (2009). Pro-proliferative FoxM1 is a target of p53-mediated repression. *Oncogene, 28*, 4295–4305.

Bartek, J., Falck, J., & Lukas, J. (2001). CHK2 kinase—A busy manager. *Nature Reviews. Molecular Cell Biology, 2*, 877–886.

Bartek, J., & Lukas, J. (2003). Chk1 and Chk2 kinases in checkpoint control and cancer. *Cancer Cell, 3*, 421–429.

Bartek, J., & Lukas, J. (2007). DNA damage checkpoints: From initiation to recovery or adaptation. *Current Opinion in Cell Biology, 19*, 238–245.

Bartek, J., & Lukas, J. (2011). Cyclin D1 multitasks. *Nature, 474*, 171–172.

Bartek, J., Lukas, C., & Lukas, J. (2004). Checking on DNA damage in S-phase. *Nature Reviews. Molecular Cell Biology, 5*, 792–804.

Baudino, T. A., & Cleveland, J. L. (2001). The Max network gone Mad. *Molecular and Cellular Biology, 21*, 691–702.

Baumann, P., & West, S. C. (1998). Role of the human RAD51 protein in homologous recombination and double-strand-break repair. *Trends in Biochemical Sciences, 23*, 247–251.

Baylin, S. B., & Jones, P. A. (2011). A decade of exploring the cancer epigenome—Biological and translational implications. *Nature Reviews. Cancer, 11*, 726–734.

Bazzoni, G., & Dejana, E. (2004). Endothelial cell-to-cell junctions: Molecular organization and role in vascular homeostasis. *Physiological Reviews, 84*, 869–901.

Behren, A., Mühlen, S., Acuna Sanhueza, G. A., Schwager, C., Plinkert, P. K., Huber, P. E., et al. (2010). Phenotype-assisted transcriptome analysis identifies FOXM1 downstream from Ras-MKK3-p38 to regulate *in vitro* cellular invasion. *Oncogene, 29*, 1519–1530.

Bektas, N., ten Haaf, A., Veeck, J., Wild, P. J., Lüscher-Firzlaff, J., Hartmann, A., et al. (2008). Tight correlation between expression of the Forkhead transcription factor FOXM1 and HER2 in human breast cancer. *BMC Cancer, 8*, 42.

Bellelli, R., Castellone, M. D., Garcia-Rostan, G., Ugolini, C., Nucera, C., Sadow, P. M., et al. (2012). FOXM1 is a molecular determinant of the mitogenic and invasive phenotype of anaplastic thyroid carcinoma. *Endocrine-Related Cancer, 19*, 695–710.

Benayoun, B. A., Caburet, S., & Veitia, R. A. (2012). Forkhead transcription factors: Key players in health and disease. *Trends in Genetics, 27*, 224–232.

Ben-Porath, I., & Weinberg, R. A. (2004). When cells get stressed: And integrative view of cellular sencence. *The Journal of Clinical Investigation, 113*, 8–13.

Ben-Porath, I., & Weinberg, R. A. (2005). The signals and pathways activating cellular sencence. *The International Journal of Biochemistry & Cell Biology, 37*, 961–976.

Benvenuti, S., Arena, S., & Bardelli, A. (2005). Identification of cancer genes by mutational profiling of tumor genomes. *FEBS Letters, 579*, 1884–1890.

Bergamaschi, A., Christensen, B. L., & Katzenellenbogen, B. S. (2011). Reversal of endocrine resistance in breast cancer: Interrelationships among 14-3-3t, FOXM1, and a gene signature associated with miosis. *Breast Cancer Research, 13*, R70.

Berger, E., Rome, S., Vega, N., Ciancia, C., & Vidal, H. (2010). Transcriptome profiling in response to adiponectin in human cancer-derived cells. *Physiological Genomics, 42A*, 61–70.

Berger, E., Vega, N., Vidal, H., & Geloen, A. (2012). Gene network analysis leads to functional validation of pathways linked to cancer cell growth and survival. *Biotechnology Journal, 7*, 1395–1404.

Besson, A., Assoian, R. K., & Roberts, J. M. (2004). Regulation of the cytoskeleton: An oncogenic function for CDK inhibitors? *Nature Reviews. Cancer, 4*, 948–955.

Besson, A., Dowdy, S. F., & Roberts, J. M. (2008). CDK inhibitors: Cell cycle regulators and beyond. *Developmental Cell, 14*, 159–169.

Bharadwaj, A. S., Bewtra, A. K., & Agrawal, D. K. (2007). Dendritic cells in allergic airway inflammation. *Canadian Journal of Physiology and Pharmacology, 85*, 686–699.

Bhat, U. G., Halasi, M., & Gartel, A. L. (2009a). Thiazole antibiotics target FoxM1 and induce apoptosis in human cancer cells. *PLoS One, 4*, e5592.

Bhat, U. G., Halasi, M., & Gartel, A. L. (2009b). FoxM1 is a general target for proteasome inhibitors. *PLoS One, 4*, e6593.

Bhat, U. G., Jagadeeswaran, R., Halasi, M., & Gartel, A. L. (2011). Nucleophosmin interacts with FOXM1 and modulates the level and localization of FOXM1c in human cancer cells. *Journal of Biological Chemistry, 286*, 41425–41433.

Bhonde, M. R., Hanski, M. L., Budczies, J., Cao, M., Gillissen, B., Moorthy, D., et al. (2006). DNA damage-induced expression of p53 suppresses mitotic checkpoint kinase hMsp1. *Journal of Biological Chemistry, 281*, 8675–8685.

Bishop, J. M. (1995). Cancer: The rise of the genetic paradigm. *Genes & Development, 9*, 1309–1315.

Björk, J. K., & Sistonen, L. (2010). Regulation of the members of the mammalian heat shock factor family. *FEBS Journal, 277*, 4126–4139.

Blagosklonny, M. V. (2006). Cell senescence: Hypermitogenic arrest beyond the restriction point. *Journal of Cellular Physiology, 209*, 592–597.

Blagosklonny, M. V., & Pardee, A. B. (2002). The restriction point of the cell cycle. *Cell Cycle, 1*, 103–110.

Blais, A., & Dynlacht, B. D. (2004). Hitting their targets: An emerging picture of E2F and cell cycle control. *Current Opinion in Genetics & Development, 14*, 527–532.

Blanco-Bose, W. E., Murphy, M. J., Ehninger, A., Offner, S., Dubey, C., Huang, W., et al. (2008). c-Myc and its target FoxM1 are critical downstream effectors of TCPOBOP-CAR induced direct liver hyperplasia. *Hepatology, 48*, 1302–1311.

Bocchetta, M., & Carbone, M. (2004). Epidemiology and molecular pathology at crossroads to establish causation: Molecular mechanisms of malignant transformation. *Oncogene, 23*, 6484–6491.

Böhm, F., Köhler, U. A., Speicher, T., & Werner, S. (2010). Regulation of liver regeneration by growth factors and cytokines. *EMBO Molecular Medicine, 2*, 294–305.

Bolte, C., Zhang, Y., Wang, I. C., Kalin, T. V., Molkentin, J. D., & Kalinchenko, V. V. (2011). Expression of Foxm1 transcription factor in cardiomyocytes is required for myocardial development. *PLoS One, 6*, e22217.

Bolte, C., Zhang, Y., York, A., Kalin, T. V., Schultz, J. E. J., Molkentin, J. D., et al. (2012). Postnatal ablation of Foxm1 from cardiomyocytes causes late onset cardiac hypertrophy and fibrosis without exacerbating pressure overload-induced cardiac remodeling. *PLoS One, 7*, e48713.

Bommer, G. T., Gerin, I., Feng, Y., Kaczorowski, A. J., Kuick, R., Love, R. E., et al. (2007). p53-mediated activation of miRNA34 candidate tumor-suppressor genes. *Current Biology, 17*, 1298–1307.

Bonet, C., Giuliano, S., Ohanna, M., Bille, K., Allegra, M., Lacour, J. P., et al. (2012). Aurora B is regulated by the MAPK/ERK signaling pathway and is a valuable potential target in melanoma cells. *Journal of Biological Chemistry, 287*, 29887–29898.

Bonner, M. Y., & Arbiser, J. L. (2012). Targeting NADPH oxidases for the treatment of cancer and inflammation. *Cellular and Molecular Life Sciences, 69*, 2435–2442.

Bordeianu, G., Zugun-Eloae, F., & Rusu, M. G. (2011). The role of DNA repair by homologous recombination in oncogenesis. *Revista Medico-Chirurgicală a Societăţii de Medici şi Naturalişti din Iaşi, 115*, 1189–1194.

Borriello, A., Cucciolla, V., Oliva, A., Zappia, V., & Della Ragione, F. (2007). p27^{Kip1} metabolism. A fascinating labyrinth. *Cell Cycle, 6*, 1053–1061.

Bose, A., Teh, M. T., Hutchison, I. L., Wan, H., Leigh, I. M., & Waseem, A. (2012). Two mechanisms regulate keratin K15 expression in keratinocytes: Role of PKC/AP-1 and FOXM1 mediated signalling. *PLoS One, 7*, e38599.

Bouchard, C., Staller, P., & Eilers, M. (1998). Control of cell proliferation by Myc. *Trends in Cell Biology, 8*, 202–206.

Boulikas, T. (1995). The phosphorylation connection to cancer. *International Journal of Oncology, 6*, 271–278.

Boura, E., Rezabkova, L., Brynda, J., Obsilova, V., & Obsil, T. (2010). Structure of the human FOXO4-DBD-DNA complex at 1.9 Å resolution reveals new details of FOXO binding to DNA. *Acta Crystallographica Section D: Biological Crystallography, 66*, 1351–1357.

Boura, E., Silhan, J., Herman, P., Vecer, J., Sulc, M., Teisinger, J., et al. (2007). Both the N-terminal loop and wing W2 of the forkhead domain of transcription factor Foxo4 are important for DNA binding. *Journal of Biological Chemistry, 282*, 8265–8275.

Bowman, A., & Nusse, R. (2011). Location, location, location: FoxM1 mediates β-catenin nuclear translocation and promotes glioma tumorigenesis. *Cancer Cell, 20*, 415–416.

Bracken, A. P., Ciro, M., Cocito, A., & Helin, K. (2004). E2F target genes: Unraveling the biology. *Trends in Biochemical Sciences, 29*, 409–417.

Braig, M., & Schmitt, C. A. (2006). Oncogene-induced senescence: Putting the brakes on tumor development. *Cancer Research, 66*, 2881–2884.

Branzei, D., & Foiani, M. (2008). Regulation of DNA repair throughout the cell cycle. *Nature Reviews. Molecular Cell Biology, 9*, 297–308.

Bräuning, A., Heubach, Y., Knorpp, T., Kowalik, M. A., Templin, M., Columbano, A., et al. (2011). Gender-specific interplay of signaling through β-catenin and CAR in the regulation of xenobiotic-induced hepatocyte proliferation. *Toxicological Sciences, 123*, 113–122.

Bravieri, R., Shiyanova, T., Chen, T. H., Overdier, D., & Liao, X. (1997). Different DNA contact schemes are used by two winged helix proteins to recognize a DNA binding sequence. *Nucleic Acids Research, 25*, 2888–2896.

Brennan, R. G. (1993). The winged-helix DNA-binding motif: Another helix-turn-helix takeoff. *Cell, 74*, 773–776.

Brent, M. M., Anand, R., & Marmorstein, R. (2008). Structural basis for DNA recognition by FoxO1 and its regulation by post-translational modification. *Structure, 16*, 1407–1416.

Brezillon, N., Lambert-Blot, M., Morosan, S., Couton, D., Mitchell, C., Kremsdorf, D., et al. (2007). Transplanted hepatocytes over-expressing FoxM1B efficiently repopulate chronically injured mouse liver independent of donor age. *Molecular Therapy, 15*, 1710–1715.

Bringardner, B. D., Bara, C. P., Eubank, T. D., & Marsh, C. B. B. (2008). The role of inflammation in the pathogenesis of idiopathic pulmonary fibrosis. *Antioxidants & Redox Signaling, 10*, 287–301.

Brown, G., Hughes, P. J., & Michell, R. H. (2003). Cell differentiation and proliferation—Simultaneous but independent? *Experimental Cell Research, 291*, 282–288.

Brumbaugh, J., Rose, C. M., Phanstiel, D. H., Thomson, J. A., & Coon, J. J. (2011). Proteomics and pluripotency. *Critical Reviews in Biochemistry and Molecular Biology, 46*, 493–506.

Buchberger, A., Bukau, B., & Sommer, T. (2010). Protein quality control in the cytosol and the endoplasmic reticulum: Brothers in arms. *Molecular Cell, 40*, 238–252.

Budd, R. C. (2002). Death receptors couple to both cell proliferation and apoptosis. *The Journal of Clinical Investigation, 109*, 437–442.

Bueno, M. J., Perez de Castro, I., & Malumbres, M. (2008). Control of cell proliferation pathways by microRNAs. *Cell Cycle, 7*, 3143–3148.

Burma, S., Chen, B. P., & Chen, D. J. (2006). Role of non-homologous end joining (NHEJ) in maintaining genomic integrity. *DNA Repair, 5*, 1042–1048.

Burtner, C. R., & Kennedy, B. K. (2010). Progeria syndrome and ageing: What is the connection? *Nature Reviews. Molecular Cell Biology, 11*, 567–578.

Butkinaree, C., Park, K., & Hart, G. W. (2010). O-linked β-*N*-acetylglucosamine (O-GlcNAc): Extensive crosstalk with phosphorylation to regulate signaling and transcription in response to nutrients and stress. *Biochimica et Biophysica Acta, 1800*, 96–106.

Butler, P. C., Meier, J. J., Butler, A. E., & Bhushan, A. (2007). The replication of β-cells in normal physiology, in disease and for therapy. *Nature Clinical Practice. Endocrinology & Metabolism, 3*, 758–768.

Buttitta, L. A., & Edgar, B. A. (2007). Mechanisms controlling cell cycle exit upon terminal differentiation. *Current Opinion in Cell Biology, 19*, 697–704.

Caestecker, K. W., & Van de Walle, G. R. (2013). The role of BRCA1 in DNA double-strand repair: Past and present. *Experimental Cell Research, 319*, 575–587.

Caino, M. C., Meshki, J., & Kazanietz, M. G. (2009). Hallmarks of senescence in carcinogenesis: Novel signaling players. *Apoptosis, 14*, 392–408.

Calderwood, S. K., Xie, Y., Wang, X., Khaleque, M. A., Chou, S. D., Murshid, A., et al. (2010). Signal transduction pathways leading to heat shock transcription. *Signal Transduction Insights*, *2*, 13–24.

Caldwell, S. A., Jackson, S. R., Shahriari, K. S., Lynch, T. P., Sethi, G., Walker, S., et al. (2010). Nutrient sensor O-GlcNAc transferase regulates breast cancer tumorigenesis through targeting of the oncogenic transcription factor FoxM1. *Oncogene*, *29*, 2831–2842.

Calvisi, D. F., Pinna, F., Ladu, S., Pellegrino, R., Simile, M. M., Frau, M., et al. (2009). Forkhead box M1B is a determinant of rat susceptibility to hepatocarcinogenesis and sustains ERK activity in human HCC. *Gut*, *58*, 679–687.

Calvisi, D. F., Simile, M. M., Ladu, S., Frau, M., Evert, M., Tomasi, M. L., et al. (2011). Activation of v-Myb avian myeloblastosis viral oncogene homolog-like 2 (MYBL2)-LIN9 complex contributes to human hepatocarcinogenesis an identifies a subset of hepatocellular carcinoma with mutant p53. *Hepatology*, *53*, 1226–1236.

Calzone, L., Gelay, A., Zinovyev, A., Radvanyi, F., & Barillot, E. (2008). A comprehensive modular map of molecular interactions in RB/E2F pathway. *Molecular Systems Biology*, *4*, 173.

Cam, H., & Dynlacht, B. D. (2003). Emerging roles for E2F: Beyond the G1/S transition and DNA replication. *Cancer Cell*, *3*, 311–316.

Campbell, S. L., Khosravi-Far, R., Rossman, K. L., Clark, G. J., & Der, C. J. (1998). Increasing complexity of Ras signaling. *Oncogene*, *17*, 1395–1413.

Campisi, J. (2000). Cancer, aging and cellular senescence. *In Vivo*, *14*, 183–188.

Campisi, J. (2001). Cellular senescence as a tumor-suppressor mechanism. *Trends in Cell Biology*, *11*, S27–S31.

Campisi, J. (2003). Cancer and aging: Rival demons? *Nature Reviews. Cancer*, *3*, 339–349.

Campisi, J. (2005a). Senescent cells, tumor suppression, and organismal aging: Good citizens, bad neighbours. *Cell*, *120*, 513–522.

Campisi, J. (2005b). Aging, tumor suppression and cancer: High wire-act! *Mechanisms of Ageing and Development*, *126*, 51–58.

Campisi, J. (2011). Cellular senescence: Putting paradoxes in perspectives. *Current Opinion in Genetics & Development*, *21*, 107–112.

Campisi, J., & d'Adda di Fagagna, F. (2007). Cellular senescence: When bad things happen to good cells. *Nature Reviews. Molecular Cell Biology*, *8*, 729–740.

Campisi, J., Kim, S. H., Lim, C. S., & Rubio, M. (2001). Cellular senescence, cancer and aging: The telomere connection. *Experimental Gerontology*, *36*, 1619–1637.

Cannell, I. G., Kong, Y. W., Johnston, S. J., Chen, M. L., Collins, H. M., Dobbyn, H. C., et al. (2010). p38 MAPK/MK2-mediated induction of miR-34c following DNA damage prevents Myc-dependent DNA replication. *Proceeding of the National Academy of Sciences of the United States of America*, *107*, 5375–5380.

Caracciolo, V., Reiss, K., Khalili, K., De Falco, G., & Giordano, A. (2006). Role of the interaction between large T antigen and Rb family members in the oncogenicity of JC virus. *Oncogene*, *25*, 5294–5301.

Carlsson, P., & Mahlapuu, M. (2002). Forkhead transcription factors: Key players in development and metabolism. *Developmental Biology*, *250*, 1–23.

Carr, J. R., Kiefer, M. M., Park, H. J., Li, J., Wang, Z., Fontanarosa, J., et al. (2012). FoxM1 regulates mammary luminal cell fate. *Cell Reports*, *1*, 715–729.

Carr, J. R., Park, H. J., Wang, Z., Kiefer, M. M., & Raychaudhuri, P. (2010). FoxM1 mediates resistance to herceptin and paclitaxel. *Cancer Research*, *70*, 5054–5063.

Carter, P. M., Heinly, T. L., Yates, S. W., & Lieberman, P. L. (1997). Asthma: The irreversible airways disease. *Journal of Investigational Allergology & Clinical Immunology*, *7*, 566–571.

Catalucci, D., Latronico, M. V., Ellingsen, O., & Condorelli, G. (2008). Physiological myocardial hypertrophy: How and why? *Frontiers in Bioscience*, *13*, 312–324.

Cates, E. C., Fattouh, R., Johnson, J. R., Llop-Guevara, A., & Jordana, M. (2007). Modeling responses to respiratory house dust mite exposure. *Contributions to Microbiology, 14*, 42–67.

Celton-Morizur, S., & Desdouets, C. (2010). Polyploidization of liver cells. *Advances in Experimental Medicine and Biology, 676*, 123–135.

Centric, G., Celton-Morizur, S., & Desdouets, C. (2012a). Hepatocytes polyploidization and cell cycle control in liver physiopathology. *International Journal of Hepatology, 2012*, 282430.

Centric, G., Celton-Morizur, S., & Desdouets, C. (2012b). Polyploidy and liver proliferation. *Clinics and Research in Hepatology and Gastroenterology, 36*, 29–34.

Cerbinskaite, A., Mukhopadhyay, A., Plummer, E. R., Curtin, N. J., & Edmondson, R. J. (2012). Defective homologous recombination in human cancers. *Cancer Treatment Reviews, 38*, 89–100.

Chakrabarty, K., & Heumann, R. (2008). Prospective of Ras signaling in stem cells. *Biological Chemistry, 389*, 791–798.

Chambard, J. C., Lefloch, R., Pouysségur, J., & Lenormand, P. (2007). ERK implication in cell cycle regulation. *Biochimica et Biophysica Acta, 1773*, 1299–1310.

Chan, D. W., Hui, W. W., Cai, P. C., Liu, M. X., Yung, M. M., Mak, C. S., et al. (2012). Targeting GRB7/ERK/FOXM1 signaling pathway impairs aggressiveness of ovarian cancer cells. *PLoS One, 7*, e52578.

Chan, D. W., Yu, S. Y. M., Chiu, P. M., Yao, K. M., Liu, V. W. S., Cheung, A. N. Y., et al. (2008). Over-expression of FOXM1 transcription factor is associated with cervical cancer progression and pathogenesis. *The Journal of Pathology, 215*, 245–252.

Chaney, S. G., Campbell, S. L., Bassett, E., & Wu, Y. (2005). Recognition and processing of cisplatin- and oxaliplatin-DNAa adducts. *Critical Reviews in Oncology/Hematology, 53*, 3–11.

Chang, H. Y., Sneddon, J. B., Alizadeh, A. A., Sood, R., West, R. B., Montgomery, K., et al. (2004). Gene expression signature of fibroblast serum response predicts human cancer progression: Similarities between tumors and wounds. *PLoS Biology, 2*, 0206–0214.

Chang, P., Steelman, L. S., Shelton, J. G., Lee, J. T., Navolanic, P. M., Blalock, W. L., et al. (2003). Regulation of cell cycle progression and apoptosis by the Ras/Raf/MEK/ERK pathway. *International Journal of Oncology, 22*, 469–480.

Chang, T. C., Wentzel, E. A., Kent, O. A., Ramachandran, K., Mulendore, M., Lee, K. H., et al. (2007). Transactivation of miR-34a by p53 broadly influences gene expression and promotes apoptosis. *Molecular Cell, 26*, 745–752.

Chang, T. C., Wentzel, E. A., Kent, O. A., Ramachandran, K., Mullendore, M., Lee, K. H., et al. (2007). Transactivation of miR-34 by p53 broadly influences gene expression and promotes apoptosis. *Molecular Cell, 26*, 745–752.

Chapman, H. A. (2011). Epithelial–mesenchymal interactions in pulmonary fibrosis. *Annual Review of Physiology, 3*, 413–435.

Chaudhary, J., Mosher, R., Kim, G., & Skinner, M. K. (2000). Role of the winged helix transcription factor (WIN) in the regulation of Sertoli cell differentiated functions: WIN acts as an early event gene for follicle-stimulating hormone. *Endocrinology, 141*, 2758–2766.

Chen, C. H., Chien, C. Y., Huang, C. C., Hwang, C. F., Chuang, H. C., Fang, F. M., et al. (2009). Expression of FLJ10540 is correlated with aggressiveness of oral cavity squamous cell carcinoma by stimulating cell migration and invasion through increased FOXM1 and MMP-2 activity. *Oncogene, 28*, 2723–2737.

Chen, Y. J., Dominguez-Brauer, C., Wang, Z., Asara, J. M., Costa, R. H., Tyner, A. L., et al. (2009). A conserved phosphorylation site within the forkhead domain of FoxM1B is required for its activation by cyclin-Cdk1. *Journal of Biological Chemistry, 284*, 30695–30706.

Chen, G., Korfhagen, T. R., Xu, Y., Kitzmiller, J., Wert, S. E., Maeda, Y., et al. (2009). SPDEF is required for mouse pulmonary goblet cell differentiation and regulates a network of genes associated with mucus production. *The Journal of Clinical Investigation, 199*, 2914–2924.

Chen, X., Müller, G. A., Quaas, M., Fischer, M., Han, N., Stutchbury, B., et al. (2013). The forkhead transcription factor FOXM1 controls cell cycle-dependent gene expression through an atypical chromatin binding mechanism. *Molecular and Cellular Biology, 33*, 227–236.

Chen, Y., & Poon, R. Y. (2008). The multiple checkpoint functions of CHK1 and CHK2 in maintenance of genome stability. *Frontiers in Bioscience, 13*, 5016–5029.

Chen, Y., & Sanchez, Y. (2004). Chk1 in the DNA damage response: Conserved roles from yeast to mammals. *DNA Repair, 3*, 1033–1037.

Chen, J., & Stubbe, J. (2005). Bleomycins: Towards better therapeutics. *Nature Reviews. Cancer, 5*, 102–112.

Chen, H. Z., Tsai, S. Y., & Leone, G. (2009). Emerging roles of E2Fs in cancer: An exit from cell cycle control. *Nature Reviews. Cancer, 9*, 785–797.

Chen, W. D., Wang, Y. D., Zhang, L., Shiah, S., Wang, M., Yang, F., et al. (2010). Farnesoid X receptor alleviates age-related proliferation defects in regenerating mouse livers by activating Forkhead Box m1b transcription. *Hepatology, 51*, 953–962.

Chen, P. M., Wu, Y. H., Li, M. C., Cheng, Y. W., Chen, C. Y., & Lee, H. (2013). MnSOD promotes tumor invasion via upregulation of FoxM1-MMP2 axis and related with poor survival and relapse in lung adenocarcinomas. *Molecular Cancer Research, 11*, 261–271.

Chen, H., Yang, C., Yu, L., Xie, L., Hu, J., Zeng, L., et al. (2012). Adenovirus-mediated RNA interference targeting FOXM1 transcription factor suppresses cell proliferation and tumor growth of nasopharyngeal carcinoma. *The Journal of Gene Medicine, 14*, 231–240.

Chen, W., Yuan, K., Tao, Z. Z., & Xiao, B. K. (2011). Deletion of Forkhead Box M1 transcription factor reduces malignancy in laryngeal squamous carcinoma cells. *Asian Pacific Journal of Cancer Prevention, 12*, 1785–1788.

Cheng, J., DeCaprio, J. A., Fluck, M. M., & Schaffhausen, B. S. (2009). Cellular transformation by simian virus 40 and murine polyoma virus T antigens. *Seminars in Cancer Biology, 19*, 218–228.

Chetty, C., Bhoopathi, P., Rao, J. A., & Lakka, S. S. (2009). Inhibition of matrix metalloproteinase-2 enhances radiosensitivity by abrogating radiation-induced FoxM1-mediated G2/M arrest in A549 lung cancer cells. *International Journal of Cancer, 124*, 2468–2477.

Chiu, S. J., Lee, Y. S., Hsu, T. S., & Chen, W. S. (2009). Oxaliplatin induced gamma-H2AX activation via both p53-depedent and -independent pathways but is not associated with cell cycle arrest in human colorectal cancer cells. *Chemico-Biological Interactions, 182*, 173–182.

Chivukula, R. R., & Mendell, J. T. (2008). Circular reasoning: MicroRNAs and cell-cycle control. *Trends in Biochemical Sciences, 33*, 474–481.

Choy, B., Roberts, S. G. E., Griffin, L. A., & Green, M. R. (1993). How eukaryotic transcription activators increase assembly of preinitiation complexes. *Cold Spring Harbor Symposia on Quantitative Biology, 58*, 199–203.

Christoffersen, N. R., Shalgi, R., Frankel, L. B., Leucci, E., Lees, M., Klausen, M., et al. (2010). p53-independent upregulation of miR-34a during oncogene-induced senescence represses MYC. *Cell Death and Differentiation, 17*, 236–245.

Chu, Y. P., Chang, C. H., Shiu, J. H., Chang, Y. T., Chen, C. Y., & Chuang, W. J. (2011). Solution structure and backbone dynamics of the DNA-binding domain of FOXP1: Insights into its domain swapping and DNA binding. *Protein Science, 20*, 908–924.

Chu, I. M., Hengst, L., & Slingerland, J. M. (2008). The Cdk inhibitor p27 in human cancer: Prognostic potential and relevance to anticancer therapy. *Nature Reviews. Cancer, 8*, 253–267.

Chu, X. Y., Zhu, Z. M., Chen, L. B., Wang, J. H., Su, Q. S., Yang, J. R., et al. (2012). FOXM1 expression correlates with tumor invasion and a poor prognosis of colorectal cancer. *Acta Histochemica, 114*, 755–762.

Chua, P. J., Yip, G. W. C., & Bay, B. H. (2009). Cell cycle arrest induced by hydrogen peroxide is associated with modulation of oxidative stress related genes in breast cancer cells. *Experimental Biology and Medicine, 234*, 1086–1094.

Ciaccia, A., & Elledge, S. J. (2010). The DNA damage response: Making it safe to play with knives. *Molecular Cell, 40*, 179–204.

Cicatiello, L., Scafoglio, C., Altucci, L., Cancemi, M., Natoli, G., Facchiano, A., et al. (2004). A genomic view of estrogen actions in human breast cancer cells by expression profiling of the hormone-responsive transcriptome. *Journal of Molecular Endocrinology, 32*, 719–775.

Cichowski, K., & Hahn, W. C. (2008). Unexpected pieces to the senescence puzzle. *Cell, 133*, 958–961.

Clark, K. L., Halay, E. D., Lai, E., & Burley, S. K. (1993). Co-crystal structure of the HNF-3/fork head-DNA-recognition motif resembles histone H5. *Nature, 364*, 412–420.

Clevers, H., & Nusse, R. (2012). Wnt/β-catenin signaling and disease. *Cell, 149*, 1192–1205.

Cole, M. D., & McMahon, S. B. (1999). The Myc oncorotein: A critical evaluation of transactivation and target gene regulation. *Oncogene, 18*, 2916–2924.

Cole, M. D., & Nikiforov, M. A. (2006). Transcriptional activation by the Myc oncoprotein. *Current Topics in Microbiology and Immunology, 302*, 33–50.

Coleman, M. L., Marshall, C. J., & Olson, M. F. (2004). RAS and RHO GTPases in G1-phase cell-cycle regulation. *Nature Reviews. Molecular Cell Biology, 5*, 355–366.

Collado, M., Blasco, M. A., & Serrano, M. (2007). Cellular senescence in cancer and ageing. *Cell, 130*, 223–233.

Collado, M., & Serrano, M. (2005). The senescent side of tumor suppression. *Cell Cycle, 4*, 1722–1724.

Collado, M., & Serrano, M. (2006). The power and promise of oncogene-induced senescence markers. *Nature Reviews. Cancer, 6*, 472–476.

Collado, M., & Serrano, M. (2010). Senescence in tumors: Evidence from mice and humans. *Nature Reviews. Cancer, 10*, 51–57.

Coller, H. A., Sang, L., & Roberts, J. M. (2006). A new description of cellular quiescence. *PLoS Biology, 4*, e83.

Colman, A., & Dreesen, O. (2009). Induced pluripotent stem cells and the stability of the differentiated state. *EMBO Reports, 10*, 714–721.

Comer, F. I., & Hart, D. W. (1999). O-GlcNAc and the control of gene expression. *Biochimica et Biophysica Acta, 1473*, 161–171.

Comer, F. I., & Hart, D. W. (2000). O-Glycosylation of nuclear and cytosolic proteins. *Journal of Biological Chemistry, 275*, 29179–29182.

Coppe, J. P., Desprez, P. Y., Krtolica, A., & Campisi, J. (2010). The senescence-associated secretory phenotype: The dark side of tumor suppression. *Annual Review of Pathology: Mechanisms of Disease, 5*, 99–118.

Corney, D. C., Flesken-Nikitin, A., Godwin, A. K., Wang, W., & Nikitin, A. Y. (2007). MicroRNA-34b and MicroRNA-34c are targets of p53 and cooperate in control of cell proliferation and adhesion-independent growth. *Cancer Research, 67*, 8433–8438.

Corney, D. C., & Nikitin, A. Y. (2008). MicroRNA and ovarian cancer. *Histology and Histopathology, 23*, 1161–1169.

Corvol, H., Flamein, F., Epaud, R., Clement, A., & Guillot, L. (2009). Lung alveolar epithelium and interstitial lung disease. *The International Journal of Biochemistry & Cell Biology, 41*, 1643–1651.

Costa, R. H. (2005). FoxM1 dances with mitosis. *Nature Cell Biology, 7*, 108–110.

Costa, R. H., Kalinichenko, V. V., Holterman, A. X. L., & Wang, X. (2003). Transcription factors in liver development, differentiation, and regeneration. *Hepatology, 38*, 1331–1347.

Costa, R. H., Kalinichenko, V. V., & Lim, L. (2001). Transcription factors in lung development and function. *American Journal of Physiology. Lung Cellular and Molecular Physiology, 280*, L823–L838.

Costa, R. H., Kalinichenko, V. V., Major, M. L., & Raychaudhuri, P. (2005). New and unexpected: Forkhead meets ARF. *Current Opinion in Genetics & Development, 15*, 42–48.

Costa, R. H., Kalinichenko, V. V., Tan, Y., & Wang, I. C. (2005). The CAR nuclear receptor and hepatocyte proliferation. *Hepatology, 42*, 1004–1008.

Courtois-Cox, S., Jones, S. L., & Cichowski, K. (2008). Many roads lead to oncogene-induced senescence. *Oncogene, 27*, 2801–2809.

Coward, W. R., Saini, G., & Jenkins, G. (2010). The pathogenesis of idiopathic pulmonary fibrosis. *Therapeutic Advances in Respiratory Disease, 4*, 367–388.

Cowling, V. H., & Cole, M. D. (2006). Mechanism of transcriptional activation by the Myc oncoproteins. *Seminars in Cancer Biology, 16*, 242–252.

Craig, E. A. (1985). The heat shock response. *CRC Critical Reviews in Biochemistry, 18*, 239–280.

Craig, E. A., & Gross, C. A. (1991). Is hsp70 the cellular thermometer? *Trends in Biochemical Sciences, 16*, 135–140.

Craig, D. W., O'Shaughnessy, J. A., Kiefer, J. A., Aldrich, J., Sinari, S., Moses, T. M., et al. (2013). Genome and transcriptome sequencing in prospective metastatic negative breast cancer uncovers therapeutic vulnerabilities. *Molecular Cancer Therapeutics, 12*, 104–116.

Crespo, P., & Leon, J. (2000). Ras proteins in the control of the cell cycle and cell differentiation. *Cellular and Molecular Life Sciences, 57*, 1613–1636.

Crystal, R. G., Bitterman, P. B., Mossman, B., Schwarz, M. I., Sheppard, L., Almasy, H. A., et al. (2002). Future research directions in idiopathic pulmonary fibrosis: Summary of a National Heart, Lung, and Blood Institute working group. *American Journal of Respiratory and Critical Care Medicine, 166*, 236–246.

Cully, M., & Downward, J. (2008). SnapShot: Ras signaling. *Cell, 133*, 1292.

Curran, D. R., & Cohn, L. (2010). Advances in mucous cell metaplasia: A plug for mucus as a therapeutic focus in chronic airway disease. *American Journal of Respiratory Cell and Molecular Biology, 42*, 268–275.

Curtin, J. C., & Lorenzi, M. V. (2010). Drug discovery approaches to target Wnt signaling in cancer stem cells. *Oncotarget, 1*, 563–566.

d'Adda di Fagagna, F. (2008). Living on a break: Cellular senescence as a DNA-damage response. *Nature Reviews. Cancer, 8*, 512–522.

Daboussi, F., Dumay, A., Delacote, F., & Lopez, B. S. (2002). DNA double-strand break repair signalling: The case of RAD51 post-translational regulation. *Cell Cycle, 14*, 969–975.

Dai, B., Gong, A., Jing, Z., Aldape, K. D., Kang, S. H., Sawaya, R., et al. (2013). Forkhead box M1 is regulated by heat shock factor 1 and promotes glioma cells survival under heat shock stress. *Journal of Biological Chemistry, 288*, 1634–1642.

Dai, B. D., Kang, S. H., Gong, W., Liu, M., Aldape, K. D., Sawaya, R., et al. (2007). Aberrant FoxM1B expression increases matrix metalloproteinase-2 transcription and enhances the invasion of glioma cells. *Oncogene, 26*, 6212–6219.

Dai, B., Pieper, R. O., Li, D., Wei, P., Liu, M., Woo, S. Y., et al. (2010). FoxM1B regulates NEDD4-1 expression, leading to cellular transformation and full malignant phenotype in immortalized astrocytes. *Cancer Research, 70*, 2951–2961.

D'Andrea, A. D. (2003). The Fanconi road to cancer. *Genes & Development, 17*, 1933–1936.

Dang, C. V. (2012). MYC on the path to cancer. *Cell, 149*, 22–35.

Dang, C. V., O'Donnell, K. A., Zeller, K. I., Nguyen, T., Osthus, R. C., & Li, F. (2006). The c-Myc target gene network. *Seminars in Cancer Biology, 16*, 253–264.

Daugaard, M., Rohde, M., & Jäättela, M. (2007). The heat shock protein 70 family: Highly homologous proteins with overlapping and distinct functions. *FEBS Letters, 581,* 3702–3710.

Davis, D. B., Lavine, J. A., Suhonen, J. I., Krautkramer, K. A., Rabaglia, M. E., Sperger, J. M., et al. (2010). FoxM1 is up-regulated by obesity and stimulates β-cell proliferation. *Molecular Endocrinology, 24,* 1822–1834.

Davisson, M. T., Schmidt, C., & Akeson, E. C. (1990). Segmental trisomy of murine chromosome 16: A new model system for studying Down syndrome. *Progress in Clinical and Biological Research, 360,* 263–280.

Dawson, M. A., & Kouzarides, T. (2012). Cancer epigenetics: From mechanism to therapy. *Cell, 150,* 12–27.

de Olano, N., Koo, C. Y., Monteiro, L. J., Pinto, P. H., Gomes, A. R., Aligue, R., et al. (2012). The p38-MAPK-MK2 axis regulates E2F1 and FOXM1 expression after epirubicin treatment. *Molecular Cancer Research, 10,* 1189–1202.

de Vries, W. V., Evsikov, A. V., Brogan, L. J., Anderson, C. P., Graber, J. H., Knowles, B. B., et al. (2008). Reprogramming and differentiation in mammals: Motifs and mechanisms. *Cold Spring Harbor Symposia on Quantitative Biology, 73,* 33–38.

Dean, J. L., McClendon, A. K., Stengel, K. R., & Knudsen, E. S. (2012). Modification of the DNA damage response by therapeutic CDK4/6 inhibition. *Oncogene, 29,* 68–80.

DeBerardinis, R. J. (2011). Serine metabolism: Some tumors take the road less traveled. *Cell Metabolism, 14,* 285–286.

DeCaprio, J. A. (2009). How the Rb tumor suppressor structure and function was revealed by the study of Adenovirus and SV40. *Virology, 384,* 274–284.

Deckbar, D., Jeggo, P. A., & Löbrich, M. (2011). Understanding the limitations of radiation-induced cell cycle checkpoints. *Critical Reviews in Biochemistry and Molecular Biology, 46,* 271–283.

DeGregori, J. (2002). The genetics of the E2F family of transcription factors: Shared functions and unique roles. *Biochimica et Biophysica Acta, 1602,* 131–150.

DeGregori, J., & Johnson, D. G. (2006). Distinct and overlapping roles for E2F family members in transcription, proliferation and apoptosis. *Current Molecular Medicine, 6,* 739–748.

Dejana, E. (2004). Endothelial cell–cell junctions: Happy together. *Nature Reviews. Molecular Cell Biology, 5,* 261–270.

Dejana, E., Orsenigo, F., & Lampugnani, M. G. (2008). The role of adherens junctions and VE-cadherin in the control of vascular permeability. *Journal of Cell Science, 121,* 2115–2122.

Dejana, E., Orsenigo, F., Molendini, C., Baluk, P., & McDonald, D. M. (2009). Organization and signaling of endothelial cell-to-cell junctions in various regions of the blood and lymphatic vascular trees. *Cell and Tissue Research, 335,* 17–25.

Dejosez, M., & Zwaka, T. P. (2012). Pluripotency and nuclear reprogramming. *Annual Review of Biochemistry, 81,* 737–765.

Delpuech, O., Griffiths, B., East, P., Essafi, A., Lam, E. W. F., Burgering, B., et al. (2007). Induction of Mxi1-SRα by FOXO3a contributes to repression of Myc dependent gene expression. *Molecular and Cellular Biology, 27,* 4917–4930.

Demirci, C., Ernst, S., Alvarez-Perez, J. C., Rosa, T., Valle, S., Shridhar, V., et al. (2012). Loss of HGF/c-Met signaling in pancreatic b-cells leads to incomplete maternal b-cell adaptation and gestational diabetes. *Diabetes, 61,* 1143–1152.

Denicourt, C., & Dowdy, S. F. (2004). Cip/Kip proteins: More than just CDKs inhibitors. *Genes & Development, 18,* 851–855.

Der, C. J., & Van Dyke, T. (2007). Stopping Ras in its tracks. *Cell, 129,* 855–857.

Desdouets, C., Sobczak-Thepot, J., Murphy, M., & Brechot, C. (1995). Cyclin A: Function and expression during cell proliferation. *Progress in Cell Cycle Research, 1,* 115–123.

Dever, S. M., White, E. R., Hartman, M. C., & Valerie, K. (2012). BRCA1-directed, enhanced and aberrant homologous recombination: Mechanism and potential treatment strategies. *Cell Cycle, 11*, 687–694.

Dhawan, S., Georgia, S., & Bhushan, A. (2007). Formation and regeneration of the endocrine pancreas. *Current Opinion in Cell Biology, 19*, 634–645.

Di Micco, R., Fumagalli, M., & d'Adda di Fagagna, F. (2007). Breaking news: High-speed race ends in arrest—How oncogenes induce senescence. *Trends in Cell Biology, 17*, 529–536.

Dibb, M., Han, N., Choudhury, J., Hayes, S., Valentine, H., West, C., et al. (2012). The FOXM1-PLK1 axis is commonly upregulated in oesophageal adenocarcinoma. *British Journal of Cancer, 107*, 1766–1775.

Diehl, A. M., & Rai, R. M. (1996). Liver regeneration 3: Regulation of signal transduction during liver regeneration. *The FASEB Journal, 10*, 215–227.

Diller, K. R. (2006). Stress protein expression kinetics. *Annual Review of Biomedical Engineering, 8*, 403–424.

Dimova, D. K., & Dyson, N. J. (2005). The E2F transcriptional network: Old acquaintances with new faces. *Oncogene, 24*, 2810–2826.

Dimri, G. P. (2005). What has senescence got to do with cancer? *Cancer Cell, 7*, 505–512.

Dominguez-Gerpe, L., & Araujo-Vilar, D. (2008). Prematurely aged children: Molecular alterations leading to Hutchinson-Gilford progeria and Werner syndromes. *Current Aging Science, 1*, 202–212.

Dorn, G. W. (2007). The fuzzy logic of physiological cardiac hypertrophy. *Hypertension, 49*, 962–970.

Down, C. F., Millour, J., Lam, E. W. F., & Watson, R. J. (2012). Binding of FoxM1 to G2/M gene promoters is dependent upon B-Myb. *Biochimica et Biophysica Acta, 1819*, 855–862.

Downward, J. (2003). Targeting Ras signalling pathways in cancer therapy. *Nature Reviews. Cancer, 3*, 11–22.

Durrani, S. R., Viswanathan, R. K., & Busse, W. W. (2011). What effect does asthma treatment have on airway remodeling? Current perspectives. *The Journal of Allergy and Clinical Immunology, 128*, 439–448.

Eddy, E. M. (2002). Male germ cell gene expression. *Recent Progress in Hormone Research, 57*, 103–128.

Eisenman, R. N. (2001a). Deconstructing Myc. *Genes & Development, 15*, 2023–2030.

Eisenman, R. N. (2001b). The Max network: Coordinated transcriptional regulation of growth and proliferation. *Harvey Lectures, 96*, 1–32.

El-Awady, R. A., Saleh, E. M., & Dahm-Daphi, J. (2010). Targeting DNA double-strand break repair: Is it the right way for sensitizing cells to 5-fluorouracil? *Anti-Cancer Drugs, 21*, 277–287.

El-Deiry, W. S., Tokino, T., Velculescu, V. E., Levy, D. B., Parsons, R., Trent, J. M., et al. (1993). WAF1, a potential mediator of p53 tumor suppression. *Cell, 75*, 817–825.

Elgaaen, B. V., Olstad, O. K., Sandvik, L., Odegaard, E., Sauer, T., Staff, A. C., et al. (2012). ZNF385B and VEGFA are strongly differentially expressed in serous ovarian carcinomas and correlate with survival. *PLoS One, 7*, e46317.

Evan, G. I., Christophorou, M., Lawlor, E. A., Ringshausen, I., Prescott, J., Dansen, T., et al. (2005). Oncogene-dependent tumor suppression: Using the dark side of the force for cancer therapy. *Cold Spring Harbor Symposia on Quantitative Biology, 70*, 263–273.

Evan, G. I., & d'Adda di Fagagna, F. (2009). Cellular senescence: Hot or what? *Current Opinion in Genetics & Development, 19*, 25–31.

Evers, B., Helleday, T., & Jonkers, J. (2010). Targeting homologous recombination repair defects in cancer. *Trends in Pharmacological Sciences, 31*, 372–380.

Ewald, J. A., Desotelle, J. A., Wilding, G., & Jarrard, D. F. (2010). Therapy-induced senescence in cancer. *Journal of the National Cancer Institute*, *102*, 1536–1546.

Ewen, M. E. (2000). Relationship between Ras pathways and cell cycle control. *Progress in Cell Cycle Research*, *4*, 1–17.

Fanciulli, M. (2006). *Rb and tumorigenesis*. Landes Bioscience. Berlin, Germany: Springer.

Faust, D., Al-Butmeh, F., Linz, B., & Dietrich, C. (2012). Involvement of the transcription factor FoxM1 in contact inhibition. *Biochemical and Biophysical Research Communications*, *426*, 659–663.

Fausto, N. (2000). Liver regeneration. *Journal of Hepatology*, *32*, 19–31.

Fausto, N. (2004). Liver regeneration and repair: Hepatocytes, progenitor cells and stem cells. *Hepatology*, *39*, 1477–1487.

Fausto, N., Campbell, J. S., & Riehle, K. J. (2006). Liver regeneration. *Hepatology*, *43*, S45–S53.

Fausto, N., Laird, A. D., & Webber, E. M. (1995). Liver regeneration 2. Role of growth factors and cytokines in hepatic regeneration. *The FASEB Journal*, *9*, 1527–1536.

Fausto, N., & Webber, E. M. (1993). Control of liver growth. *Critical Reviews in Eukaryotic Gene Expression*, *3*, 117–135.

Feder, M. E., & Hofman, G. E. (1999). Heat-shock proteins, molecular chaperones, and the stress response: Evolutionary and ecological physiology. *Annual Review of Physiology*, *61*, 243–282.

Felsani, A., Mileo, A. M., & Paggi, M. G. (2006). Retinoblastoma family proteins as key targets of the small DNA virus oncoproteins. *Oncogene*, *25*, 5277–5285.

Ferguson, D. O., Sekiguchi, J. M., Frank, K. M., Gao, Y., Sharpless, N. E., Gu, Y., et al. (2000). The interplay between nonhomologous end-joining and cell cycle checkpoint factors in development, genomic stability, and tumorigenesis. *Cold Spring Harbor Symposia on Quantitative Biology*, *65*, 395–403.

Fernandez, P. C., Frank, S. R., Wang, L., Schroeder, M., Liu, S., Greene, J., et al. (2003). Genomic targets of the human c-Myc protein. *Genes & Development*, *17*, 1115–1129.

Ferrari, S. (2006). Protein kinases controlling the onset of mitosis. *Cellular and Molecular Life Sciences*, *63*, 781–795.

Finn, K., Lowndes, N. F., & Grenon, M. (2012). Eukaryotic DNA damage checkpoint activation in response to double-strand breaks. *Cellular and Molecular Life Sciences*, *69*, 1447–1473.

Fisher, P. B., Prignoli, D. R., Hermo, H., Weinstein, I. B., & Pestka, S. (1985). Effects of combined treatment with interferon and mezerein on melanogenesis and growth in human melanoma cells. *Journal of Interferon Research*, *5*, 1–22.

Foley, K. P., & Eisenman, R. N. (1999). Two MAD tails: What the recent knockouts of Mad1 and Mxi1 tell us about the MYC/MAX/MAD network. *Biochimica et Biophysica Acta*, *1423*, M37–M47.

Forget, A. L., & Kowalczykowski, S. C. (2010). Single-molecule imaging brings Rad51 nucleoprotein filaments into focus. *Trends in Cell Biology*, *20*, 269–276.

Fossey, S. L., Liao, A. T., McCleese, J. K., Bear, M. D., Lin, J., Li, P. K., et al. (2009). Characterization of STAT3 activation and expression in canine and human osteosarcoma. *BMC Cancer*, *9*, 81.

Frame, S., & Balmain, A. (2000). Integration of positive and negative growth signals during ras pathway activation in vivo. *Current Opinion in Genetics & Development*, *10*, 106–113.

Francis, R. E., Myatt, S. S., Krol, J., Hartman, J., Peck, B., McGovern, U. B., et al. (2009). FoxM1 is a downstream target and marker of HER2 overexpression in breast cancer. *International Journal of Oncology*, *35*, 57–68.

Freund, A., Orjalo, A. V., Desprez, P. Y., & Campisi, J. (2010). Inflammatory networks during cellular senescence: Causes and consequences. *Trends in Molecular Medicine*, *16*, 238–246.

Friday, B. B., & Adjei, A. A. (2005). K-ras as a target for cancer therapy. *Biochimica et Biophysica Acta*, *1756*, 127–144.

Fridman, J. S., & Tainsky, M. A. (2008). Critical pathways in cellular senescence and immortalization revealed by gene expression profiling. *Oncogene, 27*, 5975–5987.

Fu, Z., Malureanu, L., Huang, J., Wang, W., Li, H., van Deursen, J. M., et al. (2008). Plk1-dependent phosphorylation of FoxM1 regulates a transcriptional programme required for mitotic progression. *Nature Cell Biology, 10*, 1076–1082.

Fuchs, B., & Braun, A. (2008). Improved mouse models of allergy and allergic asthma—Changes beyond ovalbumin. *Current Drug Targets, 9*, 495–502.

Fuentes, G., & Valencia, A. (2009). Ras classical effectors: New tales from in silico complexes. *Trends in Biochemical Sciences, 34*, 533–539.

Fuh, B., Sobo, M., Cen, L., Josiah, D., Hutzen, B., Cisek, K., et al. (2009). LLL-3 inhibits STAT3 activity, suppresses glioblastoma cell growth and prolongs survival in a mouse glioblastoma model. *British Journal of Cancer, 100*, 106–112.

Fujii, T., Ueda, T., Nagata, S., & Funaga, R. (2010). Essential role of p400/mDomino chromatin-remodeling ATPase in bone marrow hematopoiesis and cell-cycle progression. *Journal of Biological Chemistry, 285*, 30214–30223.

Fujimoto, M., & Nakai, A. (2010). The heat shock factor family and adaptation to proteotoxic stress. *FEBS Journal, 277*, 4112–4125.

Futreal, P. A., Coin, L., Marshall, M., Down, T., Hubbard, T., Wooster, R., et al. (2004). A census of human cancer genes. *Nature Reviews. Cancer, 4*, 177–183.

Gabai, V. L., & Sherman, M. Y. (2002). Invited review: Interplay between molecular chaperones and signaling pathways in survival of heat shock. *Journal of Applied Physiology, 92*, 1743–1748.

Gajiwala, K. S., & Burley, S. K. (2000). Winged helix proteins. *Current Opinion in Structural Biology, 10*, 110–116.

Gajiwala, K. S., Chen, H., Cornille, F., Roques, B. P., Reith, W., Mah, B., et al. (2000). Structure if the winged-helix protein hRFX1 reveals a new mode of DNA binding. *Nature, 403*, 916–921.

Garber, M. E., Troyanskaya, O. G., Schluens, K., Petersen, S., Thaesler, Z., Pacyna-Gengelbach, M., et al. (2001). Diversity of gene expression in adenocarcinoma of the lung. *Proceedings of the National Academy of Sciences, 98*, 13784–13789.

Garrido, C., Brunet, M., Didelot, C., Zermati, Y., Schmitt, E., & Kroemer, G. (2006). Heat shock proteins 27 and 70. Anti-apoptotic proteins with tumorigenic properties. *Cell Cycle, 5*, 2592–2601.

Garrido, C., Gurguxani, S., Ravagnan, L., & Kroemer, G. (2001). Heat shock proteins: Endogenous modulators of apoptotic cell death. *Biochemical and Biophysical Research Communications, 286*, 433–442.

Garrido, C., Schmitt, E., Cande, C., Vahsen, N., Parcellier, A., & Kroemer, G. (2003). HSP27 and HSP70. Potentially oncogenic apoptosis inhibitors. *Cell Cycle, 2*, 579–584.

Gartel, A. L. (2008). FoxM1 inhibitors as potential anticancer drugs. *Expert Opinion on Therapeutic Targets, 12*, 663–665.

Gartel, A. L. (2010). A new target for proteasome inhibitors: FoxM1. *Expert Opinion on Investigational Drugs, 19*, 235–242.

Gehrke, I., Gandhirajan, R. K., & Kreuzer, K. A. (2009). Targeting the Wnt/β-catenin/TCF/LEF1 axis in solid and haematological cancers: Multiplicity of therapeutic options. *European Journal of Cancer, 45*, 2759–2767.

Gemenetzidis, E., Bose, A., Riaz, A. M., Chaplin, T., Young, B. D., Ali, M., et al. (2009). FOXM1 upregulation is an early event in human squamous cell carcinoma and it is enhanced by nicotine during malignant transformation. *PLoS One, 4*, e4849.

Gemenetzidis, E., Elena-Costea, D., Parkinson, E. K., Waseem, A., Wan, H., & Teh, M. T. (2010). Induction of epithelial stem/progenitor expansion by FOXM1. *Cancer Research, 70*, 9515–9526.

Geoghegan, E., & Byrnes, L. (2008). Mouse induced pluripotent stem cells. *International Journal of Developmental Biology, 52*, 1015–1022.

Gewirtz, D. A. (1999). A critical evaluation of the mechanisms of action proposed for the antitumor effects of the anthracycline antibiotics adriamycin and daunorubicin. *Biochemical Pharmacology, 57*, 727–741.
Gharaee-Khermani, M., Hu, B., Phan, S. H., & Gyetko, M. R. (2009). Recent advances in molecular targets and treatment of idiopathic pulmonary fibrosis: Focus on TGFb signaling and the myofibroblast. *Current Medicinal Chemistry, 16*, 1400–1417.
Gialmanidis, I. P., Bravou, V., Amanetopoulou, S. G., Varakis, J., Kourea, H., & Papadaki, H. (2009). Overexpression of hedgehog pathway molecules and FOXM1 in non-small cell lung carcinomas. *Lung Cancer, 66*, 64–74.
Gidalevitz, T., Prahlad, V., & Morimoto, R. I. (2011). The stress of protein misfolding: From single cells to multicellular organisms. *Cold Spring Harbor Perspectives in Biology, 3*, a009704.
Giehl, K. (2005). Oncogenic Ras in tumor progression and metastasis. *Biological Chemistry, 386*, 193–205.
Gieling, R. G., Elsharkawy, A. M., Caamano, J. H., Cowie, D. E., Wright, M. C., Ebrahimkhani, M. R., et al. (2010). The c-Rel subunit of nuclear factor-κB regulates murine liver inflammation, wound-healing, and hepatocyte proliferation. *Hepatology, 51*, 922–931.
Gilgenkrantz, H. (2010). Rodent models of liver repopulation. *Methods in Molecular Biology, 640*, 475–490.
Gloaguen, P., Crepieux, P., Heitzler, D., Poupon, A., & Reiter, E. (2011). Mapping the follicle-stimulating hormone-induced signaling networks. *Frontiers in Endocrinology, 2*, 45.
Gong, A., & Huang, S. (2012). FoxM1 and Wnt/β-catenin signaling in glioma stem cells. *Cancer Research, 72*, 5658–5662.
Gonzales, K. A. U., & Ng, H. H. (2011). Choreographing pluripotency and cell fate with transcription factors. *Biochimica et Biophysica Acta, 1809*, 337–349.
Gopinathan, L., Ratnacaram, C. K., & Kaldis, P. (2011). Established and novel Cdk/cyclin complexes regulating the cell cycle and development. *Results and Problems in Cell Differentiation, 53*, 365–389.
Gorcynska-Fjälling, E. (2004). The role of calcium in signal transduction processes in Sertoli cells. *Reproductive Biology, 4*, 219–241.
Gordon, K. J., & Blobe, G. C. (2008). Role of transforming growth factor-β superfamily signaling pathways in human disease. *Biochimica et Biophysica Acta, 1782*, 197–228.
Goss, A. M., Tian, Y., Tsukiyama, T., Cohen, E. D., Zhou, D., Lu, M. M., et al. (2009). Wnt2/2b and beta-catenin signaling are necessary and sufficient to specify lung progenitors in the foregut. *Developmental Cell, 17*, 290–298.
Goto, H., Izawa, I., Li, P., & Inagaki, M. (2012). Novel regulation of checkpoint kinase 1: Is checkpoint kinase 1 a good candidate for anti-cancer therapy? *Cancer Science, 103*, 1195–1200.
Grandori, C., Cowley, S. M., James, L. P., & Eisenman, R. N. (2000). The Myc/Mad/Max network and the transcriptional control of cell behaviour. *Annual Review of Cell and Developmental Biology, 16*, 653–699.
Grant, G. D., Gamsby, J., Martyanov, V., Brooks, L., George, L. K., Mahoney, J. M., et al. (2012). Live cell monitoring of periodic gene expression in synchronous human cells identifies forkhead genes involved in cell cycle control. *Molecular Biology of the Cell, 23*, 3079–3093.
Green, M. R., Aya-Bonilla, C., Gandhi, M. K., Lea, R. A., Wellwood, J., Wood, P., et al. (2011). Integrative genomic profiling reveals conserved genetic mechanisms for tumorigenesis in common entities of non-Hedgkin's lymphoma. *Genes, Chromosomes & Cancer, 50*, 313–326.

Greenblatt, M. S., Bennett, W. P., Hollstein, M., & Harris, C. C. (1994). Mutations in the p53 tumor suppressor gene: Clues to cancer etiology and molecular pathogenesis. *Cancer Research, 54*, 4855–4878.

Gregory, L. G., & Lloyd, C. M. (2011). Orchestrating house dust mite-associated allergy in the lung. *Trends in Immunology, 32*, 402–411.

Grompe, M. (1999). Therapeutic liver repopulation for the treatment of metabolic liver disease. *Human Cell, 12*, 171–180.

Grompe, M. (2001). Liver repopulation for the treatment of metabolic diseases. *Journal of Inherited Metabolic Disease, 24*, 231–244.

Grompe, M. (2006). Principles of therapeutic liver repopulation. *Journal of Inherited Metabolic Disease, 29*, 421–425.

Grompe, M., Laconi, E., & Shafritz, D. A. (1999). Principles of therapeutic liver repopulation. *Seminars in Liver Disease, 19*, 7–14.

Guerra, E., Trerotola, M., Aloisi, A. L., Tripaldi, R., Vacca, G., La Sodda, R., et al. (2013). The Trop-2 signalling network in cancer growth. *Oncogene, 32*, 1594–1600.

Guha, C., Deb, N. J., Sappal, B. S., Ghosh, S. S., Roy-Chowdhury, N., & Roy-Chowdhury, J. (2001). Amplification of engrafted hepatocytes by preparative manipulation of the host liver. *Artificial Organs, 25*, 522–528.

Guney, I., Wu, S., & Sedivy, J. M. (2006). Reduced c-Myc signaling triggers telomere-independent senescence by regulating Bmi-1 and p16(INK4a). *Proceeding of the National Academy of Sciences of the United States of America, 103*, 3645–3650.

Günther, A., Korfei, M., Mahavadi, P., von der Beck, D., Ruppert, C., & Markart, P. (2012). Unravelling the progressive pathophysiology of idiopathic pulmonary fibrosis. *European Respiratory Review, 21*, 152–160.

Guo, W. J., Datta, S., Band, V., & Dimri, G. P. (2007). Mel-18, a polycomb group protein, regulates cell proliferation and senescence via transcriptional repression of Bmi-1 and c-Myc oncoproteins. *Molecular Biology of the Cell, 18*, 536–546.

Gupta, S. (2000). Hepatic polyploidy and liver growth control. *Seminars in Cancer Biology, 10*, 161–171.

Gupta, S. (2002). Hepatocyte transplantation. *Journal of Gastroenterology and Hepatology, 17*(Suppl. 3), S287–S293.

Gupta, S., Bhargava, K. K., & Novikoff, P. M. (1999). Mechanisms of cell engraftment during liver repopulation with hepatocyte transplantation. *Seminars in Liver Disease, 19*, 15–26.

Gupta, S., & Chowdhury, J. R. (2002). Therapeutic potential of hepatocyte transplantation. *Seminars in Cell & Developmental Biology, 13*, 439–446.

Gupta, S., Malhi, H., Gagandeep, S., & Novikoff, P. (1999). Liver repopulation with hepatocyte transplantation: New avenues for gene and cell therapy. *The Journal of Gene Medicine, 1*, 386–392.

Gupta, S., Rajvanshi, P., Bhargava, K. K., & Kerr, A. (1996). Hepatocyte transplantation: Progress toward liver repopulation. *Progress in Liver Diseases, 14*, 199–222.

Gupta, S., & Rogler, C. E. (1999). Lessons from genetically engineered animal models VI. Liver repopulaions systems an study of pathophysiological mechanisms in animals. *American Journal of Physiology, 277*, G1097–G1102.

Gupta, S. C., Sharma, A., Mishra, M., Mishra, R. K., & Chowdhuri, D. K. (2010). Heat shock proteins in toxicology: How close and how far? *Life Sciences, 86*, 377–384.

Gusarova, G. A., Wang, I. C., Major, M. L., Kalinichenko, V. V., Ackerson, T., Petrovic, V., et al. (2007). A cell-penetrating ARF peptide inhibitor of FoxM1 in mouse hepatocellular carcinoma treatment. *The Journal of Clinical Investigation, 117*, 99–111.

Hahn, W. C., & Weinberg, R. A. (2002a). Modelling the molecular circuitry of cancer. *Nature Reviews. Cancer, 2*, 331–341.

Hahn, W. C., & Weinberg, R. A. (2002b). Rules for making human tumor cells. *The New England Journal of Medicine, 347*, 1593–1603.

Hainaut, P., & Hollstein, M. (2000). p53 and human cancer: The first ten thousand mutations. *Advances in Cancer Research, 77*, 81–137.
Hainaut, P., Soussi, T., Shomer, B., Hollstein, M., Greenblatt, M., Hovig, E., et al. (1997). Database of p53 gene somatic mutations in human tumors and cell lines: Updated compilation and future prospects. *Nucleic Acids Research, 25*, 151–157.
Hakem, R. (2008). DNA-damage repair; the good, the bad, and the ugly. *EMBO Journal, 27*, 589–605.
Halasi, M., & Gartel, A. L. (2009). A novel mode of FoxM1 regulation: Positive autoregulatory loop. *Cell Cycle, 8*, 1966–1967.
Halasi, M., & Gartel, A. L. (2012). Suppression of FOXM1 sensitizes human cancer cells to cell death induced by DNA-damage. *PLoS One, 7*, e31761.
Halder, G., & Johnson, R. L. (2011). Hippo signaling: Growth control and beyond. *Development, 138*, 9–22.
Hall, M., & Peters, G. (1996). Genetic alterations of cyclins, cyclin-dependent kinases, and cdk inhibitors in human cancer. *Advances in Cancer Research, 68*, 67–108.
Hallstrom, T. C., Mori, S., & Nevins, J. R. (2008). An E2F1-dependent gene expression program that determines the balance between proliferation and cell death. *Cancer Cell, 13*, 11–22.
Hammad, H., Plantinga, M., Deswarte, K., Pouliot, P., Willart, M. A., Kool, M., et al. (2010). Inflammatory dendritic cells—Not basophils—Are necessary and sufficient for induction of Th2 immunity to inhaled house dust mite allergen. *The Journal of Experimental Medicine, 207*, 2097–2111.
Hanahan, D., & Weinberg, R. A. (2000). The hallmarks of cancer. *Cell, 100*, 57–70.
Hanahan, D., & Weinberg, R. A. (2011). Hallmarks of cancer: The next generation. *Cell, 144*, 646–674.
Hanna, J. H., Saha, K., & Jaenisch, R. (2010). Pluripotency and cellular reprogramming: Facts, hypotheses, unresolved issues. *Cell, 143*, 508–525.
Hannenhalli, S., & Kaestner, K. H. (2009). The evolution of Fox genes and their role in development and disease. *Nature Reviews. Genetics, 10*, 233–240.
Hansson, V., Skalhegg, B. S., & Tasken, K. (2000). Cyclic-AMP-dependent protein kinase (PKA) in testicular cells. Cell specific expression, differential regulation and targeting of subunits of PKA. *The Journal of Steroid Biochemistry and Molecular Biology, 73*, 81–92.
Hardie, W. G., Hagood, J. S., Dave, V., Perl, A. K., Whitsett, J. A., Korfhagen, T. R., et al. (2010). Signaling pathways in the epithelial origins of pulmonary fibrosis. *Cell Cycle, 9*, 1769–1776.
Hardy, K., Mansfield, L., Mackay, A., Benvenuti, S., Ismail, S., Arora, P., et al. (2005). Transcriptional networks and cellular senescence in human mammary fibroblasts. *Molecular Biology of the Cell, 16*, 943–953.
Harper, J. W., & Elledge, S. J. (2007). The DNA damage response: Ten years after. *Molecular Cell, 28*, 739–745.
Harris, T., & McCormick, F. (2010). Gene and drug matrix for personalized cancer therapy. The molecular pathology of cancer. *Nature Reviews. Drug Discovery, 7*, 251–265.
Harrison, J. C., & Haber, J. E. (2006). Surviving the break: The DNA damage checkpoint. *Annual Review of Genetics, 40*, 209–235.
Hart, G., Housley, M. P., & Slawson, C. (2007). Cycling of O-linked β-N-acetylglucosamine on nucleocytoplasmic proteins. *Nature, 446*, 1017–1022.
Hart, G. W., Slawson, C., Ramirez-Correa, G., & Lagerlof, O. (2011). Cross talk between O-GlcNAcylation and phosphorylation: Roles in signaling, transcription, and chronic disease. *Annual Review of Biochemistry, 80*, 825–858.
Hartl, F. U., Bracher, A., & Hayer-Hartl, M. (2011). Molecular chaperones in protein folding and proteostasis. *Nature, 475*, 324–332.

Hasty, P. (2008). Is NHEJ a tumor suppressor or an aging suppressor? *Cell Cycle, 7,* 1139–1145.
Havnes, C. G., & Walter, J. C. (2011). Mechanism of CRL4(Cdt2), a PCNA-dependent E3 ubiquitin ligase. *Genes & Development, 25,* 1568–1582.
Hayakawa, R., Hayakawa, T., Takeda, K., & Ichijo, H. (2012). Therapeutic targets in the ASK1-dependent stress signaling pathway. *Proceedings of the Japan Academy. Series B, Physical and Biological Sciences, 88,* 434–453.
He, X., He, L., & Hannon, G. J. (2007). The guardian's little helper: MicroRNAs in the p53 tumor suppressor network. *Cancer Research, 67,* 11099–11101.
He, L., He, X., Lim, L. P., de Stanchina, E., Xuan, Z., Liang, Y., et al. (2007). A microRNA component of the p53 tumour suppressor network. *Nature, 447,* 1130–1134.
He, L., He, X., Lowe, S. W., & Hannon, G. J. (2007). MicroRNAs join the p53 network— Another piece in the tumour-suppression puzzle. *Nature Reviews. Cancer, 7,* 819–822.
He, S. Y., Shen, H. W., Xu, L., Zhao, X. H., Yuan, L., Niu, G., et al. (2012). FOXM1 promotes tumor cell invasion and correlates with poor prognosis in early-stage cervical cancer. *Gynecologic Oncology, 127,* 601–610.
He, L., Yang, X., Cao, X., Liu, F., Quan, M., & Cao, J. (2013). Casticin induces growth suppression and cell cycle arrest through activation of FOXO3a in hepatocellular carcinoma. *Oncology Reports, 29,* 103–108.
Hedge, N. S., Sanders, D. A., Rodriguez, R., & Balasubramanian, S. (2011). The transcription factor FOXM1 is a cellular target of the natural product thiostrepton. *Nature Chemistry, 3,* 725–731.
Heineke, J., & Molkentin, J. D. (2006). Regulation of cardiac hypertrophy by intracellular signalling pathways. *Nature Reviews. Molecular Cell Biology, 7,* 589–600.
Helleday, T. (2010). Homologous recombination in cancer development, treatment and development of drug resistance. *Carcinogenesis, 31,* 955–960.
Helt, A. M., & Galloway, D. A. (2003). Mechanisms by which DNA tumor virus oncoproteins target the Rb family of pocket proteins. *Carcinogenesis, 24,* 159–169.
Hemann, M. T., & Narita, M. (2007). Oncogenes and senescence: Breaking down the fast lane. *Genes & Development, 21,* 1–5.
Hennekam, R. C. (2006). Hutchinson-Gilford progeria syndrome: Review of the phenotype. *American Journal of Medical Genetics. Part A, 140,* 2603–2624.
Henning, W., & Stürzbecher, H. W. (2003). Homologous recombination and cell cycle checkpoints: Rad51 in tumour progression and therapy resistance. *Toxicology, 193,* 91–109.
Henninghausen, L., & Robinson, G. W. (2005). Information networks in the mammary gland. *Nature Reviews. Molecular Cell Biology, 6,* 715–725.
Henriksson, M., & Lüscher, B. (1996). Proteins of the Myc network: Essential regulators of cell growth and differentiation. *Advances in Cancer Research, 68,* 111–182.
Hermeking, H. (2010). The miR-34 family in cancer and apoptosis. *Cell Death and Differentiation, 17,* 193–199.
Hermeking, H. (2012). MicroRNAs in the p53 network: Micromanagement of tumour suppression. *Nature Reviews. Cancer, 12,* 613–626.
Herr, P., Hausmann, G., & Basler, K. (2012). WNT secretion and signalling in human disease. *Trends in Molecular Medicine, 18,* 483–493.
Hesketh, R. (1997). *The oncogene and tumor suppressor gene facts book.* San Diego, CA: Academic Press.
Heyer, W. D., Ehmsen, K. T., & Liu, J. (2010). Regulation of homologous recombination in eukaryotes. *Annual Review of Genetics, 44,* 113–139.
Hightower, L. E. (1991). Heat shock, stress proteins, chaperones, and proteotoxicity. *Cell, 66,* 191–197.
Ho, C., Wang, C., Mattu, S., Destefanis, G., Ladu, S., Delogu, S., et al. (2012). AKT and N-Ras co-activation in the mouse liver promotes rapid carcinogenesis via mTORC1, FOXM1/SKP2, and c-Myc pathways. *Hepatology, 55,* 833–845.

Hochedlinger, K., & Jaenisch, R. (2006). Nuclear reprogramming and pluripotency. *Nature*, *441*, 1061–1067.

Hochedlinger, K., & Plath, K. (2009). Epigenetic reprogramming and induced pluripotency. *Development*, *136*, 509–523.

Hochegger, H., Takeda, S., & Hunt, T. (2008). Cyclin-dependent kinases and cell-cycle transitions: Does one fit all? *Nature Reviews. Molecular Cell Biology*, *9*, 910–916.

Hodgson, G., Yeh, R. F., Ray, A., Wang, N., Smirnov, I., Yu, M., et al. (2009). Comparative analyses of gene copy number and mRNA expression in GBM tumors and GBM xenografts. *Neuro-Oncology*, *11*, 477–487.

Hoeijmakers, J. H. J. (2001). Genome maintenance mechanisms for preventing cancer. *Nature*, *411*, 366–374.

Holgate, S. T. (2012). Innate and adaptive immune responses in asthma. *Nature Medicine*, *18*, 673–783.

Hollstein, M., Rice, K., Greenblatt, M. S., Soussi, T., Fuchs, S., Sorlie, T., et al. (1994). Database of p53 gene somatic mutations in human tumors and cell lines. *Nucleic Acids Research*, *22*, 3551–3555.

Hollstein, M., Sidransky, D., Vogelstein, B., & Harris, C. C. (1991). p53 mutations in human cancers. *Science*, *253*, 49–53.

Hooker, C. W., & Hurlin, P. J. (2005). Of Myc and Mnt. *Journal of Cell Science*, *119*, 208–216.

Horimoto, Y., Hartman, J., Millour, J., Pollock, S., Olmos, Y., Ho, K. K., et al. (2011). ERβ1 represses FOXM1 expression through targeting ERα to control cell proliferation in breast cancer. *American Journal of Pathology*, *179*, 1148–1156.

Hromas, R., & Costa, R. (1995). The hepatocyte nuclear factor-3/forkhead transcription regulatory family in development, inflammation, and neoplasia. *Critical Reviews in Oncology/Hematology*, *20*, 129–140.

Hua, Z. C., Sohn, A. J., Kang, C., Cado, D., & Winoto, A. (2003). A function of Fas-associated death domain protein in cell cycle progression localized to a single amino acid at its C-terminal region. *Immunity*, *18*, 513–521.

Huang, C. (2013). Wild type offspring of heterozygous prolactin receptor null female mice have maladaptive β-cell responses during pregnancy. *The Journal of Physiology*, *591*, 1325–1338.

Huang, Y. H., Li, D., Winoto, A., & Robey, E. A. (2004). Distinct transcriptional programs in thymocytes responding to T cell receptor, Notch, and positive selection signals. *Proceeding of the National Academy of Sciences of the United States of America*, *101*, 4936–4941.

Huang, W., Ma, K., Zhang, J., Qatanani, M., Cuvillier, J., Liu, J., et al. (2006). Nuclear receptor-dependent bile acid signaling is required for normal liver regeneration. *Science*, *312*, 233–236.

Huang, C., Qiu, Z., Wang, L., Peng, Z., Jia, Z., Logsdon, C., et al. (2012). A novel FoxM1-caveolin signaling pathway promotes pancreatic cancer invasion and metastasis. *Cancer Research*, *72*, 655–665.

Huang, X., & Zhao, Y. Y. (2012). Transgenic expression of FoxM1 promotes endothelial repair following lung injury induced by polymicrobial sepsis in mice. *PLoS One*, *7*, e50094.

Huen, M. S., Sy, S. M., & Chen, J. (2010). BRCA1 and its toolbox for the maintenance of genomic integrity. *Nature Reviews. Molecular Cell Biology*, *11*, 138–148.

Hui, C. C., & Angers, S. (2011). Gli proteins in development and disease. *Annual Review of Cell and Developmental Biology*, *27*, 513–537.

Hui, M. K. C., Chan, K. W., Luk, J. M., Lee, N. P., Chung, Y., Cheung, L. C. M., et al. (2012). Cytoplasmic Forkhead Box M1 (FoxM1) in esophageal squamous cell carcinoma significantly correlates with pathological disease stage. *World Journal of Surgery*, *36*, 90–97.

Hurlin, P. J., & Dezfouli, S. (2004). Functions of Myc:Max in the control of cell proliferation and tumorigenesis. *International Review of Cytology*, *238*, 183–226.

Hurlin, P. J., & Huang, J. (2006). The MAX-interacting transcription factor network. *Seminars in Cancer Biology*, *16*, 265–274.

Hurlin, P. J., Zhou, Z. Q., Toyo-oka, K., Ota, S., Walker, W. L., Hirotsune, S., et al. (2004). Evidence of Mnt-Myc antagonism revealed by Mnt gene deletion. *Cell Cycle, 3*, 97–99.

Huynh, K. M., Kim, G., Kim, D. J., Yang, S. J., Park, S. M., Yeom, Y. I., et al. (2009). Gene expression analysis of terminal differentiation of human melanoma cells highlights global reductions in cell cycle-associated genes. *Gene, 433*, 32–39.

Huynh, K. M., Soh, J. W., Dash, R., Sarkar, D., Fisher, P. B., & Kang, D. (2010). FOXM1 expression mediates growth suppression during terminal differentiation of HO-1 human metastatic melanoma cells. *Journal of Cellular Physiology, 226*, 194–204.

Hydbring, P., Bahram, F., Su, Y., Tronnersjö, S., Högstrand, K., von der Lehr, N., et al. (2009). Phosphorylation by Cdk2 is required for Myc to repress Ras-induced senescence in cotransformation. *Proceeding of the National Academy of Sciences of the United States of America, 107*, 58–63.

Iacobuzio-Donahue, C. A., Maitra, A., Olsen, M., Lowe, A. W., Van Heek, N. T., Rosty, C., et al. (2003). Exploration of global gene expression patterns in pancreatic adenocarcinoma using cDNA microarrays. *American Journal of Pathology, 162*, 1151–1162.

Iakova, P., Awad, S. S., & Timchenko, N. A. (2003). Aging reduces proliferative capacities of liver by switching pathways of C/EBPα growth arrest. *Cell, 113*, 495–506.

Iaquinta, P. J., & Lees, J. A. (2007). Life and death decisions by the E2F transcription factors. *Current Opinion in Cell Biology, 19*, 649–657.

Ingham, P. W. (2008). Hedgehog signalling. *Current Biology, 18*, R238–R241.

Ingham, P. W., Nakano, Y., & Seger, C. (2011). Mechanisms and functions of Hedgehog signalling across the metazoa. *Nature Reviews. Genetics, 12*, 393–406.

Ishikawa, K., Ishii, H., & Saito, T. (2006). DNA damage-dependent cell cycle checkpoints and genomic stability. *DNA and Cell Biology, 25*, 406–411.

Issad, T., & Kuo, M. S. (2008). O-GlcNAc modification of transcription factors, glucose sensing and glucotoxicity. *Trends in Endocrinology and Metabolism, 19*, 381–389.

Itahana, K., Campisi, J., & Dimri, G. P. (2004). Mechanisms of cellular senescence in human and mouse cells. *Biogerontology, 5*, 1–10.

Jäättelä, M. (1999). Heat shock proteins as cellular lifeguards. *Annals of Medicine, 31*, 261–271.

Jäättelä, M., & Wissing, D. (1992). Emerging role of heat shock proteins in biology and medicine. *Annals of Medicine, 24*, 249–258.

Jackson, S. P. (2009). The DNA-damage response: New molecular insights and new approaches to cancer therapy. *Biochemical Society Transactions, 37*, 483–494.

Jackson, S. P., & Bartek, J. (2009). The DNA damage response in human biology and disease. *Nature, 461*, 1071–1078.

Jackson, B. C., Carpenter, C., Nebert, D. W., & Vasiliou, V. (2010). Update of human and mouse forkhead box (FOX) gene families. *Human Genomics, 4*, 345–352.

Jaenisch, R., & Young, R. (2008). Stem cells, the molecular circuitry of pluripotency and nuclear reprogramming. *Cell, 132*, 567–582.

Janus, J. R., Laborde, R. R., Greenberg, A. J., Wang, V. W., Wei, W., Trier, A., et al. (2011). Linking expression of *FOXM1, CEP55* and *HELLS* to tumorigenesis in oropharyngeal squamous cell carcinoma. *Laryngoscope, 121*, 2598–2603.

Jeggo, P. A., Geuting, V., & Löbrich, M. (2011). The role of homologous recombination in radiation-induced double-strand break repair. *Radiotherapy and Oncology, 101*, 7–12.

Jia, R., Li, C., McCoy, J. P., Deng, C. X., & Zheng, Z. M. (2010). SRp20 is a proto-oncogene critical for cell proliferation and tumor induction and maintenance. *International Journal of Biological Sciences, 6*, 806–826.

Jiang, J., & Hui, C. C. (2008). Hedgehog signaling in development and cancer. *Developmental Cell, 15*, 801–812.

Jiang, L. Z., Wang, P., Deng, B., Huang, C., Tang, W. X., Lu, H. Y., et al. (2011). Overexpression of Forkhead Box M1 transcription factor and nuclear factor-κB in laryngeal squamous cell carcinoma: A potential indicator for poor prognosis. *Human Pathology, 42,* 1185–1193.

Jin, C., Marsden, I., Chen, X., & Liao, X. (1999). Dynamic DNA contacts observed in the NMR structure of winged helix protein–DNA complex. *Journal of Molecular Biology, 289,* 683–690.

Jin, J., Wang, G. L., Iakova, P., Shi, X., Haefliger, S., Finegold, M., et al. (2010). Epigenetic changes play critical role in age-associated dysfunctions of the liver. *Aging Cell, 9,* 895–910.

Jin, J., Wang, G. L., Salisbury, E., Timchenko, L., & Timchenko, N. A. (2009). GSK3beta-cyclin D3-CUGBP1-eIF2 pathway in aging and myotonic dystrophy. *Cell Cycle, 8,* 2356–2359.

Jin, J., Wang, G. L., Timchenko, L., & Timchenko, N. A. (2009). GSK3beta and aging liver. *Aging, 1,* 582–585.

Jirawatnotai, S., Hu, Y., Michowski, W., Elias, J. E., Becks, L., Bienvenu, F., et al. (2011). A function for cyclin D1 in DNA repair uncovered by protein interactome analyses in human cancers. *Nature, 474,* 230–234.

Jolly, C., & Morimoto, R. I. (2000). Role of the heat shock response and molecular chaperones in oncogenesis and cell death. *Journal of the National Cancer Institute, 92,* 1564–1572.

Jones, S. M., & Kazlauskas, A. (2001a). Connecting signaling and cell cycle progression in growth factor-stimulated cells. *Oncogene, 20,* 6558–6567.

Jones, S. M., & Kazlauskas, A. (2001b). Growth factor-dependent signaling and cell cycle progression. *Chemical Reviews, 101,* 2413–2423.

Jones, K., Timchenko, L., & Timchenko, N. A. (2012). The role of CUGBP1 in age-dependent changes of liver functions. *Ageing Research Reviews, 11,* 442–449.

Jones-Villeneuve, E. M., McBurney, M. W., Rogers, K. A., & Kalnins, V. I. (1982). Retinoic acid induces embryonal carcinoma cells to differentiate into neurons and glial cells. *The Journal of Cell Biology, 94,* 253–262.

Joshi, P. A., Jackson, H. W., Beristain, A. G., Di Grappa, M. A., Mote, P. A., Clarke, C. L., et al. (2010). Progesterone induces adult mammary stem cell expansion. *Nature, 465,* 803–807.

Kahlem, P. (2006). Gene-dosage effect on chromosome 21 transcriptome in trisomy 21: Implication in Down syndrome cognitive disorders. *Behavior Genetics, 36,* 416–428.

Kalin, T. V., Ustiyan, V., & Kalinichenko, V. V. (2011). Multiple faces of FoxM1 transcription factor. Lessons from transgenic mouse models. *Cell Cycle, 10,* 396–405.

Kalin, T. V., Wang, I. C., Ackerson, T. J., Major, M. L., Detrisac, C. J., Kalinichenko, V. V., et al. (2006). Increased levels of the FoxM1 transcription factor accelerate development and progression of prostate carcinomas in both TRAMP and LADY transgenic mice. *Cancer Research, 66,* 1712–1720.

Kalin, T. V., Wang, I. C., Meliton, L., Zhang, Y., Wert, S. E., Ren, X., et al. (2008). Forkhead Box m1 transcription factor is required for perinatal lung function. *Proceeding of the National Academy of Sciences of the United States of America, 105,* 19329–19334.

Kalinichenko, V. V., Gusarova, G. A., Tan, Y., Wang, I. C., Major, M. L., Wang, X., et al. (2003). Ubiquitous expression of the Forkhead Box M1B transgene accelerates proliferation of distinct pulmonary cell types following lung injury. *Journal of Biological Chemistry, 39,* 37888–37894.

Kalinichenko, V. V., Lim, L., Shin, B., & Costa, R. H. (2001). Differential expression of forkhead box transcription factors following butylated hydroxytoluene lung injury. *American Journal of Physiology. Lung Cellular and Molecular Physiology, 280,* L695–L704.

Kalinichenko, V. V., Major, M. L., Wang, X., Petrovic, V., Kuechle, J., Yoder, H. M., et al. (2004). Foxm1b transcription factor is essential for development of hepatocellular carcinomas and is negatively regulated by the p19ARF tumor suppressor. *Genes & Development, 18,* 830–850.

Kalinina, O. A., Kalinin, S. A., Polack, E. W., Mikaelian, I., Panda, S., Costa, R. H., et al. (2003). Sustained hepatic expression of FoxM1B in transgenic mice has minimal effects on hepatocellular carcinoma development but increases cell proliferation rates in preneoplastic and early neoplastic lesions. *Oncogene, 22*, 6266–6276.

Kamemura, K., & Hart, G. W. (2003). Dynamic interplay between O-glycosylation and O-phosphorylation of nucleocytoplasmic proteins: A new paradigm for metabolic control of signal transduction and transcription. *Progress in Nucleic Acid Research and Molecular Biology, 73*, 107–136.

Kampinga, H. H., Hageman, J., Vos, M. J., Kubota, H., Tanguay, R. M., Bruford, E. A., et al. (2009). Guidelines for the nomenclature of the human heat shock proteins. *Cell Stress & Chaperones, 14*, 105–111.

Kao, H. T., & Nevins, J. R. (1983). Transcriptional activation and subsequent control of the human heat shock gene during adenovirus infection. *Molecular and Cellular Biology, 3*, 2058–2065.

Karadedou, C. T. (2006). Regulation of the FOXM1 transcription factor by the estrogen receptor α at the protein level in breast cancer. *Hippokratia, 10*, 128–132.

Karadedou, C. T., Gomes, A. R., Chen, J., Petkovic, M., Ho, K. K., Zwolinska, A. K., et al. (2012). FOXO3a represses VEGF expression through FOXM1-dependent and -independent mechanisms in breast cancer. *Oncogene, 31*, 1845–1858.

Karge, H., & Borok, Z. (2012). EMT and interstitial lung disease: A mysterious relationship. *Current Opinion in Pulmonary Medicine, 18*, 517–523.

Karnoub, A. E., & Weinberg, R. A. (2008). Ras oncogenes: Split personalities. *Nature Reviews. Molecular Cell Biology, 9*, 517–531.

Kass, E. M., & Jasin, M. (2010). Collaboration and competition between DNA double-strand break repair pathways. *FEBS Letters, 584*, 3703–3708.

Kastan, M. B. (2008). DNA damage responses: Mechanisms and roles in human disease. *Molecular Cancer Research, 6*, 517–524.

Kastan, M. B., & Bartek, J. (2004). Cell-cycle checkpoints and cancer. *Nature, 432*, 316–323.

Kästner, K. H., Knöchel, W., & Martinez, D. E. (2000). Unified nomenclature for the winged helix/forkhead transcription factors. *Genes & Development, 14*, 142–146.

Katoh, M., & Katoh, M. (2004). Human FOX gene family. *International Journal of Oncology, 25*, 1495–1500.

Kaufmann, E., & Knöchel, W. (1996). Five years on the wings of fork head. *Mechanisms of Development, 57*, 3–20.

Kawabata, M., Kawabata, T., & Nishibori, M. (2005). Role of recA/RAD51 family proteins in mammals. *Acta Medica Okayama, 59*, 1–9.

Kawashita, Y., Guha, C., Yamanouchi, K., Ito, Y., Kamohara, Y., & Kanematsu, T. (2005). Liver repopulation: A new concept of hepatocyte transplantation. *Surgery Today, 35*, 705–710.

Keith, B., & Simon, M. C. (2007). Hypoxia-inducible factors, stem cells, and cancer. *Cell, 129*, 465–472.

Kelland, L. (2007). The resurgence of platinum-based cancer chemotherapy. *Nature Reviews. Cancer, 7*, 573–584.

Keller, U. B., Old, J. B., Dorsey, F. C., Nilsson, J. A., Nilsson, L., MacLean, K. H., et al. (2007). Myc targets Cks1 to provoke the suppression of p27KIp1, proliferation and lymphomagenesis. *EMBO Journal, 26*, 2562–2574.

Kelly, K., & Siebenlist, U. (1986). The regulation and expression of c-myc in normal and malignant cells. *Annual Review of Immunology, 4*, 317–338.

Kennedy, R. D., & D'Andrea, A. D. (2005). The Fanconi anaemia/BRCA pathway: New faces in the crowd. *Genes & Development, 19*, 2925–2940.

Kerkhoff, E., & Rapp, U. R. (1998). Cell cycle targets of Ras/Raf signalling. *Oncogene, 17*, 1457–1462.

Kern, F., Niault, T., & Baccarini, M. (2011). Ras and Raf pathways in epidermis development and carcinogenesis. *British Journal of Cancer, 104*, 229–234.

Khanna, K. K., & Jackson, S. P. (2001). DNA double-strand breaks: Signaling, repair and the cancer connection. *Nature Genetics, 27*, 247–254.

Kiang, J. G., & Tsokos, G. C. (1998). Heat shock protein 70 kDa: Molecular biology, biochemistry, and physiology. *Pharmacology & Therapeutics, 80*, 183–201.

Kieran, M. W., Gordon, L., & Kleinman, M. (2007). New approaches to progeria. *Pediatrics, 120*, 834–841.

Kim, I. M., Ackerson, T., Ramakrishna, S., Tretiakova, M., Wang, I. C., Kalin, T. V., et al. (2006). The Forkhead Box M1 transcription factor stimulates the proliferation of tumor cells during development of lung cancer. *Cancer Research, 66*, 2153–2161.

Kim, Y. J., Cha, H. J., Nam, K. H., Yoon, Y., Lee, H., & An, S. (2011). Centella asiatica extracts modulate hydrogen peroxide-induced senescence in human dermal fibroblasts. *Experimental Dermatology, 20*, 998–1003.

Kim, H. J., Hwang, N. R., & Lee, K. J. (2007). Heat shock responses for understanding diseases of protein denaturation. *Molecules and Cells, 23*, 123–131.

Kim, I. M., Ramakrishna, S., Gusarova, G. A., Yoder, H. M., Costa, R. H., & Kalinichenko, V. V. (2005). The Forkhead box m1 transcription factor is essential for embryonic development of pulmonary vasculature. *Journal of Biological Chemistry, 280*, 22278–22286.

King, T. E., Pardo, A., & Selman, M. (2011). Idiopathic pulmonary fibrosis. *Lancet, 378*, 1949–1961.

Kisseleva, T., & Brenner, D. A. (2008a). Fibrogenesis of parenchymal organs. *Proceedings of the American Thoracic Society, 5*, 338–342.

Kisseleva, T., & Brenner, D. A. (2008b). Mechanisms of fibrogenesis. *Experimental Biology and Medicine, 233*, 109–122.

Kiyokawa, H. (2006). Senescence and cell cycle control. *Results and Problems in Cell Differentiation, 42*, 257–270.

Kizek, R., Adam, V., Hrabeta, J., Eckschlager, T., Smutny, S., Burda, J. V., et al. (2012). Anthracyclines and ellipticines as DNA-damaging anticancer drugs: Recent advances. *Pharmacology & Therapeutics, 133*, 26–39.

Knetemann, N. M., & Mercer, D. F. (2005). Mice with chimeric human livers: Who says supermodels have to be tall? *Hepatology, 41*, 703–706.

Knight, A. S., Notaridou, M., & Watson, R. J. (2009). A Lin-9 complex is recruited by B-Myb to activate transcription of G_2/M genes in undifferentiated embryonal carcinoma cells. *Oncogene, 28*, 1737–1747.

Koh, K. P., Sundrud, M. S., & Rao, A. (2009). Domain requirements and sequence specificity of DNA binding for the forkhead transcription factor FOXP3. *PLoS One, 4*, e8109.

Koniaris, L. G., McKillop, I. H., Schwartz, S. I., & Zimmers, T. A. (2003). Liver regeneration. *Journal of the American College of Surgery, 197*, 634–659.

Koo, C. Y., Muir, K. W., & Lam, E. W. F. (2011). FOXM1: From cancer initiation to progression and treatment. *Biochimica et Biophysica Acta, 1819*, 28–37.

Korver, W., Roose, J., & Clevers, H. (1997). The winged-helix transcription factor Trident is expressed in cycling cells. *Nucleic Acids Research, 25*, 1715–1719.

Korver, W., Roose, J., Heinen, K., Weghuis, D. O., de Bruijn, D., van Kessel, A. G., et al. (1997). The human TRIDENT/HFH-11/FKHL16 gene: Structure, localization, and promoter characterization. *Genomics, 46*, 435–442.

Korver, W., Roose, J., Wilson, A., & Clevers, H. (1997). The winged-helix transcription factor Trident is expressed in actively dividing lymphocytes. *Immunobiology, 198*, 157–161.

Korver, W., Schilham, M. W., Moerer, P., van den Hoff, M. J., Lamers, W. H., Medema, R. H., et al. (1998). Uncoupling of S-phase and mitosis in cardiomyocytes and hepatocytes lacking the winged-helix transcription factor Trident. *Current Biology, 8,* 1327–1330.

Kouros-Mehr, H., Slorach, E. M., Sternlicht, M. D., & Werb, Z. (2006). GATA-3 maintains the differentiation of the luminal cell fate in the mammary gland. *Cell, 127,* 1041–1055.

Kregel, K. C. (2002). Heat shock proteins: Modifying factors in physiological stress responses and acquired thermotolerance. *Journal of Applied Physiology, 92,* 2177–2186.

Krejci, L., Altmannova, V., Spirek, M., & Zhao, X. (2012). Homologous recombination and its regulation. *Nucleic Acids Research, 40,* 5795–5818.

Kretschmer, C., Sterner-Kock, A., Siedentopf, F., Schlag, P. M., & Kemmner, W. (2011). Identification of early molecular markers for breast cancer. *Molecular Cancer, 10,* 15.

Krupczak-Hollis, K., Wang, X., Dennewitz, M. B., & Costa, R. H. (2003). Growth hormone stimulates proliferation of old-aged regenerating liver through Forkhead Box m1b. *Hepatology, 38,* 1552–1562.

Krupczak-Hollis, K., Wang, X., Kalinichenko, V. V., Gusarova, G. A., Wang, I. C., Dennewitz, M. B., et al. (2004). The mouse Forkhead Box m1 transcription factor is essential for hepatoblast mitosis and development of intrahepatic bile ducts and vessels during liver morphogenesis. *Developmental Biology, 276,* 74–88.

Kudlow, J. E. (2006). Post-translational modification by O-GlcNAc: Another way to change protein function. *Journal of Cellular Biochemistry, 98,* 1062–1075.

Kudlow, B. A., Kennedy, B. K., & Monnat, R. J. (2007). Werner and Hutchinson-Gilford progeria syndromes: Mechanistic basis of human progeroid diseases. *Nature Reviews. Molecular Cell Biology, 8,* 394–404.

Kuilman, T., Michloglou, C., Mooi, W. J., & Peeper, D. S. (2010). The essence of senescence. *Genes & Development, 24,* 2463–2479.

Kuilman, T., & Peeper, D. S. (2009). Senescence-messaging secretome: SMS-ing cellular stress. *Nature Reviews. Cancer, 9,* 81–94.

Kültz, D. (2005). Molecular and evolutionary basis of the cellular stress response. *Annual Review of Physiology, 67,* 225–257.

Küppers, M., Ittrich, C., Faust, D., & Dietrich, C. (2010). The transcriptional programme of contact-inhibition. *Journal of Cellular Biochemistry, 110,* 1234–1243.

Kurinna, S., Stratton, S. A., Coban, Z., Schumacher, J. M., Grompe, M., Duncan, A. W., et al. (2013). p53 regulates a mitotic transcription program and determines ploidy in normal mouse liver. *Hepatology, 57,* 2004–2013.

Kwak, Y. D., Wang, B., Li, J. J., Wang, R., Deng, Q., Diao, S., et al. (2012). Upregulation of the E3 ligase NEDD4-1 by oxidative stress degrades IGF-1 receptor protein in neurodegeneration. *Journal of Neuroscience, 32,* 10971–10981.

Kweekel, D. M., Gelderblom, H., & Guchelaar, H. J. (2005). Pharmacology of oxaliplatin and the use of pharmacogenomics to individualize therapy. *Cancer Treatment Reviews, 31,* 90–105.

Kwok, J. M. M., Peck, B., Monteiro, L. J., Schwenen, H. D. C., Millour, J., Coombes, R. C., et al. (2010). FOXM1 confers acquired cisplatin resistance in breast cancer cells. *Molecular Cancer Research, 8,* 24–34.

Laconi, E., & Laconi, S. (2002). Principles of heptocytes repopulation. *Seminars in Cell & Developmental Biology, 13,* 433–438.

Lacroix, B., & Maddox, A. S. (2012). Cytokinesis, ploidy and aneuploidy. *The Journal of Pathology, 26,* 338–351.

Lai, E., Clark, K. L., Burley, S. K., & Darnell, J. E. (1993). Hepatocyte nuclear factor 3/fork head or "winged helix" proteins: A family of transcription factors of diverse biological function. *Proceeding of the National Academy of Sciences of the United States of America, 90,* 10421–10423.

Lalmansingh, A. S., Karmakar, S., Jin, Y., & Nagaich, A. K. (2012). Multiple modes of chromatin remodeling by Forkhead box proteins. *Biochimica et Biophysica Acta, 1819,* 707–715.

Lambrecht, B. N. (2001a). Allergen uptake and presentation by dendritic cells. *Current Opinion in Allergy and Clinical Immunology, 1*, 51–59.

Lambrecht, B. N. (2001b). The dendritic cell in allergic airway diseases: A new player to the game. *Clinical and Experimental Allergy, 31*, 206–218.

Lange, A. W., Keiser, A. R., Wells, J. M., Zorn, A. M., & Whitsett, J. A. (2009). Sox17 promotes cell cycle progression and inhibits TGF-β/Smad3 signaling to initiate progenitor cell behaviour in the respiratory epithelium. *PLoS One, 4*, e5711.

Lanigan, F., Geraghty, J. G., & Bracken, A. P. (2011). Transcriptional regulation of cellular senescence. *Oncogene, 30*, 2901–2911.

Lanks, K. W. (1986). Modulators of the eukaryotic heat shock response. *Experimental Cell Research, 165*, 1–10.

Laoukili, J., Alvarez, M., Meijer, L. A. T., Stahl, M., Mohammed, S., Kleij, L., et al. (2008). Activation of FoxM1 during G2 requires CyclinA/Cdk-dependent relief of auto-repression by the FoxM1 N-terminal domain. *Molecular and Cellular Biology, 28*, 3076–3087.

Laoukili, J., Alvarez-Fernandez, M., Stahl, M., & Medema, R. H. (2008). FoxM1 is degraded at mitotic exit in a Cdh1-dependent manner. *Cell Cycle, 7*, 2720–2726.

Laoukili, J., Kooistra, M. R. H., Brás, A., Kauw, J., Kerkhoven, R. M., Morrison, A., et al. (2005). FoxM1 is required for execution of the mitotic programme and chromosome stability. *Nature Cell Biology, 7*, 126–136.

Laoukili, J., Stahl, M., & Medema, R. H. (2007). FoxM1: At the crossroads of ageing and cancer. *Biochimica et Biophysica Acta, 1775*, 92–102.

Larsson, L. G., & Henriksson, M. A. (2010). The Yin and Yang functions of the Myc oncoprotein in cancer development and as target for therapy. *Experimental Cell Research, 316*, 1429–1437.

Lash, J. W., Linask, K. K., & Yamada, K. M. (1987). Synthetic peptides mimic the adhesive recognition signal of fibronectin: Differential effects on cell–cell and cell–substratum adhesion in embryonic chick cells. *Developmental Biology, 123*, 411–420.

Laurendeau, I., Ferrer, M., Garrido, D., D'Haene, N., Ciavarelli, P., Basso, A., et al. (2010). Gene expression profiling of the Hedgehog signaling pathway in human meningiomas. *Molecular Medicine, 16*, 262–270.

Le Cras, T. D., Acciani, T. H., Mushaben, E. M., Kramer, E. L., Pastura, P. A., Hardie, W. D., et al. (2011). Epithelial EGF receptor signaling mediates airway hyperreactivity and remodeling in a mouse model of chronic asthma. *American Journal of Physiology. Lung Cellular and Molecular Physiology, 300*, L414–L421.

le Sage, C., Nagel, R., & Agami, R. (2007). Diverse ways to control $p27^{Kip1}$ function. miRNAs come into play. *Cell Cycle, 6*, 2742–2749.

Le, X. F., Pruefer, F., & Bast, R. C. (2005). HER2-targeting antibodies modulate the cyclin-dependent kinase inhibitor $p27^{Kip1}$ via multiple signaling pathways. *Cell Cycle, 4*, 87–95.

Ledda-Columbano, G. M., Pibiri, M., Cossu, C., Molotzu, F., Locker, J., & Columbano, A. (2004). Aging does not reduce the hepatocyte proliferative response of mice to the primary mitogen TCPOBOP. *Hepatology, 40*, 981–988.

Lee, G. H. (2000). Paradoxical effects of phenobarbital on mouse hepatocarcinogenesis. *Toxicologic Pathology, 28*, 215–225.

Lee, C., & Cho, Y. (2002). Interactions of SV40 large T antigen and other viral proteins with retinoblastoma tumour suppressor. *Reviews in Medical Virology, 12*, 81–92.

Lee, Y. S., & Dutta, A. (2009). MicroRNAs in cancer. *Annual Review of Pathology, 4*, 199–228.

Lee, E. Y., & Muller, W. J. (2010). Oncogenes and tumor suppressor genes. *Cold Spring Harbor Perspectives in Biology, 2*, a003236.

Lefebvre, C., Rajbhandari, P., Alvarez, M. J., Bandaru, P., Lim, W. K., Sato, M., et al. (2010). A human B-cell interactome identifies MYB and FOXM1 as master regulators of proliferation in germinal centers. *Molecular Systems Biology, 6*, 377.

Lehmann, K., Tschuor, C., Rickenbacher, A., Jang, J. H., Oberkofler, C. E., Tschopp, O., et al. (2012). Liver failure after extended hepatectomy in mice is mediated by a p21-dependent barrier to liver regeneration. *Gastroenterology, 143*, 1609–1619.

Leon, J., Ferrandiz, N., Acosta, J. C., & Delgado, M. D. (2009). Inhibition of cell differentiation. A critical mechanism for MYC-mediated carcinogenesis? *Cell Cycle, 8*, 1148–1157.

Leung, T. W. C., Lin, S. S. W., Tsang, A. C. C., Tong, C. S. W., Ching, J. C. Y., Leung, W. Y., et al. (2001). Overexpression of FoxM1 stimulates cyclin B1 expression. *FEBS Letters, 507*, 59–66.

Levens, D. (2002). Disentangling the MYC web. *Proceeding of the National Academy of Sciences of the United States of America, 99*, 5757–5759.

Levens, D. L. (2003). Reconstructing MYC. *Genes & Development, 17*, 1071–1077.

Levine, A. J. (2009). The common mechanisms of transformation by the small DNA tumor viruses: The inactivation of tumor suppressor gene products: p53. *Virology, 384*, 285–293.

Levine, A. J., Momand, J., & Finlay, C. A. (1991). The p53 tumor suppressor gene. *Nature, 351*, 453–456.

Lewkovich, I. P., Lajoie, S., Clark, J. R., Herman, N. S., Sproles, A. A., & Wills-Korp, M. (2008). Allergen uptake, activation, and IL-23 production by pulmonary myeloid DCs drives airway hyperresponsiveness in asthma-susceptible mice. *PLoS One, 3*, e3879.

Li, Y. Q. (2010). Master stem cell transcription factors and signaling regulation. *Cellular Reprogramming, 12*, 3–13.

Li, M. L., & Greenberg, R. A. (2012). Links between genome integrity and BRCA1 tumor suppression. *Trends in Biochemical Sciences, 37*, 418–424.

Li, Q., Jia, Z., Wang, L., Kong, X., Li, Q., Guo, K., et al. (2012). Disruption of Klf4 in villin-positive gastric progenitor cells promotes formation and progression of tumors of the antrum in mice. *Gastroenterology, 142*, 531–542.

Li, Y., Ligr, M., McCarron, J. P., Daniels, G., Zhang, D., Zhao, X., et al. (2011). Natura-alpha targets Forkhead box M1 and inhibits androgen-dependent and -independent prostate cancer growth and invasion. *Clinical Cancer Research, 17*, 4414–4424.

Li, D., Wei, P., Peng, Z., Hung, C., Tang, H., Jia, Z., Cui, J., et al. (2013). The critical role of dysregulated FoxM1-uPAR signaling in human colon cancer progression and metastasis. *Clinical Cancer Research, 19*, 62–72.

Li, S. K. M., Smith, D., Leung, W. Y., Cheung, A. M. S., Lam, E. W. F., Dimri, G. P., et al. (2008). FOXM1c counteracts oxidative stress-induced senescence and stimulates Bmi-1 expression. *Journal of Biological Chemistry, 283*, 16545–16553.

Li, J., Wang, Y., Luo, J., Fu, Z., Ying, J., Yu, Y., et al. (2012). miR-134 inhibits epithelial to mesenchymal transition by targeting FOXM1 in non-small cell lung cancer cells. *FEBS Letters, 586*, 3761–3765.

Li, S., Weidenfeld, J., & Morrisey, E. E. (2004). Transcriptional and DNA binding activity of the Foxp1/2/4 family is modulated by heterotypic and homotypic protein interactions. *Molecular and Cellular Biology, 24*, 809–822.

Li, Q., Zhang, N., Jia, Z., Le, X., Dai, B., Wie, D., et al. (2009). Critical role and regulation of transcription factor FoxM1 in human gastric cancer angiogenesis and progression. *Cancer Research, 69*, 3501–3509.

Li, L., & Zou, L. (2005). Sensing, signaling, and responding to DNA damage: Organization of the checkpoint pathways in mammalian cells. *Journal of Cellular Biochemistry, 94*, 298–306.

Liang, Y., Lin, S. Y., Brunicardi, C., Goss, J., & Li, K. (2009). DNA damage response pathways in tumor suppression and cancer treatment. *World Journal of Surgery, 33*, 661–666.

Liang, J., & Slingerland, J. M. (2003). Multiple roles of the PI3K/PKB (Akt) pathway in cell cycle progression. *Cell Cycle, 2*, 339–345.

Lieber, M. R. (1999). The biochemistry and biological significance of nonhomologous DNA end joining: An essential repair process in multicellular eukaryotes. *Genes to Cells, 4*, 77–85.

Lieber, M. R. (2010). The mechanism of double-strand DNA break repair by the nonhomologous DNA end-joining pathway. *Annual Review of Biochemistry, 79*, 181–211.

Lim, S. C., Choi, J. E., Kang, H. S., & Han, S. I. (2010). Ursodeoxycholic acid switches oxaliplatin-induced necrosis to apoptosis by inhibiting reactive oxygen species production and activating p53-caspase 8 pathway in hepatocellular carcinoma. *International Journal of Cancer, 126*, 1582–1595.

Lin, P. C., Chiu, Y. L., Banerjee, S., Park, K., Mosquera, J. M., Giannopoulou, E., et al. (2013). Epigenetic repression of miR-31 disrupts androgen receptor homeostasis and contributes to prostate cancer progression. *Cancer Research, 73*, 1232–1244.

Lin, M., Guo, L. M., Liu, H., Du, J., Yang, J., Zhang, L. J., et al. (2010). Nuclear accumulation of glioma-associated oncogene 2 protein and enhanced expression of forkhead-box transcription factor M1 protein in human hepatocellular carcinoma. *Histology and Histopathology, 25*, 1269–1275.

Lin, B., Madan, A., Yoon, J. G., Fang, X., Yan, X., Kim, T. K., et al. (2010). Massively parallel signature sequencing and bioinformatics analysis identifies up-regulation of TGFBI and SOX4 in human glioblastoma. *PLoS One, 5*, e10210.

Lindquist, S. (1986). The heat-shock response. *Annual Review of Biochemistry, 55*, 1151–1191.

Lindquist, S., & Craig, E. A. (1988). The heat shock proteins. *Annual Review of Genetics, 22*, 631–677.

Lipinski, M. M., & Jacks, T. (1999). The retinoblastoma gene family in differentiation and development. *Oncogene, 18*, 7873–7882.

Littler, D. R., Alvarez-Fernandez, M., Stein, A., Hibbert, R. G., Heidebrecht, T., Aloy, P., et al. (2010). Structure of the FoxM1 DNA-recognition domain bound to a promoter sequence. *Nucleic Acids Research, 38*, 4527–4538.

Liu, P. P., Chen, Y. C., Li, C., Hsieh, Y. H., Chen, S. W., Chen, S. H., et al. (2002). Solution structure of the DNA-binding domain of interleukin enhancer binding factor 1 (FOXK1a). *Proteins, 49*, 543–553.

Liu, M., Dai, B., Kang, S. H., Ban, K., Huang, F. J., Lang, F. F., et al. (2006). FoxM1B is overexpressed in human glioblastomas and critically regulates the tumorigenicity of glioma cells. *Cancer Research, 66*, 3593–3602.

Liu, Y., Gampert, L., Nething, K., & Steinacker, J. M. (2006). Response and function of skeletal muscle heat shock protein 70. *Frontiers in Bioscience, 11*, 2802–2827.

Liu, J., Guo, S., Li, Q., Yang, L., Xia, Z., Zhang, L., et al. (2013). Phosphoglycerate dehydrogenase induces glioma cells proliferation and invasion by stabilizing forkhead box M1. *Journal of Neuro-Oncology, 111*, 245–255.

Liu, S., Guo, W., Shi, J., Li, N., Yu, X., Xue, J., et al. (2012). MicroRNA-135a contributes to the development of portal vein tumor thrombus by promoting metastasis in hepatocellular carcinoma. *Journal of Hepatology, 56*, 389–396.

Liu, Y., Hock, J. M., Van Beneden, R. J., & Li, X. (2013). Aberrant overexpression of FOXM1 transcription factor plays a critical role in lung carcinogenesis induced by low doses of arsenic. *Molecular Carcinogenesis*, Epub: Dec 19, 2012.

Liu, X., & Marmorstein, R. (2006). When viral oncoprotein meets tumor suppressor: A structural view. *Genes & Development, 20*, 2332–2337.

Liu, Y., Sadikot, R. T., Adami, G. R., Kalinichenko, V. V., Pendyala, S., Natarajan, V., et al. (2011). FoxM1 mediates the progenitor function of type II epithelial cells in repairing alveoloar injury induced by *Pseudomonas aeruginosa*. *The Journal of Experimental Medicine, 208*, 1473–1484.

Llaurado, M., Majem, B., Casellvi, J., Cabrera, S., Gil-Moreno, A., Reventos, J., et al. (2012). Analysis of gene expression regulated by the ETV5 transcription factor in

OV90 ovarian cancer cells identifies FoxM1 over-expression in ovarian cancer. *Molecular Cancer Research, 10*, 914–924.

Lleonart, M. E., Artero-Castro, A., & Kondoh, H. (2009). Senescence induction; a possible cancer therapy. *Molecular Cancer, 8*, 3.

Locasale, J. W., Grassian, A. R., Melman, T., Lyssiotis, C. A., Mattaini, K. R., Bass, A. J., et al. (2011). Phosphoglycerate dehydrogenase diverts glycolytic flux and contributes to oncogenesis. *Nature Genetics, 43*, 869–874.

Loh, Y. H., Ng, J. H., & Ng, H. H. (2008). Molecular framework underlying pluripotency. *Cell Cycle, 7*, 885–891.

Loh, Y. H., Yang, L., Yang, J. C., Li, H., Collins, J. J., & Daley, G. Q. (2011). Genomic approaches to deconstruct pluripotency. *Annual Review of Genomics and Human Genetics, 12*, 165–185.

Lohr, D., Venkov, P., & Zlatanova, J. (1995). Transcriptional regulation in the yeast *GAL* gene family: A complex genetic network. *The FASEB Journal, 9*, 777–787.

Lok, G. T. M., Chan, D. W., Liu, V. W. S., Hui, W. W. Y., Leung, T. H. Y., Yao, K. M., et al. (2011). Aberrant activation of ERK/FOXM1 signaling cascade triggers the cell migration/invasion in ovarian cancer cells. *PLoS One, 6*, e23790.

Lorvellec, M., Dumon, S., Maya-Mendoza, A., Jackson, D., Frampton, J., & Garcia, P. (2010). B-Myb is critical for proper DNA duplication during an unperturbed S phase in mouse embryonic stem cells. *Stem Cells, 28*, 1751–1759.

Loss, E. S., Jacobus, A. P., & Wassermann, G. F. (2007). Diverse FSH and testosterone signaling pathways in the Sertoli cell. *Hormone and Metabolic Research, 39*, 806–812.

Lotterman, C. D., Kent, O. A., & Mendell, J. T. (2008). Functional integration of microRNAs into oncogenic and tumor suppressor pathways. *Cell Cycle, 7*, 2493–2499.

Louis, S. A., Rietze, R. L., Deleyrolle, L., Wagey, R. E., Thomas, T. E., Eaves, A. C., et al. (2008). Enumeration of neural stem and progenitor cells in the neural colony-forming cell assay. *Stem Cells, 26*, 988–996.

Love, D. C., Krause, M. W., & Hanover, J. S. (2010). O-GlcNAc cycling: Emerging roles in development and epigenetics. *Seminars in Cell & Developmental Biology, 21*, 646–654.

Lowe, S. W., Cepero, E., & Evan, G. (2004). Intrinsic tumour suppression. *Nature, 432*, 307–315.

Lu, Z., & Hunter, T. (2010). Ubiquitylation and proteasomal degradation of the $p21^{Cip1}$, $p27^{Kip1}$ and $p57^{Kip2}$ CDK inhibitors. *Cell Cycle, 9*, 2342–2352.

Lu, Y., Mahony, S., Benos, P. V., Rosenfeld, R., Simon, I., Breeden, L. L., et al. (2007). Combined analysis reveals a core set of cycling genes. *Genome Biology, 8*, R146.

Luedde, M., Katus, H. A., & Frey, N. (2006). Novel molecular targets in the treatment of cardiac hypertrophy. *Recent Patents on Cardiovascular Drug Discovery, 1*, 1–20.

Lukas, J., Lukas, C., & Bartek, J. (2004). Mammalian cell cycle checkpoints: Signalling pathways and their organization in space and time. *DNA Repair, 3*, 997–1007.

Lundberg, A. S., & Weinberg, R. A. (1999). Control of the cell cycle and apoptosis. *European Journal of Cancer, 35*, 531–539.

Luo, J. (2011). Cancer's sweet tooth for serine. *Breast Cancer Research, 13*, 317.

Lüscher, B. (2001). Function and regulation of the transcription factors of the Myc/Max/Mad network. *Gene, 277*, 1–14.

Lüscher, B. (2012). MAD1 and its life as a MYC antagonist: An update. *European Journal of Cell Biology, 91*, 506–514.

Lüscher, B., & Larsson, L. G. (1999). The basis region/helix-loop-helix/leucine zipper domain of Myc proto-oncoproteins: Function and regulation. *Oncogene, 18*, 2955–2966.

Lüscher, B., & Vervoorts, J. (2012). Regulation of gene transcription by the oncoprotein MYC. *Gene, 494*, 145–160.

Lüscher-Firzlaff, J. M., Lilischkis, R., & Lüscher, B. (2006). Regulation of the transcription factor FOXM1c by cyclin E/CDK2. *FEBS Letters, 580*, 1716–1722.

Lüscher-Firzlaff, J. M., Westendorf, J. M., Zwicker, J., Burkhardt, H., Henriksson, M., Müller, R., et al. (1999). Interaction of the fork head domain transcription factor MPP2 with the human papilloma virus 16 E7 protein: Enhancement of transformation and transactivation. *Oncogene, 18*, 5620–5630.

Luster, M. I., Simeonova, P. P., Gallucci, R. M., Bruccoleri, A., Blazka, M. E., & Yucesoy, B. (2001). Role of inflammation in chemical-induced hepatotoxicity. *Toxicology Letters, 120*, 317–321.

Ly, D. H., Lockhart, D. J., Lerner, R. A., & Schultz, P. G. (2000). Mitotic misregulation and human ageing. *Science, 287*, 2486–2492.

Lynch, T. P., Ferrer, C. M., Jackson, R., Shahriari, K. S., Vosseller, K., & Reginato, M. J. (2012). Critical role of O-GlcNAc transferase in prostate cancer invasion, angiogenesis and metastasis. *Journal of Biological Chemistry, 287*, 11070–11081.

Ma, Y., Lu, H., Schwarz, K., & Lieber, M. R. (2005). Repair of double-strand DNA breaks by the human nonhomologous DNA end joining pathway: The iterative processing model. *Cell Cycle, 4*, 1193–1200.

Ma, R. Y., Tong, T. H., Cheung, A. M., Tsang, A. C., Leung, W. Y., & Yao, K. M. (2005). Raf/MEK/MAPK signaling stimulates the nuclear translocation and transactivating activity of FOXM1c. *Journal of Cell Science, 118*, 795–806.

Ma, R. Y. M., Tong, T. H. K., Leung, W. Y., & Yao, K. M. (2010). Raf/MEK/MAPK signaling stimulates the nuclear translocation and transactivating activity of FOXM1. *Methods in Molecular Biology, 647*, 113–123.

MacArthur, B. D., Ma'ayan, A., & Lemischka, I. R. (2009). Systems biology of stem cell fate and cellular reprogramming. *Nature Reviews. Molecular Cell Biology, 10*, 672–681.

Mackey, S., Singh, P., & Darlington, G. J. (2003). Making the liver young again. *Hepatology, 38*, 1349–1352.

Madureira, P. A., Varshochi, R., Constantinidou, D., Francis, R. E., Coombes, R. C., Yao, K. M., et al. (2006). The Forkhead box M1 protein regulates the transcription of the estrogen receptor α in breast cancer cells. *Journal of Biological Chemistry, 281*, 25167–25176.

Major, M. L., Lepe, R., & Costa, R. H. (2004). Forkhead Box M1B transcriptional activity requires binding of Cdk-cyclin complexes for phosphorylation-dependent recruitment of p300/CBP coactivators. *Molecular and Cellular Biology, 24*, 2649–2661.

Malhi, H., & Gupta, S. (2001). Hepatocyte transplantation: New horizons and challenges. *Journal of Hepato-Biliary-Pancreatic Surgery, 8*, 40–50.

Malin, D., Kim, I. M., Boetticher, E., Kalin, T. V., Ramakrishna, S., Meliton, L., et al. (2007). Forkhead Box F1 is essential for migration of mesenchymal cells and directly induces integrin-beta3 expression. *Molecular and Cellular Biology, 27*, 2486–2498.

Malumbres, M. (2011). Physiological relevance of cell cycle kinases. *Physiological Reviews, 91*, 973–1007.

Malumbres, M., & Barbacid, M. (2003). RAS oncogenes: The first 30 years. *Nature Reviews. Cancer, 3*, 7–13.

Malumbres, M., & Barbacid, M. (2005). Mammalian cyclin-dependent kinases. *Trends in Biochemical Sciences, 30*, 630–641.

Malumbres, M., & Barbacid, M. (2009). Cell cycle, CDKs and cancer: A changing paradigm. *Nature Reviews. Cancer, 9*, 153–166.

Malumbres, M., & Pellicier, A. (1998). Ras pathways to cell cycle control and cell transformation. *Frontiers in Bioscience, 3*, d887–d912.

Manibusan, M. K., Odin, M., & Eastmond, D. A. (2007). Postulated carbon tetrachloride mode of action: A review. *Journal of Environmental Science and Health. Part C, Environmental Carcinogenesis & Ecotoxicology Reviews, 25*, 185–298.

Mannefeld, M., Klassen, E., & Gaubatz, S. (2009). B-MYB is required for recovery from the DNA damage-induced G2 checkpoint in p53 mutant cells. *Cancer Research, 69,* 4073–4080.

Marcu, K. B., Bossone, S. A., & Patel, A. J. (1992). Myc function and regulation. *Annual Review of Biochemistry, 61,* 809–860.

Markey, M. P., Bergseid, J., Bosco, E. E., Stengel, K., Xu, H., Mayhew, C. N., et al. (2007). Loss of the retinoblastoma tumor suppressor: Differential action on transcriptional programs related to cell cycle control and immune function. *Oncogene, 26,* 6307–6318.

Marsden, I., Chen, Y., Jin, C., & Liao, X. (1997). Evidence that the DNA binding specificity of winged helix proteins is mediated by a structural change in the amino acid sequence adjacent to the principal DNA binding helix. *Biochemistry, 36,* 13248–13255.

Marsden, I., Jin, C., & Liao, X. (1998). Structural changes in the region directly adjacent to the DNA-binding helix highlight a possible mechanism to explain the observed changes in the sequence-specific binding of winged helix proteins. *Journal of Molecular Biology, 278,* 293–299.

Marshall, C. J. (1996). Ras effectors. *Current Opinion in Cell Biology, 8,* 197–204.

Marshall, C. J. (1999a). How do small GTPase signal transduction pathways regulate cell cycle entry. *Current Opinion in Cell Biology, 11,* 732–736.

Marshall, C. J. (1999b). Small GTPases and cell cycle regulation. *Biochemical Society Transactions, 27,* 363–370.

Massagué, J. (2004). G1 cell-cycle control and cancer. *Nature, 432,* 298–306.

Massagué, J. (2008). TGFβ in cancer. *Cell, 134,* 215–230.

Masson, J. Y., & West, S. C. (2001). The Rad51 and Dmc1 recombinases: A non-identical twin relationship. *Trends in Biochemical Sciences, 26,* 131–136.

Masumoto, N., Tateno, C., Tachibana, A., Utoh, R., Morikawa, Y., Shimada, T., et al. (2007). GH enhances proliferation of human hepatocytes grafted into immunodeficient mice with damaged liver. *Journal of Endocrinology, 194,* 529–537.

Mathon, N. F., & Lloyd, A. C. (2001). Cell senescence and cancer. *Nature Reviews. Cancer, 1,* 203–213.

Matsumoto-Taniura, N., Pirollet, F., Monroe, R., Gerace, L., & Westendorf, J. M. (1996). Identification of novel M phase phosphoproteins by expression cloning. *Molecular Biology of the Cell, 7,* 1455–1469.

Matsuo, T., Yamaguchi, S., Mitsui, S., Emi, A., Shimoda, F., & Okamura, H. (2003). Control mechanism of the circadian clock for timing of cell division in vivo. *Science, 302,* 255–259.

Matsushima-Nishiu, M., Unoki, M., Ono, K., Tsunoda, T., Minaguchi, T., Kuramoto, H., et al. (2001). Growth and gene expression profile analyses of endometrial cancer cells expressing exogenous PTEN. *Cancer Research, 61,* 3741–3749.

Matsuzawa, A., & Ichijo, H. (2008). Redox control of cell fate by MAP kinase: Physiological roles of ASK1-MAP kinase pathway. *Biochimica et Biophysica Acta, 1780,* 1325–1336.

Mauviel, A., Nallet-Staub, F., & Varelas, X. (2012). Integrating developmental signals: A Hippo in the (path)way. *Oncogene, 31,* 1743–1756.

Mazet, F., Yu, J. K., Liberles, D. A., Holland, L. Z., & Shimeld, S. M. (2003). Phylogenetic relationships of the Fox (Forkhead) gene family in the Bilateralia. *Gene, 316,* 79–89.

McArthur, G. A., Laherty, C. D., Queva, C., Hurlin, P. J., Loo, L., James, L., et al. (1998). The Mad protein family links transcriptional repression to cell differentiation. *Cold Spring Harbor Symposia on Quantitative Biology, 63,* 423–433.

McBurney, M. W. (1993). P19 embryonal carcinoma cells. *International Journal of Developmental Biology, 37,* 135–140.

McBurney, M. W., Jones-Villeneuve, E. M., & Edwards, M. K. (1982). Control of muscle and neuronal differentiation in a cultured embryonal carcinoma cell line. *Nature, 299,* 165–167.

McClellan, K. A., & Slack, R. S. (2007). Specific in vivo roles for E2Fs in differentiation and development. *Cell Cycle, 23*, 2917–2927.

McCormick, F. (1999). Signalling networks that cause cancer. *Trends in Biochemical Sciences, 24*, M53–M56.

McCormick, F. (2011). Cancer therapy based on oncogene addiction. *Journal of Surgical Oncology, 103*, 464–467.

McDuff, F. K. E., & Turner, S. D. (2011). Jailbreak: Oncogene-induced senescence and its evasion. *Cellular Signalling, 23*, 6–13.

McGovern, U. B., Francis, R. E., Peck, B., Guest, S. K., Wang, J., Myatt, S. S., et al. (2009). Gefitinib (Iressa) represses FOXM1 expression via FOXO3a in breast cancer. *Molecular Cancer Therapeutics, 8*, 582–591.

McGowan, C. H. (2002). Checking on Cds1 (Chk2): A checkpoint kinase and tumor suppressor. *Bioessays, 24*, 502–511.

McKinnon, P. J., & Caldecott, K. W. (2007). DNA strand break repair and human genetic disease. *Annual Review of Genomics and Human Genetics, 8*, 37–55.

Medina, P. P., & Slack, F. J. (2008). MicroRNAs and cancer. *Cell Cycle, 7*, 2485–2492.

Megarbane, A., Ravel, A., Mircher, C., Sturtz, F., Grattau, Y., Rethore, M. O., et al. (2009). The 50th anniversary of the discovery of trisomy 21: The past, present, and future of research and treatment of Down syndrome. *Genetics in Medicine, 11*, 611–616.

Mehendale, H. M. (1995). Toxicodynamics of low level toxicant interactions of biological significance: Inhibition of tissue repair. *Toxicology, 105*, 251–266.

Mehendale, H. M. (2005). Tissue repair: An important determinant of final outcome of toxicant-induced injury. *Toxicologic Pathology, 33*, 41–51.

Mehendale, H. M., Roth, R. A., Gandolfi, A. J., Klaunig, J. E., Lemasters, J. J., & Curtis, L. R. (1994). Novel mechanisms in chemically induced hepatotoxicity. *The FASEB Journal, 8*, 1285–1295.

Mehendale, H. M., Thakore, K. N., & Rao, C. V. (1994). Autoprotection: Stimulated tissue repair permits recovery from injury. *Journal of Biochemical Toxicology, 9*, 131–139.

Mehta, D., & Malik, A. B. (2006). Signaling mechanisms regulating endothelial permeability. *Physiological Reviews, 86*, 279–367.

Meichle, A., Philipp, A., & Eilers, M. (1992). The functions of Myc proteins. *Biochimica et Biophysica Acta, 1114*, 129–146.

Mencalha, A. L., Binato, R., Ferreira, G. M., Du Rocher, B., & Abdelhay, E. (2012). Forkhead box M1 (FoxM1) gene is a new STAT3 transcriptional factor target and is essential for proliferation, survival and DNA repair of K562 cell line. *PLoS One, 7*, e48160.

Meng, Z., Liu, N., Fu, X., Wang, X., Wang, Y. D., Chen, W. D., et al. (2011). Insufficient bile acids signaling impairs liver repair in $CYP27-/-$ mice. *Journal of Hepatology, 55*, 885–895.

Meng, Z., Wang, Y., Wang, L., Jin, W., Liu, N., Pan, H., et al. (2010). FXR regulates liver repair after CCl_4-induced toxic injury. *Molecular Endocrinology, 24*, 886–897.

Merry, C., Fu, K., Wang, J., Yeh, I. J., & Zhang, Y. (2010). Targeting the checkpoint kinase Chk1 in cancer therapy. *Cell Cycle, 9*, 279–283.

Meuleman, P., & Leroux-Roels, G. (2008). The human liver-uPA-SCID mouse: A model for the evaluation of antiviral compounds against HBV and HCV. *Antiviral Research, 80*, 231–238.

Meuleman, P., Libbrecht, L., De Vos, R., de Hemptinne, B., Gevaert, K., Vandekerckhove, J., et al. (2008). Morphological and biochemical characterization of a human liver in a uPA-SCID mouse chimera. *Hepatology, 41*, 847–856.

Meuth, M. (2010). Chk1 suppressed cell death. *Cell Division, 5*, 21.

Meyer, N., & Penn, L. Z. (2008). Reflecting on 25 years with MYC. *Nature Reviews. Cancer, 8*, 976–990.

Michalopoulos, G. K. (2007). Liver regeneration. *Journal of Cellular Physiology, 213*, 286–300.
Michalopoulos, G. K. (2010). Liver regeneration after partial hepatectomy: Critical analysis of mechanistic dilemmas. *American Journal of Pathology, 176*, 2–13.
Michalopoulos, G. K., & DeFrances, M. C. (1997). Liver regeneration. *Science, 276*, 60–66.
Miller, J. P., Yeh, N., Vidal, A., & Koff, A. (2007). Interweaving the cell cycle machinery with cell differentiation. *Cell Cycle, 6*, 2932–2938.
Millour, J., Constantinidou, D., Stavropoulou, A. V., Wilson, M. S. C., Myatt, S. S., Kwok, J. M. M., et al. (2010). FOXM1 is a transcriptional target of ERα and has a critical role in breast cancer endocrine sensitivity and resistance. *Oncogene, 29*, 2983–2995.
Millour, J., de Olano, N., Horimoto, Y., Monteiro, L. J., Langer, J. K., Aligue, R., et al. (2011). ATM and p53 regulate FOXM1 expression in breast cancer epirubicin treatment and resistance. *Molecular Cancer Therapeutics, 10*, 1046–1058.
Minamino, T., & Komuro, I. (2006). Regeneration of the endothelium as a novel therapeutic strategy for acute lung injury. *The Journal of Clinical Investigation, 116*, 2316–23319.
Minnicozzi, M., Sawyer, R. T., & Fenton, M. J. (2011). Innate immunity in allergic disease. *Immunological Reviews, 242*, 106–127.
Minotti, G., Menna, P., Salvatorelli, E., Cairo, G., & Gianni, L. (2004). Anthracyclines: Molecular advances and pharmacologic developments in antitumor activity and cardiotoxicity. *Pharmacological Reviews, 56*, 185–229.
Mirza, M. K., Sun, Y., Zhao, Y. D., Potula, H. H. S. K., Frey, R. S., Vogel, S. M., et al. (2010). FoxM1 regulates re-annealing of endothelial adherens junctions through transcriptional control of β-catenin expression. *The Journal of Experimental Medicine, 207*, 1675–1685.
Mitra, M., Kandalam, M., Sundaram, C. S., Shenkar Verma, R., Maheswari, U. K., Swaminathan, S., et al. (2011). Reversal of stathmin-mediated microtubule destabilization sensitizes retinoblastoma cells to a low dose of antimicrotubule agents: A novel synergistic therapeutic intervention. *Investigative Ophthalmology & Visual Science, 52*, 5441–5448.
Mizuguchi, T., Mitaka, T., Katsuramaki, T., & Hirata, K. (2005). Hepatocyte transplantation for total liver repopulation. *Journal of Hepato-Biliary-Pancreatic Surgery, 12*, 378–385.
Mizuno, T., Murakami, H., Fujii, M., Ishiguro, F., Tanaka, I., Kondo, Y., et al. (2012). YAP induces malignant mesothelioma cell proliferation by upregulating transcription of cell cycle-promoting genes. *Oncogene, 31*, 5117–5122.
Moeller, A., Ask, K., Warburton, D., Gauldie, J., & Kolb, M. (2008). The bleomycin animal model: A useful tool to investigate treatment options for idiopathic pulmonary fibrosis? *The International Journal of Biochemistry & Cell Biology, 40*, 362–382.
Moens, U., Van Ghelue, M., & Johannessen, M. (2007). Oncogenic potentials of the human polyomavirus regulatory proteins. *Cellular and Molecular Life Sciences, 64*, 1656–1678.
Monroe, D. G., McGee-Lawrence, M. E., Oursler, M. J., & Wesetndorf, J. J. (2012). Update on Wnt signaling in bone cell biology and bone disease. *Gene, 492*, 1–18.
Monteiro, L. J., Khongkow, P., Kongsema, M., Morris, J. R., Man, C., Weekes, D., et al. (2012). The Forkhead Box M1 protein regulates BRIP1 expression and DNA damage repair in epirubicin treatment. *Oncogene*, Epub: Oct 29, 2012.
Mooi, W. J., & Peeper, D. S. (2006). Oncogene-induced cell senescence—Halting on the road to cancer. *The New England Journal of Medicine, 355*, 1037–1046.
Moon, A. (2006). Differential functions of Ras for malignant phenotypic conversion. *Archives of Pharmacal Research, 29*, 113–122.
Moore, B. B., & Hogaboam, C. M. (2008). Murine models of pulmonary fibrosis. *American Journal of Physiology. Lung Cellular and Molecular Physiology, 294*, L152–L160.
Morgan, D. O. (2007). *The cell cycle: Principles of control*. London, UK: New Science Press Ltd.
Morgan, D. O. (2008a). SnapShot: Cell-cycle regulators I. *Cell, 135*, 764–764.e1.
Morgan, D. O. (2008b). SnapShot: Cell-cycle regulators II. *Cell, 135*, 974–974.e1.

Morimoto, R. I. (1993). Cells in stress: Transcriptional activation of heat shock genes. *Science*, *259*, 1409–1410.
Morimoto, R. I. (1998). Regulation of the heat shock transcriptional response: Cross talk between a family of heat shock factors, molecular chaperones, and negative regulators. *Genes & Development*, *12*, 3788–3796.
Morimoto, R. I. (2008). Proteotoxic stress and inducible chaperone networks in neurodegenerative disease and aging. *Genes & Development*, *22*, 1427–1438.
Morimoto, R. I. (2011). The heat shock response: Systems biology of proteotoxic stress in aging and disease. *Cold Spring Harbor Symposia on Quantitative Biology*, *76*, 91–99.
Morimoto, R. I., Sarge, K. D., & Abravaya, K. (1992). Transcriptional regulation of heat shock genes. *Journal of Biological Chemistry*, *267*, 21987–21990.
Morris, J. R. (2010). More modifiers move on DNA damage. *Cancer Research*, *70*, 3861–3863.
Morrisey, E. E., & Hogan, B. L. (2010). Preparing for the first breath: Genetic and cellular mechanisms in lung development. *Developmental Cell*, *18*, 8–23.
Mosser, D. D., & Morimoto, R. I. (2004). Molecular chaperones and the stress of oncogenesis. *Oncogene*, *23*, 2907–2918.
Mouratis, M. A., & Aidini, V. (2011). Modeling pulmonary fibrosis with bleomycin. *Current Opinion in Pulmonary Medicine*, *17*, 355–361.
Moynahan, M. E., & Jasin, M. (2010). Mitotic homologous recombination maintains genomic stability and suppresses tumorigenesis. *Nature Reviews. Molecular Cell Biology*, *11*, 196–207.
Mucenski, M. L., Wert, S. E., Nation, J. M., Loudy, D. E., Huelsken, J., Birchmeier, W., et al. (2003). Beta-catenin is required for specification of proximal/distal cell fate during lung morphogenesis. *Journal of Biological Chemistry*, *278*, 40231–40238.
Mukhopadhyay, B., Cinar, R., Yin, S., Liu, J., Tam, J., Godlewski, G., et al. (2011). Hyperactivation of anandamice synthesis and regulation of cell-cycle progression via cannabinoid tape 1 (CB_1) receptors in the regenerating liver. *Proceeding of the National Academy of Sciences of the United States of America*, *108*, 6323–6328.
Mullan, P. B., Quinn, J. E., & Harkin, D. P. (2006). The role of BRCA1 in transcriptional regulation and cell cycle control. *Oncogene*, *25*, 5854–5863.
Mullarky, E., Mattaini, K. R., Vander Heiden, M. G., Cantley, L. C., & Locasale, J. W. (2011). PHGDH amplification and altered glucose metabolism in human melanoma. *Pigment Cell & Melanoma Research*, *24*, 1112–1115.
Müller, G. A., & Engeland, K. (2009). The central role of CDE/CHR promoter elements in the regulation of cell cycle-dependent gene transcription. *FEBS Journal*, *277*, 877–893.
Müller, I., Niethammer, D., & Bruchelt, G. (1998). Anthracycline-derived chemotherapeutics in apoptosis and free radical cytotoxicity. *International Journal of Molecular Medicine*, *1*, 491–494.
Mundle, S. D., & Saberwal, G. (2003). Evolving intricacies and implications of E2F1 regulation. *The FASEB Journal*, *17*, 569–574.
Murakami, H., Aiba, H., Nakanishi, M., & Murakami-Tonami, Y. (2010). Regulation of yeast forkhead transcription factors and FoxM1 by cyclin-dependent and polo-like kinases. *Cell Cycle*, *9*, 3233–3242.
Murray, M. M., Mullan, P. B., & Harkin, D. P. (2007). Role played by BRCA1 in transcriptional regulation in response to therapy. *Biochemical Society Transactions*, *35*, 1342–1346.
Myatt, S. S., & Lam, E. W. F. (2007). The emerging roles of forkhead box (Fox) proteins in cancer. *Nature Reviews. Cancer*, *7*, 847–859.
Myster, D. L., & Duronio, R. J. (2000). To differentiate or not to differentiate? *Current Biology*, *10*, R302–R304.
Nagai, H., Noguchi, T., Takeda, K., & Ichijo, H. (2007). Pathophysiological roles of ASK1-MAP kinase signaling pathways. *Journal of Biochemistry and Molecular Biology*, *40*, 1–6.

Nair, S. K., & Burley, S. K. (2006). Structural aspects of interactions within the Myc/Max/Mad network. *Current Topics in Microbiology and Immunology, 302*, 123–143.

Nakamura, T., Furukawa, Y., Nakagawa, H., Tsunoda, T., Ohigashi, H., Murata, K., et al. (2004). Genome-wide cDNA microarray analysis of gene expression profiles in pancreatic cancers using populations of tumor cells and normal ductal epithelial cells selected for purity by laser microdissection. *Oncogene, 23*, 2385–2400.

Nakamura, S., Hirano, I., Okinaka, K., Takemura, T., Yokota, D., Ono, T., et al. (2010). The FOXM1 transcriptional factor promotes the proliferation of leukemia cells through modulation of cell cycle progression in acute myeloid leukemia. *Carcinogenesis, 31*, 2012–2021.

Nakamura, S., Yamashita, M., Yokota, D., Hirano, I., Ono, T., Fujie, M., et al. (2010). Development and pharmacologic characterization of deoxybromophospha sugar derivatives with antileukemic activity. *Investigational New Drugs, 28*, 381–391.

Nakanishi, M., Shimada, M., & Niida, H. (2006). Genetic instability in cancer cells by impaired cell cycle checkpoints. *Cancer Science, 97*, 984–989.

Nakayama, K. I., & Nakayama, K. (2006). Ubiquitin ligases: Cell-cycle control and cancer. *Nature Reviews. Cancer, 6*, 369–381.

Nardella, C., Clohessy, J. G., Alimonti, A., & Pandolfi, P. P. (2011). Pro-senescence therapy for cancer treatment. *Nature Reviews. Cancer, 11*, 503–511.

Nelson, M., & Silver, P. A. (1986). Context affects nuclear protein localisation in *Saccharomyces cerevisiae*. *Molecular and Cellular Biology, 9*, 348–389.

Newick, K., Cunniff, B., Preston, K., Held, P., Arbiser, J., Pass, H., et al. (2012). Peroxiredoxin 3 is a redox-dependent target of thiostrepton in malignant mesothelioma cells. *PLoS One, 7*, e39404.

Ng, H. H., & Surani, M. A. (2011). The transcriptional and signalling networks of pluripotency. *Nature Cell Biology, 13*, 490–496.

Niehrs, C. (2006). Function and biological roles of the Dickkopf family of Wnt modulators. *Oncogene, 25*, 7469–7481.

Niehrs, C., & Shen, J. (2010). Regulation of Lrp6 phosphorylation. *Cellular and Molecular Life Sciences, 67*, 2551–2562.

Nigg, E. A. (1995). Cyclin-dependent protein kinases: Key regulators of the eukaryotic cell cycle. *Bioessays, 17*, 471–480.

Nigg, E. A. (2001). Mitotic kinases as regulators of cell division and its checkpoints. *Nature Reviews. Molecular Cell Biology, 2*, 21–32.

Ning, Y., Li, Q., Xiang, H., Liu, F., & Cao, J. (2012). Apoptosis induced by 7-difluoromethoxyl-5,4'-di-n-octyl genistein via the inactivation of FoxM1 in ovarian cancer cells. *Oncology Reports, 27*, 1857–1864.

Niwa, H. (2007a). How is pluripotency determined and maintained? *Development, 134*, 635–646.

Niwa, H. (2007b). Open conformation chromatin and pluripotency? *Genes & Development, 21*, 2671–2676.

Normand, G., & King, R. W. (2010). Understanding cytokinesis failure. *Advances in Experimental Medicine and Biology, 676*, 27–55.

Nowosielska, A., & Marinus, M. G. (2005). Cisplatin induces DNA double-strand break formation in *Escherichia coli* dam mutants. *DNA Repair, 4*, 773–781.

Nyberg, K. A., Michelson, R. J., Putnam, C. W., & Weinert, T. A. (2002). Toward maintaining the genome: DNA damage and replication checkpoints. *Annual Review of Genetics, 36*, 617–656.

O'Farrell, P. H. (2011). Quiescence: Early evolutionary origins and universality do not imply uniformity. *Philosophical Transactions of the Royal Society B: Biological Sciences, 366*, 3498–3507.

Obama, K., Ura, K., Li, M., Katagiri, T., Tsunoda, T., Nomura, A., et al. (2005). Genome-wide analysis of gene expression in human intrahepatic cholangiocarcinoma. *Hepatology, 41*, 1339–1348.

Obaya, A. J., Mateyak, M. K., & Sedivy, J. M. (1999). Mysterious liaisons: The relationship between c-Myc and the cell cycle. *Oncogene, 18*, 2934–2941.

Ober, C., & Yao, T. C. (2011). The genetics of asthma and allergic disease: A 21st century perspective. *Immunological Reviews, 242*, 10–30.

Obsil, T., & Obsilova, V. (2008). Structure/function relationships underlying regulation of FOXO transcription factors. *Oncogene, 27*, 2263–2275.

Obsil, T., & Obsilova, V. (2011). Structural basis for DNA recognition by FOXO proteins. *Biochimica et Biophysica Acta, 1813*, 1946–1953.

Ogrunc, M., & d'Adda di Fagana, F. (2011). Never-aging cellular senescence. *European Journal of Cancer, 47*, 1616–1622.

Ohta, T., Sato, K., & Wu, W. (2011). The BRCA1 ubiquitin ligase and homologous recombination repair. *FEBS Letters, 585*, 2836–2844.

Ohtani, N., Mann, D. J., & Hara, E. (2009). Cellular senescence: Its role in tumor suppression and aging. *Cancer Science, 100*, 792–797.

Ohtsuka, S., & Dalton, S. (2008). Molecular and biological properties of pluripotent stem cells. *Gene Therapy, 15*, 74–81.

Okabe, H., Satoh, S., Kato, T., Kitahara, O., Yanagawa, R., Yamaoka, Y., et al. (2001). Genome-wide analysis of gene expression in human hepatocellular carcinomas using cDNA microarray: Identification of genes involved in viral carcinogenesis and tumor progression. *Cancer Research, 61*, 2129–2137.

Oliver, J. D., & Roberts, R. A. (2002). Receptor-mediated hepatocarcinogenesis: Role of hepatocyte proliferation and apoptosis. *Pharmacology & Toxicology, 91*, 1–7.

Olivier, M., Hollstein, M., & Hainaut, P. (2010). TP53 mutations in human cancers: Origins, consequences and clinical use. *Cold Spring Harbor Perspectives in Biology, 2*, a001008.

Orkin, S. H., & Hochedlinger, K. (2011). Chromatin connections to pluripotency and cellular reprogramming. *Cell, 145*, 835–849.

Osborn, S. L., Sohn, S. J., & Winoto, A. (2007). Constitutive phosphorylation mutation in FADD results in early cell cycle defects. *Journal of Biological Chemistry, 282*, 22786–22792.

O'Shea, C. C., & Fried, M. (2005). Modulation of the ARF-p53 pathway by the small DNA tumor viruses. *Cell Cycle, 4*, 449–452.

Oster, S. K., Ho, C. S. W., Soucle, E. L., & Penn, L. Z. (2002). The myc oncogene: MarvelouslY Complex. *Advances in Cancer Research, 84*, 81–154.

Osterloh, L., von Eyss, B., Schmit, F., Rein, L., Hübner, D., Samans, B., et al. (2007). The human synMuv-like protein LIN-9 is required for transcription of G2/M genes and for entry into mitosis. *EMBO Journal, 26*, 144–157.

Overdier, D. G., Porcella, A., & Costa, R. H. (1994). The DNA-binding specificity of the hepatocyte nuclear factor 3/forkhead domain is influenced by amino acid residues adjacent to the recognition helix. *Molecular and Cellular Biology, 14*, 2755–2766.

Overmeyer, J. H., & Maltese, W. A. (2011). Death pathways triggered by activated Ras in cancer cells. *Frontiers in Bioscience, 16*, 1693–1713.

Özcan, S., Andrali, S. S., & Cantrell, J. E. L. (2010). Modulation of transcription factor function by O-GlcNAc modification. *Biochimica et Biophysica Acta, 1799*, 353–364.

Pan, J. S., Hong, M. Z., & Ren, J. L. (2009). Reactive oxygen species: A double-edged sword in oncogenesis. *World Journal of Gastroenterology, 15*, 1702–1707.

Pandit, B., & Gartel, A. L. (2011a). FoxM1 knockdown sensitizes human cancer calls to proteasome inhibitor-induced apoptosis but not autophagy. *Cell Cycle, 10*, 3269–3273.

Pandit, B., & Gartel, A. L. (2011b). Thiazole antibiotic thiostrepton synergizes with bortezomib to induce apoptosis in cancer cells. *PLoS One, 6*, e17110.

Pandit, B., Halasi, M., & Gartel, A. L. (2009). p53 negatively regulates expression of FoxM1. *Cell Cycle, 8*, 3425–3427.

Park, H. J., Carr, J. R., Wang, Z., Nogueira, V., Hay, N., Tyner, A. L., et al. (2009). FoxM1, a critical regulator of oxidative stress during oncogenesis. *EMBO Journal, 28*, 2908–2918.

Park, H. J., Costa, R. H., Lau, L. F., Tyner, A. L., & Raychaudhuri, P. (2008). APC/C-Cdh1 mediated proteolysis of the Forkhead Box M1 transcription factor is critical for regulated entry into S phase. *Molecular and Cellular Biology, 28*, 5162–5171.

Park, H. J., Gusarova, G., Wang, Z., Carr, J. R., Li, J., Kim, K. H., et al. (2011). Deregulation of FoxM1b leads to tumour metastasis. *EMBO Molecular Medicine, 2*, 21–34.

Park, Y. Y., Jung, S. Y., Jennings, N. B., Rodriguez-Aguayo, C., Peng, G., Lee, S. R., et al. (2012). FOXM1 mediates Dox resistance in breast cancer by enhancing DNA repair. *Carcinogenesis, 33*, 1843–1853.

Park, T. J., Kim, J. Y., Oh, P., Kang, S. Y., Kim, B. W., Wang, H. J., et al. (2008). TIS21 negatively regulates hepatocarcinogenesis by disruption of cyclin B1-Forkhead Box M1 regulation loop. *Hepatology, 47*, 1533–1543.

Park, T. J., Kim, J. Y., Park, S. H., Kim, H. S., & Lim, I. K. (2009). Skp2 enhances polyubiquitination and degradation of TIS21$^{BTG2/PC3}$, tumor suppressor protein, at the downstream of FoxM1. *Experimental Cell Research, 315*, 3152–3162.

Park, K. S., Korfhagen, T. R., Bruno, M. D., Kitzmiller, J. A., Wan, H., Wert, S. E., et al. (2007). SPDEF regulates goblet cell hyperplasia in the airway epithelium. *The Journal of Clinical Investigation, 117*, 978–988.

Park, H. J., Wang, Z., Costa, R. H., Tyner, A., Lau, L. F., & Raychaudhuri, P. (2008). An N-terminal inhibitory domain modulates activity of FoxM1 during cell cycle. *Oncogene, 27*, 1696–1704.

Parsell, D. A., & Lindquist, S. (1993). The function of heat shock proteins in stress tolerance: Degradation and reactivation of damaged proteins. *Annual Review of Genetics, 27*, 437–496.

Parsons, J. A., Brelje, T. C., & Sorenson, R. L. (1992). Adaptation of islets of Langerhans to pregnancy: Increased islet cell proliferation and insulin secretion correlates with the onset of placental lactogen secretion. *Endocrinology, 130*, 1459–1466.

Pauklin, S., Pedersn, R. A., & Vallier, L. (2011). Mouse pluripotent stem cells at a glance. *Journal of Cell Science, 124*, 3727–3732.

Pazolli, E., & Stewart, S. A. (2008). Senescence: The good the bad and the dysfunctional. *Current Opinion in Genetics & Development, 18*, 42–47.

Pei, D. (2009). Regulation of pluripotency and reprogramming by transcription factors. *Journal of Biological Chemistry, 284*, 3365–3369.

Pelengaris, S., Khan, M., & Evan, G. (2002). c-MYC: More than just a matter of life and death. *Nature Reviews. Cancer, 2*, 764–776.

Pellegrino, R., Calvisi, D. F., Ladu, S., Ehemann, V., Staniscia, T., Evert, M., et al. (2010). Oncogenic and tumor suppressive roles of Polo-like kinases in human hepatocellular carcinoma. *Hepatology, 51*, 857–868.

Penzo, M., Massa, P. E., Olivotto, E., Bianchi, F., Borzi, R. M., Hanidu, A., et al. (2009). Sustained NF-κB activation produces a short-term cell proliferation block in conjunction with repressing effectors of cell cycle progression controlled by E2F and FoxM1. *Journal of Cellular Physiology, 218*, 215–227.

Pera, M. F., & Tam, P. P. L. (2010). Extrinsic regulation of pluripotent stem cells. *Nature, 465*, 713–720.

Perego, P., Corna, E., De Cesare, M., Gatti, L., Palizzi, D., Pratesi, G., et al. (2001). Role of apoptosis and apoptosis-related genes in cellular response and antitumor efficacy of anthracyclines. *Current Medicinal Chemistry, 8*, 31–37.

Perez-Sala, D., & Rebello, A. (1999). Novel aspects of Ras proteins biology: Regulation and implications. *Cell Death and Differentiation, 6*, 722–728.

Perona, R., Moncho-Amor, V., Machado-Pinilla, R., Belad-Iniesta, C., & Sanchez Perez, I. (2008). Role of CHK2 in cancer development. *Clinical & Translational Oncology, 10*, 538–542.

Perou, C. M., Sorlie, T., Eisen, M. B., van de Rijn, M., Jeffrey, S. S., Rees, C. A., et al. (2000). Molecular portraits of human breast tumors. *Nature, 406*, 747–752.

Peterson, C. L., & Cote, J. (2004). Cellular machineries for chromosomal DNA repair. *Genes & Development, 18*, 602–616.

Petrovic, V., Costa, R. H., Lau, L. F., Raychaudhuri, P., & Tyner, A. L. (2008). FOXM1 regulates growth factor induced expression of the KIS kinase to promote cell cycle progression. *Journal of Biological Chemistry, 283*, 453–460.

Petrovich, V., Costa, R. H., Lau, L. H., Raychaudhuri, P., & Tyner, A. L. (2010). Negative regulation of the oncogenic transcription factor FoxM1 by thiazolidinediones and mithramycin. *Cancer Biology & Therapy, 9*, 1008–1016.

Phillips, B., & Morimoto, R. I. (1991). Transcriptional regulation of human hsp70 genes: Relationship between cell growth, differentiation, virus infection, and the stress response. *Results and Problems in Cell Differentiation, 17*, 167–187.

Pierrou, S., Hellqvist, M., Samuelsson, L., Enerbäck, S., & Carlsson, P. (1994). Cloning and characterization of seven human forkhead proteins: Binding site specificity and DNA bending. *EMBO Journal, 13*, 5002–5012.

Pignot, G., Vieillefond, A., Vacher, S., Zerbib, M., Debre, B., Lidereau, R., et al. (2012). Hedgehog pathway activation in human translational cell carcinoma of the bladder. *British Journal of Cancer, 106*, 1177–1186.

Pilarsky, C., Wenzig, M., Specht, T., Saeger, H. D., & Grutzmann, R. (2004). Identification and validation of commonly overexpressed genes in solid tumors by comparison of microarray data. *Neoplasia, 6*, 744–750.

Pilkinton, M., Sandoval, R., & Colamonici, O. R. (2007). Mammalian Mip/LIN-9 interacts with either p107, p130/E2F4 repressor complex or B-Myb in a cell cycle-phase-dependent context distinct from the Drosophila dREAM complex. *Oncogene, 26*, 7535–7543.

Pilkinton, M., Sandoval, R., Song, J., Ness, S. A., & Colamonici, O. R. (2007). Mip/LIN-9 regulates the expression of B-Myb and the induction of cyclin A, cyclin B, and CDK1. *Journal of Biological Chemistry, 282*, 168–175.

Pines, J. (1995). Cyclins and cyclin-dependent kinases: Theme and variations. *Advances in Cancer Research, 66*, 181–212.

Pinzone, J. J., Hall, B. M., Thudi, N. K., Vonau, M., Qiang, Y. W., Rosol, T. J., et al. (2009). The role of Dickkopf-1 in bone development, homeostasis, and disease. *Blood, 113*, 517–525.

Pipas, J. M. (2009). SV40: Cell transformation and tumorigenesis. *Virology, 384*, 294–303.

Pipas, J. M., & Levine, A. J. (2001). Role of T antigen interactions with p53 in tumorigenesis. *Seminars in Cancer Biology, 11*, 23–30.

Pirity, M., Blanck, J. K., & Schreiber-Agus, N. (2006). Lessons learned from Myc/Max/Mad knockout mice. *Current Topics in Microbiology and Immunology, 302*, 205–234.

Pitot, H. C., Dragan, Y. P., Teeguarden, J., Hsia, S., & Campbell, H. (1996). Quantitation of multistage carcinogenesis in rat liver. *Toxicologic Pathology, 24*, 119–128.

Planas-Silva, M. D., & Weinberg, R. A. (1997). The restriction point and control of cell proliferation. *Current Opinion in Cell Biology, 9*, 768–772.

Plank, J. L., Frist, A. Y., LeGrone, A. W., Magnuson, M. A., & Labosky, P. A. (2011). Loss of Foxd3 results in decreased β-cell proliferation and glucose intolerance during pregnancy. *Endocrinology, 152*, 4589–4600.

Plant, T. M., & Marshall, G. R. (2001). The functional significance of FSH in spermatogenesis and the control of its secretion in male primates. *Endocrine Reviews, 22*, 764–786.

Plopper, C. G., & Hyde, D. M. (2008). The non-human primate as a model for studying COPD and asthma. *Pulmonary Pharmacology & Therapeutics, 21,* 755–766.
Pohl, B. S., & Knöchel, W. (2005). Of Fox and Frogs: Fox (fork head/winged helix) transcription factors in *Xenopus* development. *Gene, 344,* 21–32.
Pohl, B. S., Rössner, A., & Knöchel, W. (2004). The Fox gene family in *Xenopus laevis*: FoxI2, FoxM1 and FoxP1 in early development. *International Journal of Developmental Biology, 49,* 53–58.
Polager, S., & Ginsberg, D. (2009). p53 and E2f: Partners in of life and death. *Nature Reviews. Cancer, 9,* 738–748.
Polo, S. E., & Jackson, S. O. P. (2011). Dynamics of DNA damage response proteins at DNA breaks: A focus on protein modifications. *Genes & Development, 25,* 409–433.
Polsky, D., & Cordon-Cardo, C. (2003). Oncogenes in melanoma. *Oncogene, 22,* 3087–3091.
Ponder, B. A. (2001). Cancer genetics. *Nature, 411,* 336–341.
Poss, K. D. (2010). Advances in understanding tissue regenerative capacity and mechanisms in animals. *Nature Reviews. Genetics, 11,* 710–722.
Possemato, R., Marks, K. M., Shaul, Y. D., Pacold, M. E., Kim, D., Birsoy, K., et al. (2011). Functional genomics reveal that the serine synthesis pathway is essential in breast cancer. *Nature, 476,* 346–350.
Powers, M. V., & Workman, P. (2007). Inhibitors of the heat shock response: Biology and pharmacology. *FEBS Letters, 581,* 3758–3769.
Prieur, A., & Peeper, D. S. (2008). Cellular senescence in vivo: A barrier to tumorigenesis. *Current Opinion in Cell Biology, 20,* 150–155.
Priller, M., Pöschel, J., Abrao, L., von Bueren, A. O., Cho, Y. J., Rutkowski, S., et al. (2011). Expression of FoxM1 is required for the proliferation of medulloblastoma cells and indicates worse survival of patients. *Clinical Cancer Research, 17,* 6791–6801.
Prots, I., Skapenko, A., Lipsky, P. E., & Schulze-Koops, H. (2011). Analysis of the transcriptional program of developing induced regulatory T cells. *PLoS One, 6,* e16913.
Pruitt, J., & Der, C. J. (2001). Ras and Rho regulation of the cell cycle and oncogenesis. *Cancer Letters, 171,* 1–10.
Pujadas, E., & Feinberg, A. P. (2012). Regulated noise in the epigenetic landscape of development and disease. *Cell, 148,* 1123–1131.
Pylayeva-Gupta, Y., Grabocka, E., & Bar-Sagi, D. (2011). RAS oncogenes: Weaving a tumorigenic web. *Nature Reviews. Cancer, 11,* 761–774.
Qatanani, M., & Moore, D. D. (2005). CAR, the continuously advancing receptor, in drug metabolism and disease. *Current Drug Metabolism, 6,* 329–339.
Qu, K., Xu, X., Li, C., Wu, Q., Wei, J., Meng, F., et al. (2013). Negative regulation of transcription factor FoxM1 by p53 enhances oxaliplatin-induced senescence in hepatocellular carcinoma. *Cancer Letters, 331,* 105–114.
Raassool, F. V. (2003). DNA double strand breaks (DSB) and non-homologous end joining (NHEJ) pathways in human leukemia. *Cancer Letters, 193,* 1–9.
Raassool, F. V., & Tomkinson, A. E. (2010). Targeting abnormal DNA double strand break repair in cancer. *Cellular and Molecular Life Sciences, 67,* 3699–3710.
Rabbani, A., Finn, R. M., & Ausio, J. (2005). The anthracycline antibiotics: Antitumor drugs that alter chromatin structure. *Bioessays, 27,* 50–56.
Radhakrishnan, S. K., Bhat, U. G., Hughes, D. E., Wang, I. C., Costa, R. H., & Gartel, A. L. (2006). Identification of a chemical inhibitor of the oncogenic transcription factor Forkhead box M1. *Cancer Research, 66,* 9731–9735.
Raghavan, A., Zhou, G., Zhou, Q., Ibe, J. C. F., Ramchandran, R., Yang, Q., et al. (2012). Hypoxia induced pulmonary arterial smooth muscle cell proliferation is controlled by FOXM1. *American Journal of Respiratory Cell and Molecular Biology, 46,* 431–436.

Rajalimgam, K., Schreck, R., Rapp, U. R., & Albert, S. (2007). Ras oncogenes and their downstream targets. *Biochimica et Biophysica Acta, 1773*, 1177–1195.

Ralston, A., & Rossant, J. (2010). The genetics of induced pluripotency. *Reproduction, 139*, 35–44.

Ramakrishna, S., Kim, I. M., Petrovic, V., Malin, D., Wang, I. C., Kalin, T. V., et al. (2007). Myocardium L. defects and ventricular hypoplasia in mice homozygous null for the Forkhead Box M1 transcription factor. *Developmental Dynamics, 236*, 1000–1013.

Rangarajan, A., & Weinberg, R. A. (2003). Comparative biology of mouse versus human cells: Modelling human cancer in mice. *Nature Reviews. Cancer, 3*, 952–959.

Raver-Shapira, N., Marciano, E., Meiri, E., Spector, Y., Rosenfeld, N., Moskovits, N., et al. (2007). Transcriptional activation of miR-34a contributes to p53-mediated apoptosis. *Molecular Cell, 26*, 731–743.

Raychaudhuri, P., & Park, H. J. (2011). FoxM1 a master regulator of tumor metastasis. *Cancer Research, 71*, 4329–4333.

Raymond, E., Faivre, S., Chaney, S., Waynarowski, J., & Cvitkovic, E. (2002). Cellular and molecular pharmacology of oxaliplatin. *Molecular Cancer Therapeutics, 1*, 227–237.

Reed, S. I. (2002). Cell cycling? Check on your brakes. *Nature Cell Biology, 4*, E199–E201.

Reinhardt, H. C., & Yaffe, M. B. (2009). Kinases that control the cell cycle in response to DNA damage: Chk1, Chk2, and MK2. *Current Opinion in Cell Biology, 21*, 245–255.

Ren, X., Shah, T. A., Ustiyan, V., Zhang, Y., Shinn, J., Chen, G., et al. (2013). FOXM1 promotes allergen-induced goblet cell metaplasia and pulmonary inflammation. *Molecular and Cellular Biology, 33*, 371–386.

Ren, X., Zhang, Y., Snyder, J., Cross, E. R., Shah, T. A., Kalin, T. V., et al. (2010). Forkhead Box M1 transcription factor is required for macrophage recruitment during liver repair. *Molecular and Cellular Biology, 30*, 5381–5393.

Repasky, G. A., Chenette, E. J., & Der, C. J. (2004). Renewing the conspiracy theory debate: Does Raf function alone to mediate Ras oncogenesis? *Trends in Cell Biology, 14*, 639–647.

Reuter, S., Gupta, S. C., Chaturvedi, M. M., & Aggarwal, B. B. (2010). Oxidative stress, inflammation, and cancer: How are they linked? *Free Radical Biology & Medicine, 49*, 1603–1616.

Reuther, G. W., & Der, C. J. (2000). The Ras branch of small GTPases: Ras family members don't fall far from the tree. *Current Opinion in Cell Biology, 12*, 157–165.

Rhim, J. A., Sandgren, E. P., Degen, J. L., Palmiter, R. D., & Brinster, R. L. (1994). Replacement of diseased mouse liver by hepatic cell transplantation. *Science, 263*, 1149–1152.

Ribes, V., & Briscoe, J. (2009). Establishing and interpreting graded Sonic Hedgehog signaling during vertebrate neural tube patterning: The role of negative feedback. *Cold Spring Harbor Perspectives in Biology, 1*, a002014.

Richardson, C. (2005). RAD51, genomic stability, and tumorigenesis. *Cancer Letters, 218*, 127–139.

Richter, K., Haslbeck, M., & Buchner, J. (2010). The heat shock response: Life on the verge of death. *Molecular Cell, 40*, 253–266.

Rickman, D. S., Bobek, M. P., Misek, D. E., Kuick, R., Blaivas, M., Kurnit, D. M., et al. (2001). Distinctive molecular profiles of high-grade and low-grade gliomas based on oligonucleotide microarray analysis. *Cancer Research, 61*, 6885–6891.

Rieck, S., & Kaestner, K. H. (2010). Expansion of β-cell mass in response to pregnancy. *Trends in Endocrinology and Metabolism, 21*, 151–158.

Riobo, N. A., & Manning, D. R. (2007). Pathways of signal transduction employed by vertebrate Hedgehogs. *Biochemical Journal, 403*, 369–379.

Rodriguez, J. A., Li, M., Yao, Q., Chen, C., & Fisher, W. E. (2005). Gene overexpression in pancreatic adenocarcinoma: Diagnostic and therapeutic implications. *World Journal of Surgery, 29,* 297–305.

Rohatgi, R., & Scott, M. P. (2007). Patching the gaps in Hedgehog signalling. *Nature Cell Biology, 9,* 1005–1009.

Romagnoli, S., Fasoli, E., Vaira, V., Falleni, M., Pellegrini, C., Catania, A., et al. (2009). Identification of potential therapeutic targets in malignant mesothelioma using cell-cycle gene expression profiling. *American Journal of Pathology, 174,* 762–770.

Roninson, I. B. (2003). Tumor cell senescence in cancer treatment. *Cancer Research, 63,* 2705–2715.

Rossant, J. (2008). Stem cells and early lineage development. *Cell, 132,* 527–531.

Rosty, C., Sheffer, M., Tsfrir, D., Stransky, N., Tsafrir, I., Peter, M., et al. (2005). Identification of a proliferation gene cluster associated with HPV E6/E7 expression level and viral DNA load in invasive cervical carcinoma. *Oncogene, 24,* 7094–7104.

Rottmann, S., & Lüscher, B. (2006). The Mad side of the Max network: Antagonizing the function of Myc and more. *Current Topics in Microbiology and Immunology, 302,* 63–122.

Roubertoux, P. L., & Kerdelhue, B. (2006). Trisomy 21: From chromosomes to mental retardation. *Behavior Genetics, 36,* 346–354.

Rouse, J., & Jackson, S. P. (2002). Interfaces between the detection, signalling, and repair of DNA damage. *Science, 297,* 547–551.

Rovillain, E., Mansfeild, L., Caetano, C., Alvarez-Fernandez, M., Caballero, O. L., Medema, R. H., et al. (2011). Activation of nuclear factor-kappa B signalling promotes cellular senescence. *Oncogene, 30,* 2356–2366.

Rowland, B. D., & Bernards, R. (2006). Re-evaluating cell-cycle regulation by E2Fs. *Cell, 127,* 871–874.

Roy, R., Chun, J., & Powell, S. N. (2011). BRCA1 and BRCA2: Different roles in a common pathway of genome protection. *Nature Reviews. Cancer, 12,* 68–78.

Ruiz i Altaba, A., Mas, C., & Stecca, B. (2007). The Gli code: An information nexus regulating cell fate, stemness and cancer. *Trends in Cell Biology, 17,* 438–447.

Ryan, K. E., & Chiang, C. (2012). Hedgehog secretion and signal transduction in vertebrates. *Journal of Biological Chemistry, 287,* 17905–17913.

Sadavisam, S., Duan, S., & DeCarpio, J. A. (2012). The MuvB complex sequentially recruits B-Myb and FoxM1 to promote mitotic gene expression. *Genes & Development, 26,* 474–489.

Saenz-Robles, M., & Pipas, J. M. (2009). T antigen transgenic mouse models. *Seminars in Cancer Biology, 19,* 229–235.

Saenz-Robles, M., Sullivan, C. S., & Pipas, J. M. (2001). Transforming functions of simian virus 40. *Oncogene, 20,* 7899–7907.

Sage, J. (2005). Making young tumors old: A new weapon against cancer? *Science of Aging Knowledge Environment, 2005,* pe25.

Sakurai, H., & Enoki, Y. (2010). Novel aspects of heat shock factors: DNA recognition, chromatin modulation and gene expression. *FEBS Journal, 277,* 4140–4149.

Saleh, E. M., El-Awady, R. A., & Anis, N. (2013). Predictive markers for the response to 5-fluorouracil therapy in cancer cells: Constant-field gel electrophoresis as a tool for prediction of response to 5-fluorouracil-based chemotherapy. *Oncology Letters, 5,* 321–327.

Salvatore, G., Nappi, T. C., Salerno, P., Jiang, Y., Garbi, C., Ugolini, C., et al. (2007). A cell proliferation and chromosomal instability signature in anaplastic thyroid carcinoma. *Cancer Research, 67,* 10148–10158.

Samuel, T., Weber, H. O., & Funk, J. O. (2002). Linking DNA damage to cell cycle checkpoints. *Cell Cycle, 1,* 162–168.

San Filippo, J., Sung, P., & Klein, H. (2008). Mechanisms of eukaryotic homologous recombination. *Annual Review of Biochemistry, 77,* 229–257.

Sancar, A., Lindsey-Boltz, L. A., Ünsal-Kacmaz, K., & Linn, S. (2004). Molecular mechanisms of mammalian DNA repair and the DNA damage checkpoints. *Annual Review of Biochemistry, 73,* 39–85.

Sanchez-Calderon, H., Rodriguez-de la Rosa, L., Milo, M., Pichel, J. G., Holley, M., & Varela-Nieto, I. (2010). RNA microarray analysis in prenatal mouse cochlea reveals novel IGF-I target genes: Implication of MEF2 and FOXM1 transcription factors. *PLoS One, 5,* e8699.

Santin, A. D., Zhan, F., Bignotti, E., Siegel, E. R., Cane, S., Bellone, S., et al. (2005). Gene expression profiles of primary HPV-16- and HPV18-infected early stage cervical cancers and normal cervical epithelium: Identification of novel candidate molecular markers for cervical cancer diagnosis and therapy. *Virology, 331,* 269–291.

Sargent, L., Dragan, Y., Xu, Y. H., Sattler, G., Wiley, J., & Pitot, H. C. (1996). Karotypic changes in a multistage model of chemical hepatocarcinogenesis in the rat. *Cancer Research, 56,* 2985–2991.

Sato, F., Tshuchiya, S., Meltzer, S. J., & Shimizu, K. (2011). MicroRNAs and epigenetics. *FEBS Journal, 278,* 1598–1609.

Satyanarayana, A., & Kaldis, P. (2009). Mammalian cell-cycle regulation: Several Cdks, numerous cyclins and diverse compensatory mechanisms. *Oncogene, 28,* 2925–2939.

Saucedo, L. J., & Edgar, B. A. (2007). Filling out the Hippo pathway. *Nature Reviews. Molecular Cell Biology, 8,* 613–621.

Scheper, W., & Copray, S. (2009). The molecular mechanism of induced pluripotency: A two-stage switch. *Stem Cell Reviews and Reports, 5,* 204–223.

Schmierer, B., & Hill, C. S. (2007). TGFβ-SMAD signal transduction: Molecular specificity and functional flexibility. *Nature Reviews. Molecular Cell Biology, 8,* 970–982.

Schmit, F., Cremer, S., & Gaubatz, S. (2009). LIN54 is an essential core subunit of the DREAM/LINC complex that binds to the cdc2 promoter in a sequence-specific manner. *FEBS Journal, 276,* 5703–5716.

Schmit, F., Korenjak, M., Mannefeld, M., Schmit, K., Franke, C., von Eyss, B., et al. (2007). LINC, a human complex that is related to pRB-containing complexes in invertebrates regulates the expression of G2/M genes. *Cell Cycle, 6,* 1903–1913.

Schmitt, C. A. (2003). Senescence, apoptosis and therapy—Cutting the lifelines of cancer. *Nature Reviews. Genetics, 3,* 286–295.

Schmitt, C. A. (2007). Cellular senescence and cancer treatment. *Biochimica et Biophysica Acta, 1775,* 5–20.

Schmitt, E., Gehrmann, M., Brunet, M., Multhoff, G., & Garrido, C. (2007). Intracellular and extracellular functions of heat shock proteins: Repercussions in cancer therapy. *Journal of Leukocyte Biology, 81,* 15–27.

Schmucker, D. L. (2005). Age-related changes in hepatic structure and function: Implications for disease. *Experimental Gerontology, 40,* 650–659.

Schmucker, D. L., & Sanchez, H. (2011). Liver regeneration and aging: A current perspective. *Current Gerontology and Geriatrics Research, 2011,* 526379.

Schreiber-Agus, N., & DePinho, R. A. (1998). Repression by the Mad(Mxi1)-Sin3 complex. *Bioessays, 20,* 808–818.

Schubbert, S., Shannon, K., & Bollag, G. (2007). Hyperactive Ras in developmental disorders and cancer. *Nature Reviews. Cancer, 7,* 295–308.

Schüller, U., Kho, A. T., Zhao, Q., Ma, Q., & Rowitch, D. H. (2006). Cerebellar "transcriptome" reveals cell-type and stage-specific expression during postnatal development and tumorigenesis. *Molecular and Cellular Neuroscience, 33,* 247–259.

Schüller, U., Zhao, Q., Godinho, S. A., Heine, V. M., Medema, R. H., Pellman, D., et al. (2007). Forkhead transcription factor FoxM1 regulates mitotic entry and prevents spindle defects in cerebellar granule neuron precursors. *Molecular and Cellular Biology, 27,* 8259–8270.

Scott, S. P., & Pandita, T. K. (2006). The cellular control of DNA double-strand breaks. *Journal of Cellular Biochemistry, 99*, 1463–1475.

Scott, R. E., Tzen, C. Y., Witte, M. M., Blatti, S., & Wang, H. (1993). Regulation of differentiation, proliferation and cancer suppressor activity. *International Journal of Developmental Biology, 37*, 67–74.

Scotton, C. J., & Chambers, R. C. (2007). Molecular targets in pulmonary fibrosis: The myofibroblast in focus. *Chest, 132*, 1311–1321.

Scrima, A., Fischer, E. S., Lingaraju, G. M., Cavadini, S., & Thomä, N. H. (2011). Detecting UV-lesions in the genome: The modular CRL4 ubiquitin ligase does it best! *FEBS Letters, 585*, 2818–2825.

Sears, R. C., & Nevins, J. R. (2002). Signaling networks that link cell proliferation and cell fate. *Journal of Biological Chemistry, 277*, 11617–11620.

Selman, M., King, T. E., Pardo, A., American Thoracic Society, & European Respiratory Society and American College of Chest Physicians (2001). Idiopathic pulmonary fibrosis: Prevailing and evolving hypotheses about its pathogenesis and implications for therapy. *Lancet, 378*, 1949–1961.

Sengupta, A., Kalinichenko, V. V., & Yutzey, K. E. (2013). FoxO and FoxM1 transcription factors have antagonistic functions in neonatal cardiomyocyte cell cycle withdrawal and IGF1 gene regulation. *Circulation Research, 112*, 267–277.

Serrano, M., & Blasco, M. A. (2001). Putting the stress on senescence. *Current Opinion in Cell Biology, 13*, 748–753.

Seton-Rogers, S. (2011). Metabolism: Flexible flux. *Nature Reviews. Cancer, 11*, 621.

Shafritz, D. A., & Oertel, M. (2011). Model systems and experimental conditions that lead to effective repopulation of the liver by transplanted cells. *The International Journal of Biochemistry & Cell Biology, 43*, 198–213.

Shamovsky, I., & Nudler, E. (2008). New insights into the mechanism of heat shock response activation. *Cellular and Molecular Life Sciences, 65*, 855–861.

Shao, L., Li, H., Pazhanisamy, S. K., Meng, A., Wang, Y., & Zhou, D. (2011). Reactive oxygen species and hematopoietic stem cell senescence. *International Journal of Hematology, 94*, 24–32.

Shapiro, P. (2001). Ras-MAP kinase signaling pathways and control of cell proliferation: Relevance to cancer therapy. *Critical Reviews in Clinical Laboratory Sciences, 39*, 285–330.

Sharpless, N. E. (2004). *Ink4a/Arf* links senescence and aging. *Experimental Gerontology, 39*, 1751–1759.

Sharpless, N. E., & DePinho, R. A. (2004). Telomeres, stem cells, senescence and cancer. *The Journal of Clinical Investigation, 113*, 160–168.

Shaw, A. T., Meissner, A., Dowdle, J. A., Crowley, D., Magendantz, M., Onyang, C., et al. (2007). Sprouty-2 regulates oncogenic K-ras in lung development and tumorigenesis. *Genes & Development, 21*, 694–707.

Shay, J. W., & Roninson, I. B. (2004). Hallmarks of senescence in carcinogenesis and tumor therapy. *Oncogene, 23*, 2919–2933.

Sheng, W., Rance, M., & Liao, X. (2002). Structure comparison of two conserved HNF-3/fkh proteins HFH-1 and genesis indicates the existence of folding differences in their complexes with a DNA binding sequence. *Biochemistry, 41*, 3286–3293.

Sherman, M. Y., & Goldberg, A. L. (2001). Cellular defenses against unfolded proteins: A cell biologist thinks about neurodegenerative diseases. *Neuron, 29*, 15–32.

Sherr, C. J. (1998). Tumor surveillance via the ARF-p53 pathway. *Genes & Development, 12*, 2984–2991.

Sherr, C. J. (2000). The Pezcoller lecture: Cancer cell cycles revisited. *Cancer Research, 60*, 3689–3695.

Sherr, C. J. (2001). The *INK4a/ARF* network in tumour suppression. *Nature Reviews. Molecular Cell Biology, 2*, 731–737.

Sherr, C. J. (2002). Cell cycle control and cancer. *Harvey Lectures, 96*, 73–92.
Sherr, C. J. (2004). Principles of tumor suppression. *Cell, 116*, 235–246.
Sherr, C. J. (2006). Divorcing ARF and p53: An unsettled case. *Nature Reviews. Cancer, 6*, 663–673.
Sherr, C. J., & DePinho, R. A. (2000). Cellular senescence: Mitotic clock or culture shock? *Cell, 102*, 407–410.
Sherr, C. J., & McCormick, F. (2002). The RB and p53 pathways in cancer. *Cancer Cell, 2*, 103–112.
Sherr, C. J., & Weber, J. D. (2000). The ARF/p53 pathway. *Current Opinion in Genetics & Development, 10*, 94–99.
Shi, Y., & Massagué, J. (2003). Mechanisms of TGF-β signaling from cell membrane to the nucleus. *Cell, 113*, 685–700.
Shields, J. M., Pruitt, K., McFall, A., Shaub, A., & Der, C. J. (2000). Understanding Ras: 'it ain't over 'til it's over'. *Trends in Cell Biology, 10*, 147–154.
Shinohara, A., & Ogawa, T. (1999). Rad51/RecA protein families and the associated proteins in eukaryotes. *Mutation Research, 435*, 13–21.
Siddik, Z. H. (2003). Cisplatin: Mode of cytotoxic action and molecular basis of resistance. *Oncogene, 22*, 7265–7279.
Silva, J., & Smith, A. (2008). Capturing pluripotency. *Cell, 132*, 532–536.
Silver, P. A., Brent, R., & Ptashne, M. (1986). DNA binding is not sufficient for nuclear localization of regulatory proteins in *Saccharomyces cerevisiae*. *Molecular and Cellular Biology, 6*, 4763–4766.
Silver, P. A., Keegan, L. P., & Ptashne, M. (1984). Amino terminus of the yeast GAL4 gene product is sufficient for nuclear localization. *Proceeding of the National Academy of Sciences of the United States of America, 81*, 5951–5955.
Silver, D. P., & Livingston, D. M. (2012). Mechanisms of BRCA1 tumor suppression. *Cancer Discovery, 2*, 679–684.
Sisson, T. H., Mendez, M., Choi, K., Subbotina, N., Courey, A., Cunningham, A., et al. (2010). Targeted injury of type I epithelial cell induces pulmonary fibrosis. *American Journal of Respiratory and Critical Care Medicine, 181*, 254–263.
Slawson, C., & Hart, G. W. (2011). Cross talk between O-GlcNAcylation and phosphorylation: Roles in signaling, transcription, and chronic disease. *Nature Reviews. Cancer, 11*, 678–684.
Smith, J., Tho, L. M., Xu, N., & Gillespie, D. A. (2010). The ATM-Chk2 and ATR-Chk1 pathways in DNA damage signaling and cancer. *Advances in Cancer Research, 108*, 73–112.
Somero, G. N. (1995). Proteins and temperature. *Annual Review of Physiology, 57*, 43–68.
Sommer, C. A., & Henrique-Silva, F. (2008). Trisomy 21 and Down syndrome: A short review. *Brazilian Journal of Biology, 68*, 447–452.
Song, H. K., Hong, S. E., Kim, T., & Kim, D. H. (2012). Deep RNA sequencing reveals novel cardiac transcriptomic signatures for physiological and pathological hypertrophy. *PLoS One, 7*, e35552.
Soni, M. G., & Mehendale, H. M. (1998). Role of tissue repair in toxicologic interactions among hepatotoxic organics. *Environmental Health Perspectives, 106*, 1307–1317.
Soprano, D. R., Teets, B. W., & Soprano, K. J. (2007). Role of retinoic acid in the differentiation of embryonal carcinoma and embryonic stem cells. *Vitamins and Hormones, 75*, 69–95.
Sorensen, R. L., & Brelje, T. C. (1997). Adaptation of islets of Langerhans to pregnancy: B-cell growth, enhanced insulin secretion and the role of lactogenic hormones. *Hormone and Metabolic Research, 29*, 301–307.
Sorlie, T., Perou, C. M., Tibshirani, R., Aas, T., Geisler, S., Johnsen, H., et al. (2001). Gene expression patterns of breast carcinomas distinguish tumor subclasses with clinical implications. *Proceedings of the National Academy of Sciences, 98*, 10869–10874.

Spurgers, K. B., Gold, D. L., Coombes, K. R., Bohnenstiel, N. L., Mullins, B., Mey, R. E., et al. (2006). Identification of cell cycle regulatory genes as principal targets of p53-mediated transcriptional repression. *Journal of Biological Chemistry, 281,* 25134–25142.

Stargell, L. A., & Struhl, K. (1996). A new class of activation-defective TATA-binding protein mutants: Evidence for two steps of transcriptional activation in vivo. *Molecular and Cellular Biology, 16,* 4456–4464.

Starostina, N. G., & Kipreos, E. T. (2011). Multiple degradation pathways regulate versatile CIP/KIP CDK inhibitors. *Trends in Cell Biology, 22,* 33–41.

Steer, C. J. (1995). Liver regeneration. *The FASEB Journal, 9,* 1396–1400.

Stein, G. S., & Pardee, A. B. (2004). *Cell cycle and growth, control.* Hoboken: Wiley-Liss.

Stevens, C., & LaThangue, N. B. (2003). E2F and cell cycle control: A double-edged sword. *Archives of Biochemistry and Biophysics, 412,* 157–169.

Stevenson, C. S., & Birrell, M. A. (2011). Moving towards a new generation of animal models for asthma and COPD with improved clinical relevance. *Pharmacology & Therapeutics, 130,* 93–105.

Stewart, S. A., & Weinberg, R. A. (2002). Senescence: Does it all happen at the ends? *Oncogene, 21,* 627–630.

Stewart, S. A., & Weinberg, R. A. (2006). Telomeres: Cancer to human ageing. *Annual Review of Cell and Developmental Biology, 22,* 531–557.

Stoick-Cooper, C. L., Moon, R. T., & Weidinger, G. (2007). Advances in signaling in vertebrate regeneration as a prelude to regenerative medicine. *Genes & Development, 21,* 1292–1315.

Stoyanova, T., Roy, N., Bhattacharjee, S., Kopanja, D., Valli, T., Bagchi, S., et al. (2012). p21 cooperates with DDB2 in suppression of UV-induced skin malignancies. *Journal of Biological Chemistry, 287,* 3019–3028.

Stracker, T. H., Usui, T., & Petrini, J. H. (2009). Taking the time to make important decisions: The checkpoint effector kinases Chk1 and Chk2 and the DNA damage response. *DNA Repair, 8,* 1047–1054.

Strasser, A., & Newton, K. (1999). FADD/MORT1, a signal transducer that can promote cell death or cell growth. *The International Journal of Biochemistry & Cell Biology, 31,* 533–537.

Strom, S. C., Davila, S., & Grompe, M. (2010). Chimeric mice with humanized liver: Tools for the study of drug metabolism, excretion, and toxicity. *Methods in Molecular Biology, 640,* 491–509.

Stroud, J. C., Wu, Y., Bates, D. L., Han, A., Nowick, K., Paabo, S., et al. (2006). Structure of the forkhead domain of FOXP2 bound to DNA. *Structure, 14,* 159–166.

Struhl, K. (1995). Yeast transcriptional regulatory mechanisms. *Annual Review of Genetics, 29,* 651–674.

Struhl, K. (1996). Transcriptional enhancement by acidic activators. *Biochimica et Biophysica Acta, 1288,* O15–P17.

Su, T. T. (2006). Cellular responses to DNA damage: One signal, multiple choices. *Annual Review of Genetics, 40,* 187–208.

Sudol, M., & Harvey, K. F. (2010). Modularity in the Hippo signaling pathway. *Trends in Biochemical Sciences, 35,* 627–632.

Sullivan, C., Liu, Y., Shen, J., Curtis, A., Newman, C., Hock, J. M., et al. (2012). Novel interaction between FOXM1 and CDC25A regulate the cell cycle. *PLoS One, 7,* e51277.

Sullivan, C. A., & Pipas, J. M. (2002). T antigens of SV40: Molecular chaperones for viral replication and tumorigenesis. *Microbiology and Molecular Biology Reviews, 66,* 179–202.

Sumi, Y., & Hamid, Q. (2007). Airway remodeling in asthma. *Allergology International, 56,* 341–348.

Sun, H., Teng, M., Liu, J., Jin, D., Wu, J., Yan, D., et al. (2011). FOXM1 expression predicts the prognosis in hepatocellular carcinoma patients after orthotopic liver transplantation combined with the Milan criteria. *Cancer Letters, 306,* 214–222.

Sung, P., Krejci, L., Van Komen, S., & Sehorn, M. G. (2003). Rad51 recombinase and recombination mediators. *Journal of Biological Chemistry, 278*, 42729–42732.

Swales, K., & Negishi, M. (2004). CAR, driving into the future. *Molecular Endocrinology, 18*, 1589–1598.

Symington, L. S., & Gautier, J. (2011). Double-strand break end resection and repair pathway choice. *Annual Review of Genetics, 45*, 247–271.

Taatjes, D. J., Fenick, D. J., Gaudiana, G., & Koch, T. H. (1998). A redox pathway leading to the alkylation of nucleic acids by doxorubicin and related anthracyclines: Application to the design of antitumor drugs for resistant cancer. *Current Pharmaceutical Design, 4*, 203–218.

Tagaya, E., & Tamaoki, J. (2007). Mechanisms of airway remodeling in asthma. *Allergology International, 56*, 331–340.

Takahashi, K., Furukawa, C., Takano, A., Ishikawa, N., Kato, T., Hayama, S., et al. (2006). The neuromedin U-growth hormone secretagogue receptor 1b/neurotensin receptor 1 oncogenic signaling pathway as a therapeutic target for lung cancer. *Cancer Research, 66*, 9408–9419.

Takahashi, M., Koi, M., Balaguer, F., Boland, C. R., & Goel, A. (2011). MSH3 mediates sensitization of colorectal cancer cells to cisplatin, oxaliplatin, and a poly(ADP-ribose) polymerase inhibitor. *Journal of Biological Chemistry, 286*, 12157–12165.

Takahashi, Y., Li, L., Kamiryo, M., Asteriou, T., Moustakas, A., Yamashita, H., et al. (2005). Hyaluronan fragments induce endothelial cell differentiation in a CD44- and CXCL1/GRO1-dependent manner. *Journal of Biological Chemistry, 280*, 24195–24204.

Takemura, T., Nakamura, S., Yokota, D., Hirano, I., Ono, T., Shigeno, K., et al. (2010). Reduction of RAF kinase inhibitor protein expression by BCR-ABL contributes to chronic myelogenous leukemia proliferation. *Journal of Biological Chemistry, 285*, 6585–6594.

Takizawa, D., Kakizaki, S., Horiguchi, N., Yamazaki, Y., Tojima, H., & Mori, M. (2011). Constitutive active/androstane receptor promotes hepatocarcinogenesis in a mouse model of non-alcoholic steatohepatitis. *Carcinogenesis, 32*, 576–583.

Takuwa, N., & Takuwa, Y. (2001). Regulation of cell cycle molecules by the Ras effector system. *Molecular and Cellular Endocrinology, 177*, 25–33.

Tamano, S., Merlino, G., & Ward, J. M. (1994). Rapid development of hepatic tumors in transforming growth factor a transgenic mice associated with increased cell proliferation in precancerous hepatocellular lesions initiated by N-nitrosodiethylamine and promoted by phenobarbital. *Carcinogenesis, 15*, 1791–1798.

Tan, Y., Chen, L., Yu, H., Zhu, X., Meng, X., Huang, L., et al. (2010). Two-fold elevation of expression of FoxM1 transcription factor in mouse embryonic fibroblasts enhances cell cycle checkpoint activity by stimulating *p21* and *Chk1* transcription. *Cell Proliferation, 43*, 494–504.

Tan, Y., Raychaudhuri, P., & Costa, R. H. (2007). Chk2 mediates stabilization of the FoxM1 transcription factor to stimulate expression of DNA repair genes. *Molecular and Cellular Biology, 27*, 1007–1016.

Tan, Y., Yoshida, Y., Hughes, D. E., & Costa, R. H. (2006). Increased expression of hepatocyte nuclear factor 6 stimulates hepatocyte proliferation during mouse liver regeneration. *Gastroenterology, 130*, 1283–1300.

Tapia-Alveal, C., Calonge, T. M., & O'Connell, M. J. (2009). Regulation of chk1. *Cell Division, 4*, 8.

Tarasov, V., Jung, P., Verdoodt, B., Lodygin, D., Epanchintsev, A., Menssen, A., et al. (2007). Differential regulation of microRNAs by p53 revealed by massively parallel sequencing: miR-34a is a p53 target that induces apoptosis and G_1-arrest. *Cell Cycle, 6*, 1586–1593.

Taub, R. (1996). Liver regeneration: Transcriptional control of liver regeneration. *The FASEB Journal, 10*, 413–427.

Taub, R. (2004). Liver regeneration: From myth to mechanism. *Nature Reviews. Molecular Cell Biology, 5*, 836–847.

Tavaria, M., Gabriele, T., Kola, I., & Anderson, R. L. (1996). A hitchhiker's guide to the human Hsp70 family. *Cell Stress & Chaperones, 1*, 23–28.

Teh, M. T. (2012). FOXM1 coming of age: Time for translation into clinical benefits? *Frontiers in Oncology, 2*, 146.

Teh, M. T., Gemenetzidis, E., Chaplin, T., Young, B. D., & Philpott, M. P. (2010). Upregulation of FOXM1 induces genomic instability in human epidermal keratinocytes. *Molecular Cancer, 9*, 45.

Teh, M. T., Gemenetzidis, E., Patel, D., Tariq, R., Nadir, A., Bahta, A. W., et al. (2012). FOXM1 induces a global methylation signature that mimics the cancer epigenome in head and neck squamous cell carcinoma. *PLoS One, 7*, e34329.

Teh, M. T., Wong, S. T., Neill, G. W., Ghali, L. R., Philpott, M. P., & Quinn, A. G. (2002). FOXM1 is a downstream target of Gli1 in basal cell carcinomas. *Cancer Research, 62*, 4773–4780.

Teichmann, M., Dumay-Odelot, H., & Fribourg, S. (2012). Structural and functional aspects of winged-helix domains at the core of transcription initiation complexes. *Transcription, 3*, 2–7.

Thacker, J. (2005). The RAD51 gene family, genetic instability and cancer. *Cancer Letters, 219*, 125–135.

Thannickal, V. J., Toews, G. B., White, E. S., Lynch, J. P., & Martinez, F. J. (2004). Mechanisms of pulmonary fibrosis. *Annual Review of Medicine, 55*, 395–417.

The Cancer Genome Atlas Research Network (2011). Integrated genomic analyses of ovarian carcinoma. *Nature, 474*, 609–615.

Thierry, F., Benotmane, M. A., Demeret, C., Mori, M., Teissier, S., & Desaintes, C. (2004). A genomic approach reveals a novel mitotic pathway in papillomavirus carcinogenesis. *Cancer Research, 64*, 895–903.

Timchenko, N. A. (2009). Aging and liver regeneration. *Trends in Endocrinology and Metabolism, 20*, 171–176.

Tjian, R., & Maniatis, T. (1994). Transcriptional activation: A complex puzzle with few easy pieces. *Cell, 77*, 5–8.

Tompkins, D. H., Besnard, V., Lang, A. W., Keiser, A. R., Wert, S. E., Bruno, M. D., et al. (2011). Sox2 activates cell proliferation and differentiation in the respiratory epithelium. *American Journal of Respiratory Cell and Molecular Biology, 45*, 101–110.

Torres, M., & Forman, H. J. (2003). Redox signaling and the MAP kinase pathways. *Biofactors, 17*, 287–296.

Tourneur, L., & Chiocchia, G. (2010). FADD: A regulator of life and death. *Trends in Immunology, 31*, 260–269.

Trazzi, S., Mitrugno, V. M., Valli, E., Fuchs, C., Rizzi, S., Guidi, S., et al. (2011). APP-dependent up-regulation of Ptch1 underlies proliferation impairment of neural precursors in Down syndrome. *Human Molecular Genetics, 20*, 1560–1573.

Triezenberg, S. J. (1995). Structure and function of transcriptional activation domains. *Current Opinion in Genetics & Development, 5*, 190–196.

Trimarchi, J. M., & Lees, J. A. (2002). Sibling rivalry in the E2F family. *Nature Reviews. Molecular Cell Biology, 3*, 11–20.

Tsai, H. C., & Baylin, S. B. (2011). Cancer epigenetics: Linking basic biology to clinical medicine. *Cell Research, 21*, 502–517.

Tsai, K. L., Huang, C. Y., Chang, C. H., Sun, Y. S., Chuang, W. J., & Hsiao, C. D. (2006). Crystal structure of the human FOXK1a-DNA complex and its implications on the diverse binding specificity of winged helix/forkhead proteins. *Journal of Biological Chemistry, 281*, 17400–17409.

Tsai, K. L., Sun, Y. J., Huang, C. Y., Yang, J. Y., Hung, M. C., & Hsiao, C. D. (2007). Crystal structure of the human FOXO3a-DBD/DNA complex suggests the effects of post-translational modification. *Nucleic Acids Research, 35*, 6984–6994.

Tsantoulis, P. K., & Gorgoulis, V. G. (2005). Involvement of E2F transcription factor family in cancer. *European Journal of Cancer, 41*, 2403–2414.

Tuteja, G., & Kaestner, K. H. (2007a). SnapShot: Forkhead transcription factors I. *Cell, 130*, 1160–1160.e1.

Tuteja, G., & Kaestner, K. H. (2007b). SnapShot: Forkhead transcription factors II. *Cell, 131*, 192–192.e1.

Tutt, A., & Ashworth, A. (2002). The relationship between the roles of BRCA genes in DNA repair and cancer predisposition. *Trends in Molecular Medicine, 8*, 571–576.

Tyedmers, J., Mogk, A., & Bukau, B. (2010). Cellular strategies for controlling protein aggregation. *Nature Reviews. Molecular Cell Biology, 11*, 777–788.

Uddin, S., Ahmed, M., Hussain, A., Akubaker, J., Al-Sanea, N., AbdulJabbar, A., et al. (2011). Genome-wide expression analysis of Middle Eastern colorectal cancer reveals *FOXM1* as a novel target for cancer therapy. *American Journal of Pathology, 178*, 537–547.

Uddin, S., Hussain, A. R., Ahmed, M., Siddiqui, K., Al-Dayel, F., Bavi, P., et al. (2012). Over-expression of FoxM1 offers a promising therapeutic target in diffuse large B-cell lymphoma. *Haematologica, 97*, 1092–1100.

Ueno, H., Nakajo, N., Watanabe, M., Isoda, M., & Sagata, N. (2008). FoxM1-driven cell division is required for neuronal differentiation in early Xenopus embryos. *Development, 135*, 2023–2030.

Ulloa-Aguirre, A., Zarinan, T., Pasapera, A. M., Casa-Gonzalez, P., & Dias, J. A. (2007). Multiple facets of follicle-stimulating hormone receptor function. *Endocrine, 32*, 251–263.

Ulrich, H., & Majumder, P. (2006). Neurotransmitter receptor expression and activity during neuronal differentiation of embryonal carcinoma and stem cells: From basic research towards clinical applications. *Cell Proliferation, 39*, 281–300.

Untergasser, G., Gander, R., Lilg, C., Lepperdinger, G., Plas, E., & Berger, P. (2005). Profiling molecular targets of TGF-β1 in prostate fibroblast-to-myofibroblast trans-differentiation. *Mechanisms of Ageing and Development, 126*, 59–69.

Ustiyan, V., Wang, I. C., Ren, X., Zhang, Y., Snyder, J., Xu, Y., et al. (2009). Forkhead box M1 transcriptional factor is required for smooth muscle cells during embryonic development of blood vessels and esophagus. *Developmental Biology, 336*, 266–279.

Ustiyan, V., Wert, S. E., Ikegami, M., Wang, I. C., Kalin, T. V., Whitsett, J. A., et al. (2012). Foxm1 transcription factor is critical for proliferation and differentiation of Clara cells during development of conducting airways. *Developmental Biology, 370*, 198–212.

Vakiani, E., & Solit, D. B. (2011). KRAS and BRAF: Drug targets and predictive biomarkers. *The Journal of Pathology, 223*, 219–229.

Valcourt, J. R., Lemons, J. M. S., Haley, E. M., Kojima, M., Demuren, O. O., & Coller, H. A. (2012). Staying alive. Metabolic adaptations to quiescence. *Cell Cycle, 11*, 1–17.

van den Boom, J., Wolter, M., Kuick, R., Misek, D. E., Youkilis, A. S., Wechsler, D. S., et al. (2003). Characterization of gene expression profiles associated with glioma progression using oligonucleotide-based microarray analysis and real-time transcription-polymerase chain reaction. *American Journal of Pathology, 163*, 1033–1043.

van der Heyden, M. A., & Defize, L. H. (2003). Twenty one years of P19 cells: What an embryonal carcinoma cell line taught us about cardiomyocyte differentiation. *Cardiovascular Research, 58*, 292–302.

van Dongen, M. J. P., Cederberg, S., Carlsson, P., Enerbäck, S., & Wikström, M. (2000). Solution structure and dynamics of the DNA-binding domain of the adipocyte-transcription factor FREAC-11. *Journal of Molecular Biology, 296*, 351–359.

van Gent, D. C., Hoeijmakers, J. H., & Kanaar, R. (2001). Chromosomal stability and the DNA double-strand break connection. *Nature Reviews. Genetics, 2,* 196–206.
van Riggelen, J., Yetil, A., & Felsher, D. W. (2010). MYC as a regulator of ribosome biogenesis and protein synthesis. *Nature Reviews. Cancer, 10,* 301–309.
Varjaluso, M., & Taipale, J. (2007). Hedgehog signaling. *Journal of Cell Science, 120,* 3–6.
Varjaluso, M., & Taipale, J. (2008). Hedgehog: Functions and mechanisms. *Genes & Development, 22,* 2454–2472.
Venkitaraman, A. R. (2004). Tracing the network connecting BRCA and Fanconi anaemia proteins. *Nature Reviews. Cancer, 4,* 266–276.
Ventura, A., & Jacks, T. (2009). MicroRNAs and cancer: Short RNAs go a long way. *Cell, 136,* 586–591.
Vervoorts, J., & Lüscher, B. (2008). Post-translational regulation of the tumor suppressor p27KIP1. *Cellular and Molecular Life Sciences, 65,* 3255–3264.
Vestweber, D., Broermann, A., & Schulte, D. (2010). Control of endothelial barrier function by regulating vascular endothelial-cadherin. *Current Opinion in Hematology, 17,* 230–236.
Vigh, L., Horvath, I., Maresca, B., & Harwood, J. L. (2007). Can the stress protein response be controlled by 'membrane-lipid therapy'? *Trends in Biochemical Sciences, 32,* 357–363.
Viglietto, G., Motti, M. L., & Fusco, A. (2002). Understanding p27kip1 deregulation in cancer. Downregulation or mislocalization? *Cell Cycle, 1,* 394–400.
Vispe, S., & Defais, M. (1997). Mammalian Rad51 protein: A RecA homologue with pleiotropic functions. *Biochimie, 79,* 587–592.
Voellmy, R. (1994). Transduction of the stress signal and mechanisms of transcriptional regulation of heat shock/stress protein gene expression in higher eukaryotes. *Critical Reviews in Eukaryotic Gene Expression, 4,* 357–401.
Voellmy, R. (2004). Transcriptional regulation of the metazoan stress protein response. *Progress in Nucleic Acid Research and Molecular Biology, 78,* 143–185.
Voellmy, R., & Boellmann, F. (2007). Chaperone regulation of the heat shock protein response. *Advances in Experimental Medicine and Biology, 594,* 89–99.
Vogelstein, B., & Kinzler, K. W. (2004). Cancer genes and the pathways they control. *Nature Medicine, 10,* 789–799.
Vojtek, A. B., & Der, C. J. (1998). Increasing complexity of the Ras signaling pathway. *Journal of Biological Chemistry, 273,* 19925–19928.
von Garnier, C., & Nicod, L. P. (2009). Immunology taught by lung dendritic cells. *Swiss Medical Weekly, 139,* 186–192.
Wahlstrom, T., & Henriksson, M. (2007). Mnt takes control as a key regulator of the myc/max/mad network. *Advances in Cancer Research, 97,* 61–80.
Walker, W. H., & Cheng, J. (2005). FSH and testosterone signaling in Sertoli cells. *Reproduction, 130,* 15–28.
Wallez, Y., & Huber, P. (2008). Endothelial adherens and tight junctions in vascular homeostasis, inflammation and angiogenesis. *Biochimica et Biophysica Acta, 1778,* 794–809.
Walworth, N. C. (2000). Cell-cycle checkpoint kinases: Checking in on the cell cycle. *Current Opinion in Cell Biology, 12,* 697–704.
Wan, X., Yeung, C., Young Kim, S., Dolan, J. G., Ngo, V. N., Burkett, S., et al. (2012). Identification of FoxM1/Bub1b signaling pathway as a required component for growth and survival of rhabdomyosarcoma. *Cancer Research, 72,* 5889–5899.
Wang, W. (2007). Emergence of a DNA-damage response network consisting of Fanconi anaemia and BRCA proteins. *Nature Reviews. Genetics, 8,* 735–748.
Wang, Z., Ahmad, A., Banerjee, S., Azmi, A., Kong, D., Li, Y., et al. (2010). FoxM1 is a novel target of a natural agent in pancreatic cancer. *Pharmaceutical Research, 27,* 1159–1168.
Wang, Z., Ahmad, A., Li, Y., Banerjee, S., Kong, D., & Sarkar, F. H. (2010). Forkhead box M1 transcription factor: A novel target for cancer therapy. *Cancer Treatment Reviews, 36,* 151–156.

Wang, Z., Banerjee, S., Kong, D., Li, Y., & Sarkar, F. H. (2007). Down-regulation of Forkhead Box M1 transcription factor leads to the inhibition of invasion and angiogenesis of pancreatic cancer cells. *Cancer Research, 67,* 8293–8300.

Wang, X., Bhattacharyya, D., Dennewitz, M. B., Kalinichenko, V. V., Zhou, Y., Lepe, R., et al. (2003). Rapid hepatocyte nuclear translocation of the Forkhead Box M1B (FoxM1B) transcription factor caused a transient increase in size of regenerating transgenic hepatocytes. *Gene Expression, 11,* 149–162.

Wang, I. C., Chen, Y. J., Hughes, D. E., Ackerson, T., Major, M. L., Kalinichenko, V. V., et al. (2008). FOXM1 regulates transcription of *JNK1* to promote the G_1/S transition and tumor cell invasiveness. *Journal of Biological Chemistry, 283,* 20770–20778.

Wang, I. C., Chen, Y. J., Hughes, D., Petrovic, V., Major, M. L., Park, H. J., et al. (2005). Forkhead Box M1 regulates the transcriptional network of genes essential for mitotic progression and genes encoding the SCF (Skp2-Cks1) ubiquitin ligase. *Molecular and Cellular Biology, 25,* 10875–10894.

Wang, M., & Gartel, A. L. (2012). The suppression of FOXM1 and its targets in breast cancer xenograft tumors by siRNA. *Oncotarget, 2,* 1218–1226.

Wang, B., Hikosaka, K., Sultana, N., Sharkar, M. T. K., Noritake, H., Kimura, W., et al. (2012). Liver tumor formation by a mutant retinoblastoma protein in the transgenic mice is caused by an upregulation of c-Myc target genes. *Biochemical and Biophysical Research Communications, 417,* 601–606.

Wang, X., Hung, N. J., & Costa, R. H. (2001). Earlier expression of the transcription factor HFH-11B diminishes induction of p21CIP1/WAF1 levels and accelerates mouse hepatocyte entry into S-phase following carbon tetrachloride liver injury. *Hepatology, 33,* 1404–1414.

Wang, Y., Ji, P., Liu, J., Broaddus, R. R., Xue, F., & Zhang, W. (2009). Centrosome-associated regulators of the G_2/M checkpoint as targets for cancer therapy. *Molecular Cancer, 8,* 8.

Wang, X., Kiyokawa, H., Dennewitz, M. B., & Costa, R. H. (2002). The Forkhead Box m1b transcription factor is essential for hepatocyte DNA replication and mitosis during mouse liver regeneration. *Proceeding of the National Academy of Sciences of the United States of America, 99,* 16881–16886.

Wang, X., Krupczak-Hollis, K., Tan, Y., Dennewitz, M. B., Adami, G. R., & Costa, R. H. (2002). Increased hepatic Forkhead Box M1B (FoxM1B) levels in old-aged mice stimulated liver regeneration through diminished p27Kip1 protein levels and increased Cdc25B expression. *Journal of Biological Chemistry, 277,* 44310–44316.

Wang, Z., Li, Y., Ahmad, A., Banerjee, S., Azmi, A. S., Kong, D., et al. (2010). Down-regulation of Notch-1 is associated with Akt and FoxM1 In inducing cell growth inhibition and apoptosis in prostate cancer cells. *Journal of Cellular Biochemistry, 112,* 78–88.

Wang, J. L., Lin, Y. W., Chen, H. M., Kong, X., Xiong, H., Shen, N., et al. (2011). Calcium prevents tumorigenesis in a mouse model of colorectal cancer. *PLoS One, 6,* e22566.

Wang, F., Marshall, C. B., Yamamoto, K., Li, G. Y., Plevin, M. J., You, H., et al. (2008). Biochemical and structural characterization of an intramolecular interaction in FOXO3a and its binding with p53. *Journal of Molecular Biology, 384,* 590–603.

Wang, Y., McMahon, A. P., & Allen, B. L. (2007). Shifting paradigms in Hedgehog signaling. *Current Opinion in Cell Biology, 19,* 159–165.

Wang, I. C., Meliton, L., Ren, X., Zhang, Y., Balli, D., Snyder, J., et al. (2009). Deletion of Forkhead Box M1 transcription factor from respiratory epithelial cells inhibits pulmonary tumorigenesis. *PLoS One, 4,* e6609.

Wang, I. C., Meliton, L., Tretiakova, M., Costa, R. H., Kalinichenko, V. V., & Kalin, T. V. (2008). Transgenic expression of the forkhead box M1 transcription factor induces formation of lung tumors. *Oncogene, 27,* 4137–4149.

Wang, Z., Park, H. J., Carr, J. R., Chen, Y. J., Zheng, Y., Li, J., et al. (2011). FoxM1 in tumorigenicity of the neuroblastoma cells and renewal of the neural progenitors. *Cancer Research, 71,* 4292–4302.

Wang, X., Quail, E., Hung, N. J., Tan, Y., Ye, H., & Costa, R. H. (2001). Increased levels of forkhead box M1B transcription factor in transgenic mouse hepatocytes prevent age-related proliferation defects in regenerating liver. *Proceeding of the National Academy of Sciences of the United States of America, 98,* 11468–11473.

Wang, G. L., Salisbury, E., Shi, X., Timchenko, L., Medrano, E. E., & Timchenko, N. A. (2008). HDAC1 cooperates with C/EBPα in the inhibition of liver proliferation in old mice. *Journal of Biological Chemistry, 283,* 26169–26178.

Wang, G. L., Shi, X., Salisbury, E., Sun, Y., Albrecht, J. H., Smith, R. G., et al. (2007). Growth hormone corrects proliferation and transcription of phosphoenolpyruvate carboxykinase in liver of old mice via elimination of CCAAT/enhancer-binding protein α-Brm complex. *Journal of Biological Chemistry, 282,* 1468–1478.

Wang, I. C., Snyder, J., Zhang, Y., Landar, J., Nakafuku, Y., Lin, J., et al. (2012). Foxm1 mediates a cross-talk between Kras/MAPK and canonical Wnt pathways during development of respiratory epithelium. *Molecular and Cellular Biology, 32,* 3838–3850.

Wang, H., Teh, M. T., Ji, Y., Patel, V., Firouzabadian, S., Patel, A. A., et al. (2010). EPS8 upregulates FOXM1 expression, enhancing cell growth and motility. *Carcinogenesis, 31,* 1132–1141.

Wang, Y., Wen, L., Zhao, S. H., Ai, Z. H., Guo, J. Z., & Liu, W. C. (2012). FoxM1 expression is significantly associated with cisplatin-based chemotherapy resistance and poor prognosis in advanced non-small cell lung cancer patients. *Lung Cancer, 79,* 173–179.

Wang, I. C., Zhang, Y., Snyder, J., Sutherland, M. J., Burhans, M. S., Shannon, J. M., et al. (2010). Increased expression of FoxM1 transcription factor in respiratory epithelium inhibits lung sacculation and causes Clara cell hyperplasia. *Developmental Biology, 347,* 301–314.

Warner, S. M., & Knight, D. A. (2008). Airway modeling and remodeling in the pathogenesis of asthma. *Current Opinion in Allergy and Clinical Immunology, 8,* 44–48.

Waseem, A., Ali, M., Odell, E. W., Fortune, F., & Teh, M. T. (2010). Downstream targets of FOXM1: CEP55 and HELLS are cancer progression markers of head and neck squamous cell carcinoma. *Oral Oncology, 46,* 536–542.

Weber, L. W., Boll, M., & Stampfl, A. (2003). Hepatotoxicity and mechanism of action of haloalkanes: Carbon tetrachloride as a toxicological model. *Critical Reviews in Toxicology, 33,* 105–136.

Weigel, D., & Jäckle, H. (1990). The fork head domain: A novel DNA binding motif of eukaryotic transcription factors? *Cell, 63,* 455–456.

Weigelt, J., Climent, I., Dahlman-Wright, K., & Wikström, M. (2001). Solution structure of the DNA binding domain of the human forkhead transcription factor AFX (FOXO4). *Biochemistry, 40,* 5861–5869.

Weinberg, R. A. (1995). The retinoblastoma protein and cell cycle control. *Cell, 81,* 323–330.

Weinberg, R. A. (2006). *The biology of cancer.* New York: Garland Science.

Weinberg, F., & Chandel, N. S. (2009). Reactive oxygen species-dependent signaling regulates cancer. *Cellular and Molecular Life Sciences, 66,* 3663–3673.

Weinstein, I. B. (2000). Disorders in cell circuitry during multistage carcinogenesis: The role of homeostasis. *Carcinogenesis, 21,* 857–864.

Welch, W. J. (1992). Mammalian stress response: Cell physiology, structure/function of stress proteins, and implications for medicine and disease. *Physiological Reviews, 72,* 1063–1081.

Wells, L., Vosseller, K., & Hart, G. W. (2001). Glycosylation and nucleocytoplasmic proteins: Signal transduction and O-GlcNAc. *Science, 291,* 2376–2378.

Wells, L., Vosseller, K., & Hart, G. W. (2003). O-GlcNAc: A regulatory post-translational modification. *Biochemical and Biophysical Research Communications, 302,* 435–441.

Westendorf, J. M., Rao, P. N., & Gerace, L. (1994). Cloning of cDNAs for M-phase phosphoproteins recognized by the MPM2 monoclonal antibody and determination of the phosphorylated epitope. *Proceeding of the National Academy of Sciences of the United States of America, 91*, 714–718.

Weymann, A., Hartman, E., Gazit, V., Wang, C., Glauber, M., Turmelle, Y., et al. (2009). p21 is required for dextrose-mediated inhibition of mouse liver regeneration. *Hepatology, 50*, 207–215.

White, M. K., & Khalili, K. (2004). Polyomaviruses and human cancer: Molecular mechanisms underlying patterns of tumorigenesis. *Virology, 324*, 1–16.

White, M. K., & Khalili, K. (2006). Interaction of retinoblastoma protein family members with large T-antigen of primate polyomaviruses. *Oncogene, 25*, 5286–5293.

Whitfield, M. L., George, L. K., Grant, G. D., & Perou, C. M. (2006). Common markers of proliferation. *Nature Reviews. Cancer, 6*, 99–106.

Whitfield, M. L., Sherlock, G., Saldanha, A. L., Murray, J. I., Ball, C. A., Alexander, K. E., et al. (2002). Identification of genes periodically expressed in the human cell cycle and their expression in tumors. *Molecular Biology of the Cell, 13*, 1977–2000.

Wierstra, I. (2011a). The transcription factor FOXM1c binds to and transactivates the promoter of the tumor suppressor gene E-cadherin. *Cell Cycle, 10*, 760–766.

Wierstra, I. (2011b). The transcription factor FOXM1c is activated by protein kinase CK2, protein kinase A (PKA), c-Src and Raf-1. *Biochemical and Biophysical Research Communications, 413*, 230–235.

Wierstra, I. (2013a). Cyclin D1/Cdk4 increases the transcriptional activity of FOXM1c without phosphorylating FOXM1c. *Biochemical and Biophysical Research Communications, 431*, 753–759.

Wierstra, I. (2013b). FOXM1 (Forkhead box M1) in tumorigenesis: overexpression in human cancer, implication in tumorigenesis, oncogenic functions, tumor-suppressive properties and target of anti-cancer therapy. *Advances in Cancer Research, 119*, in press.

Wierstra, I., & Alves, J. (2006a). Despite its strong transactivation domain transcription factor FOXM1c is kept almost inactive by two different inhibitory domains. *Biological Chemistry, 387*, 963–976.

Wierstra, I., & Alves, J. (2006b). Transcription factor FOXM1c is repressed by RB and activated by cyclin D1/Cdk4. *Biological Chemistry, 387*, 949–962.

Wierstra, I., & Alves, J. (2006c). FOXM1c is activated by cyclin E/Cdk2, cyclin A/Cdk2 and cyclin A/Cdk1, but repressed by GSK-3α. *Biochemical and Biophysical Research Communications, 348*, 99–108.

Wierstra, I., & Alves, J. (2006d). FOXM1c transactivates the human c-myc promoter directly via the two TATA-boxes P1 and P2. *FEBS Journal, 273*, 4645–4667.

Wierstra, I., & Alves, J. (2007a). FOXM1c and Sp1 transactivate the P1 and P2 promoters of human c-myc synergistically. *Biochemical and Biophysical Research Communications, 352*, 61–68.

Wierstra, I., & Alves, J. (2007b). The central domain of transcription factor FOXM1c directly interacts with itself in vivo and switches from an essential to an inhibitory domain depending on the FOXM1c binding site. *Biological Chemistry, 388*, 805–818.

Wierstra, I., & Alves, A. (2007c). FOXM1, a typical proliferation-associated transcription factor. *Biological Chemistry, 388*, 1257–1274.

Wierstra, I., & Alves, J. (2008). Cyclin E/Cdk2, P/CAF and E1A regulate the transactivation of the c-myc promoter by FOXM1. *Biochemical and Biophysical Research Communications, 368*, 107–115.

Wijchers, P. J. E. C., Burbach, J. P. H., & Smidt, M. P. (2006). In control of biology: Of mice, men and Foxes. *Biochemical Journal, 397*, 233–246.

Wilkinson, M. G., & Millar, J. B. A. (2000). Control of the eukaryotic cell cycle by MAP kinase signaling pathways. *The FASEB Journal, 14*, 2147–2157.

Willart, M. A., & Hammad, H. (2010). Alarming dendritic cell for allergic sensitization. *Allergology International*, *59*, 95–103.

Willart, M. A., & Lambrecht, B. N. (2009). The danger within: Endogenous danger signals, atopy and asthma. *Clinical and Experimental Allergy*, *39*, 12–19.

Willis-Martinez, D., Richards, H. W., Timchenko, N. A., & Medrano, E. E. (2010). Role of HDAC1 in senescence, aging and cancer. *Experimental Gerontology*, *45*, 279–285.

Wilson, M. S. C., Brosens, J. J., Schwenen, H. D. C., & Lam, E. W. F. (2011). FOXO and FOXM1 in cancer: The FOXO-FOXM1 axis shapes the outcome of cancer therapy. *Current Drug Targets*, *12*, 1256–1266.

Winkelman, M., Pfitzer, P., & Schneider, W. (1987). Significance of polyploidy in megakaryocytes and other cells in health and tumor disease. *Klinische Wochenschrift*, *65*, 1115–1131.

Wolberger, C., & Campbell, R. (2000). New perch for the winged helix. *Nature Structural Biology*, *7*, 261–270.

Wonsey, D. R., & Follettie, M. T. (2005). Loss of the forkhead transcription factor FoxM1 causes centrosome amplification and mitotic catastrophe. *Cancer Research*, *65*, 5181–5189.

Woodfield, G. W., Horan, A. D., Chen, Y., & Weigel, R. J. (2007). TFAP2C controls hormone response in breast cancer cells through multiple pathways of estrogen signaling. *Cancer Research*, *67*, 8439–8443.

Wu, Y. M., & Gupta, S. (2009). Hepatic preconditioning for transplanted cell engraftment and proliferation. *Methods in Molecular Biology*, *481*, 107–116.

Wu, M. Y., & Hill, C. S. (2009). TGF-β superfamily signaling in embryonic development and homeostasis. *Developmental Cell*, *16*, 329–343.

Wu, B., Hunt, C., & Morimoto, R. (1985). Structure and expression of the human gene encoding major heat shock protein HSP70. *Molecular and Cellular Biology*, *5*, 330–341.

Wu, B. J., Kingston, R. E., & Morimoto, R. I. (1986). Human HSP70 promoter contains at least two distinct regulatory domains. *Proceeding of the National Academy of Sciences of the United States of America*, *83*, 629–633.

Wu, Q. F., Liu, C., Tai, M. H., Liu, D., Lei, L., Wang, R. T., et al. (2010). Knockdown of FoxM1 by siRNA interference decreases cell proliferation, induces cell cycle arrest and inhibits cell invasion in MHCC-97H cells in vitro. *Acta Pharmacologica Sinica*, *31*, 361–366.

Wu, J., Lu, L. Y., & Yu, X. (2010). The role of BRCA1 in DNA damage response. *Protein & Cell*, *1*, 117–123.

Wyman, C., & Kanaar, R. (2006). DNA double-strand break repair: All's well the ends well. *Annual Review of Genetics*, *40*, 363–383.

Wynn, T. A. (2004). Fibrotic disease and the T(H)1/T(H)2 paradigm. *Nature Reviews. Immunology*, *4*, 583–594.

Wynn, T. A. (2007). Common and unique mechanisms regulate fibrosis in various fibroproliferative diseases. *The Journal of Clinical Investigation*, *117*, 524–529.

Wynn, T. A. (2008). Cellular and molecular mechanisms of fibrosis. *The Journal of Pathology*, *214*, 199–210.

Wynn, T. A. (2011). Integrating mechanisms of pulmonary fibrosis. *The Journal of Experimental Medicine*, *208*, 1339–1350.

Wynn, T. A., & Barron, L. (2010). Macrophages: Master regulators of inflammation and fibrosis. *Seminars in Liver Disease*, *30*, 245–257.

Xia, L., Huang, W., Tian, D., Zhu, H., Zhang, Y., Hu, H., et al. (2012). Upregulated FoxM1 expression induced by hepatitis B virus X protein promotes tumor metastasis and indicates poor prognosis in hepatitis B virus-related hepatocellular carcinoma. *Journal of Hepatology*, *57*, 600–612.

Xia, L. M., Huang, W. J., Wang, B., Liu, M., Zhang, Q., Yan, W., et al. (2009). Transcriptional up-regulation of FoxM1 in response to hypoxia is mediated by HIF-1. *Journal of Cellular Biochemistry, 106*, 247–256.

Xia, L., Mo, P., Huang, W., Zhang, L., Wang, Y., Zhu, H., et al. (2012). The TNF-a/ROS/HIF-1-induced upregulation of FoxM1 expression promotes HCC proliferation and resistance to apoptosis. *Carcinogeneis, 33*, 2250–2259.

Xia, J. T., Wang, H., Liang, L. J., Peng, B. G., Wu, Z. F., Chen, L. Z., et al. (2012). Overexpression of FOXM1 is associated with poor prognosis and clinicopathologic stage of pancreatic ductal adenocarcinoma. *Pancreas, 41*, 629–635.

Xiang, H. L., Liu, F., Quan, M. F., Cao, J. G., & Lv, Y. (2012). 7-Difluoromethoxyl-5,4′-di-n-octylgenistein inhibits growth of gastric cancer cells through downregulating forkhead box M1. *World Journal of Gastroenterology, 18*, 4618–4626.

Xie, Z., Tan, G., Ding, M., Dong, D., Chen, T., Meng, X., et al. (2010). Foxm1 transcription factor is required for maintenance of pluripotency of P19 embryonal carcinoma cells. *Nucleic Acids Research, 38*, 8027–8038.

Xu, Y., Wang, Y., Besnard, V., Ikegami, M., Wert, S. E., Heffner, C., et al. (2012). Transcriptional programs controlling perinatal lung maturation. *PLoS One, 7*, e37046.

Xu, N., Zhang, X., Wang, X., Ge, H., Wang, X. Y., Garfield, D., et al. (2012). FoxM1-mediated resistance of non-small-cell lung cancer cells to gefitinib. *Acta Pharmacologica Sinica, 33*, 675–681.

Xue, L., Chiang, L., He, B., Zhao, Y. Y., & Winoto, A. (2010). FoxM1, a Forkhead transcription factor is a master cell cycle regulator for mouse mature T cells but not double positive thymocytes. *PLoS One, 5*, e9229.

Xue, X. J., Xiao, R. H., Long, D. Z., Zou, X. F., Zhang, G. X., Yuan, Y. H., et al. (2012). Overexpression of FoxM1 is associated with tumor progression in patients with clear cell renal cell carcinoma. *Journal of Translational Medicine*, Epub: Sep 24, 2012.

Yam, C. H., Fung, T. K., & Poon, R. Y. C. (2002). Cyclin A in cell cycle control and cancer. *Cellular and Molecular Life Sciences, 59*, 1317–1326.

Yamamoto, T., Taya, S., & Kaibuchi, K. (1999). Ras-induced transformation and signaling pathways. *Journal of Biochemistry, 126*, 799–803.

Yamanaka, S. (2007). Strategies and new developments in the generation of patient-specific pluripotent stem cells. *Cell Stem Cell, 1*, 39–49.

Yamanaka, S. (2008). Pluripotency and nuclear reprogramming. *Philosophical Transactions of the Royal Society B: Biological Sciences, 363*, 2079–2087.

Yamauchi, K. (2006). Airway remodeling in asthma and its influence on clinical pathophysiology. *The Tohoku Journal of Experimental Medicine, 209*, 75–87.

Yanagida, M. (2009). Cellular quiescence: Are controlling genes conserved? *Trends in Cell Biology, 19*, 705–715.

Yanagisawa, M. (2011). Stem cell glycolipids. *Neurochemical Research, 36*, 1623–1635.

Yanagita, M. (2005). BMP antagonists: Their role in development and involvement in pathophysiology. *Cytokine & Growth Factor Reviews, 16*, 309–317.

Yang, P., Huang, S., Liu, D., Zhou, Q., Wen, Y. A., Xiang, Y., et al. (2011). Hepatic expression profile of forkhead transcription factor genes in normal Balb/c mice and their dynamic changes after bile duct ligation. *Molecular Biology Reports, 38*, 2665–2671.

Yang, E. S., & Xia, F. (2010). BRCA1 16 years later: DNA damage-induced BRCA1 shuttling. *FEBS Journal, 277*, 3079–3085.

Yang, X. H., & Zou, L. (2006). Checkpoint and coordinated cellular responses to DNA damage. *Results and Problems in Cell Differentiation, 42*, 65–92.

Yao, K. M., Sha, M., Lu, Z., & Wong, G. G. (1997). Molecular analysis of a novel winged helix protein, WIN. *Journal of Biological Chemistry, 272*, 19827–19836.

Yarden, R. I., & Papa, M. Z. (2006). BRCA1 at the crossroads of multiple cellular pathways: Approaches for therapeutic interventions. *Molecular Cancer Therapeutics, 5*, 1396–1404.

Ye, H., Holterman, A. X., Yoo, K. W., Franks, R. R., & Costa, R. H. (1999). Premature expression of the winged helix transcription factor HFH-11B in regenerating mouse liver accelerates hepatocyte entry into S-phase. *Molecular and Cellular Biology, 19,* 8570–8580.

Ye, H., Kelly, T. F., Samadani, U., Lim, L., Rubio, S., Overdier, D. G., et al. (1997). Hepatocyte nuclear factor 3/fork head homolog 11 is expressed in proliferating epithelial and mesenchymal cells of embryonic and adult tissues. *Molecular and Cellular Biology, 17,* 1626–1641.

Yeang, C. H., McCormick, F., & Levine, A. (2008). Combinatorial patterns of somatic mutations in cancer. *The FASEB Journal, 22,* 2605–2622.

Yee, A. S., Shih, H. H., & Tevosian, S. G. (1998). New perspectives on retinoblastoma family functions in differentiation. *Frontiers in Bioscience, 3,* d532–d547.

Yokomine, K., Senju, S., Nakatsura, T., Irie, A., Hayashida, Y., Ikuta, Y., et al. (2009). The Forkhead Box M1 transcription factor as a candidate target for anti-cancer immunotherapy. *International Journal of Cancer, 126,* 2153–2163.

Yoshida, K., & Miki, Y. (2004). Role of BRCA1 and BRCA2 as regulators of DNA repair, transcription, and cell cycle in response to DNA damage. *Cancer Science, 95,* 866–871.

Yoshida, Y., Wang, I. C., Yoder, H. M., Davidson, N. O., & Costa, R. H. (2007). The Forkhead Box M1 transcription factor contributes to the development and growth of mouse colorectal cancer. *Gastroenterology, 132,* 1420–1431.

Young, R. A. (2011). Control of the embryonic stem cell state. *Cell, 144,* 940–954.

Yu, E. Y., & Hahn, W. C. (2004). The origin of human cancer. *Cancer Treatment and Research, 122,* 1–22.

Yu, J., & Thomson, J. A. (2008). Pluripotent stem cell lines. *Genes & Development, 22,* 1987–1997.

Zachara, N. E., & Hart, G. W. (2006). Cell signaling, the essential role of O-GlcNAc!. *Biochimica et Biophysica Acta, 1761,* 599–617.

Zaret, K. S., & Grompe, M. (2008). Generation and regeneration of cells of the liver and pancreas. *Science, 322,* 1490–1494.

Zeng, J., Wang, L., Li, Q., Li, W., Björkholm, M., Jia, J., et al. (2009). FoxM1 is up-regulated in gastric cancer and its inhibition leads to cellular senescence, partially dependent on p27kip1. *The Journal of Pathology, 218,* 419–427.

Zetterberg, A., Larsson, O., & Wiman, K. G. (1995). What is the restriction point? *Current Opinion in Cell Biology, 7,* 835–842.

Zhang, H., Ackermann, A. M., Gusarova, G. A., Lowe, D., Feng, X., Kopsombut, U. G., et al. (2006). The Foxm1 transcription factor is required to maintain pancreatic beta cell mass. *Molecular Endocrinology, 20,* 1853–1866.

Zhang, X., Bai, Q., Kakiyama, G., Xu, L., Kin, J. K., Pandak, W. M., et al. (2012). Cholesterol metabolite, 5-cholesten-3b, 25-diol 3-sulfate, promotes hepatic proliferation in mice. *The Journal of Steroid Biochemistry and Molecular Biology, 132,* 262–270.

Zhang, Y., Lee, F. Y., Barrera, G., Lee, H., Vales, C., Gonzalez, F. J., et al. (2006). Activation of the nuclear receptor FXR improves hyperglycemia and hyperlipidemia in diabetic mice. *Proceeding of the National Academy of Sciences of the United States of America, 103,* 1006–1011.

Zhang, L., Li, T., Yu, D., Forman, B. M., & Huang, W. (2012). FXR protects lung from lipopolysaccharide-induced acute injury. *Molecular Endocrinology, 26,* 27–36.

Zhang, L., Wang, Y. D., Chen, W. D., Wang, X., Lou, G., Liu, N., et al. (2012). Promotion of liver regeneration/repair by farnesoid X receptor in both liver and intestine. *Hepatology, 56,* 2336–2343.

Zhang, N., Wei, P., Gong, A., Chiu, W. T., Lee, H. T., Colman, H., et al. (2011). FoxM1 promotes β-catenin nuclear localization and controls Wnt target-gene expression and glioma tumorigenesis. *Cancer Cell, 20,* 427–442.

Zhang, N., Wu, X., Yang, L., Xiao, F., Zhang, H., Zhou, A., et al. (2012). FoxM1 inhibition sensitizes resistant glioblastoma cells to temozolomide by downregulating the expression of DNA repair gene Rad51. *Clinical Cancer Research, 18*, 5961–5971.

Zhang, X., Zeng, J., Zhu, M., Li, B., Zhang, Y., Huang, T., et al. (2012). The tumor suppressor role of miR-370 by targeting FoxM1 in acute myeloid leukemia. *Molecular Cancer, 11*, 56.

Zhang, Y., Zhang, N., Dai, B., Liu, M., Sawaya, R., Xie, K., et al. (2008). FoxM1B transcriptionally regulates vascular endothelial growth factor expression and promotes angiogenesis and growth of glioma cells. *Cancer Research, 68*, 8733–8742.

Zhang, H., Zhang, J., Pope, C. F., Crawford, L. A., Vasavada, R. C., Jagasia, S. M., et al. (2010). Gestational diabetes resulting from impaired β-cell compensation in the absence of FoxM1, a novel downstream effector of placental lactogen. *Diabetes, 59*, 143–152.

Zhao, R., & Daley, G. Q. (2008). From fibroblasts to iPS cells: Induced pluripotency by defined factors. *Journal of Cellular Biochemistry, 105*, 949–955.

Zhao, Y. Y., Gao, X. P., Zhao, Y. D., Mirza, M. K., Frey, R. S., Kalinichenko, V. V., et al. (2006). Endothelial cell-restricted disruption of FoxM1 impairs endothelial repair following LPS-induced vascular injury. *The Journal of Clinical Investigation, 116*, 2333–2343.

Zhao, F., & Lam, E. W. F. (2012). Role of the forkhead transcription factor FOXO-FOXM1 axis in cancer and drug resistance. *Frontiers of Medicine, 6*, 376–380.

Zhao, B., Li, L., & Guan, K. L. (2010). Hippo signaling at a glance. *Journal of Cell Science, 123*, 4001–4006.

Zhao, B., Li, L., Lei, Q., & Guan, K. L. (2010). The Hippo-YAP pathway in organ size control and tumorigenesis: An updated version. *Genes & Development, 24*, 862–874.

Zhou, B. B. S., & Elledge, S. J. (2000). The DNA damage response: Putting checkpoints in perspective. *Nature, 408*, 433–439.

Zhou, Z. Q., & Hurlin, P. J. (2001). The interplay between Mad and Myc in proliferation and differentiation. *Trends in Cell Biology, 11*, S10–S14.

Zhou, J., Liu, Y., Zhang, W., Popov, V. M., Wang, M., Pattabiraman, N., et al. (2010). Transcription elongation regulator 1 is a co-integrator of the cell fate determination factor Dachshund homolog 1. *Journal of Biological Chemistry, 285*, 40342–40350.

Zhou, J., Wang, C., Wang, Z., Dampier, W., Wu, K., Casimiro, M. C., et al. (2010). Attenuation of Forkhead signaling by the retinal determination factor DACH1. *Proceeding of the National Academy of Sciences of the United States of America, 107*, 6864–6869.

Zhu, L., & Skoultchi, A. (2001). Coordinating cell proliferation and differentiation. *Current Opinion in Genetics & Development, 11*, 91–97.

Zimmerman, H. J., & Lewis, J. H. (1995). Chemical- and toxin-induced hepatotoxicity. *Gastroenterology Clinics of North America, 24*, 1027–1045.

Zou, Y., Tsai, W. B., Cheng, C. J., Hsu, C., Chung, Y. M., Li, P. C., et al. (2008). Forkhead box transcription factor FOXO3a suppresses estrogen-dependent breast cancer cell proliferation and tumorigenesis. *Breast Cancer Research, 10*, R21.

INDEX

Note: Page numbers followed by "*f*" indicate figures, and "*t*" indicate tables.

A

ACD. *See* Autophagic cell death (ACD)
Adult tissue repair, injury
  cell proliferation, 305
  CLP, 307
  endothelial repair, 307
  liver, lung and pancreas, 304
  nonproliferative functions, 306
  pancreas regeneration, 305
  PHx-induced liver regeneration, aged mice, 305–306
AEG-1. *See* Astrocyte elevated gene-1 (AEG-1)
Allergen-induced lung inflammation
  asthma, 337
  HDM exposure, 339
  myeloid inflammatory cells and Clara cells, 338
  sensitization/challenge protocol, HDM, 337
Aplasia Ras homolog member I (ARHI), 79–81
Apoptosis and cellular survival
  cancer cells, 328
  HepG2 HCC cells, 332
  IFN-β + mezerein, 331
  stimulation, 328–330
  tumor cells, 327–328
Aptamers
  advantages, 30
  description, 30
ARHI. *See* Aplasia Ras homolog member I (ARHI)
Asthma and COPD, 255–256
Astrocyte elevated gene-1 (AEG-1), 79, 80*f*
Autophagic cell death (ACD)
  atg genes and stimulation, 68–69
  bystander/double-edged sword, 69–70
  caspase-mediated proteolytic degradation, 69–70
  description, 68–69
  pathophysiological role, 69–70
  programmed cell death and NCCD, 68–69
Autophagy
  ACD (*see* Autophagic cell death (ACD))
  basal, 62
  in cancer initiation/stem cells (*see* Cancer stem cells (CSCs))
  in cancer therapy, 82–86
  cargo selection, 67
  defective mice display signs, 62
  description and classification, 63
  FOXM1 knockdown, 333
  homeostatic and regulatory mechanism, 62
  lysosomal fusion, 67–68
  macroautophagy and molecular basis, 63
  molecular events, 63–65, 64*f*
  necrosis and inflammation, 86–87
  phagophores, 65–67
  proteasome inhibitors, 333
  stress stimulates, 62
  therapy-stimulated accumulation of autophagosomes, 86–87
  in tumor dormancy, 79–81
  in tumor initiation and development, 70–73
  in tumor progression and metastasis, 73–79

B

BAL. *See* Brochoalveolar lavage (BAL)
Bax-interacting factor 1 (Bif 1), 65–66
Bif 1. *See* Bax-interacting factor 1 (Bif 1)
Biological functions, FOXM1
  B-Myb, 292
  breast cancer cells, 292
  cell proliferation and cycle progression, 283–294
  cellular senescence, 297–304
  confluent cells, 296–297
  cyclin D1/Cdk4 and D3/Cdk6, 290
  G2/M-transition and mitosis, 290

Biological functions, FOXM1 (*Continued*)
  G1/S-transition, 293
  interference, contact inhibition, 296–297
  M-phase entry and mitotic progression, 291–292
  pancreatic β-cells, 310–311
  proliferative capacity, cells, 312
  S-phase entry, 292–293
  transcriptional activity, 288–290
  transcription factor, 293–294
  xenopus laevis, 296
BMP. *See* Bone morphogenetic protein (BMP)
Bone morphogenetic protein (BMP), 259
Brochoalveolar lavage (BAL), 341

C

Cancer stem cells (CSCs)
  autophagy and brain, 82
  chronic myeloid leukemia (CML), 82
  radiation-induced autophagy in glioma, 82
  subpopulation and tumor cell heterogeneity, 81–82
  tyrosine kinase inhibitors, 82
  Wnt/Notch/Hedgehog, 81–82
Cancer targeting
  active, 7–8
  anticancer drug, 39
  bioengineering design strategies, 9, 10*t*
  biomaterials and nanotechnology, 4
  design goals, 9
  detection and treatment, 2–3
  drug uptake and release, 11–12
  dual ligand targeting, 41
  EPR, 5
  extravasation, 10–11
  ligand-conjugated particle and receptor, 8
  liposomes, 13–15
  lymphatic drainage, 6
  microbubbles, 40–41
  multidrug resistance, 12
  multifunctional delivery systems, 39–40
  nanoparticles, 5, 6*f*
  opsonization and mononuclear phagocytic system, 9–10
  particle elimination, 12–13
  passive, 5–7
  radiation and chemotherapy, 3
  signaling and delivery mechanisms, 4–5
  tumor tissue, 6–7
Cancer therapy
  ACD stimulation, 83
  protective autophagy inhibition
    Bcr-Abl-expressing hematopoietic cell transplantations, 83–84
    chemoresistance, 84–85
    chloroquine (CQ), 85–86
    ER stress inducers, 85
    estrogen-receptor-positive breast cancer cells, 84–85
    Ginsenoside F2-induced apoptosis, 83–84
    MDA-7/IL-24 and Beclin1, 83–84
    metastasis prone state, 85–86
    synergistic anticancer effects and organelle-damaging drugs, 85
    tumor-promoting effect, 86
  "type II programmed cell death," 82–83
Carbon nanotubes (CNTs)
  advantages, 23
  carboxylic acid groups, 22
  description, 22
  drug loading, 22–23
Cell cycle, FOXM1 expression
  adult liver, 210
  adult murine pancreas, 211
  farnesoid X receptor, 212
  foxm1 expression, 204–205
  gene expression analyses, human cells, 205
  induction, 208–209
  partial hepatectomy, 210–211
  proliferating cells, 204
  protein expression, 205
  quiescent cells, 207–208
  reinduction, 209–212
  tissue repair, 211
  toxic liver injury, 211
Cell proliferation
  antiproliferation signals, 283–284
  G1/S-and G2/M transition, 283
  transcription factor, 284
Cellular differentiation
  inhibition, 323–324
  lobuloalveolar differentiation, 326

mammary luminal differentiation, 325–326
versatile regulation, 326–327
Cellular senescence
  HMFs, 300
  negative-regulatory mechanisms, 300
  NIH3T3 fibroblasts, 298–299
  osteosarcoma cells, 299
  oxaliplatin, 301
  p38 activity, 302
  positive effect, FOXM1, 302–304
  premature, 302
  prevention, 297–302
  proliferation-stimulating function, 298
  ROS level, 301–302
  siRNA depletion, 301
  U2OS cells, 299
Chromatin immunoprecipitation (ChIP) assays
  direct FOXM1 target genes, 110
  in vivo occupancy, genes, 159–160
Chronic liver injury
  donor hepatocytes, 308–309, 310
  FoxM1B-overexpressing hepatocytes, 308
  repopulation efficiency, 310
  uPA/SCID, 309
Chronic myeloid leukemia (CML), 82
Chronic obstructive pulmonary disease (COPD), 255–256
Circulating tumor cells (CTCs), 79–81
CML. See Chronic myeloid leukemia (CML)
CNTs. See Carbon nanotubes (CNTs)
Conditional deletion, foxm1 mice
  adult virgin FoxM1 FL/FL, WAP-rtTA-Cre mice
    control mammary glands, 280–281
    differentiation markers, 281
    excessive cell infiltration, 280–281
    GATA-3 transcription, 281–282
    luminal differentiation, 279
    mammary stem cell pool, 278–279
    mRNA expression, 278–279
    proliferation and apoptosis, 278–279
    transcription factor GATA-3, 279–280
  airway Clara cells, $CCSP\text{-}Foxm1^{-/-}$ mice
    in distal lung saccules, 274
    in epithelial junctions, 273
    long-term maintenance, 273–274
    in naphthalene lung injury, 274
    progenitor properties, 274–275
    proliferation and differentiation, 272
    squamous, goblet, and alveolar type II cells, 272–273
    stromal tissue and peribronchiolar fibrosis, 273
    in trachea, 274
  from cardiomyocytes
    heart development, 276
    postnatal period, $a\text{-}MHC\text{-}Cre/Foxm1^{fl/fl}$ mice, 276–277
  $epFoxm1^{-/-}$ mice
    canonical Wnt/β-catenin signaling, 271–272
    cell-selective deletion, 269
    $epKras^{G12D}$, 270
    $K\text{-}Ras^{G12D}$, expression, 269, 270
    morphological and biochemical maturation, 269
    sacculation defects, 270
  FoxM1 FL/FL, WAP-Cre mice, 277–278
  SMCs, $smFoxm1^{-/-}$ mice, 275–276
Conventional transcription factor, FOXM1
  autoinhibitory N-terminus, 195–196
  cellular signals, 197
  consensus sequence, 167
  DBD (see DNA-binding domains (DBDs))
  description, 167
  ERα and E-cadherin, 167–168
  functional domains, 168–170, 193
  Gal-FOXM1c fusion proteins, 193–194
  IDs (see Inhibitory domains (IDs))
  in vitro DNA-binding studies, 194
  intramolecular interactions, 194
  NRD-C (see Negative-regulatory domain-C (NRD-C))
  NRD-N (see Negative-regulatory domain-N (NRD-N))
  TAD (see Transactivation domain (TAD))
  TRD (see Transrepression domain (TRD))
COPD. See Chronic obstructive pulmonary disease (COPD)

CSCs. *See* Cancer stem cells (CSCs)
CTCs. *See* Circulating tumor cells (CTCs)
Cycle progression
    Cdk2 and Cdk1, 284–285
    cyclin A1 and A2, 285
    cyclin/Cdk complexes, 285
    cyclin D1 and *c-myc* expression, 285–287
    protein products, 285, 286f
    S- and M-phase entry, 285

D

Damage-regulated autophagy modulator (DRAM), 70–71
DAPK. *See* Death-associated protein kinase (DAPK)
DAPK-related protein kinase-1 (DRP-1), 71
DBDs. *See* DNA-binding domains (DBDs)
Death-associated protein kinase (DAPK), 70, 71
Dendrimers
    advantages and disadvantages, 19
    description, 17–18
    drug loading, 19
    size and surface, 18–19
    structure, 17–18, 18f
Disseminated tumor cells (DTCs), 73–77, 79–81
DNA-binding domains (DBDs)
    conventional transcription factor, 169–170
    exon A4 and exon A1, 103–105
    forkhead domains, 105–106
    FoxM1B *vs*. FOXM1c, 109
    of FOXM1c (*see* FOXM1c-DBD)
    low DNA-binding affinity, 160
    TATA-boxes, 165
    winged-helix domain, 101
DNA-binding specificity, FOXM1 splice variants
    exon A4 and exon A1, 103–105
    features, FOXM1c, 106
    general mode, 105–106
DNA damage, FOXM1 expression
    anticancer drug, 221
    daunorubicin, 220
    5-fluorouracil, 221
    ionizing radiation, 221
    mRNA and protein, 220
    pancreatic cancer cells, 220
    ultraviolet radiation, 221
DNA hypermethylation, 165
DRAM. *See* Damage-regulated autophagy modulator (DRAM)
DRP-1. *See* DAPK-related protein kinase-1 (DRP-1)
Drug release
    enzyme-responsive, 36–37
    pH-responsive, 34–35
    temporal and spatial, 33–34
    temporal control, 33
    thermoresponsive, 37–39
DTCs. *See* Disseminated tumor cells (DTCs)

E

EGF. *See* Epidermal growth factor (EGF)
Electrophoretic mobility shift assays (EMSAs)
    direct FOXM1 target genes, 110, 166–167
    FoxM1B *vs*. FOXM1c, 108, 160
Elongation and multimerization, phagophores
    Atg8 and GATE-16, 66
    Atg12–Atg5 complex and WIPI1/WIPI2, 66
    Atg14L role, 66–67
    class III PI3 kinases, 65–66
    PI3P, 65–67
    ubiquitin-like conjugation systems and Vps34–Beclin1 interaction, 65–66
EMSAs. *See* Electrophoretic mobility shift assays (EMSAs)
Enhanced permeability and retention (EPR), 5
Epidermal growth factor (EGF)
    advantages, 27
    description, 26–27
    transforming growth factor, 27
EPR. *See* Enhanced permeability and retention (EPR)
Estrogen receptor α (ERα)
    as conventional transcription factor, 167–168
    and *E-cadherin*, 166
    in fluorescence anisotropy assays, 110

Index

F

Folate
- advantages, 25
- description, 24

Forkhead box M1 (FOXM1)
- activating transcription factor, 99–101, 100f
- description, 99–101, 102
- forkhead domain/box, 101
- gene regulation (see Gene regulation mechanisms, FOXM1)
- α-helices and β-strands, 101–102
- molecular function (see Conventional transcription factor, FOXM1)
- mouse models (see Mouse models, FOXM1)
- order H1-S1-H2-turn-H3-S2-W1-S3-W2, 101–102
- orthologs, 102
- splice variants (see Splice variants, FOXM1)
- target genes (see Target genes, FOXM1)

FOXM1. See Forkhead box M1 (FOXM1)

FOXM1c-DBD
- aa 235–346, 171–173
- alignment, 170–171, 171f
- DNA-binding features, 106
- DNA contacts, 171, 172f
- exon A1 in wing W2, 170–173
- vs. forkhead domains, 177–178
- H1-S1-H2-turn-H3-S2- loop (wing W1)-S3-loop (wing W2), 171
- reflections, wing W2, 179
- structure, 171
- wings W1 and W2, 178–179
- X-ray crystal structure, 173–177

FOXM1 expression
- AKT, 265
- allergen stimulation, 260
- AMPK, 262
- β-Catenin, 261
- BMP, 259
- calcium, 260
- cell cycle, 204–207
- cellular senescence, 217
- cessation, cell cycle exit, 212–217
- checkpoint kinase, 222
- c-Met, 258
- ConA/PMA, 263
- c-Rel, 261
- Cry1 and Cry2, 260–261
- cyclin D1/Cdk4 and D3/Cdk6, 225–227
- description, 197–198
- dextrose, 260
- discontinuance, 213–216
- downregulation, quiescence, 213–214
- EPS8, 260
- ERb1, 257
- factors, 200f, 205–206
- fibronectin, 262
- FOXO1+ FOXO3, 257
- GRB7, 262
- hypoxia, 257
- IGF-1, 264
- IGF-I, 262
- JNK, 266
- life span, 199–203
- LIN9, 258
- NIH3T3 fibroblasts, 215
- OGT, 228
- p130, 258
- PHGDH, 228–229
- PI3K inhibitors, 259
- p38 inhibitor, 266
- protein, 215–216, 222
- protein kinase Ca, 259
- quiescence program, 214–215
- senescent cells, 198–199
- serum starvation and contact inhibition, 215–216
- serum-starved cells, 205–206
- Shh, 262–263
- SPDEF, 260
- SRp20, 261
- STAT3, 257
- terminal differentiation, 216–217
- TGF-β1, 264
- TNF- α, 264
- ubiquitin - proteasome pathway, 223–225
- U2OS cells, 223
- Wnt3a, 227–228

G

Gene regulation mechanisms, FOXM1
  binding sites, 166–167
  ChIP assays, 166
  description, 164–165
  DNA hypermethylation, 165
  EMSAs and VEGF, 166–167
  ERα and E-cadherin, 166
  protein–protein interactions, 165–166
  TATA-boxes, 165
  transactivator, 164–165
Genomic stability
  DSBs, 316
  FOXM1 role, HR, 316
  polyploidy, 316
  SRF, 317
  target genes, 317
Global and focal DNA hypomethylation
  ChIP-seq assays, 334
  promoter methylation, 333
  RB, 335
  target genes, 334

H

Heat shock response
  FOXM1, 337
  gene expression program, 336
Homologous recombination (HR) repair, DNA damage
  DSBs, 313
  FOXM1-deficient cells, 313
  and NHEJ, 312–313
Human disease, FOXM1 expression
  asthma and COPD, 255–256
  Hutchinson-Gilford progeria, 254
  IPF, 256
  pathological cardiac hypertrophy, 255
  trisomy 21/down syndrome, 254–255
Human tumor-suppressor pathways, 244–246
Hutchinson-Gilford progeria, 254

I

Idiopathic pulmonary fibrosis (IPF), 256
IDs. See Inhibitory domains (IDs)
Inhibitory domains (IDs)
  cellular signals, 197
  description, 169–170
  Gal-FOXM1 fusion proteins, 193–194
  NRD-C, 188
  NRD-N, 189
  TRD
    of FOXM1c, 185, 190–191
    tight control, FOXM1c-TAD, 192–193
IPF. See Idiopathic pulmonary fibrosis (IPF)

K

Knockout mice, FOXM1 mouse models
  conditional deletion
    a-MHC-Cre/Foxm1$^{fl/fl}$ mice, 276–277
    from cardiomyocytes, 276
    CCSP-Foxm$^{-/-}$ mice, 272–275
    epFoxm1$^{-/-}$ mice, 269–272
    FoxM1 FL/FL, WAP-Cre mice, 277–278
    FoxM1 FL/FL, WAP-rtTA-Cre mice, 278–282
    smFoxm1$^{-/-}$ mice, 275–276
  foxm1$^{-/-}$ mouse embryos
    description, 266–267
    hearts, 267
    livers, 267
    lungs, 267
  pancreas-specific deletion, Foxm1^{Dpanc}, 268

L

Life span, FOXM1 expression
  cell cycle-dependent, 206–207
  foxm1 mRNA expression, 199
  heart, foxm1 mRNA expression, 199
  self-renewing tissues, 199
  serum refeeding, 206
Liposomes
  advantages and disadvantages, 15
  aggregation, 13–14
  drugs, 14–15
  structure, 6f, 13
Lysosomal fusion, autophagy
  H$^+$ATPase and autophagosomes, 68
  mouse cells lacking Atg5/Atg7, 68
  permeases and transporters, 68
  Rab proteins and UVRAG, 67–68

M

Maternal β cells, 312
Micelles
  advantages, 17
  description, 15–16
  drug loading, 17
  passive and active targeting, 16
  structure, 15–16, 16f
  water-insoluble drug, 16
miRNAs, FOXM1 expression
  description, 237
  RB tumor-suppressor pathway, 238–240
Mitosis
  description, 287–288
  mitotic fidelity, 288
  target genes, 288
Monoclonal antibodies (MAbs)
  advantages and disadvantages, 31–32
  and CSCs, 31
  cytokeratin, 31
  description, 31
Mononuclear phagocytic system (MPS), 9
Mouse models, FOXM1
  knockout mice (see Knockout mice, FOXM1 mouse models)
  transgenic mouse models, 282–283
  Xenopus embryos, 283
MPS. See Mononuclear phagocytic system (MPS)
Mus musculus, embryonic development
  airway Clara cells, 295–296
  foxm1 knockout mice, 294–296
  foxm1 transgenic mice, 296
Myc/Max/Mad network
  antagonistic control, 250
  IFN-b/mezerein, 249
  protein expression, 249–250
  target gene, 248–249

N

NCCD. See Nomenclature Committee on Cell Death (NCCD)
Negative-regulatory domain-C (NRD-C)
  central domain functions, 188
  functional domains, 168, 169f
  RB-independent TRD, 188
Negative-regulatory domain-N (NRD-N)
  autoinhibitory N-terminus, 189
  as central on/off switch, 189
  in trans inhibition, 189–190
  mapping, 190
  N-terminus, FOXM1c, 189
  in RB-deficient SAOS-2 cells, 190
NHEJ. See Nonhomologous end joining (NHEJ)
Nomenclature Committee on Cell Death (NCCD), 68–69
Nonhomologous end joining (NHEJ), 312–313
NRD-C. See Negative-regulatory domain-C (NRD-C)
NRD-N. See Negative-regulatory domain-N (NRD-N)

O

Opsonization and mononuclear phagocytic system
  MPS, 9
  surface hydrophobicity, 9

P

Pancreatic β-cells
  postnatal expansion and maintenance, 311
  proliferation, 310–311
Pathological cardiac hypertrophy, 255
Peptides
  advantages, 29
  arginine–glycine–aspartic acid (RGD), 29
  description, 27–28
  luteinizing hormone-releasing hormone (LHRH) receptors, 28–29
Phagophores, autophagy
  elongation and multimerization, 65–67
  formation and regulation, 65
PHGDH. See Phosphoglycerate dehydrogenase (PHGDH)
Phosphatidylinositol-3 kinase (PI3K inhibitors), 259
Phosphatidylinositol 3-phosphate (PI3P)
  Atg5–Atg12 conjugation and interaction, 64f
  description, 65–66
  omegasome and autophagosome formation, 66–67
  production, 66

Phosphatidylinositol 3-phosphate (PI3P) (*Continued*)
   rich Ω-shaped structure formed, 66–67
Phosphoglycerate dehydrogenase (PHGDH)
   protein stability, FOXM1, 228–229
   ShRNA-mediated knockdown, 229
PI3K inhibitors. *See* Phosphatidylinositol-3 kinase (PI3K inhibitors)
Polymeric particles
   advantages, 21
   description, 20
   encapsulation, 21
   nonocapsules, 21
   structure, 20
Progenitor cell renewal
   GIC differentiation, 322
   luminal differentiation, 323
Proliferation *vs.* antiproliferation signals, FOXM1 expression
   antagonistic control, 218–219
   description, 218
   p53 tumor suppressor pathway
      contradictory findings, 243–244
      daunorubicin and nutlin-3, 241–242
      E2F-1 and Sp1, *foxm1* promoter, 242–243
      MnSOD, 243
      oxaliplatin, 242
      p21 and miR34 family, 241
      target gene, 240–241

R

Radiation-induced pulmonary fibrosis
   BAL, 341
   bleomycin, 342
   ECM components, 340
   epiFoxm1-DN animals, 341
   FOXM1, 340
   lung inflammation, 341
   myofibroblasts, 342
RAS/MAPK pathway, 250–251
RB tumor-suppressor pathway
   contradictory finding, 239–240
   E2F family and pocket proteins, 238
   mRNA expression, 239
   mutant protein, 240
   siRNA, 239

S

Shh. *See* Sonic hedgehog (Shh)
Smooth muscle cells (SMCs), 275–276
Sonic hedgehog (Shh), 262–263
Splice variants
   FOXM1
      comparisons, 107–108
      description, 100*f*, 102
      DNA-binding specificity, 103–106
      expression, human and rat, 103
      FoxM1A, FoxM1B, FOXM1c and rat WIN, 103
      FoxM1B *vs.* FOXM1c, 108–109
      nomenclature, 103
      transcriptional activity, 106–107
   FoxM1B *vs.* FOXM1c
      in A2780cp and OVCA433 ovarian cancer cells, 109
      ca MEK1, 108–109
      β-catenin, 109
      with Cdk2, 109
      DNA-binding affinity, 108
      expression patterns, 109
      with p27, 109
SRF. *See* Substrate recognition factor (SRF)
Stem cell pluripotency
   alkaline phosphatase activity, 320
   P19 EC cells, 319
   somatic cells, 320
   SSEA-1, 320
   target genes, 318
   transcription factors, 320
Stem cell self-renewal
   AsPC-1 human pancreatic cancer cells, 321
   FOXM1 overexpression, 321
   MD11 GIC cells, 322
   neural cortical stem/progenitor cells, 321
   SW1783 human glioma cells, 321
Substrate recognition factor (SRF), 317

T

TAD. *See* Transactivation domain (TAD)
Target genes, FOXM1
   biological functions
      angiogenesis, 163
      in apoptosis, 163–164

Index

expression, cell cycle-related genes, 162–163
  in G2- and M-phase, 161–162
  inflammation, chemotaxis and macrophage functions, 163
  proliferative signal transduction pathways, 162–163
  transcriptome, 164
direct and possibly indirect genes
  ChIP assays, EMSAs/DNAPs, 110
  description, 110, 111*f*, 115*t*
  downregulation, 156, 157
  FoxM1B, 157
  GATA-3 transcription, 157
  KIS promoter, 110–156
  microarrays, 157
  p27 promoter, 156
  splice variants FoxM1B, 110–156
  WREs/TBEs, 110
genome-wide ChIP-Seq, 159–160
in vivo occupancy, genes, 159
oligonucleotides, murine *cdx-2* and rat genes, 159
probably only indirect genes
  Bmi-1 and *PCGF4*, 158
  Cxcl5, 158
  HELLS, LSH, and *SMARCA6*, 158
  p16 and *INK4A*, 158
  proteins, 158–159
transcription factor
  DNA-binding affinity and selectivity, 160, 161
  FoxM1B *vs.* FOXM1c, 160
  FOXM1c-DBD, 160
  genome-wide ChIP-seq assays, 160–161
  in vivo, 161
Targeting moieties
  aptamers, 30
  cancer cells, 23–24
  description, 23
  EGF, 26–27
  folate, 24–25
  peptides, 27–29
  transferrin, 25–26
  VEGF, 32–33
Target of rapamycin (TOR) kinase, 65
TGF-β1. *See* Transforming growth factor-β1 (TGF-β1)

TNF-α. *See* Tumor necrosis factor-α (TNF-α)
TNF-related apoptosis inducing ligand (TRAIL), 77, 78*f*, 79–81, 83–84
TRAIL. *See* TNF-related apoptosis inducing ligand (TRAIL)
Transactivation domain (TAD)
  FOXM1c-TAD
    coactivators, P/CAF and p300/CBP, 182–185, 183*f*
    hydrophobic residue, 181
    intramolecular interactions, 179–181, 180*f*
    with TBP, TFIIB, and TFIIEa, 182
    transactivation potential, 179–181
  functional domains, 169–170
  IDs, 190–193
  N-terminus function, 170
Transcriptional activity, FOXM1 splice variants, 106–107
Transcription factors, *Foxm1* promoter
  cancer, 237
  CREB, 229–230
  E2F family, 230–237
  miRNAs, 237–238
  p53, 230
  target gene, 230
Transferrin
  advantages, 26
  description, 25
  lipid particles, 25–26
Transforming growth factor-β1 (TGF-β1), 264
Transgenic mouse models, FOXM1
  CCSP-*Foxm1* mice, Clara cells, 283
  ep*Foxm1* mice, respiratory epithelial cells, 282–283
Transrepression domain (TRD)
  FOXM1c-TRD
    basal transcription complex, 186–187
    direct interaction, 187–188
    as ID, 185
    potency, 185
    and splice variant FoxM1B, 186
    transactivation potential, 185–186
  and NRD-C, functional domains, 168, 169–170
TRD. *See* Transrepression domain (TRD)

Trisomy 21/down syndrome, 254–255
Tumor dormancy, autophagy
   angiogenesis arrest and ARHI, 79–81
   breast cancer metastases to bone, 79–81
   DTCs and CTCs, 79–81
   β1-integrin signaling and progression, 79–81
Tumor initiation, autophagy
   AKT and Bcl-2, 71–72
   5'-AMP-activated protein kinase (AMPK), 71
   ATP:ADP ratios, 71
   BH3 receptor domain, 71–72
   Bif 1 and beclin1 identification, 72–73
   cellular senescence, 73
   DAPK and DRP-1, 71
   DRAM and PTEN, 70–71
   oncogenic and tumor-suppressive signaling pathways, 73, 74t
   p53 plays, 70–71
   tumorigenesis and decreased levels, 70
   UVRAG, 72–73
Tumor metastasis, autophagy
   anoikis, 77
   antimetastatic properties, 77
   astrocyte elevated gene-1 (AEG-1), 79, 80f
   damaged mitochondria, 77–79
   description, 73–77
   different stages, 77, 78f
   DTCs and cellular debris, 73–77
   glutamine utilization and intracellular fat storage, 77–79
   low-nutrient and low-oxygen conditions, 79
   necrosis and subsequent macrophage infiltration, 73–77
   nutrient-limited and low-oxygen conditions, 79
   organelle function, preservation and maintenance, 77–79
   primary event models, 73–77, 76f
   TRAIL, 77
   "Warburg effect" and neoplastic cells, 77–79
Tumor necrosis factor-α (TNF- α), 264

U

Ubiquitin-proteasome pathway, FOXM1 expression
   APC/C activator protein, 224
   Cdc20, 224
   serum depletion, 224
Ultraviolet radiation resistance-associated gene (UVRAG), 65–66, 67–68, 72–73
UVRAG. See Ultraviolet radiation resistance-associated gene (UVRAG)

V

Vascular endothelial growth factor (VEGF)
   advantages, 33
   description, 32
   ligands, 32–33
VEGF. See Vascular endothelial growth factor (VEGF)

W

Wnt3a, FOXM1 expression
   antagonist DKK1, 228
   signaling and receptor, 227–228

X

Xenopus embryos, FOXM1 mouse models, 283

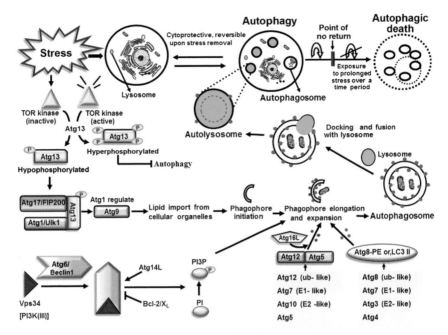

Figure 2.1, Sujit K. Bhutia *et al.* (See Page 64 of this volume.)

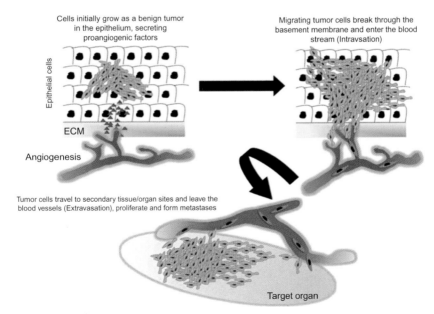

Figure 2.2, Sujit K. Bhutia *et al.* (See Page 76 of this volume.)

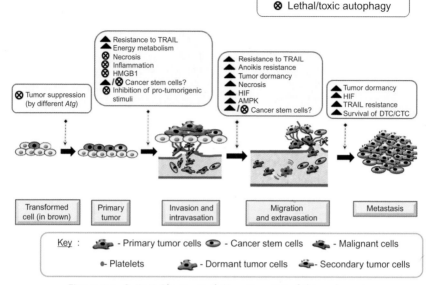

Figure 2.3, Sujit K. Bhutia *et al.* (See Page 78 of this volume.)

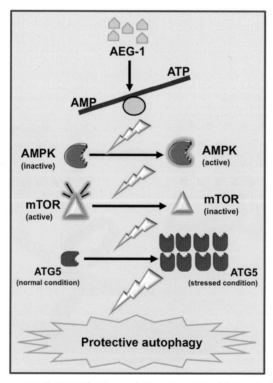

Figure 2.4, Sujit K. Bhutia *et al.* (See Page 80 of this volume.)